The Great Climate Change Deception!

The true history of the Earth's changing climate.

Volume 2. 2000BCE to 1699 CE

George Mitrovic

All sources of information are from published works and scientific studies done by experts in their fields and are in the Bibliography at the end of the book. I collate data and let the results speak for themselves.

Contents

2000 BCE	2
1800 BCE	6
1700 BCE	7
1600 BCE	12
1500 BCE. Sub-Atlantic Period or Neoglacial Period.	14
1400 BCE	20
1300 BCE. End Bronze Age. Invasion by Sea Peoples	23
1200 BCE. Late Bronze Age Collapse. Sub-Atlantic Period.	32
1100 BCE	48
1000 BCE. Sub-Atlantic Period	50
900 BCE	52
800 BCE	53
700 BCE	54
600 BCE. The Secondary Climatic Optimum or Little Optimum	55
500 BCE	56
400 BCE. Climatic Optimum. The Roman Warm Period.	58
300 BCE	60
200 BCE	62
100 BCE	63
000 BCE/CE	65
100 CE	70
200 CE	72
300 CE	74
400 CE	79
450 CE	82
500 CE	84
550 CE	94
600 CE	99

650 CE	101
700 CE	104
750 CE	106
800 CE	110
850 CE	118
900 CE The Medieval Warm Period	126
950 CE	132
1000 CE	137
1050 CE	147
1100 CE	157
1150 CE	175
1200 CE	185
1250 CE	208
1300 CE	224
1350 CE	246
1400 CE	257
1450 CE	268
1500 CE	276
1550 CE	289
1600 CE	311
1650 CE	341

Introduction

We now enter a more tumultuous period as well as a better documented period. This section covers events of the Second Millenium BCE, the Roman Warm Period which you may have not learnt of, the Antique Ice Age, the Dark Ages followed by the Medieval Warm Period and ends as the Little Ice Age peaks in the late Seventeenth Century in 1699. Yes, it looks like interminable lists of climatic events, volcanic eruptions, tsunamis, earthquakes and other events. These are proofs of change. As we advance further along and out data list becomes much closer in dates, patterns start appearing as to what may be causing what.

Why are you going to look at a long list of climatic events? Because this is my data list verifying my main hypothesis. The data also shows that the weather in the past, even until recently was a nightmare of rapidly changing climatic conditions, frequent floods and famines, enormous sea floods and frequent rises and drops of temperature that we do not see or hear very often these days. We are in the Golden Age of an interstadial. Just be thankful you did not live in the good old days.

2000 BCE

Around 2000 BCE Black Peak in southwest Alaska erupted with a VEI of 6. This would have caused a volcanic winter. This was an explosive eruption that created a caldera. Ash flow tuffs travelled down Ash Creek and Bluff Creek valleys to a depth of one hundred meters. 10 to 15 cubic kilometers of tephra was released. Other eruptions were Sheveluch in Kamchatka with a VEI of 5, Miyakejima in Japan with a VEI of 3, Nasudake in Japan with a VEI of 2, Gorely in Kamchatka with a VEI of 2, Jebel Marra in Sudan with a VEI of 4, Campi Flegrei in Italy with VEI of 4 and Raoul Island in New Zealand with a VEI of 4.

In the northernmost parts of North America, where the remnants of the ice-sheets still lingered, the warmest time was not reached until 2000 BCE. There was a major re-advance of glaciers in the Western Rockies of Alaska after 2000 BCE.

The rise of general water levels after the Ice Age was generally over by 2000 BCE where it stood between one meter and two meters higher than during the late twentieth century.

During the Second Millennium BCE there are massive discrepancies in dating. In Egypt there is a three-hundred-year gap in the Egyptian King Lists. All around the Mediterranean Sea there were civilization-destroying changes as a new dark age spread across the area. Was this all caused by the eruption of Thera at some unspecified date? What is this you say? You try and get a firm date. Even Wikipedia gives it a leeway from 1642 B to 1540 BCE. Other sources give other dates. Was it from meteoric events as evidenced by bharak stones found all over the Arabian Peninsula around the middle of the Second Millenium? There are vast areas of the Arabian Peninsula that are covered with burnt stone. In Hebrew barat, bharak, stones were described as fiery hail falling from the sky. Was there also volcanic activity in the Sinai as well as the Arabian Peninsula at the same time? There are at least ten active volcanoes on the Arabian Peninsula and many extinct ones. We have already met the "flaming potsherds" falling from the sky in Akkadian literature as the empire of Akkad fell.

And don't forget the Al Amarah meteoric impact which would have seriously affected Mesopotamia, which means Land Between the Rivers, with the two rivers being the Tigris and the Euphrates.

Dating is still a problem but let us ignore the dates of events in this period. After all, if they have several dates over 98 years for the Thera eruption plus others outside of these dates, how reliable are the dates that we have for other events in this period".

Let us start in a period where people had good opportunities across the earth to see meteorite falls.

There was a sudden change in world weather patterns. The great Pluvial or rainy period was finishing, and the drying period was starting.

Sea levels went over modern-day levels in 4,000 BCE and rose one meter more by 2000 BCE.

Around 2000 BCE it was the time of the Climate Optimum, the Altithermal, that brought favorable climatic conditions to Europe. There was ample rainfall in Europe and savage droughts in Western North America. These were millennia when high Pacific Ocean temperatures suppressed downwelling and kept marine productivity low. These Optimum Warm Periods appear to be around every thousand years.

According to W. Bruce Massie, an environmental archeologist in the period 2,350 BCE to 2000 BCE there were at least four cosmic impacts with the Earth. These were in 2,345 BCE,

2,240 BCE, 2,188 BCE and 2000 BCE with a possible fifth impact between 2,297 BCE and 2,265 BCE.

From 2000 BCE there was a spread of people building stone circles in the British Isles as well as Europe and North Africa. This might indicate increasing clear skies in this period. Were they watching for meteorites etc.?

In Arabia the Marib Dam supplied water to much of the Empty Quarter. Water was not lacking in the area until the destruction of the Marib Dam in the sixth century CE. Only a civilization capable of lifting gigantic stones like those at Baalbek in Lebanon could have built the Marib Dam, which fed a huge series of canals, which were so old that they were unknown to the Romans. The canals were made of baked clay to prevent leakage of the water into the ground and distributed the torrential rains into primary feeders and then secondary feeders and then tertiary feeders. The Marib dam irrigated about four thousand acres. In addition, a system of wells connected to the water supply furnished a small supplementary supply of water. This is civil engineering on a massive scale in supposedly primitive times. The original Great Dam of Marib was started around 2000 BCE and rebuilt several times, becoming progressively larger until it was fourteen meters tall. By the fifth century CE it had collapsed from lack of maintenance. Other important ancient dams in Yemen include the Dam of Jufaynah, the Dam of Khārid, the Dam of Aḍra'ah, the Dam of Miqrān and the Dam of Yath'ān. Historically, Yemen has been recognized for the magnificence of its ancient water engineering. From the Red Sea coast to the limits of the Empty Quarter Desert are numerous ruins of small and large dams made of earth and stone. Before 2000 BCE these were not needed. Technology and infrastructure was changing to suit changing climatic conditions.

Around 2000 BCE the weather in Australia became dryer after the wet period of the previous 2000 years. It was driest in 1200 BCE. The weather continued being dry for some centuries.

In the Baltic Sea around 2000 BCE declining salinity had changed the flora and fauna of the Litorina Sea in the Limnaea Sea stage and the water level was nowhere more than 200-200 km wide in the northern part. This is now the Baltic Sea.

Around 2000 BCE a strange people arrived in the St Lawrence Valley between New York State and Quebec and Ontario in Canada. These people were huge, rugged, very tall and with massive skulls with very broad round heads similar to the mysterious Adena People who appeared suddenly further south. The women were 170 cm tall, and the men were 180 cm tall, and they had hyperdontia or extra teeth, which is a genetic condition. Some believe that these people had lived here since 10,000 BCE.

In Egypt water was still seasonally available becoming increasingly arid with an estimated annual rainfall of four inches. There were great droughts in Egypt between 2000 BCE to 1950 BCE. These were associated with an abnormal prevalence of southerly winds from the desert and abnormally low levels of the Nile and the resultant failures of the yearly flood.

In Kashmir in India conifers were re-establishing themselves and replacing the oaks and alders that needed a warmer, moister climate. Yes, forests do move and get replaced as the climate changes. We can read climatic conditions by seeing what trees and other plants were growing at the time.

In Mesopotamia population figures of Sumerian population centers were only fifty per cent of two hundred years before due to long term drought. Down to 2000 BCE cultivation in Mesopotamia still extended fifty kilometers north of the present limit of feasibility of any such activity and a density of population grew up that could no longer be sustained in the twentieth century.

The submerged Quetzalcoatl Impact Crater in the Caribbean Sea off Veracruz State in Mexico is ten kilometers in diameter and four thousand years old. This would have been a one-kilometer cosmic body. Do you remember what a body this size can do? Impact of a 1-kilometer asteroid would probably cause significant cooling on Earth due to debris ejected in and above the atmosphere blocking sunlight, subsequent die-off of vegetation and crop failure, which then leads to mass starvation of mammals and famine. If it fell into the sea, the sea would boil at the impact point and massive tsunamis would spread out causing huge damage.

And this is what one impact can do. What about the years when there are several.

The Tver Impact Crater in Tverskaya Oblast near Moldino in Russia is five hundred meters in diameter and four thousand years old.

Already by 2000 BCE the forests had been retreating from the exposed coasts of northwest Scotland and from most of the highest places which it had reached in Scotland and Northern England. This was due to a cooling climate.

In Ireland, Wales and Cornwall, woodland still extended on the Atlantic shores and higher on the hills than any present woods, until the Bronze Age.

In southern England as early as Neolithic times around 3,000 BCE, allowing for human disturbance on the chalk downs, oak and hazel were abundant at the foot of the hills and on slopes and were used as firewood as the plateaux of the chalk downs was already grassland. By 2000 BCE the tops of the chalk downs were abandoned except for grazing and burials and for their convenience as travel routes. This meant that well into the dryer times of the Bronze Age the springs were lower, and water supply was difficult on the higher ground. The now abandoned open areas on the chalk hills became colonized by beeches which were the last trees to colonize England.

Chia Ming Lake or Meteor Lake in Taiwan is an oval meteoric impact formed lake. It was created four thousand years ago and is one hundred and twenty meters long and eighty meters wide. Meteor Lake is in the Yushan National Park in Taiwan.

Around 2000 BCE relatively few people lived along the California coast when rainfall was considerably lower than the late Twentieth Century. Before 2000 BCE most Californian societies had moved constantly, exploiting whatever foods were available. They turned to more labor-intensive foods like acorns. But the problem with acorns was that it tied women up into a daily routine of pounding and leaching of the acorns. This caused societies on the West Coast of North America to form small communities that acted as bases which they occupied for months on end instead of being totally nomadic. The Chumash were more dependent on a small area for foraging with a limited single watercourse close to their acorn trees and their mortars and pestles. By 1500 BCE the Chumash were in much larger settlements. Other sources state that the climate of California was much wetter and cooler than today. It all depends on when they dated their report. Sometimes you can only work on overall periods of phenomena to get a clearer idea.

Between 2000 BCE-1500 BCE most of the present glaciers in the Rocky Mountains south of 57 degrees North were formed.

In 1995 BCE Mount Vesuvius erupted with a VEI of 6. This would have caused a volcanic winter. It is called the Avellino eruption.

In 1990 BCE Nevado del Tolima in Colombia erupted with a VEI of 5.

In 1950 BCE Chikurachki on Paramushir Island in the Kurils erupted with a VEI of 4. Grimsvotn in Iceland erupted with a VEI of 2 as well.

Melebingoy/ Mount Parker in the Philippines erupted in 1920 BCE with a VEI of 4. Also, in 1920 BCE Sanbesan in Japan erupted with a VEI of 4 and Katla in Iceland with a VEI of 4.

In 1932 the English explorer H. St John B. Philby set off on an expedition to discover the lost city of Wabar in the Empty Quarter of Arabia, the Rub el Khali Desert. According to legend the city had been destroyed in ancient times, around 1900 BCE, by fire from Heaven and according to tradition the site was marked by a lump of iron the size of a camel. Philby only found some circular craters with rims full of great chunks of glass. Small black spherules of glass were common. Initially Philby thought that the craters were volcanic but eventually realized that they were meteoric. The glass was later found to be heavily shocked, and pressure formed forms of quartz and nickel bearing iron, namely meteoric material.

To the east of the Indus River there now lies a vast desert called the Thar Desert. The remains of towns were found here that could not apparently survive in the desert conditions. Satellite photographs showed that this was once a fertile plain that was traversed by a great river and there were even signs of canals. Now only a small part of the river remains, which is called the Ghaggar. Various scholars state that the now vanished river was the Sarasvati that was mentioned in Vedic hymns. These towns were contemporary with Harappa and Mohenjo-Daro. The vast area was the site of a very rich and cultured civilization. Sometime after 1900 BCE some great cataclysm destroyed this civilization. The Earth in the region suddenly buckled due to the pressure of tectonic plates raising the Himalayas and there were earthquakes and volcanic eruptions that basically caused the rivers to disappear into the ground.

1900 BCE

In 1900 BCE rainfall returned to its previous seasonality in Mesopotamia. People returned to Habur and Assyria. This is believed to be the period of Abraham and the beginning of Monotheism. Abraham, Avram, apparently left Ur in Chaldea, city of the Moon Goddess Nin, with his father Terah, on his trek with his tribe to the promised land of Canaan, now in Israel. These people are known as the Hapiruh from the cuneiform tablets found in Mari. The dating for this event is very inaccurate though. Dating in this entire millennium is very inaccurate.

In 1890 BCE Mount Hudson or Cerro Hudson in the Andes of southern Chile erupted with a VEI of 6. This eruption created an ash layer which it deposited about five centimeters deep as far as the shores of San Jorge Gulf on the Atlantic coast of Argentina. More ash was found in an ash layer in a peat bog in the Falkland Islands in the South Atlantic, 1200 kilometers east of Mount Hudson. This would have created a volcanic winter. More than ten cubic kilometers of tephra was ejected. Also, in 1890 BCE Mount Dana in Alaska erupted with a VEI of 5. The eruption was responsible for a pyroclastic flow that filled valleys south and west of the crater and reached the sea at Canoe Bay around seven kilometers away.

In 1880 Michoacan-Guanajuato erupted with a VEI of 3 in Mexico.

In 1870 BCE Campi Flegrei in Italy erupted with a VEI of 4.

In 1860 BCE Mount St. Helens in Washington erupted with a VEI of 6. It released 15 cubic kilometers of tephra and would have caused a volcanic winter.

In 1850 BCE Agua de Pau erupted on San Miguel Island in the Azores. It had a VEI of 5.

In 1820 BCE Rincon de la Vieja in Costa Rica erupted with a VEI of 4.

This is a large number of high VEI eruptions and please allow for the fact that VEI1 and VEI 0 eruptions are not shown in this book and neither are eruptions with no VEI.

Apparently around the time of the eruption of Thera there was a massive land displacement in Egypt. The Nilometers at Semneh, dated from the Middle Kingdom show an average rise in the waters of the Nile where the river is channeled in rock of twenty-two feet above the level at present. Had the Nile flooded much higher in early times which would be problematic for many buildings which would have been regularly inundated or did the land rise in this period? Flood levels 8 to 11 meters higher than the modern period were recorded from the reigns of Amenemhet III and his successors for a period of 27 years. Amenemhet died in 1814 BCE so this is too early for Thera/Santorini.

In 1810 BCE Villarica in Chile erupted with a VEI of 5.

1800 BCE

In 1800 BCE Reykjanes in Iceland erupted with a VEI of 2. Other eruptions were Aogashima in Japan with a VEI of 2, Miyakejima in Japan with a VEI of 3 and Mauna Loa in Hawaii with a Vei of 3.

Around 1800 BCE world sea levels were high.

In 1780 BCE Ata in Japan erupted with a VEI of 4.

Vesuvius also erupted around 1780 BCE with a Plinian eruption much larger than the eruption that buried Pompeii and Herculaneum. Vesuvius produced an early violent pumice fallout and a late pyroclastic surge that covered surroundings as far as 25 kilometers away, burying land and Bronze Age villages. Evidence showed that a sudden mass evacuation of people occurred at the beginning of the eruption. Desertification of the total habitat due to the huge eruption size caused the abandonment of the entire area for centuries. It had a VEI of 5 and released 3.9 cubic kilometers of tephra.

In 1770 BCE Mount St. Helens in Washington erupted with a VEI of 5. This is volcanic winter material. Four cubic kilometers of tephra was ejected.

Around 1750 BCE Black Peak/ Mount Veniamof in southwest Alaska erupted with a VEI of 6. This would have caused a volcanic winter. More than 50 cubic kilometers of tephra was ejected. This was a caldera forming eruption. Other eruptions were Okataina in New Zealand which erupted with a VEI of 4 and Tolbachik with a VEI of 3 in Kamchatka.

As we progress into later periods we will start getting a real estimation of climate alteration by volcanoes. In this period, unless it is stated as having occurred, then there is nothing to study as the gaps in the histories are still too large.

Around 1800 BCE a magnitude 9.5 magnitude earthquake generated tsunamis 15 to 20 meters (49 to 66 feet) in height that struck 1000 kilometers (620 miles) of the coastline of the Atacama Desert. People fled the area and did not begin to return until ca. 800 BCE; some pre-tsunami settlements were not reoccupied until between ca. 1000 and 1500 CE.

1700 BCE

Around 1700 BCE Villarica in the Chilean Andes erupted with a VEI of 5. Also, in 1700 BCE Taranaki in New Zealand erupted with a VEI of 5. Sheveluch in Kamchatka also erupted in

1700 BCE with a VEI of 4, Maruyama in Japan erupted with a VEI of 2 and Taveuni in Fiji erupted with a VEI of 2.

Now we come to one of the most influential and interesting periods in world history the influence of which still resounds around the Earth. This is also the beginning of the period of the rise of Monotheism. The belief in a single God. What were the influences in this period that might have caused such a revolutionary change of religious viewpoints from many gods to only one god?

During the Second Millennium BCE there are massive discrepancies in dating. In Egypt there is a three-hundred-year gap in the Egyptian King Lists. All around the Mediterranean Sea there were civilization destroying changes as a new dark age spread across the area. Was this all caused by the eruption of Thera at some unspecified date? What is this you say? You try and get a firm date. Even Wikipedia gives it a leeway from 1642 B to 1540 BCE. Other sources give other dates. Was it from meteoric events as evidenced by bharak stones found all over the Arabian Peninsula around the middle of the Second Millenium? There are vast areas of the Arabian Peninsula that are covered with burnt stone. In Hebrew barat stones were described as fiery hail falling from the sky. Was there also volcanic activity in the Sinai as well as the Arabian Peninsula at the same time? There are at least ten active volcanoes on the Arabian Peninsula and many extinct ones.

In 1650 BCE Arenal in Costa Rica erupted with a VEI of 4. As well Campi Flegrei erupted in Italy with a VEI of 4. Other eruptions that year were Asosan in Japan with a VEI of 3, and Sheveluch in Kamchatka which also erupted with a VEI of 3.

In 1645 BCE Aniakchak in Alaska erupted with a VEI of 6. This was a volcanic winter causing event having caused severe global cooling. More than 50 cubic kilometers of tephra was ejected. Tephra from Aniakchak was found in Greenland. The nearest coast is the north coast of Greenland which is around 2,450 miles away or around 4,000 kilometers away. This was a voluminous ignimbrite eruption that created the modern caldera and collapsed the summit of the 7,000-foot mountain. The ignimbrite extended as far as eighty kilometers from the caldera and filled adjacent valleys to the depth of 75 meters. As the ignimbrite entered the Bering Sea it generated a tsunami that reached a height of fifteen meters above mean sea level along the northern Bristol Channel coastline of Alaska. This was one of the largest volcanic events of the Late Holocene. The massive release of sulfur led to the onset of a severe volcanic winter and caused extreme environmental disruption. The effects were felt across a vast expanse of the Northern Hemisphere. It may have caused the end of the Arctic Norwegian Stone Age. During this eruption pyroclastic flows swept all the flanks of the volcano and caused a tsunami in Bristol Bay. Tephra from the eruption rained down on Alaska, with noticeable deposits being left as far as northern Europe. The eruption depopulated the central Alaskan Peninsula and caused cultural changes in Alaska. Together with other eruptions at the time, namely Thera in the Mediterranean, this eruption, which is called Aniakchak II, may have caused climatic anomalies. The present caldera was formed during this eruption. Ash was even deposited over Alaska. Tephra rained down on Alaska and even northern Europe receiving large falls.

The Dye 3 ice core from Greenland possesses an acidity level indicating obscured skies possibly from a volcanic dust veil event. This event covers a period of twenty years due to dating fuzziness. This was around 1645 BCE.

This is probably the most confusing period of world history as there are discrepancies in the Egyptian Kings Lists and they do not tally with other chronologies in the Middle East. I will use geology and cosmology to try and make sense of things.

1645 BCE is the accepted date for the end of the Middle Kingdom in Egypt, but this period is so complex you might still need to have two hundred years leeway either side.

Drastically deteriorating climatic conditions in the Levant caused the Hyksos people to overrun Egypt. The fourteenth dynasty collapsed as the Hyksos overran the Nile Delta. The historian Manetho of Sais recorded that a "blast of God" prostrated Egypt. We do not know what the "blast of God" was. Also, around this time earthquake and holocaust had destroyed Egypt. This had also occurred throughout the Middle East. Egypt was conquered by the Hyksos who came from the east when it fell in a catastrophe of nature.

Other sources state that there was a mysterious mass death at Avaris during the XIII dynasty around this time period. Other large metropolitan centers were destroyed or abandoned at this time. These included Ithtaw and Hetepsenusret. Avaris was later renamed Ramses, and the biblical Israelites were said to be enslaved in its reconstruction. Avaris was also known as Piramesse and Peru-nefer and was at the start of the overland route to Canaan and Philistia where Israel is now. A buried harbor basin was found here with a canal connected to the Pelusiac branch of the Nile River. In all Avaris had three harbors and was a major trading port. There were two islands and a tributary of the Nile running through the city as well as a second harbor and a dry dock at the Nile branch. The main harbor basin was 450 meters square.

Greek tradition states that the Great Flood of Deucalion occurred around 1630 BCE though other sources state that the Great Flood occurred much earlier and in fact there may have been more than one Great Flood. Greek myths mention several floods in their legends. The Greeks mention two World Floods. There was the Flood of Ogyges which apparently occurred between 2,137 BCE and 2,132 BCE. Another date is 2,376 BCE as well as 2,123 BCE. There is also the Flood of Deucalion which occurred apparently around 1630 BC,1528 BCE or 1460 BCE. Remember that normal calendars can have great discrepancies, not only geological calendars. The Ark of Deucalion allegedly landed on Mount Tomaros in Greece.

The Indus Valley civilization ended during a great drought. Many rivers disappeared in this period including the Sarasvati River.

In 1629 BCE a dark age descended over Babylon that lasted one hundred years. At the same time the Amorite capital is sacked and the Mittani and Hurrian overlords lose their hold of the Assyrian Empire. All of this was occurring in Mesopotamia.

During the ninth year of the reign of Amaziduga, circa 1629 BC, the Babylonian king described seeing the cosmic body Ninsianna for a period of nine months. Was this a comet? This is an exceptionally long cometary or meteoric approach. What was Ninsianna that appeared for nine months in this same period in 1629 BC? Ninsianna was called the "Holy Torch who fills the Heavens". Venus does not do this. Was the light from this object so strong that it blocked out the view of Venus as it approached from the direction of Venus towards the Earth? What was hovering over the earth for nine months? This legend of Venus, the Morning Star, as well as the Evening Star is quite familiar in Mediterranean mythology during the Second Millennium.

In 1629 BCE the Irish Annals report that there was a sudden eruption of nine lakes in Ireland.

The period of 1629 BCE to 1600 BCE is the first accepted date period for the volcanic eruption of Thera or Santorini in the Mediterranean Sea north of Crete. Controversy still surrounds the dating of the eruption as the range extends from 1629 BCE to 1220 BCE. In this text I will stick to everyone's dating so that you can make up your own minds and whilst considering this think about how would volcanic eruptions interfere with radiation levels and therefore the radiocarbon levels? Santorini had a VEI of 7. Interestingly enough we are finding more and more data indicating that Santorini might well have happened in 1629 BCE.

AMS measurements on seeds of grain found in a storage jar in the Minoan village of Akrotiri that had been buried under the pumice of the volcanic eruption of Thera-Santorini indicate a date of 1626 BCE. AMS is a type of mass spectrometry performed by converting the atoms in the sample into a bean of fast-moving ions (charged atoms). The mass of these ions is then measured by the application of magnetic and electric fields.

But as I have stated with all of these radiation releases how accurate are our dating systems? The Santorini event was much more massive than previously thought; it expelled 61 cubic kilometers of magma and rock into the Earth's atmosphere, compared to previous estimates of only 39 km^3 (9.4 cu mi) in 1991, producing an estimated 100 cubic kilometers of tephra.

1610 BCE was another date chosen for the Thera/Santorini eruption, plus or minus fourteen years. Santorini had a VEI of 7 and ejected 123 cubic kilometers of tephra which is an enormous amount. It ended the Minoan settlement on Akrotiri and the Minoan Age on Crete as well as creating a volcanic winter and massive tsunamis. The date of 1610 BCE is confirmed by the Smithsonian Global Volcanism Program. 123 cubic kilometers of tephra was ejected. The column of smoke for a VEI 7 eruption is apparently 40 to 50 kilometers high so could have been easily seen from northern Egypt and present-day Israel. Santorini is 800 kilometers from the Nile Delta in Egypt.

Akrotiri, which was a city on Thera, was buried in 1627 BCE by a massive volcanic eruption.

With everything else going on at the time, we can say pretty well that Santorini blew its top around 1627±10 years.

Only the Mount Tambora volcanic eruption of 1815, the 233 CE eruption of the Taupo Volcano, and possibly Baekdu Mountain's 946 CE eruption have released more material into the atmosphere during the past 5,000 years. In 1628 Mount Sambe in Honshu, Japan, exploded. In 1628 BCE the island of Rabaul in Papua New Guinea, north of Australia, erupted with extraordinary violence. This created a tsunami. That same year there was a simultaneous detonation of New Zealand's Taupo Valley center volcanoes which shot a two-hundred-foot-high wall of water away at speeds of more than five hundred miles per hour. The combined tsunamis scoured low lying lands in the Pacific Ocean as well as continental coastal shores. There were also massive tsunamis from Alaska and Hawaii as well. If there was anything in the Pacific Ocean it would have been drowned by the three tsunamis. Also, in 1628 BCE there was an extraordinary outgassing from Mauna Kea in Hawaii. This created a tsunami.

Tree rings in Britain and Ireland indicate enormous ashfall around 1628 BCE.

There was a comet seen in the sky over China. The Xia Dynasty ended and there was the start of the Shang Dynasty. The Chinese also recorded that at one time there were two suns in the sky, one in the east and one in the west.

In Egypt the Nile Delta suffered massive deposition, killing off most plant life across Lower Egypt.

Tree rings in England indicate enormous ashfall around 1628 BCE.

During 1628 BCE European oaks show unnaturally reduced growth.

Tree rings in Ireland indicate enormous ashfall around 1628 BCE. Growth depression occurred in European oaks in Ireland.

Tree rings in Germany indicate enormous ashfall around 1628 BCE.

Exceptionally heavy ashfall in ice cores is recorded in Greenland in 1628.

In the Babylonian Venus Tablets which was a set of astronomical records kept by Babylonian astronomers during the reign of King Amaziduga in the 1600s BCE the planet Venus disappeared from view for nine months and four days during the ninth year of his reign. Venus is one of the brightest of the celestial objects and the skies are remarkably clear over Babylon. The archeoastronomer P. J. Huber deduced that the date of this event was, amazingly enough, 1628 BCE. The planet Venus could have been obscured by a dust veil event.

What was Ninsianna that appeared for nine months in this same period in 1628 BCE? Was the light from this object so strong that it blocked out the view of Venus as it approached from the direction of Venus towards the Earth?

Mike Bailly, the dendrochronologist, looked at tree rings in the 1620s BCE. To find a definitive date for Santorini he looked at the narrowest of narrow tree rings. This was in Irish bog oaks. Some German records of tree rings also showed narrow rings around this same time. The narrowest tree ring events did not occur very often. Mike found that the narrowest tree ring events did not occur very often. At first there were only a handful of narrowest tree ring events. The Irish trees were showing the detrimental effects of large explosive volcanic eruptions. Extreme poor growth events in the Irish oaks were due to the same volcanic dust veil events which left acid layers in Greenland.

Tree rings in North America indicate enormous ashfall around 1628 BCE.

Tree rings from Anatolia in Turkey from archaeological sites indicate sudden, massive growth up to 240 per cent occurred around 240 BCE, plus or minus 25 years. This would have been due to vast amounts of magmaic water and seawater thrown up by the eruption of Santorini. This is estimated to have been around 1628 BCE. Trees had suddenly been putting on summer growth far in excess of normal due to the normally dry, desert conditions. Volcanic ash from the Santorini eruption was deposited in an easterly direction, which was Turkey. This was from research by Kuniholm on Anatolian tree rings.

Scotch pines in Sweden showed growth depression in 1628 BCE. What was responsible for this enormous ashfall in this period all over the Earth?

According to Mike Baillie, a dendrochronologist at Queens University in Belfast in Northern Ireland there was a major temperature trough worldwide around 1600BCE. Baillie had constructed an Irish bog oak dendrochronology spanning the period from 5,000 BCE to 1000 CE. The Irish bog oaks showed a pattern of narrow rings, indicating extremely minimal growth, around 1627 BCE. In fact the rings were so narrow that Baillie could barely measure them indicating that weather conditions had been catastrophic. These narrow rings covered most of the 1620s and unlike a major volcanic eruption lasted almost a decade. Even the largest volcanic eruptions only produce nuclear winter conditions for around three years. Baillie then compared his bog oak data against Greenland ice cores and a number of dates lined up indicated by ice core acid layers. These major weather events, much larger than volcanic induced nuclear winter conditions, occurred within thirty years of 210 BCE, fifty years of 1120 BCE, ninety years of 3,150 BCE and one hundred years of 4,400 BCE. When new data was published in 1987 it showed another major acid layer within twenty years of 1645 BCE.

Mr. Val LaMarche of the University of Arizona specifically mentioned the year 1627 BCE as being a year with a particularly striking frost indicating a major weather variation. This was deduced from bristle cone pine dendrochronology. LaMarche had analyzed bristlecone pines from the White Mountains along the California and Nevada borders in the United States. This was the most severe frost ring in the second millennium! LaMarche suggested that this event may have been due to the Santorini/ Thera eruption. This was according to LaMarche and Hirschboeck.

How many volcanic winters have we met so far in this book? And not just little ones. We might have even met at least one Impact Winter.

Stunted fossil tree rings were found in the Sierra Nevada that indicated that a climatic period similar to a nuclear winter had occurred around 1626 BCE.

There is a Hindu legend that states that there was a great empire that was swallowed up by the sea near the end of the "Treta Yuga", around 1621 or 1575 BCE depending on which calendar you use. This is the same period as the volcanic destruction of Thera or Santorini in the Mediterranean Sea. Mind you as with all older manuscript records the exact dates are varied and open to interpretation sometimes with great discrepancies. Thera has been dated to have occurred over a three-hundred-year period. This is around the same date of the Irish legends.

There was a volcanic winter in China around 1618 BCE during the collapse of the Xia Dynasty and its replacement by the Shang Dynasty. This was accompanied by a yellow fog, a dim sun, then three stars, frost in July, famine and the withering of all five cereals. I could not find any volcanoes erupting in this area or near it this year or in the previous few. It could have been a Kamchatkan volcano that is not listed anywhere as erupting, but Kamchatka has so many volcanoes, 300 known so far, with 29 that are active. Kamchatka is in Siberia. Ata in Japan had erupted in 1610 BCE, but it was only a VEI 3 which generally do not cause volcanic winters but can at times. King Chiah, Chieh, was the last king of the Xia Dynasty. The comet was recorded to have appeared when he was executing his faithful counselors.

Was the eruption of Santorini/Thera in 1629 BCE responsible for the volcanic winter in China?

In 1610 BCE Ata in Japan erupted with a VEI of 3.

1600 BCE.

This date of 1600 BCE is not necessarily the date of the following events. It is approximate to the century due to problems with the Kings Lists of Egypt and other Mediterranean and Middle Eastern cultures.

There are ancient ruins in southern Arabia that date back to a time when the area was fertile and well-watered. In Western Arabia, there are 28 fields of scorched and shattered stones that cover as many as 7,000 square miles each. The stones are sharp-edged, densely grouped and burned black. They are not volcanic and appear to date from the time when this area was lush and fertile and suddenly became scorched by intense heat transforming it into an instant desert. It was an intense blast of heat coupled to other catastrophic events. Under this area there is an enormous reservoir of water below the parched desert. The only source for such a large deposit of water could only have been heavy rains from the fertile times. What caused this intense heat surge and when? Could this have been in the middle of the Second Millennium BC? Or was it the Eighth Millennium BC? How accurate are our radiation decay dating procedures anyway? What had caused the heat surge onto the area? Not meteorites or comets or asteroids as they do not do this. Something else was traveling with the cosmic bodies and it was generating intense amounts of heat as well as radiation.

The Wet'suwet'en Indians of northern British Columbia remember the land of Dzilke also known as Dimlahamid which was destroyed by earthquakes and then submerged under the sea, the Pacific Ocean. This was the home of their ancestors who after the Deluge arrived on Vancouver Island around 1600 BCE. The ocean had risen in one mighty swell and submerged the city. This myth was well known throughout North America. Was this a lost Pacific Ocean civilization?

Heavy falls of ash fell on the Nile Delta in Egypt killing off most plant life in Lower Egypt around 1600 BCE.

Other sources state that there was a mysterious mass death at Avaris during the XIII dynasty around this time period. Other large metropolitan centers were destroyed or abandoned at this time. These included Ithtaw and Hetepsenusret. Avaris was later renamed Ramses and the biblical Israelites were said to be enslaved in its reconstruction.

Greenland ice cores show exceptionally heavy ash falls around 1600 BCE and Greenland entered a cooling period around 1600 BCE.

What would have caused this?

There were no recent volcanic eruptions but there was one other factor that might be taken into consideration.

Iron spherules indicating cometary collision with the Earth have been found in Tunisian peat bogs in North Africa indicating that there was a possible cometary or asteroid impact in ocean shelf sediments. This is dated to around 1600BCE or the late seventeenth century BCE.

Between 1600 BCE-1495 BCE a dark age descended over Babylon that lasted one hundred years. At the same time the Amorite capital is sacked and the Mittani and Hurrian overlords lose their hold of the Assyrian Empire.

In Iraq stone monuments from this period show a new god represented by a glyph showing a circle with a starlike design at its center with wings on either side and a tail below. These were from the reign of the Assyrian king Enlil-Nasir 1 who was a contemporary of the Mittanian king Shaushtatar who introduced the cult of the flying disc in Mitanni in Syria. This symbol was later used by the Assyrians to represent Shamash, the Sun god. Originally though this symbol was associated with a god called Antum whose name meant "from Heaven". For a brief period, the Assyrians regarded Antum as their supreme god. This supreme god was introduced by Puzur-Ashur III, who was the Assyrian king before the conquest by the Mittanian king Shaushtatar so was not an introduced cult. In this same period the same cult of the single sun disc was appearing in Hattasus in Turkiye.

The Tuatha de Danaan were a race of Seapeople who invaded Ireland around 1600BCE. The Tuatha de Danaan came from a lost land that disappeared under the Sea (Atlantic) that was called Murias. In Wales Murias is called Morvo and in Normandy in France it was called Morois. The Nemedians arrived on the south coast of Ireland in this period according to legend.

Why include legends? Legends quite often are fragments of history.

The Chott el Djerid or Shott el Djerid is a dried up marsh section of Tunisia, once a bay of the Mediterranean and later the inland Trithonis Sea, Lake Tritonia, that had a citadel island in the middle. The central plain of Tunisia was once an island.

In 1590 BCE Fuss Peak on Paramushir Island in the Kuril Islands erupted with a VEI of 3.

In 1550 BCE Hayes Volcano in Alaska erupted with a VEI of 5. Ten cubic kilometers of tephra was ejected. It erupted a series of 7 or 8 closely related tephra beds. Also, in 1550 BCE Hekla in Iceland erupted with a VEI of 4 and Vesuvius in Italy erupted with a VEI of 4.

Between 1550 BCE and 1450 BCE Mount Etna erupted in Sicily with a VEI of 5.

The Tempest Stele was created around 1540 BCE to 1517 BCE. It was found reused in the third pylon in the Temple of Karnak in Egypt tells of an extraordinarily ferocious storm that occurred during the reign of Ahmose 1, circa 1550 BCE. The text describes low level winds carrying ash towards Egypt where the sky was darkened and there were terrible storms that struck awe into the Pharoah Ahmose himself who raised a memorial pillar at Thebes. "The Gods expressed their discontent…a tempest…caused darkness in the western region. The sky was unleashed more than the roar of the crowd……was powerful……..on the mountains more than the turbulence of the cataract of Elephantine. Houses and shelters were floating on the waters like the banks of papyrus outside the royal palace for……days, with no one able to light the torch anywhere. Then his majesty descended in his boat, his council following him. The people at the east and west were silent, for they had no more clothes after the power of the god was manifested. His majesty set about strengthening the two lands, providing them with silver, gold, copper, oil, clothing, all the things they desired after which his Majesty rested in the palace-life, health, strength."

Pumice stone which is a byproduct of volcanic eruptions has been found at Tell el Dab'a in Egypt which is dated to 1540 BCE. This pumice matches the composition of pumice from Thera. Had Thera had another eruption for my research indicates that it occurred in 1629 BCE. There are no volcanoes in Egypt.

Between 1530 and 1210 BCE Witori/ Pago Volcano in Papua New Guinea erupted with a VEI of 6. This would have caused a volcanic winter. The Smithsonian Eruption list shows this as 1370 BCE± 100 years.

Some sources state that Santorini/Thera erupted in 1520 BCE.

1500 BCE. Sub-Atlantic Period or Neoglacial Period.

Eruptions in 1500 BCE were Mount Etna in Italy with a VEI of 5, Ata in Japan with a VEI of 4, and Avachinsky with a VEI of 5 in Kamchatka where more than 3.6 cubic kilometers of tephra was ejected. Other eruptions were Sheveluch with a VEI of 3 in Kamchatka and Pico de Orizaba in Mexico with a VEI of 3. In 1500 BCE Chikurachki in Kamchatka erupted with a VEI of 4.

Ata in Japan also erupted in this period with a VEI of 4. Here are four volcanic winter producing eruptions in the one year and if you include Vaca Muerta in Chile then you have five volcanic winter causing eruptions.

Here are four volcanic winter producing eruptions in the one year.

The dates for the onset of the colder regime of the Sub-Atlantic Period or Neoglacial Period, around 1500 BCE started.

Neoglacial development began to make itself felt in earnest between about 1500 BCE and 1300 BCE with advances of glaciers in Alaska and in the Alps in Europe and perhaps the first beginnings of renewal in Colorado in the Rockies, in Scandinavia and New Zealand.

Around 1500 BCE there was a sudden and sharp cooling trend in North America though not in the north.

In the British Isles and Europe no new large monoliths were created internationally after this date. Prior to this the British Isles were blessed with sunny skies, calm seas and ample but not excessive rainfall. This was a climate conducive to sea trade, cosmic observations and exterior gatherings.

1500 BCE was a rainfall maxima period in Europe. These continued at 500-year intervals with lesser periods of rain at 200 year intervals until the time of Alexander the Great (356-323 BCE).

This was the sharply cooler Neoglacial Period when glaciers advanced in Alaska again as well as the Alps in Central Europe.

There was a growth of peat bogs in Britain.

There were large fluctuations of inundation of marginal land areas around Alpine lakes. This implies increased precipitation or rain. Overall temperatures were still warmer than in the 21st Century. These were small scale fluctuations. This was during a broadscale downturn in temperature from the warmth of the Bronze Age to the chillier late Iron Age.

Around 1500 BCE world sea levels suddenly fell sixteen to twenty feet. It was getting colder.

There was a eustatic depression in the Netherlands and in Eastern Europe. There was also a sudden increase of the ice caps and the resultant chilling of the Northern Europe. The increasing ice caps resulted in lower sea levels at the same time. All of this was sudden and devastating.

Remembering that Ice Ages or glacial periods can be fully formed in only thirty to fifty years according to a lot of research, then this would be frightening.

The North Sea is bordered by Scotland, England, the Low Countries, Germany, Denmark and Norway. In this period, it did not exist. The Thames River was a tributary of the Rhine River. The mouth of the Rhine was up near Aberdeen in Scotland. Then the land level here dropped! Numerous human artifacts have been found here and analysis of pollens indicated that forests existed here up until this period or at least until 1200 BCE. This was the same time as the destruction of Lake Dwellings in Central Europe. There must have been Mesolithic and Neolithic settlements here and the English Channel might not have existed then. This was the same period as the Bronze Age in Egypt and Phoenicia. The Phoenicians were already sailing to Cornwall in this period to purchase tin.

Like the Mediterranean Sea, does the English Channel have periods of disappearing?

Investigations of the coast of Maine in the northeastern United States indicate that Gulf Stream water regularly followed the coast as far north as Maine all through the warmest post-glacial times until around 1500 BCE. Since then, it has moved out into the Atlantic (as the warm saline North Atlantic Drift) from more southerly points on the American seaboard and has never renewed its dominance so far north. In other words, the Gulf Stream had moved south and away from the mainland.

Around 1500 BCE the previously arid region around Lake Titicaca in Bolivia had rainfall that increased considerably and the lake rose more than twenty meters in two centuries. Almost immediately sedentary agricultural villages appeared along its shores. With more rain, the risk of farming the previously arid Altiplano receded somewhat. The farmers cultivated dry fields, depending entirely on seasonal rainfall which produced low crop yields. They also experimented with raised field cultivation. The lake level now fluctuated considerably from decade to decade but never returned to the arid conditions that previously existed. For a thousand years the Altiplano supported farmers and alpaca herders but as their numbers and sophistication increased raised fields dominated the agricultural economy. There is evidence of sharp cooling in Central and South America around 1500 BCE.

Around 1500 BCE the two-thousand-year-old Neolithic civilization in Britain changed completely from that of a peaceful agrarian culture to that of one that focused on building fortifications and weapons of war. The old Neolithic sites such as Stonehenge and Avebury were abandoned, and no new ones were created. There is no excavated evidence for foreign invasion or massive climate change. All Bronze Age mining and working suddenly stopped in Britain around this time as well.

In the Northwest Territories in Canada cooler summers started around 1500 BCE. This was the same time as in Greenland. From 1500 BCE the forests rapidly retreated 200 to 400 kilometers from the northern limits that they had achieved around 4,000 BCE. This withdrawal was accompanied by extremely widespread forest fires, presumably started by lightning strikes on the dead wood and within about a century the whole zone had been covered by tundra.

The Vaca Muerta Crater in Antofagasta Province in Chile is three thousand five hundred years old. The strewn field is 11.5 kilometers long and 2.1 kilometers wide. The only crater found is 10.5 meters in diameter though up to six metric tons of meteoritic stony-iron material has been found here. Vaca Muerta is near Taltal in the Atacama Desert in Antofagasta Province in Chile. The meteorite was a stony iron that fragmented into 77 pieces that scattered over 22 square kilometers. The stony irons are believed to come from the outermost part of an asteroid's metal core where molten metal iron has been forced into the overlying layer of rock.

In China the early Chinese Yang-shao and Lung-shan cultures had been overwhelmed by a horse-riding people, invading from central Asia and possibly signaling an early stage of increasing difficulty there. This was probably from increasing aridity or colder winters or both. Bamboo retreated from northern parts of China, and the dates of rice and fruit harvests became later. The Han River, 33° North in China, froze. The climate and temperature were similar to those of the late twentieth century.

Around 1500 BCE earthquake and holocaust had destroyed Cyprus. This had also occurred throughout the Middle East.

Around 1500 BCE a flow of basaltic lava erupted from a fissure in the ground in Sinai in Egypt and set fire to forests leaving a desert behind. What created the fissure?

Around 1500 BCE there were regrowth phases in the peatbogs and fluctuations of the levels of lakes in and around the Alps in Europe. This affected the human settlements at their margins.

Pollen analysis from the fens, swamps, of Germany show massive and sudden climate change.

Though many people had settled northern Greenland including the Inuit, Eskimos, by 2,500 BCE, by 1500 BCE human activity declined. At the same time there was a general southward movement of early peoples traceable along the shores of Hudson Bay and Labrador. The climate was getting colder.

Around 1500 BCE there was the creation of an impact crater in the Great Rann of Katch in Western India. The crater is one kilometer in diameter and ages vary from four thousand years ago to 1500 years ago. The Luna Crater is near the village of Luna in the Great Rann of Katch on the Banni Plains in Gujarat in India.

Around 1500 BCE the Indian text the Rig Veda was composed shortly after the alleged Aryan invasion of the Indus Valley and the sudden destruction of its civilization. The Rig Veda is a series of hymns to the gods and the first hymn is devoted to Agni, the name of which means fire, who was the chief deity at the time, and he spanned the gulf between heaven and Earth as well as being the envoy of all of the other gods. The alleged Aryan invaders of Mohenjo-Daro and Harappa in Pakistan who destroyed the cities and slaughtered all of the inhabitants practiced human sacrifice to Agni who was described as the charioteer of the skies and the one who shines in heaven. It was alleged that when Agni was born, he filled the space between heaven and Earth with light and his head was in the sky and he had long flaming hair that resembled the tail of a horse. This might well be a description of a comet or asteroid. Was it describing this?

In Iran, then Persia, the cult of Zoroastrianism appeared around 1500 BCE. Ahura Mazda was depicted as the one god, the wise lord, who was represented by the Faravahar, a glyph depicting a human figure standing in a winged disc with a fan shaped tail spreading out from underneath it. The earliest depictions do not include the human figure inside the winged disc. Around this time earthquake and holocaust had destroyed Persia. This had also occurred throughout the Middle East.

In Iraq stone monuments from this period show a new god represented by a glyph showing a circle with a starlike design at its center with wings on either side and a tail below. These were from the reign of the Assyrian king Enlil-Nasir 1 who was a contemporary of the Mittanian king Shaushtatar who introduced the cult of the flying disc in Mitanni in Syria. This symbol was later used by the Assyrians to represent Shamash, the Sun god. Originally though this symbol was associated with a god called Antum whose name meant "from Heaven". For a brief period, the Assyrians regarded Antum as their supreme god. This supreme god was introduced by Puzur-Ashur III, who was the Assyrian king before the conquest by the Mittanian king Shaushtatar so was not an introduced cult. In this same period the same cult of the single sun disc was appearing in Hattasus in Turkiye.

Around 1500 BCE all Bronze Age mining and working suddenly ceased in Ireland.

Around 1500 BCE there was a massive lava flow in the Jezreel Valley in Israel that filled it in. The volcanoes in this area were supposed to have been extinct since prehistoric times. A Phoenician vase was found buried in the lava which helped date it.

Around this time earthquake and holocaust had destroyed Palestine. This had also occurred throughout the Middle East.

Around 1500 BCE the previously peaceful Olmec civilization was supplanted by a warlike Olmec culture. What had caused it to suddenly change? We have also seen other cultures revert from peaceful to warlike in this same period. What was agitating them?

During the mid-Second Millenium there was a massive climate disturbance at the end of the Bronze Age. Incidentally there are two dates for this, 1500 BCE and 1200 BCE. Opulent plenty was replaced by striking poverty. Pollen studies showed that there was a climatic catastrophe. This was called a Fimbul-Winter where snow continued through winter and summer uninterrupted for years. This was in Scandinavia and was probably the origin of Viking mythology.

Around 1500 BCE all Bronze Age mining and working suddenly ceased in Spain. The Morra, Motilla and Castellijo communities of the La Mancha bronze working culture were suddenly abandoned. Numerous bronze-working cultures suddenly disappeared in this period. Was there a disruption to the copper supply or was there a disruption to the cultures that supplied the copper? Isn't it odd that the European and Middle Eastern Bronze Age cultures suddenly stopped at the same time that all work stopped in the Copper mining regions around the Great Lakes in North America.

Around 1500 BCE records indicate that a new Supreme God had appeared in the pantheon of the Gods of the Mittani in Syria. Unlike the rest of the Mittanian gods who were depicted in anthropomorphic form like combination humans and animals, this new god was called "Ir", "the Bringer". The symbol for Ir was a pair of bird wings and tail surmounted by a circle. Within this circle there is an eight-pointed star indicating that the new god was identified with some form of heavenly body. This was at the same time as the cult of Aten suddenly appeared in Egypt with its aerial fiery disc.

Is the sudden appearance of religions based on flying discs actually refer to cometary or meteoric bodies? Cometary bodies can be seen for a long time and meteorites tend to be over in a few seconds.

Were all of these Mediterranean and Middle Eastern cultures hallucinating?

On the Syrian coast as well as inland there is a stratigraphic and chronological rupture between the strata of the Late Bronze Age at Qalaat-er-Rous, Tell Simiriyan, Byblos and the necropolises of Kafer-Djarra, Qraye and Majdalouna which suddenly ceased to be used. The site of Hama was interrupted at the same moment that the Middle Kingdom in Egypt suddenly ended.

Early legends state that the Dardanelles in Turkiye were opened up by an earthquake during the time of the Pelasgian king Phoroneus when the waters of the Mongolian Sea rushed through the depression of the Caspian and Black Seas. This was the flood of Deucalion, son of Prometheus. Phoroneus was the first man created, an Adam. This was allegedly around 1500 BCE. Like the Pillars of Hercules at the Gibraltar end of the Mediterranean which appears to have several openings and closings, the same appears to have occurred at the Bosphorus end of the Mediterranean.

At Alaca Huyuk in Turkiye the transition from the Middle Bronze Age to the Late Bronze Age was marked by upheaval and destruction.

Tarsus in Cilicia in southern Anatolia in Turkiye was destroyed and there was a vast reduction in population. Between the strata of the brilliantly developed civilization and the late Bronze Age is a layer of earth that is five foot thick without a sign of habitation in it.

At Alisar in Turkiye all life disappeared in this period. There was a period of extreme poverty and no trade.

Troy in Turkiye was destroyed and there was a vast reduction in population.

Around 1500 BCE all copper mining and working suddenly ceased in the United States. Some sources say that this was 1200 BCE There appears to be a 300-year error factor here. The same as in Europe and the Middle East.

What is this about copper mining and smelting and their cessation around 1500 BCE on both sides of the Atlantic Ocean coincidentally at the end of the Bronze Age. Bronze was created by using copper and a small quantity of tin to harden it occasionally with other materials such as arsenic.

There may have been several cataclysms, both localized and international.

In a Silk text called the "Mawangdui Silk Almanac" there is a reference to a huge comet seen in the sky over China in 1486 BCE approximately that had ten tails and to the Chinese this indicated the worst of portents. This incidentally was the first comet recorded by the Chinese who plotted all cometary positions to occur thereafter. The comet was named "Lao-Tien-Yeh", the

"Great God" and became the Supreme God of China for a very turbulent and warlike period thereafter. This was by King Tang in the beginning of the Shang Dynasty. Here is another version of the same event. Coincidentally with this appearance a new god appeared in China. Zi Cheng, the third king of the Shang Dynasty, overthrew their rulers the Xiaw with the help of this new supreme God called Lao-Tien-Yeh, the "Great God". This became the new ruler of the Universe, and its symbol was a circle with a series of straight lines radiating in a fan shape beneath it. This is very similar to the new Supreme God symbols that suddenly appeared in the Middle East and Egypt bringing a new monotheism into previously pantheistic cultures.

An odd coincidence from this period is that Atenism, the worship of Aten, the Sun Disc, which later became the official religion of Egypt, started in 1486 BCE. If it was the sun that was suddenly being worshipped as the supreme Deity, why did the Egyptians not just refer to it as Ra, the sun's traditional name and use the winged disc, the traditional symbol for it? This Atenist disc was a heavenly body that shone from above and sent down light. The literal translation of Aten is "Disc".

Other sources state that there was a mysterious mass death at Avaris during the XIII dynasty around this time period. Other large metropolitan centers were destroyed or abandoned at this time. These included Ithtaw and Hetepsenusret. Avaris was later renamed Ramesses, and the biblical Israelites were said to be enslaved in its reconstruction.

Various authorities such as Hevelius, Calvisius, Herlicus and Rockenbach place the date of the Exodus as 1495 B.C. Others place it around 1200 BCE.

These authors are cited by Hevelius in his Cometographia of 1668 CE where using original manuscripts he wrote that in 1495 BCE a comet was seen in Syria, Babylonia, India, in the sign Jo (Capricorn), in the form of a disc.

Incidentally in 1486 BCE the Egyptians recorded in the Tulli Papyrus that a great celestial disc of fire appeared in the sky over the Temple of Karnak and that the priests prostrated themselves in fear. This fiery disc was so unique that it was recorded as never having been seen before and was one rod in length by one rod in width and it extended to the limits of the hour supports of the heavens. The Egyptian rod was 170 feet. The Egyptians used to say that the Full Moon was five cubits wide and the rod was one hundred cubits long. This was enormous. The Tulli Papyrus also stated that the aerial object broke up after several days had passed and became more numerous in the sky. "In the year 22, in the third month of winter, in the sixth hour of the day, the scribes of the House of Life noticed that there was a disc of fire coming from the sky". "The hearts of the scribes became terrified and confused, and they lay themselves flat on their bellies". The fiery disc was described as " a marvel never before known since the foundation of this land." The Tulli papyrus states that the fiery disc divided up into smaller fiery discs until after a few days they became more numerous in the sky. The fragments then grew in size and shined in the sky more than the brightness of the sun. Fish and birds were also said to have fallen from the sky at this time. Ancient cultures often referred to meteors as birds or fiery fish, the firebird being a name for them amongst the ancient Greeks.

The disc of fire was recorded as appearing early in the year that Tuthmosis overthrew his mother around 1486 BCE. This was the year before the advent of Atenism.

An almost identical glyph as that depicting Antum of the Assyrians showed up in Hattasus in Turkiye around 1486 BCE. This symbol depicts a star surrounded by a circle with wings each side and fan-tail shaped tail beneath.

In 1460 BCE Taupo volcano in New Zealand erupted with a VEI of 6. Seventeen cubic kilometers of tephra was released. This is definitely volcanic winter material. This was from Horomatangi Reefs.

In Iraq stone monuments from 1458 BCE show a new god represented by a glyph showing a circle with a starlike design at its center with wings on either side and a tail below it. These were from the reign of the Assyrian king Enlil-Nasir 1 who was a contemporary of the Mittanian king Shaushtatar who introduced the cult of the flying disc in Mitanni in Syria. This symbol was later used by the Assyrians to represent Shamash, the Sun god. Originally this symbol was associated with a god called Antum whose name meant "from Heaven". For a brief period, the Assyrians regarded Antum as their supreme god. This supreme god was introduced by Puzur-Ashur 111, who was the Assyrian king before the conquest by the Mittanian king Shaushtatar so was not an introduced cult. In this same period the same cult of the single sun disc was appearing in Hattasus in Turkiye.

In 1450 BCE Arenal in Costa Rica erupted with a Vei of 4. Taveuni in Fiji also erupted that year with a VEI of 2.

Around 1450 BCE the fields of northern Crete were rendered barren for over a decade by a fall of volcanic ash. Some researchers put the eruption of Thera at 1450 BCE.

Coincidentally the sacred cave of Arkalochori in Crete had its roof collapse in 1450 BCE. Masses of Bronze Age bronze, silver and gold artifacts were found here including many weapons such as labrys axes or double-sided axes. Current consensus is that Santorini erupted around 1629 BCE.

There were eight volcanic eruptions in Japan in 1450 BCE but none recorded near Crete or anywhere in the Mediterranean though of this we are not sure as there. There is an unnamed source shown for 1454 BCE and another unknown source for 1459 BCE.

Was there an unfound 1450 BCE Mediterranean eruption?

Some researchers stated that massive tidal waves came from the north, the direction of Thera (Santorini), and devastated many towns on the north cost of Crete. At various sites in Crete the remains of Minoan plaster, pottery and food as well as fossilized seashells and marine animals were found in layers of residue up to seven meters higher than sea level. The shells and pebbles could have only been scooped up from the seabed and then dumped on the shores. The Palace of Knossos was destroyed suddenly. The port of Palakaistro on the east of Crete had its town walls, which originally faced the sea, destroyed or missing completely. Parts of these walls were pushed fifteen meters above sea level and several hundred meters inland. Indications are that the tidal wave that hit the north coast of Crete in 1450 BCE was 26 meters high.

All the palaces except Knossos in Crete were burned. Other sources state that they were all destroyed by fire, including Knossos. Others state that it wasn't.

Welcome to the puzzles of history and geology. Another date for the Santorini or Thera eruption! Why not try 1450 BC? With corroborating evidence? Who is right? Which dates are right? What do we do with Thera? Were there several eruptions of this volcano in this period?

The explosion of Thera is regarded as one of the greatest volcanic eruptions in history and allegedly changed the nature of civilization around the Mediterranean. The main cone of the circular island was thrown up into the air to an immense height leaving a crater four hundred meters deep. It was estimated that one hundred and fifty cubic kilometers of stone was blasted into the air.

In 1956 two prominent geologists, Dragoslav Ninkovic and Bruce Heezen, surveyed the seabed around the island of Santorini from their survey ship the "Vema" and ascertained that the exact size of the volcanic crater, thirty square miles, indicated that a massive explosive volcanic event had occurred. The explosion was 6,000,000 kilotons sending a mass of material that was seventy cubic miles in size that erupted skywards. The famous Krakatoa eruption of August 1883 was only 1000,000 kilotons. The atomic bomb that destroyed half of the Japanese city of Nagasaki was only 20 kilotons.

The "Vema" survey indicated that pumice and volcanic debris covered the seabed only to the southeast of the island of Santorini, in the direction of Egypt. The winds would have carried the ash and debris cloud five hundred miles to Egypt where it could have been responsible for the plagues of Egypt and the other phenomena reported in this period.

Some historians state that the time of the Biblical Exodus was around 1450 BCE. Seeing it took forty years to get from Egypt to the Promised Land, though it should have taken eleven days, this date would be anything from 1490 BCE to 1410 BCE. This was during the reign of Amenhotep III who lived from 1479 BCE to 1425 BCE.

In 1440 BCE Katla in Iceland erupted with a VEI of 4. That same year Nasudake in Japan erupted with a VEI of 2.

In 1430 BCE Vesuvius erupted in Italy with a VEI of 4.

1400 BCE

In 1400 BCE Karymsky in Kamchatka erupted with a VEI of 2.

There was a massive eruption of lava on the Sinai Peninsula around 1400 BCE which burned down forests leaving a desert behind. Haven't we met this same disaster before but with a different date?

Between 1400 BCE and 1300 BCE in Palestine the cities of Beth-Shan and Ai were suddenly laid waste. At Beth-Shan there was an accumulation of debris one meter thick. The same was also found at Tell el Hesy. At Beth-Mirsim in Palestine there was a noticeable interruption to the habitation at the site.

Between 1400 BCE and 1300 BCE the cities of Ugarit, Byblos, Chagar Bazar, Tell Brak and Tepe Gawra were suddenly laid waste in Syria (including what is now Lebanon).

At Ras-Shamra, Ugarit, in Syria a massive sea tide or tsunami broke onto the land bringing destruction in its wake.

In Anatolia in Turkiye the cities of Alaca Huyuk, Tarsus and Alisar were suddenly laid waste between 1400 BCE and 1300 BCE.

What was causing this mass destruction?

At Jericho there had been earth tremors, and the city had been destroyed. The great wall surrounding it fell in an earthquake.

Crete in the fourteenth century BCE was the home of a seafaring nation called the Minoans who were one of the wealthiest cultures in the Mediterranean. The eruption of Thera was enough to destroy this culture completely as Crete was only seventy miles south of Thera. The firestorm that was heading southeast towards Egypt would have passed over Crete destroying everything in its path. This would not be the only danger to pass over though. In the 1930s the Greek archeologist Spyridon Marinates discovered a villa on the north coast of Crete where the walls had been strained outwards in a curious way. Large upright stones were prised out of position as if being attacked by a massive external force. This suggested that this event was caused by a massive backwash from a tidal wave that had indeed drowned the harbor town. This was in Amnisos, the harbor town of the Minoan capital of Knossos. In the ruins Marinatos found an Egyptian artifact that dated the eruption and tidal wave to the reign of the Egyptian Pharoah Amenhotep III who ruled from around 1385 BCE to 1360 BCE. Numerous towns and villages on the north coast of Crete had been totally destroyed. This artifact was an alabaster jar discovered in the 1930s. In 1999 the Greek archeologist Kristos Vlachos examined this jar and discovered that the inscription on it referred to the thirty-third year of the reign of the father of Amonhotep, Tuthmosis IV. It was the last year of Tuthmosis reign which would have been around this period. The problem being that the Egyptian Kings List is notoriously difficult to ascertain as to actual dates though many place this date to be around 1400 BCE. The eruption of Thera then must have been later. Or there might have been several of them!

In 1370 BCE Witori/Pago in New Britain in Papua New Guinea erupted with a VEI of 6. Thirty cubic kilometers of tephra was released. This would cause a volcanic winter.

1360 BCE. Welcome to the puzzles of history and geology. Another date for the Santorini or Thera eruption. With corroborating evidence. Who is right? Which dates are right?

What do we do with Thera?

Both sets of dates are accepted. Were there two eruptions of Thera or an island nearby? They can disappear under the sea pretty quickly at times.

What else occurred in this period?

This is another date for the Santorini or Thera eruption. With corroborating evidence. Who is right? Which dates are right? The volcanic island of Thera, otherwise known as Santorini in the Aegean Sea erupted around 1360 BCE. Previously it was determined that Thera had been buried around 1675 BCE as well as 1450 BCE. Were there two eruptions of Thera or an island nearby? They can disappear under the sea pretty quickly at times. What else occurred in this period? Other sources state that Thera erupted in 1220 BCE contemporary with the events depicted on the temple of Medinet Habu in Egypt. 1629 BCE seems to be the leading date so far.

Several scholars place the date of the Exodus of the Hebrews to 1360 BCE. The dating of Exodus is equally as controversial as the dates in the Egyptian Kings List which is still controversial as is also the dating of the Thera eruption.

Akhenaten, the so-called heretic pharaoh of Egypt who introduced the monotheistic cult of Atenism, ruled from 1352 BCE to 1360 BCE. Both sets of dates are accepted. Though there are controversies over the Egyptian Kings Lists as well with wide discrepancies in dates. We might have to allow for overlap periods of several years?

This general 200 year period at the end of the Bronze and into the Dark Ages is almost impossible to decipher. Civilization in the eastern Mediterranean had crashed literally.

1352 BCE was when Amenhotop III, the father of Akhenaten, erected numerous states of the Goddess Sekhmet, the Goddess of Destruction, throughout Egypt. The Temple of Maat was also reconsecrated to Sekhmet who was always depicted as a fiery Lion-headed body flying through the sky. Was this a comet or a large asteroid or meteorite?

In 1352 BCE the explosion of Thera is regarded as one of the greatest volcanic eruptions in history and allegedly changed the nature of civilization around the Mediterranean. The main cone of the circular island was thrown up into the air to an immense height leaving a crater four hundred meters deep. It was estimated that one hundred and fifty cubic kilometers of stone was blasted into the air. Other sources state that Thera erupted in 1220 BCE contemporary with the events depicted on the temple of Medinet Habu in Egypt.

Interestingly enough, in 1351 BCE Amenhotep III had thousands of statues of Sekhmet, the lion-headed goddess of devastation installed across Egypt in this period. The reign of Amenhotep ended around the beginning of that of Akhenaton. It was at the end of the reign of Amenhotep III that Avaris, later called Ramesses, was being rebuilt by the Israelites. Sekhmet was the daughter of Ra the Sun god, and she had almost annihilated mankind in the remote past and natural disasters in Egypt were attributed to her. Sekhmet had obscured the sun and rained down fire from heaven. Humanity was saved only through the personal intervention of her father Ra, the Sun God. What sort of natural disaster occurred that resulted in Amenhotep creating more statues of Sekhmet than any other Egyptian god or goddess?

In the last year of Amenhotep III's reign a new temple to the Chief Goddess Mut was suddenly reconsecrated into a temple to the Goddess of Destruction named Sekhmet. Inscriptions also stated that by decree Sekhmet should replace Mut as the principal Goddess. Statues of the goddess Sekhmet virtually became the most common statue of a goddess in Egypt. No other deity of Egypt was represented by so many large-scale statues almost all of whom were erected by Amenhotep III.

Amenhotep III started a massive new building project in Avaris in which the people of Habiru were forced to work. The Habiru had been an enslaved Semitic people since the reign of Tuthmoses III who regarded them as enemy aliens and thusly they were enslaved and put into work gangs. Were the Habiru actually the Hebrews? Tuthmoses III was fighting Semitic tribes in Canaan at the time. Avaris was much later named Ramesses around 1290 BCE. Amenhotep III died around 1349 BCE though there is controversy over this date due to irregularities in the Egyptian Kings List.

In the middle of all of this Amenhotep IV ruled from 1353 BCE to 1336 BCE. Amenhotep IV changed his name to Akhenaten and established a brief period of monotheism in Egypt after the fifth period of his reign. Aten established a religion based on the Aten. Not to be confused with the Sun, which was called Ra by the Egyptians. The Aten was a glowing mass in the sky and was not even represented by the sun symbol. What was the Aten then?

In 1350 BCE Mount Fuji or Fujisan in Honshu, Japan, erupted with a VEI of 5. Also, in 1350 BCE Avachinsky in Kamchatka erupted with a VEI of 5. More than 1.2 cubic kilometers of tephra was ejected. Other eruptions were Asosan with a VEI of 3 and Tokachidake with a VEI of 3 in Japan, and Krasheninnikov with a VEI of 3 in Kamchatka.

In 1345 BCE Quetrapillan in Chile erupted with a VEI of 3.

In 1330 BCE Sheveluch in Kamchatka erupted with a VEI of 3.

In 1320 BCE Yucamane in Peru erupted with a VEI of 5.

In 1996 radiocarbon tests at the center for Isotope Research at Groningen University in the Netherlands determined from the study of six separate samples of cereal grains found in the burned layer of the citadel of Jericho that it was destroyed around 1320 BCE. Was this the date of Joshua's alleged destruction of the city? We have already had an earlier date for the destruction in 1630 when it was destroyed by an earthquake. Is our dating in this period interfered with by massive radiation surges from the Aten Comet?

In 1310 BCE Mount Tarawera in New Zealand erupted with a VEI of 5. This would have caused a volcanic winter. 5 cubic kilometers of tephra was ejected. Cerro Bravo in Colombia erupted with a VEI of 4 and Soufriere Guadeloupe in the Lesser Antilles erupted with a VEI of 3.

1300 BCE. End Bronze Age. Invasion by Sea Peoples

In 1300 BCE Vulcano in Italy erupted with a VEI of 3 along with Taveuni in Fiji with a VEI of 2.

Between 1300 BCE and 300 BCE Yantani Volcano erupted in southwest Alaska with a VEI of 5.

Around 1300 BCE a sharp cooling phase arrived in California as witnessed by bristlecone pines. This time there was no full and lasting recovery at any time since. Most if not all of the glaciers present in the late twentieth century in the United States Rockies south of the Canadian border are believed to have formed since 1500 BCE.

Between 1300 BCE and 900 BCE there were increases in temperature by 1° C overall though with an increase of 2° C in Europe on average.

During the thirteenth and twelfth centuries BCE there were onslaughts by people called the Sea Peoples in the Mediterranean.

In Central Europe there were catastrophic floods that drowned lake villages that had been there since 6,000 BCE. The sites were abandoned after the flooding.

Between 1300 BCE and 1200 BCE the Mycenean city of Tiryns in the Argolid of Greece was devastated by a massive flood. Parts of the lower town were buried under several feet of mud. After this event the inhabitants built a dam above the citadel as well as converted the river so that this would never happen again. Was this a local flood or part of something much larger in this period? The rest of the Mycenean, which was a late Bronze Age culture was imploding at this same time. The flood debris was up to sixteen feet thick! Later in this same period Tiryns was destroyed by an earthquake as was Mycenae, the center of Mycenean culture in Greece. The flash flood at Tiryns came not from a tsunami but was transported downhill by a stream on the slopes east of the citadel. This was probably caused by an earthquake. After this event the inhabitants of the city built a thirty-three-foot-high dam built of cyclopean blocks and then dug a one-mile-long channel to divert the river away. This Achaean dam is still standing over three thousand years later.

There was a great eruption of Bronze Age people from the Hungarian plains soon after 1300 BCE. Was this due to flooding? This was the same time that the Phrygians charged into Asia Minor. Was there a climatic disturbance further east at this time?

At the ancient city of Hazor which is now Tell-el-Qedah in Israel, which is nine miles north of the Sea of Galilee, the remains of an ancient fortified palace were found that had been destroyed by fire around 1300 BCE. Broken Mycenean or early Greek pottery was found in the level of destruction from this period. This type of pottery was not exported to the Middle East after the thirteenth century BCE. The destruction of the city had been intentional as statues and temple decorations had been deliberately defaced.

Around 1300 BCE there was an influx of people from the north into Italy. These were probably the ancestors of the Etruscans and the Romans.

Or was there an impact event?

Remember, that dating in this period is extremely confusing.

Between 1300 BCE and 300 BCE Yantani Volcano erupted in southwest Alaska with a VEI of 5.

In 1290 BCE Agua de Pau erupted on San Miguel Island in the Azores with a VEI of 4.

There was a recession of hazel and lime trees and the simultaneous spreading of hornbeam in southwestern Poland between 1270 BCE and 1160 BCE.

In 1253 BCE Aguilera in Chile erupted with a VEI of 5.

In 1250 BCE Arenal in Costa Rica erupted with a VEI of 4. Other eruptions were Iwatesan with a VEI of 3 and Miyakejima with a VEI of 3 in Japan. Taupo in New Zealand also erupted with an ejection of 0.05 cubic kilometers of tephra from Te Kohaiakatu Pt.

In 1250 BCE a Hittite queen from Anatolia in Turkey wrote to the Egyptian king Ramesses II stating that she had no grain in her lands. There must have been famine there for shortly after this a Hittite trade embassy arrived in Egypt in order to procure barley and wheat for shipment to Anatolia.

Around this time the Hittites abandoned the Anatolian plateau.

Between 1250 BCE to 1175 BCE a catastrophic earthquake ripped much of the North African shoreline, tumbling cities into the Mediterranean Sea. Diodorus Siculus had written about this, and it was later confirmed by geology. It has been proposed that a ripple of earthquakes had also occurred in Greece, Egypt and the Middle East and then into Turkey.

On top of the Sacred Way which was built by Ramesses II from his palace in Luxor to the Temple of No Amun in Karnak, Egypt, and along which his funeral procession passed in 1232 BCE traces have been found of a layer of black volcanic ash three to four millimeters thick and above that a layer of red volcanic ash several millimeters deep. This volcanic ash had fallen here shortly after 1232 BCE. Where is the nearest volcano to Egypt? Santorini or Arabia? Had a still undiscovered undersea volcano in the Mediterranean appeared, erupted and then disappeared?

Or as I asked earlier, was this because of an impact event?

In 1230 BCE Villarica in Chile erupted with a VEI of 4.

In 1229 BCE Pharoah Merneptah of Egypt defended Egypt against the Hanebu, or Sea People. The Hanebu returned forty years later during the reign of Ramesses III, Ramses III.

Who were the Hanebu also known as the Sea People?

In the Karnak manuscript from Egypt, it states that in the fifth year of the reign of the Pharoah Merneptah, 1232-1222 BCE, Libya, North Africa, had become a barren desert and the Libyans came to Egypt to seek sustenance for their bodies. This would be 1227 BCE.

From 1225 BCE to 1175 BCE much of the Aegean and eastern Mediterranean was rocked by a series of earthquakes. This is called an earthquake storm in which a seismic fault keeps unzipping by unleashing a series of earthquakes over years or decades until all the pressure along the fault line has been released.

The remains of the city at Ugarit, now Ras Shamra, on the Syrian coast are nine thousand years old, dating from at least 7,000 BCE. It is situated only 12 kilometers north of the modern port of Latakia. The ruins of Ugarit have been vitrified. This is where the brick or stone has been fused and develops a glasslike glaze. It requires such extreme heat that its origins are unknown. The People of the Sea finally destroyed the city in 1200 BCE. A baked clay text from Ugarit in Syria stated that the star Anat fell onto the Syrian land, set it on fire and confusing the two twilights. "The star Anat has fallen from heaven: he slew the people of the Syrian land and confused the two twilights and the seats of the constellations". Ugarit was the capital of Syria and was destroyed by violent earth tremors and sudden outbreaks of fire around 1200 BCE. Underneath the layer of ruins was a long sword bearing the seal of the Egyptian pharoah Merneptah, who died about 1222 BCE, while above the burnt remains was Philistine pottery. This indicates that the final destruction occurred here between 1222 BCE and 1200 BCE. This catastrophe stratum has been found across Syria as well as the Middle East as well as in Asia Minor. Numerous cities and towns were destroyed by earthquakes as well as intense heat. There is no mention of wars or invaders though.

In 1220 BCE Katla erupted in Iceland with a VEI of 3.

The level of the Caspian Sea suddenly subsided around 1220 BCE. Russian scientists have found the ruins of a mysterious submerged city here that would date from before this period.

James Mavor of the Woods Hole Oceanographic Institute reckoned that there were three waves with each having a height of two hundred feet even when they were hitting the shores of Crete when Thera erupted in 1220 BCE. W. Brandstein deduced that the eruption of Thera produced such an enormous wave that the flood reached the capital Knossos eight kilometers inland and at an elevation of forty meters above sea level and destroyed it. Brandstein also stated that a catastrophe that could destroy Crete with all of its towns and cities must have also caused damage at Athens in Greece as the epicenter was midway between Athens and Crete.

Around 1220 BCE there were numerous sudden volcanic eruptions in the Sinai Peninsula in Egypt.

Also, around 1220 BCE there were major earthquakes in Egypt and the Temple of Karnak was damaged.

At this point there is controversy. Several dates have been proposed from 1486 BCE to 1232 BCE. This is confounded even more by the equally great controversy over the Egyptian King Lists and the eternal disputes over their dates as well.

In Exodus 1:11 it was recorded that the Children of Israel built for Pharoah treasure cities Pithom and Ramesses. Both of these cities were built under Ramesses II, the Great. Exodus then states that the king of Egypt died, and it was Ramesses II who died after a seventy-two-year reign in 1213 BCE. After his death the plagues of Egypt occurred in which the Israelites made their escape. Was Ramses II actually the king of Egypt that they are referring to?

The Papyrus Ipuwer states that the land turns around as does a potter's wheel, and the Earth was turned upside down. There was terrible devastation. The Ipuwer Papyrus according to Egyptologists S. Morenz, J. Leiden and C. Baux was dated to between 1220 BCE and 1205 BCE.

The book of Exodus as well as the writings of Ipuwer, an Egyptian eyewitness of the events, records that the river, the Nile, is blood. These ancient witnesses stressed that the water appeared bloodlike, not dust covered, as they would well have known the difference. Ipuwer also wrote that plague is throughout the land and blood is everywhere. Exodus states that the river (Nile) stank, and the water could not be drunk. The skin of men and animals was irritated by the dust which caused boils, sickness and the death of cattle. Wild animals that were frightened by the events came closer to the towns which would account for the plagues recorded in this period as well. The red dust covered everything.

Was this red dust part of a cometary shower of rusty pigment? Many meteorites are iron chondrites but how would they rust in space and then fall as dust?

After the fall of red dust there was a small dust like the ashes of a furnace coming down. This was followed by the fall of a grievous hail, such as had not been seen in Egypt since its foundations. The Universal Cataclysm? Stones of barad, barak, translated as hail, fell. Jewish Mishradic and Talmudic sources state that the stones were hot and, in the Scriptures, they were described as mingled with fire. Their fall was accompanied by loud noises, *kolot*, that was rendered as thunder. The word for thunder though is *raam*, which was not used here. The din of the falling stones terrified the people as much as the stones themselves. The red dust had frightened the people and there was a warning to keep men and cattle under shelter as the hailstones would come down and kill them. This is referred to in Exodus. Ipuwer wrote that the falling stones and fire made the frightened cattle flee. Trees are perished, and no fruits or herbs are found. Grain has perished on every side. In one day, the fields were turned into wasteland. The book of Exodus wrote that the hail, stones of *barad*, smote every herb of the field and brake every tree of the field. Ipuwer stated that the gates, columns and walls are consumed with fire. The fire was to the end of heaven and to the end of the earth. Ipuwer stated that the southern ship, Upper Egypt, is adrift. The towns are destroyed. Great and small say "I wish I were dead!" The papyrus mentions that the residence is overturned in an hour and there are long lists of places and buildings that have been destroyed. At the end it states that Mankind is destroyed… all these years are confusion… thou hast kept alive a few among them but they cover their faces for fear of the morrow. The eighth plague of Exodus was of *barad* and fire mixed together and there were loud noises and *barad* and the fire ran along the ground. The papyrus Ipuwer states that gates, columns and walls are consumed by fire. The sky is in confusion. The fire almost exterminated mankind.

The Jewish Midrashim in a number of texts states that naphtha, together with hot stones poured upon Egypt. The Egyptians refused to let the Israelites go and He (God) poured out naptha, petroleum, over them, causing burning blisters. It was a stream of hot naptha.

According to the Apocrypha and the Pseudepigrapha of the Old Testament, the water that quencheth all wrought the fire more mightily.

Numerous Rabbinical sources state that an exceedingly strong wind endured for seven days and all the time the land was shrouded in darkness so dense that people could not stir from their place. The darkness was such that artificial light or flame could not be seen. Their eyes were blinded by it and their breath choked. Rabbinical tradition states that the vast majority of the Israelites perished. After the hurricane had finished, the Pharaoh pursued the evildoers, Israelites, to the place called Pi-Khiroti. In Exodus the Egyptians pursued them and overtook them encamping by the sea at Pi-ha-khiroth. Here the Pharaoh perished in the whirlpool or sea returning.

Ipuwer states that the children of princes are dashed against the walls. The children of princes are cast out in the streets. The prison is ruined.

The tenth plague was a great earthquake that destroyed the palaces, prisons and houses of Egypt. Each house received a fatality. The houses were smitten.

All the while according to Artapanus, quoted by Eusebius, where he states that the hail still fell so that those who fled from the earthquake were killed by the hail. All of the houses fell and most of the temples. The population fled and lived in fields. The Israelites as well as others fled as the lightning lightened the world and the earth trembled and shook. There was a portent in the sky which looked like a stretched arm. Was this a Comet?

Exodus states that the lightnings lighted the world. The earth trembled and shook. Thou leddest thy people like a flock by the hand of Moses and Aaron. They followed a portent in the sky that looked like an outstretched arm.

According to the Midrashim the seventh plague was the plague of *barad, barak*, which was a plague of earthquake, fire and meteorites. The structures that were erected by the Israelites in Pithon and Ramesses collapsed or were swallowed up by the earth. An inscription dating from the beginning of the New Kingdom refers to a temple of the Middle Kingdom that was swallowed by the ground at the end of the Middle Kingdom.

At midnight all the houses were shook, smitten, and there was not a house where there was not one dead. According to the Midrashim the last night in Egypt was as bright as noon on the day of the summer solstice. This happened on the night of Passover, the fourteenth of Aviv, the first month as it was celebrated originally. The first month of the Egyptians was called Thout and the thirteenth of Thout was a day of sadness and a very bad day for it was the day that Horus waged war with Seth. The Egyptians reckoned the beginning of the day from sunrise whereas the Hebrews count the new day from sunset. This means that the 13th of Thout and the 14th of Aviv were the same day. Interestingly enough in the Aztec calendar the thirteenth day of the month is called olin, motion or earthquake, when a new sun initiated another world age.

The population fled. Tents are what they make like the dwellers of the hills according to Ipuwer. The Book of Exodus describes a hurried flight from Egypt on the night of the tenth plague. A mixed multitude of non-Israelites left Egypt together with the Israelites who spent their first night in *Sukkoth* or huts.

Later in Exodus we have the description of the sea being rent apart for the escape of the Israelites. "The pillars of the Heavens trembled….and He divideth the sea with his power. He divided the sea and caused them to go through and he made the waters to stand as a heap. Then the Great Sea (Mediterranean) broke into the Red Sea in an enormous tidal wave."

Was this a tsunami retreating back into the sea before returning to the shore?

On the way from Egypt to Palestine the Israelites were at the site of Lake Serbon where Typhon and Zeus-Jupiter were said to have warred and where Typhon descended to Earth into the lake in Greek mythology. Was Typhon the pillar of cloud by day and the pillar of fire by night that the Israelites were following to the Promised Land?

Apollodorus states that the summits of mountainous Thrace in Greece received the name Haemus and that there was a tradition that the summit was so named because of the stream of blood which gushed out on the mountain when the heavenly battle was fought between Zeus, the hurler of thunderbolts, and Typhon was struck by a thunderbolt.

When Pharaoh and his army were destroyed by the rush of water many Israelites were destroyed as well by the onrush. At the same time God sent against the Djorhomites who inhabited the Tihama, the thousand-mile-long coastal region of the Red Sea in Saudi Arabia, swift clouds, ants and other signs of his rage and many of them perished and in the land of Djohaina an impetuous torrent carried off all of them in a night at a place called Idom which means Fury. In the Kitab Alaghani it is mentioned that at this same time a tribe was forced to migrate due to a plague of ants from Hedjas, in the Tehama, to their native land where they were destroyed by Toufan, which is a deluge.

The pharaoh according to the Shrine of El Arish pursued the evil-doers to the place called Pi-Khiroti. Exodus states that the Egyptians pursued the Israelites, all the horses and chariots of Pharaoh……and overtook them encamping by the sea beside Pi-ha-kiroth. The shrine of El Arish also stated that the Pharaoh fought with the evil-doers in this pool, the place of the whirlpool. The evil-doers prevailed not over his Majesty who leapt into the place of the whirlpool. Exodus states that the horse of Pharaoh went in with his chariots and with his horsemen into the sea and the Lord brought again the waters of the sea upon them. On the shrine of El Arish, it was written that the land was in great distress and misfortune fell upon the whole earth. There was a terrible uproar in the capital and for nine days none could leave the palace. During these nine days of storm there was such a tempest that neither men nor gods could distinguish the faces around them. Incidentally on the shrine of El Arish the king Typhon is called Tawi-Thom who reigned only for a short time after the death of Merneptah, 1213 BCE to 1203 BCE, and how in following rebels, some believe these to be the children of Israel, was drowned in Lake Serbonis, which is now called Sebchat-Berdawil, east of Port Said in Egypt. In Egypt the El Arish inscribed shrine stated that there was a deep prolonged darkness and a pharaoh had died. An heir of this pharaoh was also injured in this period by heavenly fire that had killed his entourage.

Incidentally on the shrine of El Arish on the Mediterranean coast east of Lake Serbonis, the king Typhon is called Tawi-Thom who reigned only for a short time after the death of Merneptah, 1213 BCE to 1203 BC, and how in following rebels, some believe these to be the children of Israel, was drowned in Lake Serbonis, which is now called Sebchat-Berdawil, east of Port Said in Egypt. This lake is also called Lake Serbon. Lake Serbon is also called the Serbonian Bog, an area of wetland in a lagoon lying between the eastern Nile delta, the Isthmus of Suez, Mount Casius or Ras Kasaroun, and the Mediterranean Sea in Egypt. Ras Kasaroun or Mount Casius according to Herodotus marked the boundary between Syria and Egypt. Remember that Israel was not a country in this time.

The mire at the bottom of the Sea of Passage was heated by the Pillar of Fire to boiling point. The pillar of fire and smoke that the people of Israel were following was also capable of leveling mountains.

Approximately seven weeks after the night of the Exodus there was the day of Revelation at the Mount of the Law Giving, possibly Mount Sinai. Exodus describes more fiery hail and coals of fire coming down from the sky as well as earthquakes and shakings and trumpetings coming from the sky. The mountain is described as burning though it is not volcanic. Mount Sinai is around 300 miles south of the Serbonian Bog and 49 days is six miles per day.

The Mount of the Law Giving was quaking so badly according to the Midrashin and the Talmud that it appeared as if it were lifted up and shaken above the heads of the people. The Pillar of the Lord was directly overhead. Then the Earth shook and trembled and the foundations of the hills moved and were shaken because the Lord was wroth. The Lord bowed the Heavens and came down and darkness was under his feet. At the brightness that was before him thick clouds passed as well as hail stones and falling coals of fire. The Lord thundered in the heavens and shot out lightnings, then the channels of water were seen, and the foundations of the world were discovered. The Lord caused the deeps to tremble and the spheres were alarmed.

The big controversy here is that no one actually knows where Mount Sinai was with candidates ranging from Mount Sinai on the Sinai Peninsula to Jebel al Lawz in Saudi Arabia with its strangely blackened peak. Jabal al Lawz is in Midian in Saudi Arabia. Jabal Maqla in Saudi Arabia is also suggested as Mount Sinai as it has a distinct blackened peak the cause of which is unknown there are clear marks here of the top of the mountain being scorched.

The Sinai Desert is today covered with blackened stones that are scattered across the landscape and for which there is no known reason unless it is the Biblical one. At Har Karkom, Jebel Ideid, a sacred mountain community dating back to 3,000 BCE, there are many boulders here several feet in size that have been mysteriously blackened and on which numerous ancient travelers have etched signs and symbols. The rocks are blackened only on the surface. Har Karkom means Mountain of Saffron and is in the southwest Negev Desert, halfway between Petra and Kadesh.

On the large mountain plateau of Har Karkom, also in Sinai, the top is covered in an expanse of black stone fragments known to the locals as hamada. In some places in ancient times the hamada layer has been cleared to form what are called hut circles. The blackened rocks in the Sinai resemble volcanic rocks but there are no volcanoes in the area.

Exodus mentions that the Israelites went through the land of Midian after crossing the Red Sea. There are volcanic fields here called the Older Harrats such as Harrat Khaybar and Harrat Rahat that cover parts of the western Arabia Plate. The land of Midian is on the right side of the Gulf of Aqaba.

As the Israelites went from Mount Sinai into the desert they were covered by clouds and the light of Noga was upon them. Noga is the planet Venus, and it must have been extraordinarily bright to be seen through the cloud cover. This is corroborated by the Ipuwer Papyrus which stated that men look like gem-birds, blackbirds, and squalor is throughout the land. There are none whose clothes are white in these times, and all are laid low by terror.

Three days journey away from the Mountain of the Law Giving, Mount Sinai, and it happened that the fire of the Lord burnt among them and consumed them that were there in the uttermost parts of the Israelite camp. The remaining Israelites continued on their way. Then came the revolt of Korah and his confederates and the earth opened her mouth and swallowed them all up. All Israel that was around them fled at the cry of them. There came a fire from the Lord and consumed the two hundred and fifty men who had offered incense. When they kindled the fire of incense, the vapors which rose out of the cleft in the rock caught the flame and exploded. Unaccustomed to handling this substance, which was volatile, the Israelite priests also fell victim to it. The two elder sons of Aaron, Nadab and Abihu died before the Lord when they offered strange fire before the Lord in the wilderness of Sinai. The fire was called strange because it had not been known before and because it was of foreign origin. Should we be looking for an area of the Middle East where natural gas rises to the surface? This could also apply to the burning bush of Moses on Mount Sinai. There is natural gas all over Saudi Arabia.

In the Hermitage Papyrus III6b it asks "how fareth this land (Egypt)? The sun is veiled and will not shine that men may see. None will live when the storm veileth it. All men are dulled through the want of it and the sun separateth himself from men. None will know that it is midday and the sun's shadow cannot be distinguished on the sundial. The sun is in the sky like the moon".

The Ipuwer Papyrus records that there was fiery destruction across Egypt.

The Egyptians told Solon that the fall of Phaethon was at this time and that the Greeks called the fiery comet Phaethon but to the Egyptians it was known as Sekhmet. In the texts of Sethos II, 1215 BCE to 1210 BCE, Sekhmet was a circling star which spread out her fire in flames.

Numerous Rabbinical sources state that an exceedingly strong wind endured for seven days and all the time the land was shrouded in darkness so dense that people could not stir from their place. The darkness was such that artificial light or flame could not be seen. The eyes were blinded by it and their breath choked. Rabbinical tradition states that the vast majority of the Israelites perished. Some sources state that forty-nine out of fifty Israelites were killed by it.

The tenth plague was a great earthquake that destroyed the palaces, prisons and houses of Egypt. Each house received a fatality. The houses were smitten. All the while according to Artapanus, quoted by Eusebius, where he states that the hail still fell so that those who fled from the earthquake were killed by the hail. All of the houses fell and most of the temples. The population fled and lived in fields. The Israelites as well as others fled as the lightning lightened the world and the earth trembled and shook. There was a portent in the sky which looked like a stretched arm. A Comet?

During the tenth plague of earthquake the Pharaoh rose up in the night, he, and all his servants, and all the Egyptians. There was a great cry in Egypt. There was not a house where there was not one dead. Exodus states that there was not a house where there was not one dead. Houses fell, smitten by one great blow. The Angel of the Lord passed over the houses of the children of Israel in Egypt and smote the houses of the Egyptians. The Passover Haggadah states that the firstborn of the Egyptians didst thou crush at midnight. One interpretation of first born, *bkhor*, in the text of the plague is a corruption of chosen.

Ipuwer states that the children of princes are dashed against the walls. The children of princes are cast out in the streets. The prison is ruined.

Artapanus describes the last night before Exodus which is quoted by Eusebius. There was hail and earthquake by night so that those who fled the earthquake were killed by the hail and those who sought shelter from the deadly hail were killed by the earthquake. At that time all the houses fell in and most of the temples.

At midnight all the houses of Egypt were smitten. There was not a house where there was not one dead. This happened on the fourteenth of the month Aviv. This was the night of Passover. Aviv according to Exodus was the first month. Thout was the name of the first month of the Egyptians. The thirteenth day of the month Thout is a very bad day. Thou shalt do nothing on this day. It is the day of the combat which Horus waged with Seth. The Hebrews counted the beginning of the day from sunset. The Egyptians reckoned from sunrise.

Around 1220 BCE the region of the now vanished Sarasvati River uplifted decimating the cities dependent on the Sarasvati River for survival.

The Babylonians in the valley of the Euphrates River referred to a rain of fire from the sky in this period.

The Quiche Manuscript of the Quiche Maya stated that the population of Mexico perished in a rain of bitumen. There descended from the sky a rain of bitumen and a sticky substance. The earth was obscured, and it rained day and night. Men ran hither and thither and were as if seized by madness. They tried to climb to the roofs and the houses crashed down. They tried to climb the trees, and the trees cast them far away. When they tried to escape in caves and caverns these were suddenly closed. In the Annals of Cuauhtitlan from Mexico there was an Age which ended in a rain of fire which was called Quiauh-tonatiuh which means the sun of fire-rain.

In Siberia the Voguls recalled an age when God sent a sea of fire upon the earth. The cause of the fire they called the fire-water. The Voguls stated that for seven winters and summers the fire has raged. It has burnt up the earth.

Tribes in the Sudan which is just south of Egypt remember when the night would not come to an end.

This was the period of the Achaean Demise when there was widespread destruction of cities in mainland Greece, the Levant and on the island of Cyprus.

Here we have Thera again! Is Thera being used as the excuse for any odd events in the Mediterranean in the Second Millenium BCE?

When the volcano on Thera apparently erupted in 1220 BCE Professor G. Marinos of the University of Salonika discovered that a layer of lava was washed up onto the island of Anaphe. This layer of lava was five meters thick and lay at a height of two hundred and fifty feet above sea-level. Anaphe is only twenty-five kilometers from Thera, and these are the remains of amongst the highest tsunami levels ever found.

A. G Galanopoulos stated that Santorini's volcanic cone collapsed and formed a huge caldera which sucked in billions of gallons of seawater which created a tsunami between three hundred and six hundred feet high which moved out from Thera crashing over the Cycladic Islands and Crete as well as the shores of the Aegean Sea and the Mediterranean. Many islands in the Western Mediterranean sank into the sea.

The Hittite Old Kingdom of Hattasus in Turkiye collapses. Though there are indications that Hattasus was not destroyed until 1200 BCE. In 1200 BCE the ruins were vitrified by an unknown source during an invasion by the people of the Sea who had fled their now destroyed island home in the west.

In 1213 BCE Ramesses II died.

Sometime between 1213 BCE and 1203 BCE, Merneptah, the son of Ramesses II, sent grain to the land of Hatti. He wrote that he had "caused grain to be taken in ships, to keep alive this land of Hatti". This confirms another famine in Anatolia in Turkey. Additional correspondence between Hatti and Egypt confirms a continuing famine during the following decades. Merneptah is also recorded to have sent consignments of grain to Ugarit in northern Syria to relieve the famine. There are many records from this period in regard to aid shipments from Egypt being shipped to countries that needed it

Three years later in 1210 BCE the Hittite king Tudhaliyas, Tudhaliya, IV died.

David Kaniewski found that from the late thirteenth century BCE to the ninth century BCE evidence from Tell Tweini (ancient Gibala) in northern Syria indicates climate instability and a severe drought episode. Pollen retrieved from the site suggested that drier climatic conditions occurred in the Mediterranean belt of Syria.

Kaniewski also found that there was a drought in Cyprus at the same time. Major environmental changes occurred in the area during the late Bronze Age. This was from 1250 BCE to 850 BCE. The area around Hala Sultan Teke, the Larnaca salt lake complex, turned into a drier landscape and the precipitation and groundwater probably became insufficient to maintain sustainable agriculture. This had been a major Bronze Age port.

Brandon Drake added that oxygen-isotope data from mineral deposits within Soreq Cave in northern Israel indicate that there was low annual precipitation during the transition from the Bronze Age to the Iron Age.

Brandon Drake also found that stable carbon isotope data in pollen cores in from Lake Voulkaria in Western Greece show that plants were adapting to arid environments. This indicates that the arid environment was ongoing over quite a period for plants to start adapting to it.

Also, in this period Brandon Drake found from sediment cores in the Mediterranean that there was a drop in temperature of the surface of the sea. This would have caused a reduction in precipitation on land by reducing the differential between land and sea. This event must have started around 1250 BCE.

Drake also mentions that in the Northern Hemisphere at the same time there was a sharp increase in temperatures immediately before the collapse of Mycenaean palatial centers, possibly causing droughts, but after the collapse there was a sharp decrease in temperature during the abandonment of the Mycenaean centers. This means that it first got colder and then suddenly got even more colder during the Greek Dark Ages.

Israel Finkelstein and Dafna Langgut noted that fossil pollen particles from a twenty-meter-long core drilled through sediments at the bottom of the Sea of Galilee also indicate a period of severe drought beginning around 1250 BCE. In the southern Levant, a second core drilled on the Western shore of the Dead Sea provided similar results. The two cores also indicated that the drought in this region may have ended by 1100 BCE.

The level of the Caspian Sea suddenly subsided around 1220 BCE. Russian scientists have found the ruins of a mysterious submerged city here that would date from before this period.

1200 BCE. Late Bronze Age Collapse. Sub-Atlantic Period.

In 1200 BCE Raoul Island in the Kermadec Islands near New Zealand erupted with a VEI of 4, Hakoneyama in Japan erupted with a VEI of 2, and Taveuni in Fiji erupted with a VEI of 2.

The Sub Atlantic Period is the current climate of the Holocene that is ongoing. Average temperatures are lower than during the preceding Subboreal and Atlantic Periods.

Interestingly enough if you divide 9,000 years by 13 lunar months which is the number of lunar months in a year for agricultural calendars and deduct it from when Plato wrote about Atlantis, you mysteriously end up in this date period. Interesting, isn't it?

Summer temperatures in the Sub Atlantic Period were generally cooler than the modern period. Up to 1° Celsius. At the same time the winter precipitations were augmented by up to 50%. Overall, the climate was wetter and cooler.

There was a sudden climatic decline around 1200 BCE in Europe in which the average temperatures dropped by 3° to 4° Celsius. Had there been large volcanic eruptions or meteorite impacts within the previous two years?

There is a strong belief that the Late Bronze Age collapse coincided with the onset of a 300-year long drought that began around 1200 BCE. This climate shift caused crop failures, death and famine which precipitated socio-economic crises and forced regional human migrations in the eastern Mediterranean and southwest Asia.

International trade in the Eastern Mediterranean and the Fertile Crescent suddenly stopped all over the vast area. There would be roughly 400 years of extreme poverty and desolation from Greece to the Middle East. There was a great reduction in the human population and in other places nomadic living replaced settled living.

The Bronze Age in Europe and the Middle East suddenly ended.

Between 1200BCE to 1000 BCE there was considerable traffic between Scandinavia and Ireland indicating a minimum of storminess.

A sequence of temperature increases in the Northern Hemisphere, followed by temperature decreases and increased aridity during the early Iron Age resulted in a hydrological anomaly or less available water between 1200 BCE and 850 BCE.

The civilizations of the eastern Mediterranean met fiery ends including Troy which was destroyed, rebuilt and then destroyed again between 1200 BCE and 1150 BCE. The Bronze Age ended in the region with the destruction of every major city in Anatolia, Greece and the Middle East. The three main cities in Cyprus were destroyed as well as Ugarit in Syria. Along the Levant coast from Lebanon to what is now Israel major cities were destroyed and in Greece all of the palaces of Late Helladic Greece such as Mycenae were destroyed. Egypt escaped destruction but resisted hordes of armed refugees who attempted to invade the country. These were the People of the Sea who came from the West stating that their home island had been destroyed by a massive cosmic object.

There was widespread drought around 1200BCE.

Lago Cardiel in Western Argentina suddenly rose by a massive amount in 1200 BCE.

At Lake Van at the headwaters of the Tigris and Euphrates Rivers in Armenia the water level rose 250 feet in only two years. What is with the suddenly rising lakes in 1200 BC? Lake Van is the largest soda lake on earth. This would have required 150 inches of rainfall.

Water levels in Lake Titicaca in Bolivia suddenly rose by a very large level in 1200 BCE.

Incidentally the same thing that had happened in southern Germany, Armenia and Lago Cardiel in Argentina, had happened at Lake Titicaca in Bolivia where the strand lines are still visible rising up at an angle from the water surface.

The Panela impact Crater in Pernambuco in Brazil is elliptical being six hundred meters long and five hundred meters wide. Breccia, ejecta, shatter cones, spherules and impact melt have been found here. This was in 1200 BCE.

The Benin and Yoruba of West Africa remember Mangala who was deliberately left behind when his home island in the Atlantic Ocean sank beneath the sea. Mangala survived in a watertight vessel that had been built for himself and his followers and arrived in West Africa where an earlier flood survivor, a female called Amma had installed herself as the ruler. After Amma's death Mangala laid claim to the throne but was opposed by Amma's twin brother Pemba who was eventually banished and Mangala took the throne.

The Benin and Yoruba remember Nana Buluku who was a royal personage who belonged to a race called the Sea Peoples who conquered the West African kingdom of Aja in West Africa. This according to tradition was around 1200 BCE. Nana Buluku and her husband Wulbari came from an island in the Atlantic that had been destroyed in a cataclysm. Were these the same Sea People who attempted to invade Egypt in the same period? Or were these some of the Sea People?

Most of Britain is depopulated suddenly around 1200 BCE and development is stopped at Stonehenge and the site is abandoned.

Around 1200 BCE the island of Gran Canaria in the Canary Islands was devastated by a massive volcanic eruption. The island was covered with a thick layer of lava and ash which covered a man-made wooden post and a pine trunk. The Canary Islands were covered in lava and ash and the trees had died. Did the island shelfs drop here?

Around 1200 BCE lake levels across Central Africa rose significantly.

There was a climatic decline around 1200 BCE in Europe in which the average temperature dropped by three to four degrees Celsius. There was dramatic climate regression all over Central Europe.

During the Bronze Age there were no glaciers in the Eastern Alps, but glaciers developed after 1200 BCE and reached far distances.

Many of the lakes of Central Europe sunk and made their dried-out shores habitable. Large segments of the population then built pile dwellings around them with the piles going down to seven meters due to the falling water table. A massive drought gripped Central Europe. Lakes were dropping in level in Central Europe, yet were rising in Africa, Armenia, Bolivia and Brazil.

Around 1200 BCE many lake pile villages or lake villages which were villages raised on piles standing in lakes were swamped by rapidly rising water levels. The villages were covered with mud, sand and calcareous deposit. This occurred in Scandinavia, Germany, Switzerland and northern Italy. Villages and fortified camps were destroyed by huge fires around 1200 BCE. Great masses of cinders were found and walls and fortifications were vitrified by the intense heat.

The Bronze Age suddenly stopped and coastal regions of Europe were evacuated.

Excavated oracle bones from 1200 BCE indicate a national obsession in this period with catastrophes in China.

Arid conditions thoroughly desiccated Crete. Herodotus the Greek historian states that after the Trojan War, Crete was so beset by famine and pestilence that it was virtually uninhabited. These conditions point to drought.

The three principal cities in Cyprus in the Mediterranean were suddenly destroyed by earthquake and fire between 1200 BCE and 1150 BCE.

James Mavor of the Woods Hole Oceanographic Institute reckoned that there were three waves with each having a height of two hundred feet even when they were hitting the shores of Crete when Thera erupted in 1220 BCE. Here is Thera again.

W. Brandstein deduced that the eruption of Thera produced such an enormous wave that the flood reached the capital Knossos eight kilometers inland and at an elevation of forty meters above sea level and destroyed it.

There was a general destruction by fire of palaces in Crete in 1200 BCE. The Minoan palace at Knossos was abandoned after earthquake damage and fire.

Brandstein also stated that a catastrophe that could destroy Crete with all of its towns and cities must have also caused damage at Athens in Greece as the epicenter was midway between Athens and Crete.

Brandstein quoted legends that stated that Poseidon in his rage overwhelmed the fertile plains of Eleusis and all of Attica. Athens was flooded as well and only those who fled to the mountains escaped with their lives. Pausanias stated that circa 150 CE during his lifetime there still existed a broad cleft in the rock north of the Ilissus Valley near the Temple of Athenian Zeus through which the flood waves were said to have rushed in. If this were indeed true, then the tsunami that hit Athens would have been seventy meters high.

In the Eastern Mediterranean house construction changed from flat to pitched gable roofs due to the earthquakes that had occurred.

Around 1200BCE in the reign of Ramesses III, an inscription on the Temple of Medinet Habu in Egypt describes a tsunami. "The might of Nun (Ocean) broke out and fell on our towns and villages in a great wave… the head of (the Sea Peoples) cities went under the sea: their land is no more."

What caused all of the tidal waves and destruction?

The answer is unknown and further confused by the modern acceptance of 1500 BCE being the date of the Thera/Santorini eruption. The Smithsonian Global Volcanism dates the eruption of Thera/Santorini at 1610 BCE.

Egypt found itself threatened by an invasion of people from the northeast and bands of Hittites and Syrian raiders as well as other migrants coming from Libya. All this possibly points to drought in the invaders' homelands. There was widespread drought around the eastern Mediterranean at this time.

The Bronze Age in Europe and the Middle East suddenly ended.

The Kalevala, the sacred book of the Finns of Finland tells of a time when hailstones of iron fell from the sky and the sun and moon disappeared because they were stolen from the sky and did not appear again. After a while a new sun and moon were placed in the sky. Most meteorites are iron chondrites and were used as sources of iron in ancient times.

Between the archipelago of Heligoland in the North Sea and Eiderstedt in Schleswig-Holstein on the German mainland, is a large area of what were coastal lands that were flooded around 1200 BCE when sea levels in the area rose on average of twenty meters. Originally large rivers flowed through here. These being the Weser, the Elbe, the Eider and the Hever which has since disappeared and originally flowed into the Eider Estuary about twenty kilometers west of Heligoland. These old river courses were still on old sea charts. A vast stretch of marshland stretched far out into the North Sea. Immediately east of Heligoland was an island which was flooded over in this same period but was still listed as being in existence until its final disappearance during the major flooding of the thirteenth and fourteenth centuries CE. Copper was mined here in Neolithic times. Jurgen Spanuth has postulated that this was the site of Atlantis, the capital of the People of the Sea who came from the far North around this time and are depicted on the Temple of Medinet Habu in Egypt.

There was a drastic rise in lake levels in Germany unequalled in history. This is especially so in the Saalachsee, the Ammersee and the Federsee in the south of the country. There was widespread abandonment of lake-pile villages. New lakes were formed near Memmingen, Munich, Ravensburg and Toelz.

Some sources state that this occurred in the middle of the Second Millenium at the end of the Bronze Age. The Bronze Age has two ends then, 1500 BCE and 1200 BCE.

Also, in this period lakes called the Ammersee and Wurmsee in the Bavarian Alps as well as other lakes in the Alpine foothills were suddenly tilted so one end was left up and the other end was left down by some gigantic tectonic movement. The same had occurred with the Bodensee, Lake Constance, in Switzerland. The old strand lines are still visible running obliquely to the present lake levels.

The Black Forest in Bavaria was incinerated suddenly. Burn strata lying between the pollen maxima of fir trees and beech trees in the Black Forest contain mountain pines. After a long period of warm and favorable climate during which the mountains were covered in beech woods which was followed by a time of dryer weather in which the beech trees were replaced by mountain pines which were the trees that were consumed by the massive fires. These were followed by coniferous forests indicative of a colder and wetter climate.

Villages and fortified camps were destroyed by huge fires around 1200 BCE in Central Germany. Great masses of cinders were found, and walls and fortifications were vitrified by the intense heat.

Around 1200 BCE Dorian tribes were able to move into Greece from the north as the country was virtually deserted. Greece continued only sparsely populated until 850 BCE.

In Greece, civilizations collapsed completely. Arid conditions thoroughly desiccated the Mycenaean Peloponnese. The palaces at Mycenae were destroyed and the culture collapsed around 1200 BCE. When the Dorians arrived one hundred years later, they found the area depopulated.

Major earthquake damage and fire occurred in Athens, Mycenae and Tiryns.

With the collapse of the Greek Bronze Age, writing came to an end. The scripts employed in Minos in Crete and Mycenae, later dubbed Linear A and Linear B, fell into disuse. This was the start of 300 years of lack of civilization after the collapse of the Bronze Age.

In 1200 BCE the Mycenean civilization suddenly imploded. One theory by Rhys Carpenter is that there was a northward shift of dry desert winds from the Sahara Desert that resulted in arid conditions arising in Mycenae and the Peloponnese as well as Crete and Anatolia. The arid conditions brought down the Mycenaean civilization.

The major palaces of Mycenae and Late Helledic Greece were destroyed at the same time. They were all destroyed by fire. This was the same time as the same was occurring in Crete.

Around 1200 BCE there began an abnormally hot and dry period in Greece which led to a catastrophic drought and to the disappearance of forests. Prior to this Greece was fertile and thickly forested.

There was the discovery of tree ring patterns and a destructive earthquake and flood in the Argive Plain in Greece.

Of 320 settlements in Greece in the thirteenth century BCE only forty were inhabited in the twelfth century BCE. The population had shrunk suddenly to one hundredth of what it had been.

There was a massive exodus of refugees from the Greek mainland as well as Cyprus. The Greek mainland became desolate and abandoned. Literacy vanished for the next three hundred years after the fall of Achaean or Mycenean civilization.

The huge hydraulic works at Lake Copais were suddenly abandoned around 1200 BCE resulting in widespread flooding of the Copais Basin. Early myth states that the lake was originally dug by Hercules and drained into the sea via underground channels that drained into the sea to the northeast. These subterranean channels were built long before 1400 BCE. Lake Copais was eventually drained in the 1930s. Another legend of Lake Copais is that it was the source of the Ogygian Flood that was regarded as a world flood.

Around 1200 BCE an extensive area of the Hungarian Plain where several large rivers converge was totally submerged by catastrophic flooding. Also, villages and fortified camps were destroyed by huge fires. Great masses of cinders were found, and walls and fortifications were vitrified by the intense heat.

Around 1200 BCE there was extraordinarily high volcanic activity in the volcanoes of Iceland in the North Atlantic. Great lava masses covered the whole of the island.

An undated report according to the "Bahman Yast" stated that at the end of a world age in Eastern Iran or India the sun remained visible for ten days in the sky. This was around 1200 BCE.

The Mahabharata describes a celestial battle or war around this date.

In the Persian-Russian borderland there was a cultural hiatus as there was no continuity between the Middle Bronze Age and the Late Bronze Age. In the Caucasus not an archaeological vestige was found of the centuries in between.

The eleventh tablet of the Epic of Gilgamesh states that from out of the horizon rose a dark cloud and it rushed against the earth. The land was shriveled by the heat of the flames. Desolation stretched to heaven. All that was bright was turned into darkness. Nor could a brother distinguish his brother. For six days the hurricane, deluge and tempest continued sweeping the land and all humanity back to its clay was returned.

The Iranian Holy book the Anugita states that a threefold day and a threefold night concluded the end of a great age.

The Bundahis, another Iranian Holy book, mentions that the world was dark at midday as though it were in deepest night. According to the Bundahis this was caused by a war between the stars and planets.

The city of Assur in Assyria was laid waste around 1200 BCE and the origin of the intense heat is inexplicable as it was so strong that it melted together hundreds of fragments of brick and reddened and vitrified the entire core of the great ziggurat.

Around 1200 BCE annual growth rings in Irish bogs suddenly declined and lake levels suddenly rose in Northern Ireland, especially at Loughbashade.

In the traditions of the pre-Prophetic period of Israel there are numerous stories of before the arrival of the Northerners, Ha Saponi, whose leading tribe was the Philistines, when fire fell from the heavens and burnt up the entire country. It is believed that the Philistines were involved with other peoples who formed the Sea Peoples.

In Italy around 1200 BCE many lake pile villages or lake villages which were villages raised on piles standing in lakes were swamped by rapidly rising water levels. The villages were covered with mud, sand and calcareous deposit. This also occurred in Scandinavia, Germany, Switzerland and northern Italy.

In Macedonia villages and fortified camps were destroyed by huge fires around 1200 BCE. Great masses of cinders were found, and walls and fortifications were vitrified by the intense heat.

There was a catastrophic earthquake that devastated the Atlantic coast of North Africa. The Greek scholar Diodorus Siculus whilst doing research in the library in Caesarea, the capital of Mauretania, found texts that stated that this earthquake had occurred in this same period.

Study of foraminifera in the Mediterranean Sea showed that the temperature of the water had risen since 5,000 BCE until it reached a peak between 3,200 and 2,400 years ago when it was similar in temperature to the Caribbean Sea. Suddenly these warm water foraminifers were covered with layers of volcanic ash from the most violent volcanic eruptions since the end of the last Ice Age. The volcanoes of the eastern and western Mediterranean were active at the same time. Above the ash strata were now the remains of cold-water foraminifera that showed that after the eruptions a period of cold set in which lasted for several centuries, and the temperature of the Mediterranean had fallen more than at any other time in the last seven thousand years. If Thera/Santorini was not responsible for this period of destruction, then what was? Were there other volcanoes erupting in the Mediterranean or were there as yet unfound bolide impacts?

Between 1200BCE and 700 BCE the population was low in Mexico due to water shortage.

Sedimentary evidence from North Africa shows abrupt general cooling with excessively low tree growth. There was extremely low tree growth and global widespread flooding as well.

In North America Utah's Great Salt Lake and the Waldsea Basin in Canada reached abnormally high-water levels and the western and central United States grew suddenly cooler and more moist.

Franzen and Larsson state that relatively large extraterrestrial bodies hit the eastern North Atlantic Ocean.

The present North Sea was still not with the same configuration that it is today. The Atlantic Ocean sent its waters to the shores of Scotland and Norway as well as to the recently formed English Channel. Human artifacts and remains of land animals have now been found under where the North Sea is now.

Forty-five miles from the coast of Norfolk at a depth of thirty-six meters a spearhead carved from a deer antler was found embedded in a block of peat which can only be formed on land. Analysis of pollens confirmed that forests existed there before the sudden sinking of the land around 1200 BCE.

The Dogger Bank in the middle of the North Sea was still covered with forests as the stumps of trees still with their roots attached were found there.

When the average temperature dropped by three to four degrees the snow line in the Norwegian Mountains in Norway which had previously been at 1900 meters above sea level sank by four hundred meters to fifteen hundred meters. This indicated sudden and extremely cold temperatures. Was this a mini-Ice Age?

In Palestine, at Beth Mirsim, there was an interruption in habitation at the site after the fall of the Middle Kingdom in Egypt. At this time Egypt ruled Palestine.

In Beth-Shan between the layers of the Middle Bronze and Late Bronze Ages there was an accumulation of debris one meter thick. This indicates that the transition from the Middle to Late Bronze Ages was a time of upheaval. At Tell el Hesy there was the same hiatus.

Earth tremors played havoc in Jericho, Megiddo, Beth-Shemesh, Lachish, Ascalon and Tell Taanak. In Jericho the city had been repeatedly destroyed. The great wall surrounding it fell shortly after the end of the Middle Kingdom in Egypt.

In the manuscripts of Avila and Molina who collected the traditions of the Indians of the New World it was related that the sun did not appear for five days. A cosmic collision of stars preceded the cataclysm. People and animals tried to escape to mountain caves. Scarcely had they reached there when the sea, breaking out of bounds following a terrifying shock, to rise on the Pacific coast. But as the sea rose, filling the valleys and the plains around, the mountain of Ancasmarca in Peru rose, too, like a ship in the waves. During the five days that the cataclysm lasted the sun did not show its face and the earth remained in darkness.

Sedimentary materials from Scandinavia show abrupt general cooling with excessively low tree growth. There was extremely low tree growth and global widespread flooding.

During the Bronze Age it was the climatic optimum in Scandinavia and the country was thickly forested up to the Arctic Circle. There were warmth loving deciduous trees up the northern coasts. Suddenly around 1200 BCE these huge forests burnt up and the weather abruptly changed.

The huge warmth loving deciduous forests that had covered Scandinavia to the North Coasts vanished around 1200 BCE and were replaced by the present coniferous forests.

Long established burial customs were abandoned, tumuli or ceremonial mounds are no longer constructed, and the display of rich funeral objects is drastically reduced.

The lower levels of glaciers in Scandinavia descended during the Sub-Atlantic by one hundred to two hundred meters.

There was a sudden drop in temperatures in Scotland after the eruption of Hekla 3 in Iceland. The climate deteriorated.

Sudden climate deterioration occurred all over South America closely corresponding with a neoglacial interval at the same time.

There is a drastic rise in lake levels in Sweden unequalled in history. There was widespread abandonment of lake-pile villages.

Prior to the cold period of 1200 BCE wheat and grape vines were grown north of Stockholm in Sweden. This has been impossible ever since. Millet, which had been cultivated far to the north of Stockholm would only thrive in the extreme south.

Around 1200 BCE many lake pile villages or lake villages which were villages raised on piles standing in lakes were swamped by rapidly rising water levels. The villages were covered with mud, sand and calcareous deposits. This occurred in Scandinavia, Germany, Switzerland and northern Italy.

There is a drastic rise in lake levels in Switzerland unequalled in history. This is apparent on the Bodensee, also known as Lake Constance. There was widespread abandonment of lake-pile villages. The Bodensee suddenly rose thirty feet, and the bed was suddenly tilted by some massive tectonic movement. Numerous lake beds tilted in this area.

Many bronze objects were found in the passes in Switzerland especially the St Bernard Pass. Mines that were worked in the Swiss Alps were suddenly abandoned. It also appeared that the passes were no longer travelled in this period.

On the Syrian coast and the interior there is a stratigraphic and chronological rupture between the strata of the Middle Bronze Age and the Late Bronze Age. This was found at Qalaat-er-Rous, Tell Simiriyan, Byblos (now in Lebanon) and in the necropolises of Kafer-Djarra, Qraye and Majdalouna. All of the necropolises in the upper valley of the Orontes River were no longer used.

The city of Hama was interrupted and destroyed at the moment that the Middle Kingdom in Egypt went down.

This catastrophe stratum has been found across Syria as well as the Middle East as well as in Asia Minor. Numerous cities and towns were destroyed by earthquakes as well as intense heat. There is no mention of wars or invaders though.

Ras Shamra shows a marked gap between the horizons of the Middle and the late Bronze Ages.

At Alalakh, also known as Tell Atchana, in Northern Syria the royal buildings had been burned so completely that the very core of the walls had bright red crumbling mud bricks. The mud and lime wall plaster had become vitrified and basalt wall slabs in some areas had actually melted. What caused this extreme level of heat? This was around 1200 BCE.

When the volcano on Thera erupted in 1220 BCE Professor G. Marinos of the University of Salonika discovered that a layer of lava was washed up onto the island of Anaphe. This layer of lava was five meters thick and lay at a height of two hundred and fifty feet above sea-level. Anaphe is only twenty-five kilometers from Thera, and these are the remains of amongst the highest tsunami levels ever found. Once again, was there a second Thera/Santorini eruption?

A. G Galanopoulos stated that Santorini's volcanic cone collapsed and formed a huge caldera which sucked in billions of gallons of seawater which created a tsunami between three hundred and six hundred feet high which moved out from Thera crashing over the Cycladic Islands and Crete as well as the shores of the Aegean Sea and the Mediterranean.

Around 1200 BCE many islands in the Western Mediterranean sank into the sea.

Frost occurred in Tunisia in North Africa which normally never has frost.

Around 1200 BCE arid conditions thoroughly desiccated Anatolia in Turkiye.

Life vanished in Troy, Bogazkoy, Tarsus, and Alisar Huyuk. This is one of the dates for the fall of Troy in Turkiye. This was the beginning of a Dark Age that lasted until 750 BCE.

In Tarsus between the strata of the of the brilliantly developed Middle Bronze Age and that of the Late Bronze Age, there is a layer of earth five feet thick which was found without a sign of habitation. This is called a hiatus.

The city of Hattasus, capital of the Hittites, or Hatti, was built around 5,000 BCE. In 1200 BCE the ruins were vitrified by an unknown source during an invasion by the People of the Sea. There are many subterranea under the city, some up to 70 yards long. The architecture was megalithic. The capital of the Hittites, Hattasus, in Turkey is destroyed in an unknown type of conflagration. The Hittite state collapsed at the same time. The empire of Hatti suffered severe famine at the same time. The Hittite Old Kingdom collapses. Though there are indications that Hattasus was not destroyed until 1200 BCE. The ruins of Hattasus are near Corum in Turkey.

The Hittites seem to have suddenly abandoned the Anatolian Plateau in Turkey. Drought appears to have caused these migrations.

The Hittite Old Kingdom of Hattasus in Turkey collapses. Though there are indications that Hattasus was not destroyed until 1200 BCE. In 1200 BCE the ruins were vitrified by an unknown source during an invasion by the People of the Sea who had fled their now destroyed island home in the west. The island home of the People of the Sea was said to have been an island in the Western Ocean, Atlantic, that was destroyed by the Goddess Sekhmet, the Egyptian Goddess of Destruction, who appeared as a lion in the sky with a fiery mane.

At Alaca Huyuk the transition from Middle Bronze Age to Late Bronze Age was marked by upheaval and destruction. Virtually every excavated site in Asia Minor suffered the same fate followed by the same hiatus.

In North America Utah's Great Salt Lake and the Waldsea Basin in Canada reach abnormally high-water levels.

The Western and central United States grew suddenly cooler and moister.

Alaska suddenly also grew cooler and moister.

Our next site has three names. They are Isle Royale, Isla Royale and Royale Island in Keweenaw County in Michigan. Another strange civilization that considerably changed the ecology of the area that they were in but of which no traces of their homes remain was an industrial society on Isla Royale, a large island in Lake Superior. On Isle Royale there are the extensive remains of a large copper mining center which over a long period of time mined and refined around two million pound of copper but of which there are no traces of who did the mining, refining, organizing and trading. Using normal trading logic, one would assume that this was organized from a center on a major trading artery which in this case would be Lake Superior. But was lake Superior always around? Geological theory is that the Great lakes were formed when the last Ice Age ended therefore before that they were river valleys. We would have to look at the river courses themselves to see where the ancient copper trading centers were as there are no traces of these on the shores of the Great Lakes, including Lake Superior, or on Isle Royale itself or any of the numerous other islands in the Great Lakes. This would have been a long-settled civilization and of reasonable size. Wealth creates growth in any society whether Neolithic or modern so we can safely assume that the mining centers are now submerged by the Great Lakes themselves.

Vast copper mining works were found reaching to a depth of sixty feet. The richest veins were followed up even when interrupted and the excavations were drained by underground drains. In one place the excavations extended in a nearly continuous line for two miles. There are no remains of the dead and no mounds near these mines indicating that the miners came from a distance. There are unusual formations called the "Garden Beds of Michigan", which were the fields from which the miners drew their food. Still no townships though.

The Chippeway Indians have traditions of the origin of the manufacture of copper implements along the shores of Lake Superior. Incidentally pure copper was found in large amounts on the shores of Lake Superior where the remains of ancient mines have been found from thousands of years ago.

Some researchers propose that the areas copper mines were active from 3,000 BCE to 1200 BCE. Others date the mines as going back to 6,000 BCE.

An interesting anomaly here is that the ingots of copper were often made in the shape of an object that resembled the reel of a kite, thereby having these ingots named reels. This was the shape of the ingots found in North America as well as in the Mediterranean area where they resembled cured ox hides which looked the same. Neither design, exactly the same, had any practical purpose. Where the market was for the mined North American copper has always been a mystery. Where the Old World got the large masses of copper needed for bronze technology that it needed for weapons superiority was also a mystery. Was there someone exporting copper to the Old World from the New? Ask the Phoenicians and the Minoans.

As an added mystery half socket hafting of copper objects, which was common in Europe, was only found around the Copper culture on the Western Great Lakes of North America.

Millions of tons of copper were excavated and smelted here. But where are they? Some bronze objects and numerous copper ones were found in North America but there must be millions of bronze or copper objects that have never been found. We are talking about industrial quantities. One of the ways that the ancient miners removed the copper was by building large bonfires on top of the copper bearing ores. When the rock was glowing from the heat water was poured over it and the ore split releasing the pure copper. This is one of the few areas on Earth where pure copper has been found. Remember you only need to add tin in low proportions, around ten per cent, and you have bronze.

Nobody knows what happened to the Copper Mining Culture of the Great Lakes as their tools were found in position waiting for them to return but they never did. They disappeared around 1200 BCE. Some say 1500BCE

This was the only place in the world with a large lode of native copper that is in the form of pure metal in nuggets and chunks. A long time ago someone dug 5,000 pit mines extending for more than 100 miles along the South shore of lake Superior as well as islands in the lake. Every copper mine functioning now has been worked in primitive times. There are no indications of settlements and no human or animal bones in the area or traces of burials. The Indians did not know or care of this place. 100 million to 500 million pounds of copper may have been extracted here in ancient times.

The water levels of the Great Lakes were much lower then. Is this where we would find the population, industrial-smelting and trading centers that were needed for such a vast enterprise?

Some of these mines are dated back to 6,000 BCE. Who needed copper then?

The Menomonie Indians of Canada state that seafaring men of fair complexion were excavating massive amounts of copper here in the Upper Great Lakes. From around 3,000 BCE to around 1200 BCE when mining suddenly stopped five hundred million pounds weight of high-quality copper were removed from around here. Tin was also mined along the Michigan shores of the Great Lakes and tin and copper together create bronze. Incidentally the Bronze Age in Europe was around this same period before suddenly ending. Egyptian legends mention massive amounts of copper being taken from a subdued nation that formerly lived on an island in the Atlantic Ocean. Was there traffic going both ways?

Masse states that there was a local terrestrial impact around 1200 BCE in the Badlands of northern Montana. This is a one-mile-wide impact crater caused by an explosion similar to a 120-megaton blast. Glassy plates and small, spherical blobs rich in iron and titanium have been found around it. The massive impact was west of Broken Bow in Custer County in Nebraska.

In 1198 BCE the New Kingdom suddenly ends. The Harris Papyrus states that there were prodigious clouds of ash coming from the west during Ramesses III coronation in 1198 BCE. Shortly after Ramesses III stopped the invasion of the Sea Peoples who had told his scribes that a shooting star had burned their homeland before it sank into the Western Sea. The Harris Papyrus is a document circa 1180 BCE that summarizes all of the accomplishments of Ramesses III. Ramesses III had defended Egypt from invading Sea Peoples, also mentioned on the walls of the temple of Medinet Habu. After the defeat of the Peoples of the Sea Ramesses III sent an expedition to the land of Ataka for the great copper foundries that were there. The copper was piled in stores when it got back to Egypt where it was in the hundreds of thousands the colors of gold, which would indicate very high-grade copper. Ataka-Atika was not the home of the Peoples of the Sea though as it had been destroyed by an aerial object crashing into it. Incidentally the Michigan copper mines in North America were abandoned at this same time. Was there a connection? Was this Ataka? Ataka was also called Atika. The trade with Atika was done by galley and in the Harris Papyrus, Ramesses III states that he sent his messengers to the country of the Atika. To the great copper mines that were in this place. There galleys carried the copper. The mines were found abounding in copper which was loaded by ten-thousands into their galleys. In this period Cyprus was a large area for copper mining. So was Timna in southern Israel in southwestern Arabia. There were even prehistoric copper mines in Cornwall such as the Great Orme Mine.

There are massive copper reserves in Spain and in Cornwall. Were there also large reserves of copper on a lost land mass in the Atlantic Ocean, the Western Sea?

The Medinet Habu inscriptions mention the great darkness which prevailed before the beginning of his reign. Inscriptions on the walls of the Victory Temple of Ramesses III at Medinet Habu in the Upper Nile Valley tell how the Atlantean or Hyperborean invaders of Egypt were destroyed. "The shooting star was terrible in its pursuit of them" before their island disappeared under the Atlantic Ocean. When was this though or was this the destruction of the last portion of Atlantis in Neolithic times?

The Victory Temple of Medinet Habu was completed around 1180 BCE.

The temple's exterior walls are covered with lengthy descriptions of the war between Ramesses III as well as incised carvings of all of the combatants. The temple records the testimony of captured warrior Sea-Peoples who stated that they came from an island called Netero that was in the far Western Sea (the Atlantic). Netero was set ablaze by the fiery goddess Sekhmet and then sunk into the sea. Sekhmet was identified as a threatening comet or shooting star and was a celestial phenomenon.

This invasion by the Sea Peoples was in 1190 BCE according to scribes at Medinet Habu which was eight years after the loss of their Atlantic Island home. The Sea People were called the Meshwesh or Hanebu. The Medinet Habu inscriptions also mention catastrophes occurring the same as Ipuwer did. The Nile dried up, the land was parched, the house of the thirty chief nobles is destroyed, the people are starving, and all water is useless. What which hath never happened before hath happened.

Ramesses III at Medinet Habu declared that the whole delta of the Nile is suddenly flooded by the sea. Was this a tsunami?

The "Peoples of the Sea" were also called the "Peoples of the Isles". These people wandered and conquered around the Mediterranean Sea and the Mycenean Greeks, the Hittites and many lesser kingdoms were swept out of existence by them. The "Peoples of the Sea" were also called the Tjeker, the Shekelesh, the Teresh, the Weshesh and the Sherden or Sardan.

The walls of the temple of Medinet Habu mention that the foreign countries (the Sea People) had made a conspiracy in their islands. All at once the lands were removed and scattered in the fray. No land could stand before their arms. From Khatte, Qode, Carchemich, Arzawa, and Alashiya on being cut off one at a time. A camp was set up in one place in Amarru. They desolated its people, and its land was like that which has never come into being. They were coming forward towards Egypt, which the flame was prepared before them. Their confederation was the Peleset, Tjekker, Shekelesh, Dananu and Weshesh lands united. Khatte was the land of the Hittites, in Anatolia in Turkey. Qode was probably Kizzuwatna near the Gulf of Iskenderun in Anatolia. Carchemich was an important ancient capital in northern Syria. Arzawa was another land in Anatolia in Turkey. Alashiya was Cyprus or a land in Cyprus and was a regional source for copper.

Some sources state that the "Peoples of the Sea" were the mercenaries for people called the Pereset who were richly clad warriors.

The Denien were referred to by Ramesses III as Peoples of the Islands. Some sources state that this invasion of Egypt was by Greeks and people from Asia Minor.

In the Hermitage Papyrus III6b it asks how fareth this land (Egypt)? The sun is veiled and will not shine that men may see. None will live when the storm veileth it. All men are dulled through the want of it and the sun separateth himself from men. None will know that is is midday and the sun's shadow cannot be distinguished on the sundial. The sun is in the sky like the moon.

At the end of the Nineteenth Dynasty, around 1197 BCE, Pharoah Seti I saw Sekhmet which was a circling star that spat flames and was seen across the known world.

The narrative of "the Destruction of Men by the Gods" in the tomb of Seti I states that the Gods delegated the goddess Hathor in her form as Tefnut or Sekhmet to punish mercilessly the rebellious people who did not submit to the will of the gods. This goddess went out and killed the men on earth...and lo! Sekhmet waded with her feet through many nights in their blood down to the city of Heracleopolis, known as Henen-nesut, the House of the Royal Child, and eventually as Ehnasya. During the First Intermediate Period it was the capital of Lower Egypt. The noted archaeologist Professor Edouard Naville presented strong arguments that the Sphinx was actually a depiction of Hathor as Tefnut or Sekhmet as Hathor's murderous aspect. Hathor was the feminine personification of Horus, and her name means House of Horus. One of the names of the Sphinx was Harmachis or Hor-em-akhet, Horus of the Horizon or Horus of the Necropolis. Was Horus of the Horizon actually Sekhmet on the Horizon? Was the Exodus myth about the killing of the first-born sons, actually a referral to the destruction of Henen-nesut, the House of the Royal Child.

Seti I was the son of Ramesses I and father of Ramesses II. This was not long after the reign of Akhenaten.

The Ipuwer Papyrus records that there was fiery destruction across Egypt.

This was the period of the invasion of the mysterious Peoples of the Sea. Possibly affected by the sudden climatic change, and possibly meteoric phenomena, they traveled by land and sea besieging ports and inland cities around the Mediterranean Sea looking for somewhere to settle. They had sacked Hattasus in Turkiye and Ugarit as well as a host of other cities and then headed for Egypt. In 1200 BCE an alliance of People of the Sea and Libyans attacked Egypt from Syria by land as well as by sea. This horde comprised men, women and children and travelled by oxcart as well as by ship. Hundreds of ships sailed alongside those who were travelling by land. This horde was confronted by the Egyptian navy at an eastern mouth of the Nile and during the attack archers poured volleys of shots into the attacking ships. Ramesses eventually won and captured vast numbers of cattle and killed more than two thousand attackers. Piles of severed heads were placed in front of him, and the numbers were checked by scribes against a tally of severed penises. As you do.

Inscriptions on the walls of the victory temple of Ramesses III at Medinet Habu in the Upper Nile Valley tell how the Atlantean or Hyperborean invaders of Egypt were destroyed. "The shooting star was terrible in its pursuit of them" before their island disappeared under the Western Sea or Atlantic Ocean. When was this though or was this the destruction of the last portion of Atlantis in Neolithic times? Was this the actual legend of Atlantis?

The goddess Hathor was the Egyptian Goddess who was described at Medinet Habu as a flaming planet or comet that destroyed the island of the Sea People who had mysteriously arrived from the West.

The Medinet Habu texts state that the forests and fields of the People of the Sea were burnt up with fire and the heat of him, Sekhmet, has burnt their countries. The fire of Sekhmet has burnt the land of the Nine Bows. Jurgen Spanuth believed that this referred to Heligoland in the North Sea based on older Egyptian divisions of the Earth into bows.

According to Herodotus the Garamantes were a chariot people who came from the west to try and invade the Mediterranean around the time of the Trojan War. The Garamantes wore the same armored vests and crested helmets as the Sea People who tried to invade Egypt in 1198 BCE.

In the Medinet Habu texts Ramesses III reported that Libya, North Africa, had become a desert. A terrible torch hurled flame from heaven to destroy their souls and lay waste their land. Their bones burn and roast within their limbs. This same text also mentioned that the Nile River dried up and the land of Egypt fell victim to drought.

According to the Deuteronomy 2:23, Amos 9:7 and Jeremiah 47:4 the Philistines, the great trading race, appeared from the island of Caphtor to Canaan. The Philistines were the remnants of the country of Caphtor. Why are they regarded as remnants? Were they survivors? This was only a few years before the Israelites reached Canaan. Several sources state that Caphtor was either Crete or Cyprus. Was Caphtor actually the same island that the "Peoples of the Sea" had fled from? Were the Pereset the Philistines or the Libyans? The Philistines were regarded as one group of the Peoples of the Sea.

One of the carvings has an inscription identifying a group of prisoners in the clothes of the Pereset as Tjeker. Another group was identified as Denien. The third and largest group in identical dress and headgear are designated as Pereset.

On the murals of the Temple of Medinet Habu the Pereset and their allies the "People of the Sea" are easily recognized by their clothing. The Pereset wear crownlike helmets on their heads and rich garments. The soldiers of the Peoples of the Sea have horned helmets sometimes with a ball or disc between the horns. These people do not look like a conquering horde but as a well-organized state and armed forces.

Originally as recorded on the temple walls the Peoples of the Sea and the Pereset made war as allies of the Egyptians against the Libyans. They are shown slaying the Libyans. In the next part of the temple inscriptions the Pereset are shown as the main foes of the Egyptians and the People of the Sea were still allied to Ramesses III. They are seen parading with the Pharaoh and their helmets, shields, spears and swords are carefully reproduced where they march to the sound of an Egyptian trumpeter. They are also seen marching swiftly in military array.

Eventually in the great battle at the mouth of the Nile River, the Peoples of the Seas wearing horned helmets, without discs between them, appear on hostile vessels. The Egyptian fleet routs the Peoples of the Sea and the Pereset alike. When the Peoples of the Sea are shown on Egyptian vessels they are depicted as fettered captives.

After the battle of the Nile there are depictions of captives with arms and necks in stocks and bound by ropes. These captives are easily recognizable as Peoples of the Sea and Pereset by their costume.

Ramesses III wrote on the walls of Medinet Habu that the invaders were coming whilst the flame was prepared before them. The Confederation was the Peleset, the Theker, the Shekelesh, the Denyen and Weshesh lands united. Ramesses prepared his frontier in Zahi and caused the Nile mouth to be prepared like a strong wall with warships, galleys and coasters equipped. Walls were raised at the mouths of the Nile. The bas-relief shows five vessels of the invading fleet engaged with four Egyptian ships. Ramesses wrote that now the northern countries penetrated the channels of the Nile mouths. Ramesses was gone like a whirlwind amongst them. Ramesses wrote on that those who came on land were overthrown and slaughtered and those that entered the Nile mouths were like birds ensnared in the net. They that entered into the midst of the Nile mouth were caught, fallen into the midst of it, pinioned in their places, butchered and their bodies hacked up. There is an inscription showing Ramesses standing on a rostrum before a fortress built at the mouth of the Nile as his officials present him with captives. Ramesses wrote that the leaders fled wretched and trembling.

The Peleset were the Philistines who had originally come from Cyprus. The Denyen or Danuna apparently came from the Aegean Sea. The Shekelesh came from Sicily. Another group of Sea People were the Shardana who were renowned as pirates. Eventually most of the remaining Sea People were allowed to settle in Canaan by Ramesses III after they were defeated.

The Sea People were camped at Amaru or Amurru in Northern Lebanon/ northwestern Syria when Ramesses III decimated and defeated them. Ramesses III recorded himself as leading a glorious procession of Sea People prisoners on his return journey. Amaru, incidentally, was also the name of the lost Egyptian homeland in the Western Ocean.

Around 1198 BCE work at Stonehenge suddenly stopped and was never resumed again with the advent of the new cold period that lasted 500 years. This also occurred at other megalithic observatories as well. After the passing of 500 years, no one around had any idea how to use the stone circles let alone what they were for.

Brandon Drake found that there was a decline in the surface temperature of the Mediterranean Sea around 1190 BCE that resulted in less snow and rainfall. This would have dramatically affected the Mycenaean centers in Greece that were dependent on high levels of agricultural.

David Kuniholm compiled tree ring sequences that indicated that climate deterioration peaked in Turkiye in the period of 1185 BCE to 1141 BCE.

In 1180 BCE Merapi in Indonesia erupted with a VEI of 4.

Around 1180 BCE population centers in Greece such as Pylos, Gla, Midea, Prosymna and Berbati were destroyed and not rebuilt. This could be the consequences of civil war as several ancient authors recorded that civil war raged through Greece after the Trojan War. During this period many of the mainland people fled to the Greek Islands to areas of greater security. The population density on the Greek mainland dropped dramatically. The quality of Mycenaean pottery had dropped as had building construction. Within a century of this many cultural and technological achievements of the Late Bronze Age vanished, central political control broke down, the elaborate administration disintegrated, and the art of writing disappeared.

This collapse of mainland Greek civilization was during the invasion of the Sea Peoples.

Eric. H. Kline wrote that prior to 1177 BCE the Sea Peoples had landed and set up camp in Syria and then proceeded down the coast of Canaan (including parts of Syria, Lebanon and Israel) and then went into the Nile Delta of Egypt. The Sea Peoples did not wear uniforms or polished outfits. One group wore feathered headdresses, another wore skull caps, others wore horned helmets or went bareheaded. Some had short, pointed beards and dressed in short kilts, either bare-chested or with a tunic. Others had no facial hair and wore longer garments, almost like shirts. They came on boats, wagons, oxcarts, and chariots. The Sea Peoples came in waves.

Kline continued. No country was able to oppose this invading mass of humanity. The great powers of the day-The Hittites, the Mycenaeans, the Canaanites, the Cypriots, and others, fell one by one. Some joined the Sea Peoples who had migrated eastward from their lands in the West.

1177 BCE was the eighth year of the Pharaoh Ramesses III's reign when he was victorious over the Sea People many of whom he settled in Canaan.

In 1171 BCE in the Baltic Sea, a tsunami with wave heights of at least 10 meters (33 feet) had run-up heights in Sweden of up to 14.5 to 16.5 meters (48 to 54 feet).

Ice cores from Camp Century in Greenland show that a global catastrophe threw several thousand cubic kilometers of ash into the atmosphere around 1170 BCE. Furnas in Iceland erupted with a VEI of 4.

Were the eruption of Furnas and the Baltic Sea tsunami related?

In 1160 BCE Galeras in Colombia erupted with a VEI of 2.

In Anatolia the Hittite Empire collapsed. Again. As I keep saying, the dates are quite confusing in this period.

In 1159 BCE the volcano Hekla 3 in Southern Iceland erupted causing a twenty-year long climate anomaly over Europe. Around this period there was extraordinarily high volcanic activity in the volcanoes of Iceland in the North Atlantic. Great lava masses covered the whole of the island. This eruption had a VEI of 4.

Greenland ice cores show an extremely high acid content indicative of unnatural weather patterns around 1159 BCE. This was the same time as the eruption of Hekla 3 in Iceland. There was a sudden drop in temperatures in Iceland after the eruption of Hekla 3. The climate deteriorated.

According to Mike Baillie, a dendrochronologist at Queens University in Belfast in Northern Ireland there was a major temperature trough worldwide around 1159 BCE. Baillie had constructed an Irish bog oak dendrochronology spanning the period from 5,000 BCE to 1000 CE.

Around 1159 BCE there were unusual weather conditions and ash rains in China.

In China a comet was seen and there was the collapse of the Shang Culture.

Also, in 1159 BCE there was a sudden drop in temperatures in England after the eruption of Hekla 3 in Iceland. The climate deteriorated. Evidence in this period shows wholesale abandonment of uplands in England followed by an upsurge in the construction of fortified defensive sites.

The 1159 BCE Hekla eruption would have triggered crop failures and hunger over a wide part of northern Europe. In Europe there were colder wetter conditions as well as crop failures and famines. In Europe there were unusual tree ring markings.

Tree-rings indicate climatic upheaval in Greece in the period 1159 BCE to 1141 BCE. This was the period of the sudden abandonment of the Mycenean Civilization.

Tree rings in Ireland indicate enormous ashfall around 1159 BCE. This is the same time as the eruption of Hekla 3 in Iceland. Tree rings virtually disappeared in this period. This was evidenced by high acidity in ice-cores from Greenland. There was the virtual disappearance of tree rings in Ireland indicating a volcanic winter.

An acid layer in Greenland indicated a massive volcanic eruption for the period 1120+/- 50 BCE.

Were there several volcanic eruptions?

Annual growth rings in Irish bogs suddenly declined in 1159 BCE.

Radiocarbon dates for peat in Ireland confirmed that the peat had been buried by the Hekla eruption.

Baillie mentions that extremely narrow tree rings were from 1159-1145 BCE.

Around 1159 BCE was the collapse of the Bronze Age in the Mediterranean region.

There was the sudden collapse of palace economies of the Aegean and Anatolia which were replaced after a hiatus by the isolated village cultures of the Dark Age period of the history of the Middle East.

Starting in 1159 BCE, the time of the eruption of Hekla 3 in Iceland, there was a marked increase in temperature and aridity in Mesopotamia.

There was a sudden drop in temperatures in Scotland after the eruption of Hekla 3 in Iceland. The climate deteriorated. There was upland abandonment followed by an upsurge in construction of defensive sites.

A conspicuous ash layer from the Hekla eruption is marked in Swedish bogs by peat layers signaling colder and wetter conditions.

David Kuniholm compiled tree ring sequences that indicated that climate deterioration peaked in Turkiye in the period of 1185 BCE to 1141 BCE. Around 1159 BCE oaks in Turkiye grew extremely fast as the climate in the area became unusually wet from its usual very dry weather.

Around 1153 BCE there was a massive famine in Egypt in this period and Ramesses III died that same year. In 1153 BCE trees showed sudden defoliation.

In 1150 BCE Cuicocha in Ecuador erupted with a VEI of 5 and Izu-Toba in Japan erupted with a VEI of 4 that same year.

There was an enhanced growth anomaly in Turkiye in this period. There was sudden tree growth in 1150 BCE.

There is a Babylonian reference from the twelfth century BCE that there was a comet sighted during the reign of Nebuchadnezzar I who reigned from 1126 BCE to 1103 BCE. This comet rivalled the sun in brightness. For this to be it must have been very close to the earth.

Excavated oracle bones from this period indicate a national obsession with catastrophes in China. In 1122 BCE during a battle between the last Shang Emperor Chou Hsin and Wu Fang, founder of the Chou Dynasty, it was recorded that there was a battle in the sky between comets during the battle. It was also reported that a single comet appeared with its tail pointing towards the people of Yin. At this time dust had been falling for ten days. Note that they say falling and not blowing.

In 1120 BCE Kusatsu-Shiranesan erupted in Japan with a VEI of 4.

1100 BCE

In 1100 BCE Hekla volcano in Iceland erupted with a VEI of 5. The other eruption that year was Taveuni in Fiji with a VEI of 2.

Between 1100 BCE and 800 BCE there was a recurrence of temperatures approaching the warmest post-glacial level.

Baillie stated that in 1100 BCE there was a major temperature trough in Europe, the Americas, the Near East and the Antarctic. This was a sudden cold period. Which one is correct. All indicators so far are of a cold period.

Between 1100 BCE and 900 BCE there were periods of increased warmth though with frequent dry, blocked spells of anticyclonic weather.

In China between 1100 BCE and 800 BCE the warmth of the post-glacial times came to an end, never to be restored. Does this mean that the feared temperatures of the late twentieth century that are approaching are actually lower than the post-glacial temperatures?

In 1090 BCE Kikai in Japan erupted with a VEI of 2.

In 1059 BCE the conquering King Wu was thwarted apparently by the appearance of a comet which aided the people of Yin and their king Chou Hsin.

Around 1100 BCE to 700 BCE the North Greenland Norse settlements were again abandoned and the same general southward movement of 1500 BCE was repeated.

In 1059 BCE Cerro Bravo in Colombia erupted with a VEI of 4. That same year Cotopaxi in Ecuador erupted with a VEI of 4.

In 1050 BCE Mount Pinatubo in Luzon in the Philippines erupted with a VEI of 6. This level of eruption would have caused a volcanic winter over much of the planet. 10 to 16 cubic kilometers of tephra was released.

Also, in 1050 BCE Taupo in New Zealand erupted with a VEI of 4 and an ejection of 0.1 cubic kilometers of tephra from Motutaiko Island. Khodutka in Kamchatka also erupted with a VEI of 4. Also, in 1050 BCE Taupo in New Zealand erupted with a VEI of 4 as did Khodutka in Kamchatka with a VEI of 4. Other eruptions were Buzzard Creek in Alaska with a VEI of 2, Cerro Bravo in in Colombia with a VEI of 4, Cotopaxi in Ecuador with a VEI of 4 and Apoyeque in Nicaragua with a VEI of 4. Just one of these could create a volcanic winter. Here were five! What was happening in 1050 BCE?

The volcanic winter seems to tie in with reports of global cooling for a period.

The shells of plankton collected from sediments points to reduced river flow around 1050 BCE. There was an overall increase of salinity in the Bay of Bengal.

There are three meteorite craters at Aundha Nagnath near Aundha Village near Talni in the Hingoli District of the Marathwada Region of Maharashtra State in India. The largest is 1.5 kilometers across, the second is two hundred and fifty meters across and the third is twenty meters across. They were created around three thousand one hundred years ago. These date from 1100 BCE. Incidentally there is a major temple here called the Aundha Nagnath Temple which is dedicated to Lord Shiva and claimed to be the eighth of the twelve Jyotirlingas in India. The Jyotirlingas are where Lord Shiva appeared as an endless pillar of light. Was this a meteor?

The Trimbakeshwar Impact Crater in the Nashik district of Maharashtra State in India is three thousand one hundred years old and is two kilometers in diameter. Trimbakeshwar is also one of the twelve Jyotirlingas created by Lord Shiva. The Jyotirlingas were endless pillars of light created by Lord Shiva.

In 1050 BCE Alaska and the Central and Western United States suddenly grew cooler and moister.

Also, in 1050 BCE in South America there was massive climate deterioration that closely corresponded with a near-glacial interval that was taking place at the same time.

Records of vegetation from the Nile River Delta point to a series of regional droughts including droughts in 2050 BCE and 1050 BCE.

In 1045 BCE the Shang Dynasty suddenly collapsed in the Yellow River in China.

In 1040 BCE Brennisteinsfjoll in Iceland erupted with a VEI of 2. Mount Redoubt also erupted.

In 1030 BCE Fujisan in Japan erupted with a VEI of 4.

Between 1030 BCE and 830 BCE Khodutka erupted in Kamchatka in Siberia. It had a VEI of 5. Indications are that this eruption was actually around 800 BCE deposited tephra over much of Southern Kamchatka.

Some researchers place the famine of King David of Israel to be around 1021 BCE. These are parts of Psalm 18. During the reign of David, the earth shook and trembled… foundations of the hills moved and were shaken…smoke…fire…darkness…dark waters… thick clouds of the skies… hail stones and coals of fire … thundered in the heavens…hail stones and coals of fire again… arrows…shot out lightenings…. Channels of water were seen…the foundations of the world were discovered. *Barakh* in early Hebrew meant fiery hail or fiery stones.

In 1020 BCE Taveuni in Fiji erupted with a VEI of 2.

In 1010 BCE Tongurahua erupted in Ecuador with a VEI of 5. This would have produced a volcanic winter. Also, in 1010 BCE Taupo in New Zealand erupted with a VEI of 4 and Te Kohaiakatu Pt. shot out 0.4 cubic kilometers of tephra.

What is the tally of large meteorite impacts in the Second Millenium? The Vaca Muerta Crater in Antofagasta Province in Chile, the Luna Crater in the Sind of India, the Panela Crater in Pernambuco in Brazil, Broken Bow in Nebraska, Aundha Nagnath in Maharashtra State in India as well as the Trimbakeshwar Crater in the Nashik district of Maharashtra State in India. There are also legends of fiery disaster coming from the sky onto Libya in North Africa and the legendary destruction of the island home of the People of the Sea by a cosmic body.

1000 BCE. Sub-Atlantic Period

In 1000 BCE Kujusan in Japan erupted with a VEI of 4, and Krasheninnikov erupted with a VEI of 3 in Kamchatka, and Snaefellsjokull also erupted in Iceland with a VEI of 2.

1000 BCE was a rainfall maxima period in Europe. These continued at 500-year intervals with lesser periods of rain at 200-year intervals until the time of Alexander the Great (356-323 BCE).

Over the whole period of the Second Millenium BCE there was major glacier regrowth and falling sea-levels.

Extraterrestrial bodies hit the eastern North Atlantic Ocean according to Palmer and Bailey from 1000 BCE to 950 BCE. This mainly affected the Mediterranean parts of Africa and Europe.

Franzen and Larsson state that relatively large extraterrestrial bodies hit the eastern North Atlantic Ocean.

Glacier advances, changes in the composition of the forests and retreat of the forest from its previous northern and upper limits, indicate significant cooling of world climates, its start being detectable in some places from as early as 1500 BCE. In Europe the most marked change seems to have been from 1200BCE to 700 BCE. By 700-500 BCE prevailing temperatures must have been 2° Celsius lower than they had been half a millennium earlier and there was a great increase of wetness everywhere north of the Alps. It was a period of mild winters and great windiness: cooling of the summers was presumably one the most notable features as well.

The climate became colder in the early Iron Age.

Between 1000 BCE and 000 BCE a drying trend shrank forests in Central Africa. Drying climate created the Dahomey Gap where savanna cut through the rainforest in West Africa. Oil palms spread out and proliferated in the gap. Human collection of and use of wood for smelting iron could have contributed to the decline of the forests.

The Dalgaranga Meteor Crater is only twenty-four meters in diameter and is three thousand years old. Fragments of mesosiderite stony-iron meteorite around the crater confirm the impact origin. The assymetrical crater structure and the ejecta blanket indicates that the falling body did so at a very low angle. This was near Mount Magnet in Western Australia.

There was a dust veil event recorded in China in 1000 BCE.

In the First Millenium BCE during the wet period wooden trackways were laid across the fens and marshlands in Somerset, England and elsewhere in England. They were later abandoned in favor of the use of boats. Lake villages were built at Glastonbury and Meare, possibly to take advantage for defensive purposes of the wet conditions then prevailing in the Somerset levels. There were shallow lakes or meres there again in the sixteenth century CE which have since disappeared.

Around 1000 BCE there was a drying period and farming extended to the flat lowlands of Holderness near the east coast of Yorkshire and after, though the region ultimately became marshy again.

Around 1000 BCE Alpine lakes in Europe shrank to a minimum. This was a drier period. In this time the chief settlements were in moist places and agriculture was carried on above the forest level, even above Alpine passes that are now glaciated.

The European peat bogs dried out in a warmer interval around 1000 BCE. New lake settlements were built after that, and farming activity was renewed even above the Alpine forest limit.

Glaciers advanced for five hundred years from 1000 BCE in the Alps in Europe. In this period the upper forest limit was lowered. Because of this the previously busy mines in the Alps, as well as traffic routes, were abandoned.

In the German Rhineland a vast majority of oak trunks show evidence of massive flooding around 1000 BCE.

Between 1000 BCE-400 BCE spruce trees spread across all central Sweden indicating a colder climate.

In 990 BCE Kujusan in Japan erupted with a VEI of 4.

In 960 BCE Antillanca Volcanic Complex erupted in Chile with a VEI of 5.

In 950 BCE Miyakejima in Japan erupted with a VEI of 4. Sheveluch in Kamchatka erupted with an explosion of VEI 5 and Sand Mountain Field erupted in Oregon with a VEI of 4.

In 930 BCE Mount Fuji, Fujisan, in Honshu, Japan, erupted. It was a VEI 5 eruption. Another eruption that year was Khodutka in Kamchatka with a VEI of 5. Azufral in Colombia erupted with a VEI of 4.

There was a recession of hazel and lime trees and the simultaneous spreading of hornbeam in the western reaches of the lower Oder Valley in Germany between 930 to 830 BCE.

In 920 BCE Sollipulli in Chile erupted with a VEI of 5.

900 BCE

In 900 BCE the eruptions were Izu-Shima in Japan with a VEI of 2, and Sheveluch in Kamchatka with a VEI of 3.

From 900 BCE there was a warming period and from 900 BCE to 500 BCE there were warmer sea temperatures.

The period of 900 BCE to 200 BCE may have been influenced by the expansion of the Polar Vortex that was expanding/cooling, therefore both increasing the jet stream strength with greater vigor of depressions and nudging the whole weather pattern southwards.

After 3,000 BCE there had been an increased frequency of cooler and stormier periods that peaked in 900 BCE to 650 BCE. There was a marked increase in storm frequency. In Britain in the west and north of the islands there was increased wetness, and it was cooler, more unsettled and stormier.

In Europe and England there was much evidence of raised trackway building indicating a very wet landscape. The wet climate lingered on. Wooden trackways in Somerset in England were built, apparently in an effort to keep open established routes across the Somerset Levels when it was becoming increasingly marshy.

There was growth of peat bogs.

There was woodland decay with the treeline moving both south in latitude and lower in altitude and retreating from exposed /west facing coastal areas.

The tendency to build hill forts was a response to wetter conditions overall according to some sources.

In the east and south of Britain other than being notably dryer in the east the increased storminess implied some effects such as coastal inundation.

People throughout highland Britain retreated to lower levels due to the sudden cold change. There were major vegetational changes as woodland gave way to grassland. Cooler temperatures and ample rainfall had the effect of fostering agricultural productivity resulting from new farming methods, especially the use of the plough and iron tools. The carrying capacity of good soil rose sharply as the land was enclosed and kept permanently cleared of regenerated woodland.

During the ninth century BCE the climate of the Eurasian steppe suddenly became colder and drier. Within generations standing water supplies had dried up. The drought played havoc with seasonal movements of flocks and herds.

The Chinese author Chin Li-Hsiang (1252-1303 AD) studied the evidence of pattern of cultivation in China in the Chou, Chin and two Han Dynasties in the period between 900 BCE and 220 CE and that that the climate was much warmer than in his own times.

Between 900 BCE to 300 BCE in Lebanon there was a lowered snow line in the high mountains. This also happened in the Near East and in Equatorial Africa.

In 880 BCE Vesuvius in Italy erupted with a VEI of 4.

The First Book of Kings mentions a drought in King Ahab's reign that lasted three years.

In 850 BCE Nevado del Ruiz in Colombia erupted with a VEI of 4. Also, in 850 BCE Katla in Iceland erupted with a VEI of 4. Still in 850 BCE Zavaritsky in Kamchatka erupted with a VEI of 4. Other eruptions were Karymsky with a VEI of 3 and Krasheninnikov in Kamchatka with a VEI of 4.

There were four major eruptions with a VEI of 4 which can cause stratospheric weather problems. How much can four VEI 4s make?

A sharp cold snap occurred simultaneously over a wide area in 850 BCE. This coincided with a sudden reduction in sunspot activity, an increase in cosmic flux and a much higher production of Carbon 14 in the atmosphere. This was a reduction in solar activity. The sun actually shone less for a few centuries. Solar activity seems to be the forming mechanism behind the change to cooler and wetter conditions in higher and middle latitudes.

In 830 BCE Tengger Caldera in Indonesia erupted with a VEI of 4.

In 820 BCE Soufriere in Guadeloupe erupted with a VEI of 3.

800 BCE

In 800 BCE Yantarni in the Aleutians erupted with a VEI of 5 as well as Momotombo in Nicaragua which erupted with a VEI of 4. Taupo in New Zealand also erupted from Ouaha Hills ejecting 0.23 cubic kilometers of tephra, Zavaritsky in Kamchatka with a VEI of 2, Belknap in Oregon with a VEI of 2, and Three Sisters in Oregon with a VEI of 2.

There was flooding of Central European Lake Dwelling villages around 800 BCE. The new lake settlements that were built around 1000 BCE and farming activity had been renewed even above the Alpine forest limit. All of this renewal collapsed in the wetter, colder climate that ruled after 800 BCE.

The climate around Halstatt in Austria built up around the salt mines and trade of the area near Salzburg collapsed catastrophically.

There were variations in moisture levels and forest growths in the valleys of Mexico and the Yucatan, as well as Cambodia, that hinted at serious flooding.

During the eighth century BCE the drought on the Asian steppe sent nomads pouring into China. They were repulsed, setting in motion a domino effect of population movements that brought horse-using nomads to the Danube Basin in Europe and the eastern frontier of the Celtic world. Weather around 800 BCE became warmer than the late twentieth century.

In Norway in 800 BCE the glaciers had been advancing and had reached positions almost as far forward as in the worst periods of recent centuries.

There was a boost in precipitation in Nazca in Peru between 800 BCE to 650 CE.

Around 800 BCE the spruce tree arrived from the east into Scandinavia. This was also the time that the spruce tree became general in Finland and crossed Central Sweden and migrated into Hedmark in eastern Norway.

Around 800 BCE to 400 BCE there was unmatched wetness in the west of Britain that at the great bog at Tregaron in west Wales nearly one meter of peat was added in only four centuries. This was as much as the thickness in the succeeding two thousand years.

In 780 BCE Sheveluch in Kamchatka erupted with a VEI of 4 and Pico de Orizaba erupted in Mexico with a VEI of 3.

In 770 BCE Concepcion in Nicaragua erupted with a VEI of 4.

In 776 BCE climatic catastrophes struck Egypt. This period lasted until 687 BCE.

In 750 BCE Towada on Honshu in Japan erupted with a VEI of 4. Another eruption was Izu-Toba with a VEI of 3.

Between 750 BCE and 500 BCE there was a notable wet climate in northwest Europe.

In Norway and the European Alps, the glaciers were advancing, perhaps more rapidly than at any other stage since the warmest post-glacial times. They came forward from their minimal extent to produce moraines almost as far forward as those of the recent cold centuries of the twentieth century. In some places they even overstepped this.

Between 750 BCE to 350 BCE Mount Tongariro erupted in New Zealand. It had a VEI of 5.

In 749 BCE Kie Matubu in Indonesia erupted with a VEI of 3.

In 730 BCE Cerro Bravo in Colombia erupted with a VEI of 4.

700 BCE

In 700 BCE the eruptions were Niigata-Yakeyama with a VEI of 3, Ata with a VEI of 3 and Nasudake with a VEI of 3, all in Japan,

Measurements taken from European peat bogs show a decrease in solar radiation. This decline in solar irradiance led to increased precipitation and rain, which reduced deserts and expanded steppe grasslands in Central Asia and southern Siberia. This produced more plentiful forage which would have increased the nomadic population of the regions as their horses and livestock would have more to eat.

The eruptions in 700 BCE were Niigata-Yakeyama in Japan with a VEI of 3, and Nasudake in Japan erupted with a VEI of 3.

In 700 BCE Mesopotamia experienced dry conditions. An Assyrian court astrologer wrote that due to the difficult conditions "No harvest was reaped". Around this time the Assyrian Empire collapsed when Medes and Babylonians attacked it.

In 690 BCE Pululahua in Ecuador erupted with a VEI of 5. This can cause a volcanic winter.

In 680 BCE Taveuni in Fiji erupted with a VEI of 2.

In 670 BCE Villarica in Chile erupted with a VEI of 4.

In 665 BCE Ljosufjoll in Iceland erupted with a VEI of 2.

In 660 BCE Alney-Chashakondzha erupted in Kamchatka with a VEI of 3.

In 650 BCE Krafla in Iceland erupted with a VEI of 4 along with Ata in Japan which erupted with a VEI of 4, Sheveluch with a VEI of 4 and Alney-Chashakondzha with a VEI of 3 in Kamchatka.

Around 650 BCE there was a general increase in rainfall and snowfall with the temperatures falling. In Britain the Western and highland areas were notably wet. There was the abandonment of upland settlements and the retreat of the treeline to lower altitudes. The east of Britain became wetter suggesting a tendency to cyclonic activity over the British Isles rather than passing to the northwest.

In 640 BCE Irazu in Costa Rica erupted with a VEI of 3.

In 610 BCE Nevado del Tolima in Colombia erupted with a VEI of 3.

600 BCE. The Secondary Climatic Optimum or Little Optimum

In 600 BCE Aogishima in Japan erupted with a VEI of 4. Other eruptions were Izu-Oshima with a VEI of 3, Miyakejima with a VEI of 3 in Japan, Taveuni in Fiji with a VEI of 2 and Vesuvius with a VEI of 3 in Italy.

In the middle of the millennium there were sharp changes and very severe weather. This seems to be every thousand years in the middle of the millennium.

There was a gradual, fluctuating recovery of warmth and a tendency toward a drier climate in Europe over the 1000 years after 600 BCE. Particularly after 100 BCE, leading to a period of warmth and apparently of high sea level around 400 CE. After some reversion to colder and wetter climes in the next three to four centuries, sharply renewed warming from about 800 CE led to an important warm epoch which seems to have culminated around 900 CE-1200 CE in Greenland and 1100AD to 1300 CE in Europe. In these few centuries the climates in the countries concerned evidently became nearly as warm as in the post glacial warmest times. This was in the European Dark Ages and in the Early Middle Ages.

Around 600 BCE there were frequent sandstorms in Jutland in North Germany and Denmark. A raised peat bog at Fugslo Mose was found to contain layers of mineral dust attributed to frequent sandstorms. The greatest were from 600 BCE to 00 BCE and in the later Middle Ages.

Around 600 BCE there was a series of blowing sand incidents in the Netherlands. The sea levels were lower again. Every time the sea levels lowered there was an increase in blowing sand incidents.

In Denmark in this same period, study of a peat bog has shown the intrusion of layers of blown sand between 600 BCE and the time of Christ. Lower sea levels caused this.

In France the Rhone Glacier started to melt. Many other glaciers started to melt as well.

In Babylon in Iraq there is a comparison of barley harvests between Old Babylon between 1800 to 1600 BCE with later Babylon around 600 BCE. J. Neumann of the Hebrew University of Jerusalem found that the mean date of the beginning of the barley harvest had become later by more than a month, shifting from late March to early May. This indicates a gradual cooling of the climate.

Around 600 BCE and for several centuries there were cooler conditions in lower latitudes in Africa and in Hither Asia. Hither Asia was an area encompassing Assyria, Babylonia, Persia, Elam, Syria, Palestine and the Hittite world.

In Israel there was more frequent snowfall that lay in longer into the summer on the slopes of Lebanon possibly through the year.

In the eastern and southern Mediterranean in this colder climatic regime, the people were perhaps helped by the increased winter rains which brought a greater degree of fertility to Greece and the northern fringe of Africa. This made it possible for the later Carthaginian and Roman croplands in North Africa which were to become the Roman granary in later periods.

In Greece in this period the inhabitants had gone over to warmer clothing and pitched gable roofs on their houses instead of the previous flat ones as the climate got colder. The previous Mycenean and Minoan Crete and other Greek cultures went around in semi-nudity with the warmer climate previously.

In 596 BCE GRIP ice cores from Greenland indicate an extraordinary amount of ammonium released into the atmosphere. The next peak would be 1312-1315 CE. Grip is Greenland Ice Core Project.

In 590 BCE Apagado in Chile erupted with a VEI of 4 and Mont Pelee in Martinique erupted with a VEI of 4. Adatarayama in Japan also erupted with a VEI of 3.

In 580 BCE Soufriere Guadeloupe in the Lesser Antilles erupted with a VEI of 2.

In 550 BCE Shikotsu in Japan erupted with a volcanic winter causing VEI of 5. Also, in 550 BCE Tongariro in New Zealand erupted with a VEI of 5. Tongariro ejected 1.2 cubic kilometers of tephra. Other eruptions were Bakening in Kamchatka with a VEI of 2, Kusatsu-Shiranesan in Japan with a VEI of 3 and Mount Adams in Washington with a VEI of 2.

In 530 BCE Mount St. Helens in California erupted with a VEI of 5.

500 BCE

In 500 BCE Mount Tarumae in Hokkaido, Japan, erupted with a VEI of 5 and Sheveluch in Kamchatka erupted with a VEI of 4.

Between 500 BCE and 100 BCE Popocatepetl erupted in Mexico with a VEI of 5.

Around 500 BCE from what is known of the thermal patterns over the northern hemisphere, suggests that a cooling Arctic had pushed the cyclonic activity south over northern Europe. This was probably from the development of cold westerlies across the Atlantic from Canada and Greenland to Britain and central Europe in winter and the prevalence of cyclonic northwesterly winds in summer over the British Isles and across Europe to the Mediterranean. This was a time of sharp climatic cooling and glacier growth in most parts of the world.

There were exceptionally high waters and flooding from North Sea storm surges, when the winds veered to the northwest and north behind some of the most intense cyclonic depressions.

There was some outstanding storminess in this period. This is indicated by the activity of sand dunes around the coasts of northwest Europe from south Wales to Denmark. Certain spits of land on the east coast of Scotland in the Firth of Forth appear to have formed as sand dunes, or sandbanks, in northerly storms and have been dated to 500 BCE. A slight lowering of the sea level due to the build-up of glaciers may have contributed by exposing greater expanses of sand in the estuaries and along the coast.

Ancient harbor works from this period and now generally submerged at Naples in Italy and in the Adriatic Sea suggest a mean sea level about one meter lower than the late twentieth century. Allowing for Mediterranean Sea instability this suggests that sea levels dropped in the Mediterranean by about two meters or more from 2000 BCE to 500 BCE. What is certain is that world sea levels rose during the Roman Empire, finally reaching a high stand around 400 CE comparable with or slightly higher than the late twentieth century.

Around 500 BCE the climate in England became much wetter than before in eastern parts so that now all parts of England, Wales and Ireland were affected by the notably wet weather. This implies that the winds were no longer predominantly from the west and that cyclonic activity frequently passed over or near the southern part of the British Isles. This extended further south and east than it had done in the immediately preceding centuries. This resembled Europe in the fifteenth century CE.

The ancient ridge routes which had already been established across England, such as the Icknield Way from near Stonehenge to Norfolk, the Cotswold ridgeway from near Bristol to Lincoln, and the route later called the Pilgrim's Way from Winchester to Canterbury, avoided not only the thicker forest but the often-swampy lowland areas.

In Central America the cold period lasted from 500 BCE to 100 BCE, which was contemporary with Europe and much of North America. Notably the United States east of the Rockies and the central Canadian Arctic. They were probably moister than in North America than in Central America, if the summer rains did not move so far north.

500 BCE was a rainfall maxima period in Europe. These continued at 500-year intervals with lesser periods of rain at 200 year intervals until the time of Alexander the Great (356-323 BCE).

In 500 BCE a periodic comet that ushered in a low temperature phase returned for a closer passage, raining down a barrage of meteoric destruction around the Northern Hemisphere. It has been postulated that the great megalithic observatories could not be used due to the cloud cover that lasted generally until around 700 BCE. This is the beginning of the Sub Atlantic Period though other sources state that it started in 450 BCE or 625 BCE.

World sea levels were lower after 500 BCE. This as we know indicates a lowering of temperatures and the growth of the ice masses.

One aspect of the colder climate around 500 BCE in northwest Europe was the storminess and the frequency of blowing sand near the coasts. This was probably caused by the lowered sea level that resulted from the build-up of glaciers.

The overall decline of Holocene forests around 500 BCE favored the spread of the oil palm in Central Africa and produced a shift from landscapes of evergreens to savannas.

Antarctic zones had entered a colder period by 500-300 BCE than in the preceding millennia.

The cooling stage in the Arctic continental region lowered the prevailing summer temperatures by as much as 3° to 4° C.

In Australia around 500 BCE the climate started to get increasingly wet until 500 CE when it went dry again. The dry conditions lasted until 1200AD.

In China temperatures continued increasing from the higher than late twentieth century temperatures that had been achieved around 800 BCE. It was now possible to grow two crops of millet a year in the southern part of Shantung at 36° North.

In the European Alps about 500 BCE the level of the lakes rose. In the Bodensee (Lake Constance) the rise exceeded thirty feet, most of the lake villages were destroyed and settlement in the Alps reached a minimum and concentrated in the lowest valleys. This climatic fluctuation had the appearance of a catastrophe. There was a sharp sinking of the forest limit in the Alps in Hallstatt times. The formerly thriving mining industry and traffic in the high Alps ceased. The lake settlements around the fringe of the Alps were finally abandoned. From this wettest period recovery followed. The rainfall by Roman times was very little above the present. Within the whole Aline region population seems to have fallen to a minimum, largely confined to the warmest valleys.

Around 500 BCE the glaciers were advancing in the Alps and lakes flooded the surrounding settlements. The glaciers brought to an end the previously flourishing high-level mining: for instance, gold mining in the Hohe Tauern. The glaciers stopped the traffic over the Alpine passes. The upper limits of the forests fell sharply, and their composition altered. The oaks and other broad-leafed trees lost a great deal of ground to firs and pines. Since 1200 BCE spruce had been coming from the southeast.

In Tierra del Fuego in South America, about 54° south, there was a spread of forest to cover the whole island on both sides on both sides of the watershed.

In 479 BCE during the Persian siege of the maritime city of Potidaea, Greece, Herodotus reports how Persian attackers attempting to take advantage of an unusual retreat of the water were suddenly surprised by "a great tide, higher, as the locals say, than any one of many that had been before". Herodotus attributes the cause of the flash flood to Poseidon's wrath.

In 476 BCE a massive meteorite crashed to earth in Thracia (Thrace). This was a thunderstone the size of a chariot.

In 450 BCE Pululahua erupted in Ecuador with a VEI of 4 and Bezymianny in Kamchatka erupted with a VEI of 4. Mutnovsky in Kamchatka erupted with a VEI of 2 and Mount Cameroon in Cameroon erupted with a VEI of 3.

In 440 BCE Mont Pelee erupted in Martinique with a VEI of 4.

In 426 BCE a tsunami struck the gulf between the northwestern tip of Euboea and Lamia in Greece. The Greek historian Thucydides (3.89.1–6) described how the tsunami and a series of earthquakes affected the Peloponnesian War (431–404 BC) and, for the first time, associated earthquakes with waves in terms of cause and effect.

In 410 BCE Mount Meager Massif erupted in British Columbia with a VEI of 5.

400 BCE. Climatic Optimum. The Roman Warm Period.

Around 400 BCE, give or take fifty years, the Mount Meager Massif in British Columbia erupted with a VEI of 5. This is probably the eruption that created the Bridge River Vent around 400 BCE which caused lahars, lava dome extrusions and pyroclastic flows. Eastward migration of the eruption column spread material across Western Canada. Ash fell as far east as the Fraser River which is 110 kilometers to the east. This eruption would have caused a volcanic winter. In

400 BCE Cotopaxi in Ecuador erupted with a VEI of 4, Taveuni in Fiji erupted with a VEI of 2, Mount Adams in Washington with a VEI of 2, Sheveluch in Kamchatka with a VEI of 3 and Reykjanes in Iceland with a VEI of 2.

This was a period of warm and stable climate from roughly 400 BCE to 200 CE. Rome in Italy and the Han Dynasty in China flourished during this period.

Around 400 BCE the great steppes of Asia were entering a large drought period. The steppes extended from the eastern margins of Europe to east Asia, bounded by desert to the south and cold boreal forest to the north. The steppes boundary had shifted constantly since the Ice Age, expanding and contracting north and south as rainfall patterns shifted. Like the Sahara Desert the Eurasian steppes acted like a pump, sucking in nomadic people during periods of high rainfall, pushing them out to the margins and onto neighboring lands when drought came.

Between 400 BCE to 300 BCE there was a possible peak of stormy, wet conditions in northwest Europe including the British Isles. Rainfall was as much as forty per cent higher than the late 20th Century. The summers were frequently cool and unsettled.

In the British Isles there was significant evidence of wooden trackway construction across lowlands to avoid long detours around flooded land and also to enhance security. Hill forts may have been built to stand up out of the bogs, swamps and fenlands.

Between 400 BCE-200 CE in Egypt. During the climate optimum flooding in the Nile River led to a ration of good harvests, boosting agricultural productivity. Egypt was the major grain supplier to Rome.

Evidence collected from the Po River delta, the Adriatic Sea and Alps indicate rising temperatures in Italy. This was a detectable warm period. This caused population growth.

Between 400 BCE-200 CE in Libya in North Africa there were Roman fortified farms called Gsur, or castles, that stand in areas today that receive scant rainfall and are inhabited by pastoralists who grow crops in valley floors following rain. There were huge Roman cities in Libya that could not survive in the deserts there now.

In Rome between 400 BCE-200 CE during the Climatic Optimum Roman authors observed the shift in range of trees, including birch and chestnut, and of the cultivation of olives and grapes. The author Colomella wrote that formerly the unremitting severity of winter could not safeguard any shoot of the vine or olive planted in them. Now the coldness had abated, and the weather was becoming more clement, producing olive harvests and the vintages of Bacchus in the greatest abundance. Olives were being grown in new regions including Gaul and France. Roman agriculture also expanded the area for growing grapevines to the north.

Between 400BC-200 BCE residents of Palmyra in Syria stored water in cisterns and used it for agriculture in an area that is now desert. Was there more rain in this period to fill these cisterns?

In the winter of 398 BCE the Tiber River in Rome was frozen. It was frozen again in 396 BCE.

In 396 BCE Corbetti in Ethiopia erupted with a VEI of 5. This was enough for a volcanic winter.

In 380 BCE Arenal in Costa Rica erupted with a VEI of 4.

In 373 BCE a comet was seen at the time of the great earthquake and tidal wave that destroyed the cities of Helice and Bura in Greece. Helice was some twelve stades, about two kilometers, from the beach. The Greek writer Ephorus claimed that the comet was seen to split into two parts. Was the earthquake caused by the impact of debris of the comet? There is a

Greenland GRIP record of an ammonium spike at this same time. The writer Aelian recorded that five days before the city of Helice was destroyed, all the wild creatures from centipedes to martyns got on the road and left town. The fate of the city, which remained permanently submerged, was often commented on by ancient writers.

In 360 BCE Vulcano Island in the Aeolian Islands in Italy erupted with a VEI of 2.

In 350 BCE Yakedake in Japan erupted with a VEI of 4 and the Three Sisters in Oregon erupted with a VEI of 4. Other eruptions were Hachijojima with a VEI of 3, Akita-Komagatake with a VEI of 3 in Japan, Gorely with a VEI of 3 in Kamchatka, and Stromboli in Italy with a VEI of 2.

The period around 350 BCE was a period of massive increased rainfall. A great number of towns and communal settlements were submerged under enormous depths of flood silt. Some villages were often completely rebuilt on the ruins of their predecessors.

In Scandinavia it was so wet around 350 BCE that boats were used to travel from one village to another.

The British Isles were so marshy and swampy that the earlier settlers had to use wooden trackways to cross the fens and marshlands in Essex and other flat lowlands. Later these trackways had to be abandoned in favor of boats. Villages were actually built up high in the middle of lakes to minimize the effects of further storm floods.

By 350 BCE it had gotten so wet that trackways were abandoned on the Somerset Levels in England and replaced by boats.

In 340 BCE Merapi erupted in Indonesia with a VEI of 3.

In 330 BCE Taveuni in Fiji erupted with a VEI of 3.

In 320 BCE Atacazo in Ecuador erupted with a VEI of 5. This is volcanic winter causing.

300 BCE

In 300 BCE Sheveluch erupted in Kamchatka with a VEI of 4. Other eruptions were Heidarspordar in Iceland with a VEI of 2, Vulcano in Italy with a VEI of 2, and the Wapi Lava Field in Idaho with a VEI of 2.

By 300 BCE the ecotone between continental and Mediterranean climatic zones had moved north, this was at least as far as modern-day Burgundy. This brought a much more Mediterranean climate, with warm dry summers and wet winters to the more southerly Celtic domains. Roman agriculture was based on extensive production of a few crops such as wheat and millet for large urban populations and was much better suited to the semi-arid southern European environment. As the ecotone moved north Rome gained power rapidly. The warm conditions persisted through the next five centuries. The northern frontier of the Mediterranean zone now lay far to the north. This lengthened the growing season for cereal crops that supported Rome's garrisons and cities.

In Europe mean temperatures were at least 1°C below those of the warmest post-glacial periods. Possibly it might have been 2°C.

Between 300 BCE and 300 CE conditions in West Africa were stable and dry. As they were also in Southeast Asia and the Amazon Basin, with rainfall below modern levels.

For some time after 300 BCE forests covered the whole of Tierra del Fuego.

Due to the warmness, the glaciers in the Alps were in retreat from 300 BCE to 400 CE. That the Alpine glaciers were not so extensive as they are now is indicated by Roman gold mines high up in the Alps in the Sonnblick area in Austria. Some of these are probably still under ice since others have only come to light recently as the ice receded. Traffic over the Alpine passes continued even in wintertime the climate was so mild.

After 300 BCE the Sahara Desert started completely drying.

The Tiber in Rome froze in the winter of 271 BCE. There were a few severe winters in fact when the Tiber froze in this period and snow lay around for many days yet the general weather was warmer.

Beech trees also grew around Rome.

In 270 BCE Ata in Japan erupted with a VEI of 4 and Arenal in Costa Rica also erupted with a VEI of 4.

Between 270 BCE to 70 BCE Masaya in Nicaragua erupted with a VEI of 5.

In 259 BCE Methana in Greece erupted with a VEI of 3.

In 250 BCE Raoul erupted in the Kermadec Islands near New Zealand with a VEI of 6. More than ten cubic kilometers of tephra were released. Denham caldera was formed during a major dacitic eruption and truncated the western side of the island and is 6.5 x 4 kilometers wide. This was definitely volcanic winter causing. Anything over VEI 4 is also volcanic winter creating.

Mount Rainier in Washington also erupted in 250 BCE with a VEI of 4 and Nasaduke in Japan with a VEI of 2 .

Around 250 BCE lake villages were built on piles at Glastonbury and Meare in England the water levels were so high.

In 230 BCE Antillanca Volcanic Complex in Chile erupted with a VEI of 5. Cotopaxi in Ecuador also erupted with a VEI of 4 as well in 230 BCE.

In 226 BCE an earthquake destroyed the Colossus of Rhodes and the city of Kameiro in the eastern Mediterranean.

In 217 BCE Vesuvius in Italy erupted with a VEI of 3. During a year that was wracked by earthquakes, a shower of glowing stones appeared suddenly from the south, had fallen at Praeneste near Rome, while at Capua there was the appearance of the sky on fire. There were sun darkenings and sky-glows and according to the poet Silius Italicus, the cause was Mount Vesuvius. The problem is that Praeneste, now Palestrina, is over one hundred miles northwest of Mount Vesuvius. "Vesuvius…thundered, hurling flames worthy of Etna from her cliffs; and the fiery crest, throwing rocks up to the clouds, reached the trembling stars." This indicates that the Romans knew that Vesuvius was the same as Etna and was a volcano. There were no more eruptions until 79 CE.

Livy wrote that the sun's disc seemed to be diminished in 217 BCE. Could this be Vesuvius though as a VEI 3 eruption does not send its smoke column more than fifteen kilometers into the sky?

In 212 BCE Livy reported on the fall of stones in Italy.

In 210 BCE Stromboli erupted in Italy with a VEI of 2.

Dendrochronology indicated that there was a dramatic narrow tree ring event from 208 to 204 BCE.

In 208 BCE in China the stars were lost from view for three months and there were famines and dynastic change.

In 207 BCE the Hudson Mountains erupted in Antarctica with a VEI of 4.

In 207 BCE a comet or asteroid crashed into modern-day Germany. This unleashed energy equivalent to thousands of atomic bombs. The 1.1-kilometer diameter rock whacked into southeast Bavaria, leaving an exceptional field of meteorites and impact craters that stretch from the town of Altoetting to an area around Lake Chiemsee. The original space rock possibly broke up seventy kilometers above the earth and the largest chunk hit the earth with the power of 8,500 Hiroshima atomic bombs. The forest beneath the blast would have ignited suddenly, burning until the impact's blast wave blew it out. Dust must have been blown into the stratosphere where it would have been transported around the earth, and the region must have been devastated for decades. The biggest crater is Lake Tuttensee, a circular lake measuring 370 meters, 1200feet, across. There are scores of smaller craters and other meteoric impacts in an elliptical field, inflicted by other debris. Celtic artifacts were found that were scorched on one side. Tree ring evidence from Irish oaks showed a slowdown in growth around 207 BCE. This may have been caused by a dust-veil event which filtered out sunlight. This is called the Chiemgau Impact.

Around the same time as the Chiemgau Impact, Roman authors wrote about showers of stones falling from the skies and terrifying the population. Might this impact been a couple of years earlier?

This 207 BCE event in the tree rings stood out as the narrowest ring event in the first millennium BCE. A possible dust veil event created crop failures, famines and epidemics as well as drought. China and the Mediterranean were affected at the same time yet again.

The area of deserts decreased during the period of Western Han dynasty from 206 BCE to 24 CE. The Western Han dynasty enjoyed better grain harvests in this period. After this period the state was more likely to suffer fiscal problems due to decreased harvests.

There were significant frost rings on both Sheep and Campito Mountains in California that date back to 206 BCE. A frost ring is found in dendrochronology.

Livy wrote that in the year 206 BCE two suns were seen over Alba (in Italy) and added that light appeared in the night at Fregellae, also in Italy.

Livy wrote that in 205 BCE the Romans consulted the Sybilline Books and were told to bring the image of Cybel, the Magna Mater, to Rome from Asia Minor. Cybele was brought to Rome to save the city. Livy wrote that this was due to showers of stones that were seen in 205 BCE.

Livy wrote that in 204 BCE a halo had encircled the sun.

And nothing was happening?

200 BCE

What was happening in 200 BC? Nevado del Ruiz in Colombia erupted with a VEI of 4, Mont Pelee in Martinique with a VEI of 4, Popocatepetl in Mexico with a VEI of 5, Yufu-Tsurami with a VEI of 4, Hakusan with a VEI of 4 in Japan and Taupo in New Zealand with a VEI of 4. Taupo ejected 0.28 cubic kilometers from Te Kohaiakatu Pt, Taveuni in Fiji had a VEI of 2, and Mutnovsky in Kamchatka had a VEI of 2.

Ksudach in Kamchatka had a VEI of 3. A VEI of 5 always causes a volcanic winter but even lower ranked eruptions do cause atmospheric disturbances as well. H. H. Lamb stated that any eruption including and above VEI 2 can do this dependent on the sulfur content released.

Around 200 BCE was the beginning of a notable upturn in temperature levels. A steady recovery (after the cooler late Iron Age) from now, right through the 'Roman-British' period, only petering out somewhere around 350 CE. Mean temperature levels based on 'middle' England would peak at around, or a shade below those of the 'Climatic Optimum' of the Early Middle Ages & precipitation amounts *overall* were declining - though adequate. The climate would become generally 'benign' by the time of the early years of the 1st Century CE, and eventually villas became bigger and built on hillier sites in areas we would not normally construct 'high status' buildings (at least until very recent times). Southern Britain in particular was self-sufficient in wine (a staple of Romano-British life, not a luxury), and there is evidence of the export of the same, implying very good conditions for growth and ripening. However, there are records (principally from the Roman occupation period), of severe and snowy winters. This is a warning to us nowadays not to assume that a 'warmer' climate necessarily means we will not have "severe" winters. Sea levels were thought to have risen between 1 and 2 meters compared with before and after.

Mean temperature levels in Middle England peaked at or around those of the Climatic Optimum.

In China temperatures started dropping from the 500 BCE highs. Chinese weather seems to run in opposition to that of Europe and North America at times.

Irish bog-Oak chronology suddenly ended in 200 BCE. This implied that some environmental alteration had made bogs incapable of supporting oak trees at this time. Trees disappear when the climate becomes unsuitable and then recolonize areas when it is suitable again

In 197 BCE Thera/Santorini erupted in the Mediterranean with a VEI of 3.

In 190 BCE Krysuvik-Trolladyngja erupted in Iceland with a VEI of 2.

In 183 BCE Vulcano in Italy erupted with a VEI of 4.

In 180 BCE Changbaishan in China/Korea erupted with a VEI of 4. Cayambe in Ecuador also erupted with a VEI of 4.

The Tiber River in Rome froze in the winter of 177 BCE.

In 170 BCE Masaya in Nicaragua erupted with a VEI of 5. Arenal in Costa Rica also erupted with a VEI of 4.

In 150 BCE Krasheninnikov in Kamchatka erupted with a VEI of 4. Other eruptions were Izu-Oshima in Japan with a VEI of 3 and Sheveluch in Kamchatka with a VEI of 3.

Between 150 BCE and 50 CE Apoyeque in Nicaragua erupted with a VEI of 6. This would have caused a major volcanic winter. This is regarded as one of the largest explosions known in history. What affect would it have had on the cultures around it?

In 122 BCE Mount Etna in Sicily erupted with a VEI of 5. This would have caused a volcanic winter.

Between 120 BCE-114 BCE in Jutland in northwest Germany and Denmark. This was the period of the Cymbrian/ Cimbrian/Kymbrian Flood where there was a large-scale incursion of the sea in the region of the Jutland Peninsula that resulted in a permanent alteration of the coastline with much land lost. The Jutland Peninsula forms part of the continental portion of Denmark and part of northern Germany. There was a great storm or a great series of storms in the North Sea Basin. These sea floodss affected the coastlines of Denmark, the Netherlands and Germany and probably the east coast of England. These changes are consistent with the change of temperature regime as it implies alteration and or intensification of jet stream patterns which often accompany major changes of climatic type. It is believed that one of the results of the Cymbrian Flood of the coasts of the German Bight was responsible for setting off a migration of

Celtic tribes. This event apparently caused the Cimbri to leave what are now the Dutch Polderlands and they migrated south with the Ambrones and Teutons and eventually into conflict with the Romans, precipitating the Cimbrian War, 113-101 BCE.

100 BCE

In 100 BCE Kujusan in Japan erupted with a VEI of 4. Other eruptions were Mutnovsky in Kamchatka which erupted with a VEI of 2 and Shikotsu in Japan with a VEI of 2.

From the first century BCE water receded worldwide due to higher temperatures until 200 CE, followed by a major high stand and incursion of the sea around 300 CE to 400 CE. Sea level was again rather lower in the seventh and eighth centuries but was higher again in the late thirteenth to fifteenth centuries.

There was the start of colder and drier climate starting in Greenland around 100 BCE.

The water level of the Bodensee (Lake Constance) was near its present level after rising thirty feet or ten meters 500 years before.

In 91 BCE Vulcano Island with a VEI of 3 erupted in Italy.

In 90 BCE Taveuni in Fiji erupted with a VEI of 2.

In 80 BCE Tenerife in the Canary Islands erupted with a VEI of 4. Ata in Japan also erupted with a VEI of 4 as well as El Misti in Peru with a VEI of 4.

In 60 BCE an earthquake of intensity IX and an estimated magnitude of 6.7 caused a tsunami on the coasts of Portugal and Galicia.

What happened in 50 BCE?

In 50 BCE the volcano Apoyeque on the Chiltepe Peninsula in Nicaragua erupted. This was one of the largest eruptions known in history. With this error factor could this eruption actually be 50 AD? Eighteen cubic kilometers of tephra were released. The VEI was 6. Also, in 50 BCE Akan in Japan erupted with a VEI of 4 and Rungwe in the Great Rift Valley of Tanzania erupted with a VEI of 4. Other eruptions were Tungurahua with a VEI of 3 and Cotopaxi with a VEI of 3 in Ecuador, Raoul Island in New Zealand with a VEI of 3, Miyakejima with a VEI of 3, Norikuradake with a VEI of 3, and Shikotsu with a VEI of 3, in Japan, and Three Sisters in Oregon with a VEI of 3. . Krafla in Iceland with a VEI of 2, Grimsvotn in Iceland with a VEI of 2, Akita-Komagatake in Japan with a VEI of 2, and Stromboli in Italy with a VEI of 2.

I am wary of lumps of eruptions appearing at the fifty year or hundred-year points in centuries. This is lazy editing by other authorities and the Smithsonian Global Vulcanism Program.

In 50 BCE Greenland ice cores indicate an ice-core acidity peak.

On December 31st 46 BCE Santorini erupted with a VEI of 3.

In March 44 BCE Mount Etna erupted with a VEI of 3.

Pliny wrote that there were portentous and protracted eclipses of the sun that occurred around 44 BCE. There were not solar eclipses visible from anywhere in the Roman Empire between February 48 BCE to December 41 BCE. There was a spectacular daylight comet in 44 BCE and a dust veil event occluded the sky over Italy in the spring. This was attributed to an unconfirmed eruption of Mount Etna.

Virgil wrote that after the death of Caesar in 44 BCE, "How often we saw Etna flooding out from her blast furnaces, boiling over the Cyclopean Fields, and whirling forth balls of flame and molten stones". And the Romans, we are told, did not know about volcanoes? They even named an island Vulcano because it was one. The Romans believed that Vulcano was the chimney of Vulcan, the God of Fire.

Sulfate deposits were found in Greenland ice cores for this year and there was also tree-ring data from North America pointing to a climate change in the late 40s. What hit and where it hit is unknown. Or is our date out by a year and the Okmok Caldera, dated as 43 BCE, in the Aleutians might be the culprit?

In 44 BCE there was a Dust Veil event in China. Chinese histories record a red daylight comet in May and June. The comet's color can be attributed to dust in the atmosphere.

Between 44-42 BCE six consecutive grain harvests failed in China over three years.

Scuderi's Foxtail pine chronology from the Sierra Nevada in California, showed that there was a large decrease between 44 and 43 BCE and continued decreased ring width values in both 42 and 41 BCE.

About 43-42 BCE Cassius requested aid from Cleopatra in Egypt but she responded that she could not supply any as Egypt was in the grip of famine and pestilence. Paul Handler has proposed there is the possibility that a dust veil event can reduce rainfall in the sources of the Nile River by altering monsoon patterns by cooling the surface of the Pacific.

In 43 BCE the Okmok Caldera erupted in the Aleutians. It was a VEI 6 eruption that spewed out 50 cubic kilometers of tephra that covered about one thousand square kilometers of Umnak Island. Some of the tephra was carried away as far as Greenland. This eruption led to crop failures and famine around the Mediterranean Sea. This potentially influenced events leading to the fall of the Roman Republic and the end of Pharaonic rule in Egypt. This volcanic winter is regarded as the event that might have changed the history of Egypt. Humans abandoned a village on Carlisle island 141 kilometers west of Okmok. There was impact on the island of Four Mountains west of Carlisle as well to the inhabitants. The eruption released about 15-16 teragrams of sulfur into the stratosphere causing a volcanic winter with 0.7°-7.4° (1.3°-13.3° F) cooling of the Northern Hemisphere. In the Mediterranean cooling reached about 1-4° C (1.8°-7.2° F). 43 BCE. 43 BCE and the following two years were among the coldest recorded during the last 2,500 years with the following decade being the fourth coldest.

In 43 BCE there were frost rings in bristlecone pines in the Sierra Nevada in California according to LaMarche and Hirschboeck. This was a worldwide cooling event.

Between May 43 BCE and March 42 BCE there was no evidence of a summer as there was no summerlike change in chemical composition of the atmosphere between the two acid peaks.

In 43 BCE Chinese records state that in April it snowed, and frost killed the mulberries. The sun was pale blue and cast no shadows. There was a famine in China.

Between 43 and 42 BCE there was still weather disruption in China. By October 43 BCE the sun appeared to have regained its color, but by spring 42 BCE the sun, moon and stars again appeared veiled and indistinct.

In 43 BCE the eruption of Okmok led to cold weather, snowfall, famines and a failure of floods on the Nile in Egypt. This led to the final collapse of the Ptolemaic Dynasty and the Roman Republic after the 31 BCE Battle of Actium leading to the Roman Empire.

In 43 BCE there were epidemics that were linked to Okmok.

Between 20 BCE and 180 CE Furnas on San Miguel island in the Azores erupted with a VEI of 5.

In 10 BCE Sheveluch erupted in Kamchatka with a VEI of 3.

The Roman author Strabo wrote in his *Geography*, about ten BCE, that olives were now being cultivated in southern France and its northern limit was about the same as in modern times.

The Roman agricultural writers Saserna, father and son, wrote that in the last century BCE cultivation of the olive and vine were spreading further north in Italy where in the previous century winters had been too cold for transplants to survive.

000 BCE/CE

Around the time of Christ world sea levels were very high. This indicates that the weather was very warm.

Southern Britain became self-sufficient in wine and was even exported, which implied that there were very good conditions for growth and ripening. Southern Britain was too cold normally to grow grapes before this period.

There were Roman records though of severe snowy winters at the time though in England. You can still have a warm climate and still have severe snowy winters.

Sea levels rose between one and two meters compared with before and after.

In the first century CE the Roman author Pliny writes that there were elephants native to North Africa including an extended area at the southern foot of the Atlas Mountains where numerous herds wandered in the forests and emerged in winter to roam over rich pastures. They seem to have finally died out in the third century CE possibly through increased aridity of the landscape.

From studying rock drawings of a wide range of animals and finds of skeletons, including elephants, there was enough surface water for animals to pass across a terrain that is now the Western and Central Sahara Desert. There was also human occupation.

There are unmistakable indications of formerly more extensive water supply and irrigation in Sinkiang in China. In the Tarim Basin in Central Asia, which was crossed by the Silk Road, and was used by trading caravans in Roman times, there was a chain of cities and settlements as well as the remains of forests. This area is now desert.

There were rivers in the Thar Desert, Rajasthan, of India and Pakistan that were still flowing during the Indus Valley Civilization which are now gone. This includes the Sarasvati River which was only rediscovered a few years ago and was formerly known only from legends. There were unmistakable signs of former more extensive water supply and vegetation, and this also includes in southern Asia crossed by Alexander the Great's army on expeditions in Iran and the march to the Indus between 330 and 323 BCE.

Other areas that were conquered by the desert in the last two thousand years are Palmyra in Syria, Petra on the fringe of Palestine and the Syrian desert in Jordan, and parts of the Anatolian Plateau in Turkiye. Never think now that the deserts that you see now, were there in the past.

In 1939 and 1940 on the shores of the Bering Sea in Alaska the remains of an ancient city of 800 houses was found, north of 68 degrees, 130 miles within the Arctic Circle. The age of the city is unknown but is at least 2000 years old. Ivory carvings, unlike Eskimo or American Indian, have been found here. Bodies in wooden log tombs had artificial eyeballs of ivory inlaid with jet. Numerous delicately carved and engraved implements resembling northern Chinese work of one thousand BCE were found. They were very sophisticated people and more advanced than the Eskimos and were clearly derived from central Asia. The city was designed on a grid pattern with wide avenues and like the cities of the Indus Valley was designed from the start. This city was called Ipiutak and according to legend was built by a fair-haired and blue-eyed race. To support such a large community there must have been a favorable agricultural climate and seeing that at present Ipiutak is beyond the Arctic Circle this then is technically impossible. Modern archaeology states that the Eskimos built the city which the Eskimos deny stating that the city was built by a white race when the climate here was warmer. Tilt the earth a bit and you get warmer in some places and colder in others. I tend to believe the Inuit or Eskimos.

There are ruins of a city at Point Barrow in North Slope, Alaska, that are the same age as the ruins of Ipiutak on Point Hope. There are oriented shallow depressions or lakes in the permafrost near Point Barrow that are aligned northwest to southwest. These are the same age and the same direction and shape as the Carolina Bays and are from the same period. Is there a connection to the Awak Meteor crater nearby?

In the early 1950's Dr J. Louis Giddings, Jnr, of the University of Pennsylvania conducted a study of ancient Arctic ruins encircling the entire top of the world in prehistoric times. After an expedition to northern Canada, he brought back grooving tools and side blades of a distinctive type thought previously to only come from Europe and Siberia.

On the California coast the Chumash had larger villages but for several centuries warmer temperatures had reduced the natural upwelling of the Pacific which had reduced the number of anchovies that had sustained growing village communities. Fisheries were generally less productive than in earlier times. The problem was that the populations of Chumash on the Channel Islands and the mainland were still rising steadily. In good years there were actually no surpluses so bad years were horrendous.

In 9 CE a great flood in the Thames valley in southern England killed men, women, children and cattle.

Also, in 9 CE there was a great overflow of the River Humber in England, flooding the country all around. The Humber is a large tidal estuary and North Sea inlet, separating Yorkshire from Lincolnshire.

Also, in 9 CE the town of Pickering in North Yorkshire, England, was burnt by lightning.

We shall see more of these towns that were often burnt down due to lightning. Was it because of the thatched roofs? Or was it lightning at all?

In 10 CE Mont Pelee on Martinique erupted with a VEI of 4.

Also, in 10 CE a famine that lasted six years hit Ireland. There was a general fruitlessness of poor harvest, that gave rise to famine and great mortality.

In 14 CE there was a great overflow of River Severn in England, causing great damage.

From 14-17 CE heavy rains and severe flooding killed millions of people in China in the first century CE. Flood control projects on the Yangtze River or Yellow River, Huang He, plains expanded to hundreds of miles of levees. The intense farming that sustained the Han Dynasty's large and growing population increased erosion and led to further construction of levees. In 14-17 CE the levee system broke down in a series of massive floods and caused many deaths and crisis in the empire. The river changed course and forged a new path to the sea, a hundred miles away from its former mouth. Repair work took several decades.

In Rome, Italy, in 15 CE, the Tiber River overflowed and did such serious damage that it was proposed in the Roman Senate to diminish its waters by diverting some of the chief tributaries.

In 17 CE an earthquake destroyed 13 cities in Asia Minor, primarily in Lydia in Turkey.

Eruptions starting in 19 CE, repeating in 49 CE and 60 CE, created the island of Theia, "Divine", now called Palaea Kameini in the caldera of Santorini/Thera. Philostratus reported the unusual detail that in the 60 CE eruption that the sea receded one kilometer from the south coast of the island before returning. What better a tsunami description.

In 20 CE Merapi in Indonesia erupted with a VEI of 4.

The Roman author Pliny the Elder, 23 CE to 79 CE, stated in his *Natural History* that the beech tree which he regarded as a mountain tree formerly grew within the precincts of Rome but did so no longer in his day. The climate seeming too hot there now, as it was also for chestnut trees. Theophrastus had also mentioned that beech trees had grown in Rome, but it was now too hot for this. This indicates that the climate of Rome was getting much warmer after the period of 500 BCE that was much colder.

In 24 CE the city of Caerleon in south Wales was burnt by lightning.

In 29 CE there was a great overflow of the River Trent in England.

In 30 CE the eruption was Ata in Japan with a VEI of 3.

In 35 CE Quetrupillan in Chile erupted with a VEI of 4.

Columella, a Roman writer about 30 CE to 60 CE in *De De Rustica* wrote that due to increasing warmth the vine and the olive were still slowly working their way northwards up the leg of Italy. Columella wrote that the position of the heavens has changed… regions which on account of the regular severity of the weather could give no protection to any vine or olive stock planted there, now that the former cold has abated and the weather is warmer, produce olive and vintages in greatest abundance. (The grape vine had been cultivated in Rome about 150 BCE. Author)

Temperatures in Southern England in this period were warmer than in the 21st Century.

In 40 CE Turrialba in Costa Rica erupted with a VEI of 4. Other eruptions were Pico de Orizaba in Mexico with a VEI of 3, Ischia in Italy with a VEI of 3, and The Three Sisters in Oregon with a VEI of 2.

In the year 43 CE, a violent storm almost destroyed Emperor Claudius near the islands of the southern coast of France. Claudius sailed from Rome to visit England. He was almost shipwrecked twice. First, off the Ligurian coast and then near Isles d'Hyères. The storms were caused by the penetrating cold wind, known as the Mistral.

On December 31st 46 CE Thera/ Santorini erupted in the Mediterranean again with a VEI of 3.

In 48 CE the River Thames in southern England overflowed. The water extended through four counties. 10,000 people drowned and there was much damage to property. Much cattle in four counties were drowned as well.

In 50 CE the volcano Ambrym in Vanuatu erupted. It is regarded as the third highest explosive force in the Smithsonian Institution's Volcanic Explosivity Index. The VEI was 6. Sixty to eighty cubic kilometers of tephra was released and the caldera was formed. Other eruptions in 50 CE were Tolbachik in Kamchatka with a VEI of 4, Mont Pelee on Martinique with a VEI of 4 and Gorely in Kamchatka with a VEI of 3. Other eruptions were Stromboli in Italy with a VEI of 2, Kusatsu-shiranesan in Japan with a VEI of 2, Fujisan with a VEI of 2, and Mutnovsky in Kamchatka with a VEI of 2.

In 50 CE there was a severe winter in England and all rivers and lakes froze from November to April.

In 52 CE the city of Winchester in England was burnt by lightning. Here is that lightning again.

In California the unpredictably drought-prone world came to an end around 55 CE. This was replaced by permanent drought.

In 60 CE part of the city of Edinburgh, Scotland was burnt by lightning.

On February 5th 62 CE there was an earthquake in the Bay of Naples in Italy that brought down a large part of Pompei and caused severe damage in Herculaneum and Nuceria.

In 63 CE Mount Churchill in eastern Alaska erupted producing a tephra field extending over 250 miles. It had a VEI of 6 and ejected 25 cubic kilometers of tephra. This was one of two White River Ash eruptions which covered vast expanse of Alaska and Western Canada and has been found as far as Europe. There is evidence that the Athabascan people migrated out of the region and into the present-day United States as a consequence of the eruption.

In 66 CE in the Oise region of France, the winter was very cold. Was this caused by Mount Churchill three years before?

In 67 CE a hurricane blew down 15,000 houses in England and killed a multitude of people.

In 68 CE the Isle of Wight was said to have separated from England. It was reported that in England, there was a volcanic eruption followed by an inundation of the sea or a tsunami. The Isle of Wight separated from Hampshire. "By a Flood and Earthquake, the Isle of Wight was torn from Hampshire". The erupting volcano is unknown though we have to allow for dating variations. There are no volcanoes in England or in the British Isles. Jad a dyke been destroyed then that went from the Isle of Wight to the mainland? During this period wagons were used to transport agricultural goods from the island to the mainland as there were no towns on the Isle of Wight, only large Roman farming enterprises. Off Yarmouth the Solent between the island and the mainland is less than a few meters at times depending on the tides. The Shingles bank can sometimes be above sea level. Where the volcano was that the Romans reported is a mystery but possibly was some sort of fiery phenomenon out in the Atlantic Ocean. The Romans knew what volcanoes were contrary to popular opinion.

In the year 69 CE, part of London, England was burnt by lightning.

In 69 CE the Island of Ischia which is near Naples in Italy erupted.

There were three major eruptions in 70 CE. These were Guagua Pichincha with a VEI of 4 and Cotopaxi with a VEI of 4, both in Ecuador as well as Tacana in Mexico-Guatemala which erupted with a VEI of 4 in 70 CE.

Tacitus reports that an unprecedented drought took place in the year 70 CE. There was no water in the north of Gaul in western Europe and the Rhine River in western Germany was barely seaworthy because of the low water level.

On October 24th 79 CE Vesuvius erupted in Italy. It was a Plinian eruption with a VEI of 5. The two cities of Pompei and Herculaneum were buried in lava and over one thousand people died. 2.8 to 3.8 cubic kilometers of tephra was ejected. Interestingly enough, Pliny the Younger witnessed a small tsunami in the Bay of Naples during the eruption of Mount Vesuvius.

There was a terrible period of suffering from 79 CE to 88 CE. when the Roman world seemed to be shaken to its physical foundations. A devastating drought and famine swept over the Italian peninsula. It is said that 10,000 citizens died in a single day at Rome during its height. Tacitus left a grim picture of the distress and suffering. Houses were filled with dead bodies and the streets with funerals. Did the eruption of Vesuvius cause this?

The following year, 80 CE, Furnas in Iceland erupted with a VEI of 5 and Rausudake in Japan erupted with a VEI of 3. Furnas would have also caused a Volcanic Winter, and that combined with the Volcanic Winter caused by Vesuvius would have made it last longer. Possibly up to the ten years as described by Tacitus.

In 89 CE a major earthquake struck Baekje and Seoul in Korea. Houses were broken and lots of people died.

In 90 CE Sete Cidades in the Azores erupted with a VEI of 4 as well as Pico de Orizaba in Mexico with a VEI of 3.

In the year 94, the city of Bangor, Wales was burnt by lightning.

In the year 98 CE, Camelon, Scotland, the Picts chief town, was burnt by lightning. What is it with these lightning attacks burning down towns?

100 CE

In 100 CE Raoul Island in the Kermadec Islands in New Zealand erupted with a VEI of 4. Sheveluch in Kamchatka erupted with a VEI of 4, Tungurahua in Ecuador with a VEI of 3, Tenchozan in Japan with a VEI of 3, and Fujisan in Japan with a VEI of 2.

In England in 107 CE it rained 9 months, washed corn wheat out of the Earth, and drowned cattle. A famine then struck England.

In 110 CE an earthquake caused the administrative center of the Dian Kingdom in Yunnan in China to be flooded. Probably thousands of people died.

In the year 111 CE, Chester, England was burnt by lightning.

In 115 CE the River Severn in England overflowed and drowned 5,000 head of cattle and people in their beds.

On December 13th 115 CE a huge 7.5 magnitude earthquake hit Antioch in Turkey killing 260,000 people. Underwater geoarchaeological excavations on the shallow shelf (around 10 meters depth) at Caesarea, Israel, documented a tsunami hitting the ancient port. Talmudic sources record a tsunami on 13 December 115 CE that affected Caesarea and Yavneh. The tsunami was likely triggered by an earthquake that destroyed Antioch and was generated somewhere along the Cyprian Arch fault system.

In 119 CE a famine struck Britain "after a pillar of fire was seen for several nights in the air".

In 120 CE Sheveluch in Kamchatka erupted with a VEI of 3.

In the year 125 CE in England there was a remarkable snowstorm that produced "great loads, and smothered much cattle".

In the year 128 CE, part of Edinburgh, Scotland was burnt by lightning.

In 130 CE Ata in Japan erupted with a VEI of 4. Mont Pelee in Martinique also erupted with a VEI of 4.

In England in 130 CE, there were a hailstorm with hailstones 12 inches "about", fatal to people and cattle.

In 131 CE in Dorsetshire in southwest England, there was an inundation of the sea, which came 20-miles inland. There was great loss of life and property. All of Dorset would have been flooded.

In 134 CE a severe winter struck England and the River Thames was frozen for 2 months.

In 139 CE, the River Thames in England was dry for 2 days.

In 140 CE Pico de Orizaba erupted in Mexico with a VEI of 3.

In 141 or 142 CE a magnitude 8 earthquake triggered a sever tsunami that caused an inundation at Rhodes. This was called the Lycian Earthquake and also affected Licia, Caria and the Dodecanese in the easter Mediterranean. The 141 Lycia earthquake occurred in the period CE 141 to 142. It affected most of the Roman provinces of Lycia and Caria and the islands of Rhodes, Kos, Simi and Serifos. It triggered a severe tsunami which caused major inundation. The epicenter for this earthquake is not well constrained, with locations suggested at the northern end of Rhodes, on the Turkish mainland north of Rhodes near Marmaris and beneath the sea to the east of Rhodes.

Between 144 CE to 146 CE the population of parts of Egypt were reduced massively due to famine.

Between 145 CE and 345 CE Mount Churchill in eastern Alaska erupted with a VEI of 6 which would have caused volcanic winters.

The greatest eruption in 150 CE was Masaya in Nicaragua which erupted with a VEI of 5. Cotopaxi in Ecuador also erupted with a VEI of 4, Mashu in Japan erupted with a VEI of 4, Ata in Japan with a VEI of 4, Bardarbunga in Iceland with a VEI of 2, Torfajokull with a VEI of 3, Stromboli in Italy with a VEI of 2, Taranaki in New Zealand with a VEI of 3, Krummel-Garbuna-Welcker in Papua New Guinea with a VEI of 2, and Izu-Oshima with a VEI of 3.

In 153 CE England experienced three months of frost and the River Thames froze. All the rivers in England were frozen for three months.

In Rome, Italy in 154 CE, during the 16th year of the reign of the emperor Antoninus, the city suffered from the following calamities. First the Tiber River overflowed its banks. Then a fire destroyed a greater part of the city. Then a famine swept away a great number of its citizens.

In 160 CE Calbuco in Chile erupted with a VEI of 4. Also, Liamuiga in St Kitts and Nevis in the Caribbean erupted with a VEI of 4.

In 166 CE a plague brought from Macedonia reached Rome and spread around the empire.

In 170 CE Sheveluch erupted in Kamchatka with a VEI of 3.

In 171 CE to 174 CE the population of Egypt was greatly reduced. Between this plage and the previous one in Egypt in 144 CE, the population was reduced by a third!

In 171 CE there was a general plague in Europe?, which was fatal to most nobility. It was attended with a famine followed by floods.

In 172 CE Vesuvius erupted again in Italy with a VEI of 3.

In 173 CE in England, there was three month's frost followed by dearth. A dearth is a scarcity and dearness of food or a famine. There had been a great snowstorm that produced "a heavy load, lay 13 weeks, and frost".

In 180 CE Cotopaxi erupted in Ecuador with a VEI of 4.

In 188 CE swarms of locusts filled the air and covered the ground in the Roman province of Apulia in southeastern Italy. These locusts destroyed the crops and ushered in a famine. Sicinius was dispatched with an army to try to battle the winged pests. Thousands of peasants lay down to die on the highroads, and so dire was the pestilence, which accompanied the famine, that even the vultures refused to feed upon the fallen.

In Rome, Italy in 188 CE, a fire caused by lightning utterly destroyed a great part of the Capitol, a famous library, and several contiguous buildings. Eusebius says it consumed whole quarters of the city, and in them the libraries.

Mount Wrangell in Alaska erupted with a VEI of 4 in 190 CE. So did Tengger Caldera in Indonesia which erupted with a VEI of 2.

There were strong summer monsoons in China between 190 CE to 530 CE. This steadily weakened to 850 CE. This is from stalagmite records.

In 192 CE a famine struck Ireland. Bad harvest caused general scarcity, mortality and immigration – "so the lands and houses, territories and tribes, were emptied."

200 CE

In 200 CE Harra of Ahrab erupted in Yemen with a VEI of 2, Gorely in Kamchatka with a VEI of 3, Fujisan in Japan with a VEI of 2, Glacier Peak in Washington with a VEI of 4, and Mount Adams in Washington with a VEI of 2.

In the first and second century CE, El Mirador, Edzna and the other growing Maya population centers collapsed and were abandoned by the Maya people. This was due to a great drought.

There was a drying period in the third century CE in the Western Roman Empire, that came at the same time that Rome almost collapsed. Good harvests became more infrequent.

In the year 202 CE, the city of Bath in southwestern England was partly burnt by lightning.

In 203 CE Vesuvius erupted with a VEI of 4. Also to erupt that year was Chichinautzin in Mexico with a VEI of 3.

Between 210 CE to 410 CE Witori/ Pago in Papua-New Guinea erupted with a VEI of 5.

In the year 212 CE at Leicester in the east Midlands of England, there was a thunder and lightning storm that was fatal to people and cattle.

In 214 CE the River Trent in England, flooded and overflowed its banks 20 miles (32 kilometers) on each side and drowned many people. Where did this volume of water come from?

In 218 CE in Northumberland, northern England, there was a great flood of the River Tweed. The River Tweed flows through the Borders region of Great Britain. The River Tweed had a sudden inundation and destroyed a considerable number of the inhabitants on its banks. A great flood in Tweed, destroyed very much people and cattle.

In 220 CE Pico de Orizaba in Mexico erupted with a VEI of 3 and Fujisan in Japan erupted with a VEI of 2.

The winter was very severe in England in 220 CE with a frost lasting five months.

In 222 CE Vesuvius erupted in Italy with a VEI of 2.

In 228 CE a famine struck Scotland and thousands were starved.

In 230 CE Sheveluch in Kamchatka erupted with a VEI of 3.

In 230 CE there was a great frost in England. The River Thames at London was frozen over for 6 weeks.

On March 15th 233 CE the Hetepe eruption of the Taupo volcano occurred in New Zealand. This was New Zealand's largest eruption within the last 20,000 years. This is one of the largest eruptions in the last 5,000 years equaling the Thera/ Santorini eruption around 1600 BCE. The VEI was 7. It released 120 cubic kilometers of tephra. The main extremely violent pyroclastic flow travelled at close to the speed of sound and devastated the surrounding area, climbing over 1500 meters to overtop the nearby Kainamawa Ranges and Mount Tongaririo while covering the land within eighty kilometers with ignimbrite. Particles were found in

Greenland and Antarctic ice cores. There would have been a severe volcanic winter afterwards. This definitely would have caused a volcanic winter.

In 233 CE in Scotland, it rained 7 months together causing a famine.

In 234 CE in Canterbury in southeast England, a storm threw down 200 houses and killed several families.

In 240 CE Ksudach volcano in Kamchatka, Russia, erupted. This was the largest historical event on the Kamchatka Peninsula. It was as large as the 1883 eruption of Krakatoa in Indonesia and released 20 to 26 cubic kilometers of tephra and 3 to 4 cubic kilometers of pyroclastic flows. This eruption formed the caldera. The VEI was 6. This is volcanic winter forming.

In England in the year 242 CE, there was a great snowstorm. "Very deep, Northampton and neighbouring Shires, much Cattle and Sheep loft lost in it."

In 245 CE there was a great sea floods over Lincolnshire. Thousands of acres of Lincolnshire were flooded and have never been recovered. In Lincolnshire in northeast-central England, an eruption of the sea laid underwater many thousands of acres, which have not been recovered to this time.

Sea floods of the North Sea are coastal floods associated with extratropical cyclones crossing over the sea. The severity is affected by the shallowness of the sea and the orientation of the shoreline relative to the storm's path as well as the timing of the tides. The water level can rise to more than five meters or 17 feet above the normal tide as a result of storm tides.

In 250 CE Sheveluch in Kamchatka erupted with a VEI of 4, Mutnovsky in Kamchatka with a VEI of 2, Gorely in Kamchatka with a VEI of 3, Nasudake in Japan with a VEI of 3, Fujisan in Japan with a VEI of 3, Izu-Oshima in Japan with a VEI of 3 and Stromboli in Italy with a VEI of 2.

I know, the lists are boring, but I am presenting evidence and there is a lot of it. As our dates get closer together, you will see repetition of phenomena indicating far more chaotic weather patterns than you ever realized.

In 250 CE in southeast England, the River Ouse now called the Great Ouse in Bedfordshire overflowed and drowned many people and cattle.

The winter of 250 CE was very cold and the River Thames in England was frozen for approximately nine weeks.

The transgression of the sea over the previous coastline of Flanders in Belgium and the Netherlands between 250 CE to 275 CE had caused a depopulation of the coastal plain there.

Between 250 CE and 750 CE spruce trees spread over southern Sweden.

From 251 CE to 268 CE there was an epidemic in Italy and Africa which claimed 5,000 lives a day in Rome alone.

In 252 CE on February 1st Mount Etna erupted in Italy with a VEI of 3.

In 253 CE a hurricane blew down 900 houses in London, England.

In 255 CE Quetrupillan in Chile erupted with a VEI of 3.

In 259 CE a famine struck Wales. Thousands were "pined to death". This indicated that people wasted away.

In 260 CE Miyakejima in Japan erupted with a VEI of 4 and Nevado del Tolima in Colombia with a VEI of 3. In New Zealand Taupo erupted from Horomatangi Reef.

In 262 CE a famine at Rome, Italy, was attended by a plague.

In southwest Anatolia in Turkey in 262 CE many cities were inundated by the sea, with cities in what was then called Roman Asia reporting the worst tsunami damage. In many places fissures appeared in the earth and filled with water; in others, towns were inundated by the sea.

In the year 268 CE, part of Worcester, England was burnt by lightning.

In 270 CE Ata in Japan erupted with a VEI of 4 and Katla erupted in Iceland with a VEI of 3.

In 272 CE there was so grievous a famine in Britain, that people were forced to eat the bark from trees and roots. Was this caused by Katla erupting in 272 CE? Icelandic eruptions do produce a lot of sulfur.

In the year 276 CE, the climate in Britain was significantly warmer than present. Wines were first made in Britain in this year. Other sources state that grape vines were first grown in Southern England in the previous century.

In 290 CE Parinacota in Chile-Bolivia erupted with a VEI of 4 and Mono-Inyo Craters in California with a VEI of 3.

In the winter of 290 to 291 CE the winter was very severe and the River Thames in England was frozen for approximately six weeks. Most of the rivers in Britain were frozen for six weeks. The frost lasted 6 weeks and was severe over all of Britain.

The winter of 291 to 292 CE was very severe. The rivers in northern France froze.

In 298 CE in Wales, there was a great drought. A famine struck Wales after a comet was seen.

In the winter of 298 to 299 CE the winter was very harsh in the north of Gaul. During the time of Ancient Rome, Gaul was a region of Western Europe encompassing present day France, Luxembourg and Belgium, most of Switzerland, the western part of Northern Italy, as well as the parts of the Netherlands and Germany on the left bank of the Rhine River.

300 CE

In 300 CE Zaozan in Japan erupted with a VEI of 4 along with Mont Pelee in Martinique which erupted also with a VEI of 4.

Between 300 CE and 400 CE world sea levels were high.

Around 300 CE the Caspian Sea was at a very low level. The Caspian Sea is an inland sea so not affected by world sea levels.

Around 300 CE the barbarian movements out of Asia which troubled the Roman Empire over a long period, can be associated with times of drought in central and western Asia.

In the Mediterranean, North Africa, and far to the east into Asia, there were severe droughts in 300 CE and 400 CE.

Starting in 300 CE there was a 36-year drought in Cyprus in the Mediterranean. Due to a famine the inhabitants forsook the island and fled.

In the Eastern Roman Empire there was a period of higher rainfall and humidity that increased agriculture in the Eastern Mediterranean and Anatolia in Turkiye. In the Eastern Empire settlements flourished with cultivation of grain, olives, walnuts and fruit including the interior of Anatolia.

In the late Roman Imperial era the Nile floods in Egypt became less reliable for farming due to the decrease in water flowing down the Nile. Other reports state that the summer monsoonal rains in Ethiopia which feed the seasonal high flood of the Nile were substantial. Was one report from early in the century and the other the later part of the century? Which one?

By the end of the fourth century climatic conditions had become colder and wetter in Europe and the Mediterranean zone started retreating far southward.

Between 300 CE and 400 CE Europe had a warm and dry climate which became colder and wetter.

There is evidence of greater warmth in that grapes were being grown in Paris in the fourth century CE. There were also grapes still being grown in southern England as well.

In California and other parts of Western North America in the period of 300 CE to 1150 CE there seemed to be increased warfare. Skeletons found from this period had a high incidence of head injuries, apparently inflicted by clubs or axes, peaking in the centuries before 1150 CE when the incidence of wounds declined sharply. There were also numerous projectile or arrow wounds from this period as well.

Starting in 300 CE there were moister periods in the Sahel of Africa. The zone in which millet was cultivated moved north into what previously was the Sahara Desert but by 1000 CE had fallen back. After 300 CE rainfall increased by 125 per cent to 150 per cent of modern rates until 700 CE. In this time Lake Chad expanded dramatically.

In 303 CE Vesuvius in Italy erupted with a VEI of 2.

In the year 305 CE the town of Dunbarton in Scotland was burnt by lightning.

A famine prevailed in Scotland beginning in 306 CE and lasted four years. Thousands died; "most grievous and fatal.

In 307 CE, a famine prevailed in Cappadocia, Turkey.

In 310 CE Witori/Pago in Papua New Guinea erupted with a VEI of 5. This, yes you have heard it lots of times before, would have created a volcanic winter.

Swedish Geologist Jens Ormo believes that a meteorite crater found in the Appenine Mountains in Italy crashed to earth around 312 CE though it might be later due to contamination of the site. The meteorite had a one kiloton impact equivalent to a very small nuclear blast and produced shock waves, earthquakes and a mushroom cloud. There has been a legend of a falling star in the Apennine Mountains since Roman times. This was the same year that Constantine saw a celestial vision in the sky that led him to victory over the barbarians at the gates of Rome and the conversion to Christianity eventually. This was also the time when Constantine defeated his fellow Emperor Maxentius at the Milvian Bridge over the Tiber.

The Sirente Impact Crater in Abruzzo in Italy is recorded to have fallen in 312 CE. The small circular "Cratere del Sirente" in central Italy is clearly an impact crater and is surrounded by numerous smaller, secondary craters. The date has been confirmed by radiocarbon dating. The Sirente crater is sixty miles east of the Milvian Bridge and is in the Appenine Mountains.

The Sirente Crater is 13 kilometers from the village of Secinaro and it is ten kilometers across. This was part of the Roman municipium of Superaequum which was a local village that had been suddenly abandoned, possibly as a consequence of a fire during the fourth century CE. Cristian catacombs dating to the same period reveal bodies were piled up hurriedly in a manner indicating a public calamity. A story, taken from the oral traditions of Abruzzo, concerning the region's religious conversion from Paganism to Christianity possibly records the impact event in the 5th century AD.

It was in the afternoon...an uproar hit the mountain and quartered the giant oaks announcing the violent arrival of the Goddess. A sudden and intense heat overwhelmed the people, and a shout echoed all around, splitting the air with its trail of violence ... Suddenly, over there, in the distance, in the sky, a new star, never seen before, bigger than the other ones, came nearer and nearer, appeared and disappeared behind the top of the eastern mountains. Peoples' eyes looked at the strange light growing bigger and bigger. Soon the star shone as large as a new sun. An irresistible, dazzling light pervaded the sky. The oak leaves shuddered, discolored, and curled up. The forest lost its sap.

The Sirente was shaking. In a tremendous rumble the statue sank into a sudden chasm. The satyrs and the Bacchantes fell down senseless. A huge silence fell. It seemed as if time had stopped in the ancient wood near the temple at the foot of the Sirente, and it looked like the mountain had never existed. The entire valley became dumb. Not a breath of wind could be heard, nor a sheep bleating from the numerous herds, nor a rustle from the strong trees, nor a human sound. After an endless period of time, when stars shone in the sky without the moon, a new breeze came to stir the leaves; sheep were heard again, and the Mountain was dressed in the light of a new dawn.

Faint stars disappeared, blue sky slowly came back and the Sirente became a golden mountain in the first rays of the new sun. It looked like the Valley was full of roses. Newly awake, men listened closely to the death rattle of the Goddess at the foot of the wood; and then they saw the statue of the Madonna with the Holy Child in her arms who was sitting on a throne of light and was surrounded by light.

Could this have also been the symbol in the sky that the Emperor Constantine saw over the Milvian Bridge? The Milvian Bridge and Sirente are only 63 miles apart east to west.

In 317 CE in southeast England, on the Isle of Thanet (Kent), there was a flood with loss of life and property. The Isle of Thanet lies at the most easterly point of Kent, England. While in the past it was separated from the mainland by the nearly 2000 feet (610 m) wide River Wantsum, it is no longer an island. Thanet was flooded and people and cattle lost.

In 320 CE Miyakejima in Japan erupted with a VEI of 3 and Taveuni in Fiji had a VEI of 2.

In 323 CE in northeast England, the inhabitants of Ferne Island off the coast of Northumberland were destroyed by an inundation of the sea. Ferne Island in now called Farne Island. Ferne Island is 7 miles southwest of Holy Island. Holy Island is now called Lindisfarne.

In 329 CE the winter was severe in England. Most rivers were frozen for 6 weeks and there was deep snow in Wales and there was a frost in Britain. Most rivers were frozen for six weeks. In Wales there was a great snowstorm "so much cattle fmothered, smothered, in Wales as caufed, caused, a Dearth."

In 330 CE Tengger Caldera in Indonesia erupted with a VEI of 3 and Nasudake in Japan had a VEI of 2.

Also, in 330 CE in northwest England, there was an irruption of the sea in Lancashire. In a volcanic "eruption," lava, ash, and other things burst forth, or out. But in an "irruption," the explosion occurs the other way, into something. It can also mean the disappearance into something. Could this also refer to the receding of the sea before it comes back during a tidal wave or tsunami?

In 331 CE a famine struck Antioch, Turkey. This city was afflicted by so terrible a famine that a bushel of wheat was sold for 400 pieces of silver. During this grievous distress, Constantine sent to the Bishop 30,000 bushels of corn grain, besides an immense quantity of all kinds of provisions, to be distributed among the ecclesiastics, widows, orphans, etc.

In 333 CE a sea floods struck Friesland in the Netherlands and many villages were lost.

In 333 CE, there was a great famine and pestilence in Syria.

In 336 CE famine and plague depopulated Syria and Cilicia in Southeastern Anatolia.

In 338 CE there was a famine in Britain and Wales. 40,000 people starved to death.

In 340 CE Akutan in the Aleutians erupted with a VEI of 5 and Izu-Oshima off the coast of Honshu in Japan had a VEI of 4.

In 345 CE it rained for five months continuously and was followed by a dearth in England.

In 349 CE in northwest England, 420 houses in Carlisle now Carlisle-Cumbria, were blown down by a storm and many people were killed.

On November 15th 350 CE Asamayama in Japan erupted with a VEI of 4.

Mont Pelee in Martinique had a VEI of 4, Nevado del Ruiz in Colombia had a VEI of 3, Tungurahua in Ecuador had a VEI of 3, Taveuni in Fiji had a VEI of 2, Mashu in Japan had a VEI of 2, Ksudach in Kamchatka had a VEI of 2 and Fujisan in Japan had a VEI of 3.

Between 350 CE and 500 CE the water levels of Lake Titicaca on the Altiplano of Bolivia were higher than those of the late twentieth century.

On 5 August 352 CE, it snowed on the Esquiline (seven hills of Rome, Italy), in the heat of the Roman summer.

In 353 CE, there was an inundation in Cheshire, England by which 3,000 persons and an innumerable quantity of cattle perished.

In 354 CE in northern Gaul, the spring rains, were more frequent than usual, causing the streams to swell. The harsh winter of 355/356 CE in northern Gaul caused a large number of people to freeze to death.

During the winter of 356-357 CE in northern Gaul, the Meuse River was frozen during the months of December and January. The winter had been preceded by a hot, dry summer.

The summer drought of 357 CE allowed individuals to ford and cross the Rhine River in Germany.

The winter of 357-358 CE was very hard. The Seine River in France carried ice the size of "blocks of marble". The winter in Paris, France was extraordinarily cold.

In 358 CE in Cheshire England, there was an irruption of the sea; several thousand (about 5,000) people drowned, and there was much damage.

The winter of 358-359 CE was very hard. Ice formed on the Seine River in France.

Also, that winter severe frost lasted 14 weeks in England and Scotland.

In 360 CE tree ring records from north-central China show that Central Asia endured three multidecade droughts. This coincided with invasions of the Western Roman Empire by people from the Asian steppes such as the Huns and Avars who must have been propelled westward by the mega-droughts.

These Chinese mega-droughts may have been caused by The El Nino-Southern Oscillation or ENSO. These ENSO conditions fluctuate between the El Nino (warm phase), expressed as warm ocean waters in the eastern equatorial Pacific Ocean and the La Nina (cold) phase with cooler waters. During El Nino, (warm) phases a weakening of the trade winds allows warm waters to migrate eastward along the equatorial Pacific. The warmer surface temperatures fuel stronger convection cells in the atmosphere and lower atmospheric pressures in the Central Pacific. This alters atmospheric and oceanic circulation patterns in such a way to produce positive feedback when even more warm water piles up on the eastern side of the Pacific and upwelling of cold waters along Peru's coast weakens. The opposite conditions-strong trade winds, increased upwelling and colder surface waters in the eastern Pacific characterized La Nina (cold) phases. Shifts between these two phases currently occur every two to seven years. El Nino and La Nina conditions affect weather and climate worldwide.

ENSO cycles shift the jet stream and thus storm tracks, in midlatitude regions. The shifting of the Jetstream during El Nino winters increases precipitation in California while a more northerly track of the Jetstream during La Nina years produces drier conditions across the southern United States and wetter winters in the Pacific northwest. Indonesia, northern Australia, parts of South America and the southeast coast of South America are drier. Central Asia experiences dry conditions during La Nina and wetter conditions during El Nino. Tree ring records indicate prevailing La Nina conditions during this time.

In 362 CE, a prodigious drought and heat killed all the fruits of the earth. Hence people were forced to eat the flesh of uncommon and filthy beasts. Another source indicated there was a great drought that was universal over all the world in the year 360.

In 363 CE on May 18th there was the Galilee earthquake that hit Syria, the Holy Land and Petra in Jordan. Thousands of people died. This was a pair of earthquakes based on the Dead Sea Transform fault system between the Dead Sea and the Gulf of Aqaba.

In 365 CE in Egypt, there was an inundation consequent upon an earthquake tsunami that destroyed many of the inhabitants. On the morning of 21 July 365 CE, an earthquake triggered a tsunami more than 100 feet (30 meters) high, devastating Alexandria and the eastern and southern shores of the Mediterranean, killing thousands, and throwing ships nearly two miles inland. This tsunami also devastated many large cities in what are now Libya and Tunisia. The anniversary of the disaster was still commemorated annually in the late sixth century in Alexandria as a "day of horror."

Researchers at the University of Cambridge recently carbon dated corals off the coast of Crete that were raised 10 meters and out of the water during the earthquake, indicating that the tsunami was generated by an earthquake on a pronounced fault in the Hellenic Trench. Scientists estimate that such an uplift is likely to only occur once every 5,000 years; however, the other segments of the fault could slip on a similar scale every 800 years or so. The island of Crete was uplifted by nine meters. The 365 Crete earthquake occurred at about sunrise on 21 July 365 in the Eastern Mediterranean, with an assumed epicenter near Crete. Geologists today estimate the undersea earthquake to have been a moment magnitude 8.5 or higher. It caused widespread destruction in the central and southern Diocese of Macedonia (modern Greece), Africa Proconsularis (northern Libya), Egypt, Cyprus, Sicily, and Hispania (Spain). On Crete, nearly all towns were destroyed. The earthquake was followed by a tsunami which devastated the southern and eastern coasts of the Mediterranean, particularly Libya, Alexandria, and the Nile Delta, killing thousands and hurling ships 3 km (1.9 miles) inland. The quake left a deep impression on the late antique mind, and numerous writers of the time referred to the event in their works.

The winter was extremely harsh in January 366 CE in northern Gaul.

A shower of hail fell at Constantinople Istanbul, Turkey on July 2nd 367 CE. The hailstones were so large that it filled a man's hand and each as solid as a stone. The hail killed many people and cattle.

In 368 CE in Sicily Italy, there was an irruption of the sea and great destruction.

In 370 Kujusan in Japan erupted with a VEI of 3.

In 374 CE in England, there was a drought that was followed by a famine.

Also, in 374 CE in Caesarea in Palestine, there was a great drought followed by a famine. During the Byzantine era Palestine was region between the Mediterranean Sea and the Jordan River and various adjoining lands. Caesarea is located mid-way between Tel Aviv and Haifa, Israel.

In 375 CE a most grievous famine afflicted Phrygia in Turkey, so the inhabitants were obligated to shift their habitations elsewhere.

In 379 CE Vesuvius in Italy erupted with a VEI of 2.

In 380 CE Sete Cidades in the Azores erupted with a VEI of 4 and Sheveluch in Kamchatka with a VEI of 3.

In 381 CE a famine struck Antioch, Turkey. During the reign of Theodosius, the Great, the country was again visited by a famine; which was accompanied by grievous plagues. There was also a terrible famine amongst the Goths, East Germanic tribes.

In 382 CE at Cape St. Vincent in Portugal an earthquake and tsunami sank two islets according to Ammianus Marcellinus.

In 387 CE in Cheshire, England, there was an overflowing of the River Dee, and great destruction. drowned much people and cattle.

In 390 CE Kikai erupted in Japan with a VEI of 3 and Veer in Kamchatka erupted with a VEI of 2.

In 393 CE in Egypt, there was a great inundation of the Nile River, which threatened ruin to Alexandria and Libya.

In 399 CE Chichinautzin erupted in Mexico with a VEI of 3.

The winter in 399-400 CE was very severe, even in Provence in southeastern France. The Rhône River froze along its entire length.

In the late 300s tree ring studies on larches which grew on the upper tree line near Zermatt in Switzerland indicate a gradual buildup of warmth with only small variations from year to year in the late 300s, followed by rather sharp variations between about 400 CE and 415 CE and a marked cold period thereafter.

400 CE

In 400 CE Arenal in Costa Rica erupted with a VEI of 4. Tolbachik in Kamchatka also had a VEI of 4 as well as Raoul Island in the Kermadec Islands near New Zealand, Fujisan in Japan had a VEI of 2, Taveuni in Fiji had a VEI of 2, and Pacaya in Guatemala had a Vei of 2.

The warmth in Roman times in Europe seems to have continued and perhaps reached its maximum and greatest consistency in the fourth Century CE. There was an increasingly warm, dry tendency of the summer climate up to 400 CE. There were cold winters, but these seemed to be insufficiently severe to have any lasting effects.

The existence of pre-Norman conquest salterns, saltpans or sandacres, over which the tide washed and from which the salt-saturated sand was then taken, outside the later sea-dykes in the English fenlands, on the Lincolnshire coast, may or may not point to a period of slightly lowered sea level between the late Roman and the Medieval high-water marks.

By 400 CE the rising sea levels produced a notable incursion of the sea from the Wash into the English Fenland and maintained estuaries and inlets that were navigable by small craft on the continental shore of the North Sea from Flanders to Jutland. This circumstance may have helped the Anglian and Saxon migrants launching out across the North Sea from their previous continental homelands. The Wash is a shallow, natural rectangular bay on the east coast of England. It is an inlet of the North Sea and the largest bay in England. It is partly in Lincolnshire and partly in Norfolk.

There was a bout of storminess between 400 CE and 440 CE which accounted for a quarter of all sea floods known from that millennium, with coastal changes in the south of England and losses of life on the Dutch coast.

Around 400 CE at West Stow near Bury St Edmunds in Suffolk, a Saxon village was established on the edge of what is now marshy land in a shallow valley. Wheat, barley, oats, rye and flax were grown for a time but in the seventh century the site was abandoned. The site may have become too marshy following the very wet years of the sixth century and later, especially in the 580s and 600s. Archaeologists suggested that the marsh was encroaching.

The early monastic institution at Glastonbury in Somerset in this period drained the marshes there and the monks of Glastonbury became regarded as the leading experts of the time in land drainage. There appeared to be a preoccupation among the Saxon population in England at the time with draining marshlands in river valleys and the monks were given such land to drain elsewhere.

The level of the Caspian Sea changed around 400 CE and the remains of intermittent rivers and lakes and abandoned settlements in Sinkiang and central Asia indicated that drought had developed on such a scale that traffic along the Silk Road stopped completely. Old shorelines and old harbor installations indicate a very low level of the Caspian Sea. It has been suggested that it was the drying up of pastures used by nomads in central Asia that set off a chain reaction of barbarian tribes and unsettled peoples migrating westwards into Europe. They ultimately undermined the Roman Empire.

Between 400 CE and 600 CE there was the start of a colder, dryer period in Greenland. This was paralleled in Europe. There was a far greater downput of snow than since.

The isotope record from northwest Greenland where the colder regime of the 400 ADs appears as a very minor development, though somewhat prolonged, and was followed by warmth as early as the 600s, which continued and built up to a maximum of warmth in the 1200s. This may be attributed to recurrent anticyclones near and over northern Greenland, repeatedly giving southerly winds over the whole of western Greenland and the regions around the Davis Strait.

There was a sharp cooling in the Alps around 400 CE followed by great advances of glaciers reaching maximum positions in the seventh century quite similar to those about 1820 CE.

During the Winter of 400-401 CE the Pontus Sea (Black Sea) was frozen over, also the sea between Constantinople (now Istanbul) and Scutari (Üsküdar) inlet to the Sea of Marmara from the Black Sea in Turkey. The Pontus Sea was entirely frozen over for the space of 20 days, and the sea between Constantinople and Scutari, Turkey. Parts of the Bosphorus were also frozen. The winter was very cold in Asia Minor. Men crossed over from Asia Minor to the Crimea. When the thaw came such mountains of ice passed by Constantinople that they frightened the citizens. This is the first record that I could find of icebergs in the Black Sea.

In 401 CE the Rhone River in France was frozen firm across its entire width. On January 28th passengers on foot and horseback went on the ice without running any risk between Dauphine in the Alps and Vivarais.

Also, in 401 CE the river Thames in England was frozen over for two months.

In 410 CE Merapi in Indonesia erupted with a VEI of 3.

In 410 CE in Rome, Italy, there was a famine followed by a plague. Under the Emperor Honorius (who reigned from 395 to 414), so great was the scarcity and dearth of victuals in Rome, Italy, that in the open marketplace, this voice was heard – set a price on man's flesh. St. Jerome alluding to this plague, says: the rage of the starved with hunger broke forth into abominable excess, so as people mutually devoured the members of each other. Nay, even the tender mother spared not the flesh of her sucking child, but received him again into her bowels whom she had brought forth a little before. In Rome, Italy, when Lucius Minutius was first made overseer of the grain, many commoners left so that they should not be tortured with a long famine, covered their faces and cast themselves headlong into the Tiber River.

In 416 CE in Essex in England, a great part of Colchester was destroyed, and several people were killed by a storm. Colchester is located in Essex, in the east coast of England. Part of Colchester was burnt by lightning and people were burnt by it.

In 419 CE Flanders in Belgium was hit by a major sea floods at the same time that Kent in England was hit.

Also, in 419 CE the coast of Hampshire was overwhelmed by the sea. There was an inundation of the sea and great destruction, near Southampton that drowned many people.

The Solent between the Isle of Wight and southern England was devastated by a sea floods as well as Southampton Water which is a tidal estuary north of the Solent and the Isle of Wight, was devastated by a sea floods.

The Goodwin Sands in the North Sea were submerged.

What sort of sea floods was this?

In 420 CE Osorno in Chile erupted with a VEI of 4.

During 421-422 CE in the provinces of Pontus and Paphlagonia currently located in northeastern and north central Turkey respectively on the coast of the Black Sea, because of the severity of the famine, many parents had their children castrated and sold as eunuch slaves.

In 430 CE Irazu erupted in Costa Rica with a VEI of 3.

In 430 CE there was a plague in Britain, in which so many people died that the living were scarcely sufficient to bury the dead.

In 431 CE Ilopango erupted in El Salvador. It had a VEI of 7 and ejected 106.5 cubic kilometers of tephra. This would have created a volcanic winter. Fallout from the eruption column blanketed an area of at least ten thousand kilometers with pumice and ash to a depth of 50 centimeters. An area of nearly two million kilometers was covered to a depth of 0.5 centimeters which would have stopped all agricultural production in the most severely affected areas for decades. The eruption possibly caused the abandonment of Teotihuacan in Mexico. This was one of the most powerful eruptions in the last 7,000 years. The dating was established from volcanic shards taken from ice cores from Greenland and levels of sulfur recorded in ice cores from Antarctica. It is believed to have led to the abandonment of Teotihuacan by its inhabitants.

In 435 CE hundreds died during a sea floods that hit Friesland in the Netherlands.

In 439 CE in England, there was a famine after the comet.

In 439 CE the Three Sisters in Oregon erupted with a VEI of 2.

In 440 CE the eruptions were Tutuila in American Samoa with a VEI of 3 and Asosan in Japan with a VEI of 3.

In 441 CE in Wales, the sea made great inroads, on both north and south coasts, many people and much cattle drowned.

In 443 CE there was an extraordinary severe winter in England. There was so much snow that covered the ground for such a long time (scarcely dissolved in six months after) that it caused great destruction of people and cattle.

450 CE

In 450 CE Harunasan erupted with a VEI of 3 in Japan and Tolbachik with a VEI of 3 in Kamchatka

After 450 CE world sea temperatures cooled for some ten centuries, resulting in a strong upwelling and an abundance of fish.

In 450 CE in Italy, there was a severe famine – so severe that parents ate their children. During this severe famine the Roman emperor decreed that parents who sold their children into slavery had the right to purchase them back with a 20% surcharge. So many people sold their children to buy food. This was followed by a plague.

In 450 CE during the reign of Turgina there was a great famine in the Kashmir region of India. This famine was attributed to frost.

Between 450 to 700 CE glaciers in Switzerland advanced their positions due to a cold period before the Medieval Warm Period.

Glaciers in north Norway advanced to prominent proxima between 450 CE and 850 CE.

There were glacial maxima on Baffin Island and Alaska between 450 CE and 850 CE. This was a very cold period.

On the California coast the ocean's temperatures cooled and upwelling intensified after 450 CE. The inshore fisheries improved dramatically. Unfortunately, with cooling came drought and there were many more mouths to feed. Some areas were over-fished. For eight centuries the droughts intensified as the climate became more unpredictable. Periodic El Ninos brought violent storms and floods, suppressed upwelling, and uprooted inshore kelp beds where fish abounded. In the interior away from the coast where the droughts interfered with groups that relied on nut harvests, grasses and game. Now there were more people, more fixed territorial boundaries and intensified competition for oak groves. Chiefs vied with each other for more territory and resources and communities fought with each other for food. At the same time permanent water supplies shrunk dramatically. These fixed communities did not have the option to move.

In 451 CE Mont Pelee on Martinique erupted with a VEI of 4.

The years 451 and 452 CE in Britain were terrible drought years. They were followed by floods.

In 454 CE in what is now present-day Turkey, in the former regions called Phrygia, Galatia, Cappadocia, and others, there was a great drought, followed by famine and then the plague struck. From January to September a famine and a plague of locust from 2 to 5 years. Under Martianus, the Emperor of the East, happened a great drought in both Phrygia, in both Galatias, in Cappadocia and in Cilicia, followed by a famine. This compelled men to eat uncommon and hurtful food. From this drought and bad food, ensued a plague. It caused inflammation for the first two days, so as the bodies of the sick swelled, they lost their eyes, had a cough at the same time which killed them the third day. No cure could be found; delirium and watchings attended it. This calamity laid waste Palestine and many other provinces; for famine and pestilence overspread the earth.

In 460 CE the Ardèche and the Durance rivers in France were entirely frozen. The winter was very severe. The Ardèche River is located in southcentral France and the Durance River is located in southeastern France.

In the winter of 461 CE, it was so cold that the Danube River was frozen. Theodomir (King Theodomir of the Ostragoths Amal) with his army crossed the ice on the frozen Danube River to avenge his brother's death.

Also, in the winter of 461 CE, the Var River in southeast France was frozen.

That same year the winter in Swabia (currently a region of Bavaria, Germany) and Provence (a region of southeastern France) was very severe.

In 462 CE, the Var River in southeast France also froze completely.

As well in 462 CE the Black Sea froze completely. This is very rare but was becoming more common.

In 462 CE the Rhône River in France was frozen across its width.

In 466 CE a grievous famine prevailed in Britain; and a pestiferous smell in the air plague killed both man and beast.

In the winter of 468 CE in France was very severe and there was an unusual reversal of the seasons. The extreme rigor of the year 468 CE in Gaul was due solely to the complete reversal of the four seasons and their weather.

In 469 CE in Constantinople (Istanbul, Turkey), there was much flooding, consequence of four days of incessant rain. Terrible rains fell in Constantinople and Bythinia, which ceased not for four days. Floods turned mountains to a plain. Towns were drowned. Bythinia is the Sea of Marmara region south of Istanbul, Turkey.

In 470 CE Fujisan erupted in Japan with a VEI of 3.

In 470 CE in Scotland there were ten months of rain altogether. This caused the death of beasts, livestock, and a dearth.

Mount Vesuvius in Italy erupted on November 5th in 472 CE with a VEI of 5. This eruption covered the whole surface of Europe with a fine dust according to the historian Marcellinus. The ash fell on Constantinople in Turkiye, over 1200kilometers away. Also, in Turkiye the sky over Constantinople appeared in flames. The ash that fell on Constantinople was one palm deep.

In the winter of 473-474 CE in north and south Wales, there was a great snowstorm. Snow lay 4 months and caused the destruction of much cattle. The winter of 474 in Britain was very cold. There was 4 months of frost and great snow.

In 475 CE famine oppressed the Gallicans, Rhaetians, Noricans, and other Northern Nations, most of Europe. The Gallicans refers to the people from Gaul. Rhaetians refers to the people from an ancient Roman province that included present-day eastern Switzerland and western Austria. Noricans or Noricum refers to the people from mostly modern-day Austria. There was a famine in the Northern Nations of Europe partly caused by locusts. Under certain weather-related conditions, solitary grasshoppers can undergo a physical transformation and develop swarming behavior. Swarms of locusts can travel great distances, rapidly stripping fields and greatly damaging crops.

It is recorded that in 476 CE at I-hsi and Chin-ling in China thundering chariots fell to the ground "like granite" and vegetation was scorched.

Also, in 465 CE in Rome, Italy, there was a plague from rains, thunder and lightning.

In 477 CE there was a famine in Britain from locusts.

In the year 478 CE, great damage was done to the city of Winchester in England from lightning.

In 480 CE Taveuni in Fiji erupted with a VEI of 2 and Belknap in Oregon erupted with a VEI also of 2.

In 480 CE the Tiber River in Italy froze over.

In 480 CE there was drought followed by famine in Scotland after the appearance of a comet.

In 484 CE in Africa, there was a famine caused by drought. There was such a drought as dried up all springs and rivers. Rational and brute animals strove for the withered grass roots in the open fields. So great was the famine; that men died on heaps. All roads were lined with their dead carcasses, without anybody to bury them. This laid waste to Africa and the Vandals. There was neither dew nor rain. The earth was parched. There was no corn grain, vines, olives, or other fruits, nor leaves on any trees. Hence there came a grievous plague. During this period of time, Africa generally referred to North Africa (north Tunisia and eastern Algeria). Vandals were Germanic people who crossed into North Africa from Spain in the year 429.

In 490 BCE Newberry in Oregon erupted with a VEI of 4.

In 492 CE Stamford in England was burnt down from lightning.

500 CE

In 500 CE Sheveluch in Kamchatka erupted with a VEI of 4 and Miyakojima erupted with a VEI of 3 in Japan.

Between 500 CE and 1200AD ocean temperatures displayed marked instability.

Around 500 CE the storminess of the latter part of the 5th century had re-arranged some coastal alignment in East Anglia. This may have been because of the increased frequency of inland storm-driven surges rather than a general world-wide sea level rise.

There was a significant rise in peat bog deposits indicating greater wetness and cyclonicity.

From 1500 to 500 years before the present the temperature of the sea went into a colder cycle after which the temperature warms until modern times.

The coldest sea surface temperatures since the Ice Age fell between 1500 and 500 years before the present.

In 500 CE the Australian climate started getting dryer after the one-thousand-year wet period. These dry conditions lasted until 1200AD when it went wet again.

Early Irish monk-explorers in remote waters around Iceland and near Greenland in the sixth to eight centuries CE encountered sea ice. This was almost unrecorded from the ninth century to 1200s CE by the Vikings.

Between 500 CE and 1300 CE there were multidecade cold periods with droughts that interrupted warm and wet periods in the Americas.

In northern Quebec and in the North-West Territories west of Hudson Bay, extensive pollen research indicated some recovery of the forest associated with warming of the summers from 500 CE to some time around 1000 CE to 1200 CE or 1250 CE.

In the sixth century CE exceptionally, heavy rains caused by an El Nino Event destroyed generations worth of irrigation canals in riverbeds along Peru's North Coast.

Off the California coast Chumash Communities abandoned Santa Cruz Island, the largest of the Channel Islands, during this dry period. This was probably due to insufficient surface water.

Climatic conditions changed and the Mediterranean ecotone had moved further south. Conditions were cooler and wetter throughout the west and made large-scale cereal production very much harder over Gaul. The frontier between the continental and Mediterranean climatic zones once again lay across North Africa.

Around 500 CE there was a sudden lessening of the population in Mexico.

On July 8th 505 CE Vesuvius erupted with a VEI of 4. Other sources say November 9th.

In the year 507 CE the frost in Britain was the most severe for two months.

In the winter of 507-508 CE the Danube River was frozen over and more or less all the rivers of Europe were frozen as well.

In 508 CE there were possibly severe winters as rivers were frozen for two months in the British Isles. All the rivers in Britain were frozen for over 2 months.

On July 8th 512 CE Vesuvius erupted again in Italy with a VEI of 4.

In 514 CE, during the reign of Cissa, King of the West Saxons, reigned so severe a famine, that both men and women in great flocks and companies cast themselves from the rocks into the sea. In the Anglo-Saxon Chronicle, Cissa is identified as one of the three sons of Ælle, who arrived in Britain in the year 477, at Cymenshore which is traditionally thought to have been in the Selsey area of Sussex.

In 516 CE a massive sea floods hit Friesland in the Netherlands.

517 CE was the start of a five-year long period of drought and pestilence in Palestine.

In 520 CE Karkar Island in the Bismarck Sea near Papua New Guinea erupted with a VEI of 4.

On June 1st 520 CE Harunasan in Japan erupted with a VEI of 4.

Fujisan in Japan erupted with a VEI of 2, and Taveuni in Fiji had a VEI of 2.

In 520 CE in Venice, Italy, there was a famine. The city received relief from Theodoric the Great.

Cantref y Gwaelod or the Lowland Hundred was permanently flooded by a sea floods in 520 CE. Much cultivable land was lost to the sea. The Lowland Hundred was between Ramsey Island and Bardsey Island in what is now Cardigan Bay. It could have stretched up to twenty miles west from the coast. The flooding occurred when the sea breached a dike in a storm. In 520 CE there was a major storm surge in Cardigan Bay.

The legendary city of Lyonesse was apparently lost permanently to a sea floods around 520 CE in Cornwall. There were massive storm surges in this period.

Around this same time there was widespread abandonment of land and cultivation in the relatively low-lying Jaeren coastal region of southwest Norway.

On June 1st 520 CE Harunasan in Japan erupted with a VEI of 4.

In 522 CE a drought started in the Levant in the Middle East that lasted several decades and affected the water supply in Persia and Constantinople.

Wilhelm Klinkerfues wrote that in the year (524 after the birth of Christ) though there occurred also much running of the stars from evening quite to daybreak, so that everybody was frightened, and we know of no such event besides…for 20 days there appeared a comet, and after some time there occurred a running of the stars from evening till early, so that people said that all the stars were falling.

De Visser mentions that in the sixth month of the fifth year of the P'u t'ung era (524 AD) dragons fought in the pond of the King of K'uh o. They went westward as far as Kien ling ch'ing. In the places they passed all the trees were broken. The divination was the same as in the second year of the T'ien kien era (503 AD). Namely that their passing Kien ling and the trees being broken indicated that there would be calamity of war for the dynasty.

From 525 CE to 557 CE there were many earthquakes recorded at Constantinople and vicinity.

In 525 CE the Thames River was frozen for six weeks indicating that it was a severe winter. The river Trent flooded and drowned 6,000 cattle.

In The Roman Empire, the period of 526 to 550 CE had the highest number of recorded famines for the winter period of 100 BCE to 800 AD.

In 526 CE Vulcano erupted in Italy with a VEI of 3.

In 526 CE the city of Antioch in what is now Turkey was destroyed by fire falling from the sky, killing 250,000 people. It was described by John Malalas describing that buildings were incinerated, and sparks of fire appeared out of the air and burned everyone they struck like lightning. The surface of the earth boiled, and foundations of buildings were struck by thunderbolts thrown up by earthquakes and were burned to ashes by fire…it was a tremendous and incredible marvel with fire belching out rain, rain falling from tremendous furnaces, flames dissolving into showers. There was an earthquake at Antioch and the city was greatly damaged.

There might have been an outgassing event at Antioch. John of Ephesus wrote that in addition to the fire, "Moist dust bubbled up from the depths of the earth, and the sea gave off a great stench: and the dust could be seen bubbling up in the water as it threw up seashells…"

Another description of the Antioch event described liquid mud (sea sand as it were) boiling and bubbling up from the nether regions. There was "a rank stench of the sea, and water seemed to flow out, just as if sea water were coming up with the hot mud". Was this a massive methane gas explosion bubbling out of the sea that ignited?

Also, in 526 CE there was a dry fog in England. This was accompanied by an earthquake and volcanic eruptions. What are these volcanic eruptions that are occasionally recorded as occurring in England?

In 530 CE eruptions were Sheveluch in Kamchatka with a VEI of 3 and Fujisan with a VEI of 3.

Halley's comet was recorded from China and Byzantium in 530 CE. In Byzantium it was regarded as a tremendous great star in the western region, sending a white beam upwards; its surface emitted flashes of lightning. … it continued shining for twenty days. During 530 CE many large shooting stars were seen from China.

In 530 CE the town of Colchester in Essex, England was burnt by lightning.

In 531 CE a famine struck in South Wales and a small plague.

In 531 CE there was a great meteor shower over the Mediterranean.

In 531-532 CE John Malalas recorded that there was a great shower of stars from dusk until dawn "so that everyone was astounded". This was in the region of Constantinople.

In 532 CE a shower of stars was seen over China. Theophanes also refers to the same.

In 533 CE there was the sighting of a comet.

For three days there was a north-westerly storm, and a castle was washed away completely in Friesland in the Netherlands.

In 534 CE a sore famine struck in Italy.

Ice cores from Antarctica show substantial sulfate deposits around 534+ 2 CE. Probably 535 AD.

In 535 CE the Rabaul Caldera erupted on New Britain in Papua New Guinea. It possibly caused the volcanic winter of 536 as it was regarded as a large eruption. It is unknown what the possible VEI was for this eruption.

In 535 CE Krakatoa, Krakatoa, in Indonesia erupted with a VEI of 4.

In 535 CE to 536 CE there is believed to have been a volcanic eruption in the East Indies at around 4 degrees south that was estimated to have put around 300 megatons of aerosols into the stratosphere. This would have brought about an abrupt drop in world-wide temperature and concomitant changes in atmospheric and perhaps oceanic circulation. One of the effects would be famine which was experienced all over the known world. There was a severe plague in the years 541 CE to 544 CE. Was the dating of Ilopango in 539 CE wrong?

There is some belief that El Chichon in Mexico might have been responsible for the events of 536 CE, but it does not show in any volcanic eruption lists of VEI 4 or above so may be unlikely. El Chichon is now shown in the eruption list Global Volcanism Program | Eruption Search Results (si.edu) from the Smithsonian. Mind you, this list is not infallible I have found.

The event of 535-536 was the single most abrupt climatic occurrence during the past two thousand years. It was possibly caused by a volcano, or volcanoes, with an intensity greater than that of Tambora in Indonesia in 1815. Ilopango is still looking like the firm favorite.

How exact are our volcanic eruption dates?

Radiocarbon dating of tree ring records has allowed scientists to construct a reliable record of the concentration of carbon-14 in the atmosphere through time.

In principle, this composite record allows eruptions to be dated by matching the wiggly trace of carbon-14 in a tree killed by an eruption to the wiggly trace of atmospheric carbon-14 from the reference curve ("wiggle-match" dating).

Scientists presently use wiggle-match dating as the method of choice for eruption dating, but the technique is not valid if carbon dioxide gas from the volcano is affecting a tree's version of the wiggle.

This could indicate that 550 CE eruptions might well have been between 536 and 540 AD.

In 535 CE there was the sighting of a comet.

In 535 CE a famine struck Ireland. It was caused by the destruction of food and scarcity. The famine last four years.

An extreme weather event took place in 535-536 CE. The effects were widespread. It caused unseasonable weather, crop failures and famines worldwide. The Byzantine historian Procopius recorded of 536, in his report on the wars with the Vandals, "during this year a most dread portent took place. For the sun gave forth its light without brightness...and it seemed exceedingly like the sun in eclipse, for the beams it shed were not clear." There were low temperatures during the summer. Snow reportedly fell in August in China delaying the harvest. There was a dense dry fog in the Middle East, China and Europe.

In 536 CE droughts occurred in Peru, which affected the Moche culture.

The 536 CE volcanic winter was the most severe and protracted episode of climatic cooling in the Northern Hemisphere in the last 2000 years. The current theory is that possibly three, at present unknown, volcanoes erupted and ejected massive amounts of sulfate aerosols into the atmosphere which reduced the solar radiation reaching the earth's surface and cooled the atmosphere for several years. Summer temperatures in Europe fell 2.5° Celsius below normal in Europe. The lingering impact of the volcanic winter of 536 was augmented in 539-540 CE when another volcanic eruption caused summer temperatures in Europe to fall by 2.7° Celsius below normal in Europe. This was probably Ilopango. The volcanic eruptions caused crop failures and were accompanied by the Plague of Justinian, famine and millions of deaths and initiated the Late Antique Little Ice Age which lasted from 536 to 560.

536 CE was noted as the second coldest summer in 1500 years in a temperature sensitive record from Fenno-Scandian pine trees. This occurred in 541 CE as well. Overall global temperatures were reduced until 550 CE.

Scuderi following a temperature reconstruction from his chronology cited 535, 536 and 541 as his second, third and fourth coldest years in the last two thousand years.

Bristle-cone pines in the Sierra Nevada show 536 CE as the third coldest year.

Baillie mentions that European oak growth suddenly dropped by 85 per cent in 536 CE. 537 and 538 seemed to recover a bit until 540 CE when oak growth dropped by 25 per cent. It took until 545 CE for recovery to complete.

Pentti Zetterberg working with Finnish pines found a really dramatic event across 540 CE. Following exceptionally wide growth in 535 CE, mean ring width in 536 CE drops by 67 per cent. By 542 CE growth is a mere 13 per cent of that in 535 CE.

Douglass in the American Southwest working on Douglas Firs and junipers in 536 CE as marked by a ring so stressed that it was specifically recorded as "often microscopic and sometimes absent. This was basically a no growth period due to extreme cold.

In 536, both dates have been mentioned, the densest, the most persistent dry fog in recorded history covered Europe, southwestern Asia and China. There was widespread famine, hunger and bubonic plague that followed. The historian Procopius, writing from Carthage in North Africa stated, "the sun gave forth its light without brightness, like the moon during this whole year, and it seemed increasingly like the sun in eclipse, for the beams it shed were not clear nor such as is accustomed to shed". Recent research indicates that the dry fog started on March 24th 536. The reports of the haze only come from North of 35°.

John of Ephesus wrote that the sun was dark, and its darkness lasted for eighteen months: each day it shone for about four hours: and still this was only a feeble shadow…the fruits did not ripen, and the wine tasted like sour grapes.

The star Canopus was difficult to see. Canopus is the brightest star in the southern constellation of Carina and the second brightest star in the Northern Hemisphere night sky. It is unknown how Polaris, the North Star, was affected as it was used by sailors to navigate their courses because it appears to be fixed in the direction of the North.

Britain experienced its worst weather in a century.

In the years 536 CE and 537 CE there was a persistent dry mist on the Mediterranean Sea. This caused rotten cold summers and snowy cold winters.

In 536-537 China suffered a major drought, and dust storms, yellow dust rained like snow and snow fell the following August, ruining the crops. There were also summer frosts and hail.

In 536-538 severe droughts hit northern China and extended into Mongolia and Siberia where tree rings show some of the coldest conditions in the last fifteen hundred years. Drought settled over the grassland steppe. This forced the nomadic Avars to move westward toward Europe, skirting the northern shores of the Caspian Sea and eventually finding themselves in the fertile grasslands north of the Caucasus Mountains. Eventually they ended up in Hungary. In 536

CE the horse-based economy of the war-like Avars foundered, and their vassals, the cattle-herding Turks, overthrew them. Driven from the steppes, the Avars joined forces with the Slavs in Hungary, on the borders of the Roman Empire.

Ice cores from Greenland show substantial sulfate deposits around 534+ 2 CE.

Both Greenland and Antarctica ice ores indicate that in 536 CE massive layers of sulfuric acid of volcanic origin fell.

Greenland ice cores linking layers of ammonium to high-energy atmospheric interactions with objects coming from space were in 539 CE, 626 CE, 1014 CE and 1908, the Tunguska Event. Did the high-energy object impact with the Earth and where?

Greenland ice cores suggest North America as the area of the volcano involved due to tephra deposits. Ice cores from Greenland and Antarctica show record levels of sulfuric acid of volcanic origin during the sixth century CE which lasted some years. The acid could only come from either a huge volcanic eruption that shot millions of tons of fine volcanic ash into the atmosphere or from a comet hitting one of the earth's oceans. The candidate is El Chichon volcano in Chiapas, Mexico.

The Smithsonian Global Volcanism Program writes that El Chichon erupted in 480 CE ± 200 years so it could be eligible. There was a massive ashfall deposit rich in lithics and large carbonized tree trunks.

In 536 Krakatoa in Indonesia erupted with a VEI of 4. Had Krakatoa erupted twice in two years or was there a misdate somewhere? I would follow the misdate allowing for wiggle room.

In Persia, Iran, there was a severe drought.

Crops failed throughout southern Iraq.

Snow fell in Mesopotamia.

Tree ring analysis by the dendrochronologist Mike Baillie showed abnormally little growth in Irish oak. In 542 CE there was another sharp drop after a partial recovery. Old Irish annals recorded that in 536 CE to 539 CE there was a failure of bread.

Crops failed throughout Italy.

In Western North America drought occurred in 536 and 542 to 543 CE.

Ice cores from the Andes in South America show extreme aridity, or dryness, also occurred, affecting the Moche Culture in Northern Peru in 536 CE. Other sources say that between 534 and 540 CE there was a drought in the Moche Empire in Peru.

Other sources state that there was a forty-year long drought in Nazca in Peru in this period. The drought may have been related to changes in the Humboldt Current and may be related to the events around 540 CE.

Tree rings from Scandinavia and Western Europe show an abrupt slowing in tree growth, and therefore rings, between 536 and 545. There was an abrupt drop in temperature followed by an incomplete recovery. The only tree rings to show poor weather were from 50° to 70° north. The cooler temperatures coincided with a period when atmospheric pressure was high over Greenland and the North and low over the Azores in the middle of the Atlantic Ocean. The prevailing Westerlies slowed, and bitter, dry weather settled over Europe. A widespread drought followed, penetrating deep into Eurasia.

Tree rings indicate that in 536 CE the climate cooled by as much as 3° C. This was from Swedish and Siberian trees. The temperature recovered over the next three years. The implied North Atlantic Oscillation Index for this period would have been highly negative with well-above average pressure over Greenland/Iceland sector and lower values around the Azores. The NAOI is based on the surface sea-level pressure difference between the Subtropical Azores High and the Subpolar Low.

In the Swiss Alps in 2018 tephra fragments with an Icelandic signature were found in Switzerland in a glacier from the Alps. There were no signs in Iceland of this eruption though. Katla in Iceland did erupt in 540 CE but there is no record of its VEI as none is listed. Hekla in 550 CE also erupted and once again there is no recorded VEI. Eyjafjallajokull also erupted in 550 CE, once again with no VEI. How much "wiggle room" has affected these results? Were they only a few years earlier if one of them was sending tephra fragments capable of reaching Switzerland?

In the Roman Provinces of Bulgaria and Scythia drought caused major suffering in 538 AD.

In 538-539 CE in the month of December there was a great and terrible comet that appeared in the sky at evening-time for one hundred days (elsewhere, several days) according to Zacharias of Mythylene, now Mytilene, on Lesbos. This was in the eleventh year of Justinian.

In 538 CE: the land of Italy lay uncultivated last year, hence a great famine. Such as dwelt in the region of Emilia in northern Italy left their seats and goods and went into the region of Picenum in east-central Italy and even there no less than 50,000 died of famine. Then the starved throwing off all humanity killed and ate one another. Delicate mothers eat their tender babes. Two women killed 17 men and ate them. A woman in Milan in northern Italy ate her dead son. People kneeling down on their knees and hands to eat grass and herbs, fell down with weakness and died. Nor was there any to bury them. Others eat dogs, mice, cats and the vilest animals. The Tuscans from Tuscany in north-central Italy were also starved, but bread made of earthnuts was a help to them. Far greater still were the numbers of starved beyond the Ionian borders. When they had nothing to eat, they became extenuated and pale, their flesh withered away and became black. The disease spread as among great herds of cattle. Their bile was redundant, there was no juice left in their bodies. Their skin was hardened, and became dried like leather, and clave to the bones. Their livid color became black. Men looked like charcoal wood, their countenance was senseless and stern. They died everywhere, partly from hunger and partly from too great satiety. Having been burnt up within, after the natural heat was extinguished. For having been starved, if they had any opportunity to feed freely, being not able to digest their food, they died so much sooner. The famine was so great in the region of Liguria in the coastal region of northwestern Italy that many mothers ate their own dearest children. The west coast region of Campania in southern Italy also suffered. Nor did Picenum's being a seacoast save it. In the following year, 539, the grain sprang up by themselves, without the labor of farmers and oxen. They shook in the wind because there was no one left to reap them.

In 538 there was famine in twelve provinces of China. This drought started in 535. In Northern China there was a mortality of 70 to 80 per cent among the population and cannibalism also occurred.

In 538 CE in Poland and Western Ukraine famine affected local farmers who went raiding their Roman neighbors for food.

In 539 CE the volcano Ilopango in El Salvador erupted. It produced widespread pyroclastic flows and devastated Mayan cities. The ash cloud fallout blanketed an area of 10,000 square kilometers waist-deep in pumice and ash which would have stopped all agricultural endeavor in the area. The eruption and subsequent weather events and agricultural failures directly led to the abandonment of Teotihuacan in Mexico. This was one of the largest volcanic eruptions in the last seven thousand years. It devastated an area up to one hundred kilometers in radius from the volcano. Thousands of people died, and it ended the Maya presence in the highlands. Allowing for dating variance this could have been a few years before. Ignimbrite surrounded the volcano up to thirty meters thick which extended into several countries. Most of the tephra went south and southeast. The VEI was 6 and 84 cubic kilometers of tephra was ejected. As Central American volcanoes had a high sulfur content of 1000 ppm or more meaning that Ilopango would have had forty per cent than Tambora, though Ilopango was a smaller eruption. There would have been no survivors within 50 kilometers of the eruption based on the depth of the tephra. All Mayan culture in the area was destroyed. The mystery eruption of 540 CE deposited more sulphate in ice cores than that of Tambora in 1815. There was an enormous eruption somewhere to be larger than Tambora.

Greenland ice cores indicate that there was an acid kill in this period simultaneously with an ammonium spike.

In 539 there was a failure of bread in Ireland only three years after a previous event. This indicates that the harvest crashed again.

In 540 CE there was a VEI 6 eruption of the Rabaul volcano in Papua New Guinea that ejected 30 cubic kilometers of tephra. Did this cause the volcanic winter of 536? We can allow for dating irregularities. This is also listed as occurring in 535 AD. Remember that wiggle. The column would have been over 20 kilometers high and described as Ultra-Plinian which indicates a dispersal index of 50,000 square kilometers or 19,000 square miles. The dispersal index is the surface area covered by an ash or tephra fall.

There were several eruptions in 540 CE but they do not have VEIs so they could have done anything. These were Katla in Iceland, Ischia in Italy, Merapi in Indonesia, Colima in Mexico, Kilauea in Hawaii, and Alamagan in the Northern Mariana Islands.

Ice cores from Greenland showed a large acid layer caused by an explosive volcanic event that footnoted a large acid layer at 540-+/- CE that corresponded with exceptionally narrow tree rings in Irish oaks.

Ice cores from Greenland and Antarctica indicate that there were two major volcanic eruptions in 540 CE. One in the Northern Hemisphere and the other in the Southern Hemisphere.

In California foxtail pines in the Sierra Nevada of California studied by Louis Scuderi showed that there were several very cold years just around 540 CE.

The Byzantine Empire and other areas experienced a phase of abrupt cooling after a possible volcanic eruption in 540 CE. There was a veil of dust and tree ring records confirm the sudden cooling. Volcanic eruptions with the same results had occurred before in 536 CE and later in 547 CE. This was called the Antique Little Ice Age. This ended the most protracted period of climatic cooling in the Northern Hemisphere in the last 2000 years.

A severe plague in the years 541 CE to 544 CE affected up to 25 per cent of the populations of Africa, Europe and Asia.

John of Ephesus recorded that by 541-542 CE up to 230,000 people in the city of Constantinople had died of the pestilence. The streets were full of corpses as there were neither litters to carry them off for burial nor diggers for graves and the corpses were piled up in the streets.

The Justinian Plague, believed to be Bubonic Plague, lasted from 541 CE to 549 CE. The pestilence ravaged Europe, reducing the population of the Roman Empire by a third, reaching as far east as China and as far northwest as Great Britain.

There was a dry period around Lake Titicaca on the Altiplano of Bolivia between 540 CE and 610 AD,

In 540 CE according to Gildas the whole island of Britain was on fire from sea to sea… until it had burned almost the whole surface of the island and was licking the Western Ocean with its fierce red tongue. Every now and again we have strange reports from England. About volcanoes that are not there, burning seashores and now an all-consuming fire.

In the same period a pestilence nearly destroyed the whole nation of Britain. It was called the Yellow Pestilence because it caused all persons who were seized by it to be yellow and without blood. "It appeared to men a column of a watery cloud, having one end trailing along the ground, and the other above proceeding in the air, and passing through the whole country like a shower going through the bottom of valleys. Whatever living creatures it touched with its pestiferous blast, either immediately died or sickened for death. So greatly did the aforesaid destruction rage throughout the nation, that it caused the country to be almost deserted". This was recorded by St. Teilo, Bishop of Llandaff Cathedral in Morganwg, South Wales who left South Wales for Brittany to escape the pestilence. Interestingly the Red Dragon on the Welsh flag, Draig Goch, in one interpretation Draig is translated as Mellt Distaw (sheet lightning) and also Mellt Didaranau, (Lightning unaccompanied by thunder). The earliest usage of the word is Maen Mellt, which refers to a meteorite. Maen means stone and melt translates to lightning.

Tree rings indicate that there was an abrupt drop in temperatures with an extremely slow recovery with trees not recovering until 550 CE. There was a temperature drop of 3° in 540 to 541 which lasted ten years. These tree rings were replicated worldwide. Stratospheric sulphate lasts at most three years before it is returned to the earth and during this time the depth of the haze steadily reduces. Was this 3° colder than the previous temperature falls? There had been a previous drop of 3° in 536 CE as well. This was then a total temperature drop of 6° in only four years. Tree rings after 540 narrowed worldwide and temperatures dropped significantly with several very cold years.

In 540 CE the Great Dam of Marib, which had been built in the seventh century BCE, though other sources disagree and believe it was built earlier, collapsed. The dam was a central part of the south Arabian Civilization and by 550 CE the dam was a complete loss and thousands of people migrated to Medina which then was another oasis on the Arabian Peninsula. The Arab tribes began thinking of conquest for survival and in 610 CE a new leader unified them by the name of Muhammad.

In 540 CE the Gupta Dynasty in India suddenly collapsed.

A legend from Japan found in the Enoshime Engi, states that sometime around 540 CE there was an apparition of the bright goddess Benzaiten. She was seen to look like an autumn moon enveloped in mist. She was adorned with a long jade pendant and her descent was accompanied by a strumming or slapping sound. She was accompanied by a myriad spirits of dragons, fire, thunder and lightning that made great boulders descend from above the clouds. She had arrived after a period when dark clouds covered the sky, and the earth quaked continuously for eleven days. What a wonderful description of a meteorite shower.

In 541 CE Roger of Wendover wrote that a comet was seen in Gaul, France, so that the whole sky seemed on fire. There rained real blood from the clouds … and a dreadful mortality ensued.

In 541 CE the sea advanced on Thrace by four miles and covered it in the territories of Odyssos and Dionysopolis and also Aphrodison. Many were drowned in the waters. By God's command the sea then retreated to its own place. Odyssos is now Varna in Bulgaria, Dionysopolis is now in the town of Balchik in Bulgaria and Aphrodison's present whereabouts are unknown. Not in Google or in my collection of atlases of the Greco-Roman world. Varna and Balchik are on the Black Sea coast.

Tree rings studied by Michael Baillie show abnormally little growth in Ireland in 542 CE after a partial recovery after 536 CE.

In 542 to 543 Bubonic plague which seemed to have come from Egypt or Ethiopia during the reign of Justinian and spread far and wide over the Roman world and beyond, reached Persia and the Indies and the ports of Europe. Between 542 CE and 565 CE half the population of the Byzantine Empire and of Europe died. This was apparently around one hundred million people.

All of this from only two volcanic eruptions, and possibly a bolide impact, as yet unfound.

Yes, I know. You weren't taught about this, were you?

The Winter of 543-544 CE in Gaul, Western Europe, was so severe because of the ice and snow, that the birds and other wild animals could be caught by hand.

In 544-545 CE there was an intensely cold winter in London and the south of England. It might have been over a wider area as well.

The Winter of 544-545 CE was very cold in France similar to the winter of 544.

During the winter of 545 CE in England the cold was so intense that birds allowed themselves to be caught by hand.

In 545 CE there was the greatest famine of grain, wine and oil. Then came the terriblest and greatest plague over all the world that ever was paralleled or recorded in history. It spared neither age, sex, rank, nor place. God only could afford the least help, not man or art. It began among the Egyptians at Pelusium, Egypt; thence it spread over the globe, not missing one corner, nor did it seize the same person twice. It began thus: Demons in human shape appeared to many, and when they fell upon them, they imagined themselves struck by some man, and disease quickly fell on them. Some from the beginning, as they were able, prayed that the distemper might be removed; and as if agitated by some evil spirit, did not hear their friends calling on them. They were shut up in close places. The same happened to some in their sleep, for they were quickly taken with a fever, both heat and color of the body continuing the same, nor was there any inflammations, as is common with feverish people, but a cough from the first evening of the fever. No medicines were given, none being suspicious of the danger. The same day in some, although in others later, a tuber appeared in one place or another on their body. Moreover, some were lethargic, or comatose. Others were foolish, some lost all memory, neglecting even their food, and they died. In their foolishness, they imagined themselves caught by someone and cried out they were assaulted and turning from they fled. Their servants and nurses suffered severe and intolerable things from them. So that they as well as the sick challenged compassion, not that they were affected with the disease, for that at present hurt none by contagion. But being furious, they either leaped out of bed, or hurried to the rivers to quench their thirst. They could hardly be restrained by force. Some died the same day, others several days after. This plague raged three months in Constantinople, Istanbul, Turkey. At first only a few died. After five or ten thousand were carried out daily. Many rich men, having all their servants dead, died rather from want of assistance than of the disease, and laid unburied. Pelusium is located in the easternmost mouth of the Nile River. Today it is called Tell el-Farama, located in the extreme northwestern Sinai not really very far to the east of Port Said, Egypt.

The winter of 547 CE was very cold, like the year 544. The frozen rivers of France could be traversed as if it was on dry land. This was the time in Gaul that was so very cold that the ice on the frozen rivers carried the weight of people.

The Byzantine Empire and other areas experienced a phase of abrupt cooling after a possible volcanic eruption in 547 CE. There was a veil of dust,and tree ring records confirm the sudden cooling. Volcanic eruptions with the same results had occurred before in 536 CE and later in 540 CE.

In 549-549 CE a severe gale or storm hit London. Many houses were damaged, and 250 persons were killed.

550 CE

On June 1st 550 CE Harunasan in Japan erupted with a VEI of 5. There were two large explosive eruptions. Also, during the same year Rausudake in Japan erupted with a VEI of 4. Arenal in Costa Rica erupted with a VEI of 4. Raoul Island in the Kermadec Islands erupted off the coast of New Zealand with a VEI of 4. Other eruptions were Stromboli in Italy with a VEI of 3, Rausudake with a VEI of 4 and Asosan with a VEI of 3 in Japan, Tolbachik with a VEI of 3, Kikhpynch with a VEI of 3 and Gorely with a VEI of in 3 Kamchatka. Ata in Japan erupted with a VEI of 2.

How exact are our volcanic eruption dates? This is more like sloppy dating.

Radiocarbon dating of tree ring records has allowed scientists to construct a reliable record of the concentration of carbon-14 in the atmosphere through time.

Between 550 CE to 1150 CE there was a long cold drought period in which the population of Cantona, west of Mexico City, dropped from a high of 70,000 people in 900 CE to 5,000 people in 1050 CE.

Teotihuacan, the largest city in Central America at the time and with a population of one hundred thousand people, was suddenly abandoned and set on fire, apparently by its own people. How exact is this date? Could it have been 540 when the Mayan Hiatus occurred further south? Were sulfuric acid clouds that would kill plant life from the Ilopango eruption in El Salvador responsible?

In Central America there was a period starting in 550 CE when building work ceased and some cities were abandoned. This is called the Maya Hiatus and it lasted one hundred years and its start was shortly after the 540 CE eruption of Ilopango in El Salvador. It took decades to centuries before some of the old sites were reoccupied.

In the area of the Maya in Central America there were relatively wet times between 550 CE and 750 CE.

From 550 CE to 750 CE there were relatively wet times in the Carioco Basin in Venezuela.

In 552 CE in Greece, there was an inundation from the sea; part submerged. This inundation was likely caused by a tsunami from a massive earthquake. Thomas Short has the following entry for the year 552 A.C.: "There was a great Earthquake in Greece which overturned many towns, as Naupaictum, Petra, Corona, and others. The sea also broke in and overflowed many places in Greece, and on its going back left innumerable unknown fishes on the shore." Robert A. Juhl believes this event occurred in 551. An earthquake generated tsunami struck

Greece on 7 July 551. It destroyed and temporarily submerged the Temple of Olympus in Etolia, Greece. A second tsunami struck Beirut, Lebanon two days later on 9 July.

The earthquake of 9th July 551 CE in and around Lebanon was one of the largest earthquake events during the Byzantine Period. This earthquake was associated with a tsunami along the Lebanese coast and a local landslide near Al-Batron. A large fire in Beirut in Lebanon continued for two months. All the cities on the Phoenician coast from Tyre to Tripoli were reduced to ruins and many ships were sunk. Around 30,000 people died in Beirut. The earthquake was felt from Alexandira in Egypt to Antioch in Syria.

There was a warm period between 551 CE to 760 AD.

In 553 CE the rainstorms were violent in Scotland where it rained five months incessantly in Scotland producing a dearth.

In the winter of 553-554 CE there was a severe winter though both winters were notably cold. The winter was so severe with frost and snow that the birds and wild animals became so tame as to allow themselves to be taken by hand.

In 554 CE there was a severe winter worldwide. Tree-ring data suggests a period of reduced growth for Western Europe.

In Ireland in 554 CE, it was so cold and severe with ice and snow that birds and wild animals became so tame as to allow themselves to be taken by hand.

In 554 CE it was so cold that the rivers of France could be traversed on foot as if on dry land.

In 558 CE there was a dreadful plague that extended all over Europe, Asia and Africa, and did not cease for many years.

In the winter of 558 CE the Danube River was frozen over and more or less all the rivers of Europe were frozen.

In the year 559 CE the Bulgarians crossed the frozen Danube River, spread over the region of Thrace, and were close to the suburbs of Constantinople.

Both July and August of 559 CE in Western Europe, were terribly agitated from east to west by an overflow of the sea, and by storms and earthquakes.

Between 560 CE and 660 CE Opala in Kamchatka erupted with a VEI of 5.

In 563 CE at Lake Geneva in Switzerland and France there was an underwater mudslide probably generated by a landslide that triggered a collapse of sediments at the mouth of the River Rhone, the tsunami traveled the length of Lake Geneva, reaching a height of 16 meters (52 feet) in some places. The wave probably killed hundreds, or even thousands, of people.

Between 564 CE and 594 CE there was a three-decade drought cycle in the Moche Empire in Peru. Rainfall was thirty per cent less than normal.

In 564 CE in England there were great rain floods.

After long continued rains, followed a great inundation of the Tiber River in Italy, which overflowed the whole low country. Then came a sweeping epidemic. There was a plague in Rome, Italy from the rains and floods of 564 CE.

In 566 CE there was a great storm affecting the eastern and mid coasts of southern England. There was serious damage.

The winter of 567 CE was called the "Winter of the Comet". In France there was a huge abundance of ice and snow. Wild animals could be taken by hand. It snowed for five months. There were many dead birds. The winter of 566 was very rigorous in southern France. The large amount of snow covered the earth for more than five months. The intensity of the cold destroyed many animals.

In 575 CE the coastal regions of southeastern England, parts of Essex, Suffolk, and Norfolk were inundated from the sea. The sea drowned much people and cattle.

In 580 CE Sheveluch in Kamchatka erupted with a VEI of 4 as well as Izu-Oshima with a VEI of 3 in Japan.

In September 580 CE it was unusually warm and trees blossomed in France. In October, the Rhône and Saône rivers flooded and rose much higher than usual. In 580 in Western Europe, there was an earthquake, large hail, fierce storms and rains.

In 580 CE in Anglesea, Wales, there was much damage by the sea. A great many people and cattle drowned in Anglesea.

In the year 580, a great flood occurred in east-central France. This was the oldest recording of an overflow of the Rhône River. The plain of Brotteaux in Lyon, was changed into a huge lake, and the damage was considerable. The Rhône and the Saône rivers, which formed a junction, towards St. Nizier; waters rose to such heights that many of the walls of the city (Lyon) was taken and many buildings destroyed. The water, after four days of flooding, seemed finally to begin retreating, when the sky again was covered with dark clouds and heavy rain fell. All the inhabitants of the plain before this calamity, fled with their wives, their children and their most valuable property to the hills of Saint-Just and Saint Sebastian. There, night and day they spent in prayer. The people of France, lying near the Liger now Loire River and Phadan now Rhône River, were almost swallowed up by great rains, which poured down continuously for 20 days. Huge rains swelled prodigiously all the rivers of France. Terrible floods followed, especially in Lyon and Limagne large plain in the Auvergne region of France. The violence of the waters submerged the herds, destroyed crops, and ruined many homes. In Auvergne in south-central France, they could not sow the land. In Lyon, where the Rhône and Saône rivers joined together, the rivers overflowed their banks and destroyed many buildings, and even overturned a portion of the city walls. The terrified inhabitants, fearing a new flood, took refuge with their wives, their children and what they value most in the hills of Saint-Just and Saint Sebastian. Hail, earthquakes, explosions of lightning and a terrible storm came and add to the spectacle of desolation. This upheaval broke out towards the beginning of autumn. Once the rain had ceased, the trees flowered a second time during the year. The rain fell in torrents for twelve consecutive days in the Auvergne, and for twenty days in Lyon.

Italy suffered prodigiously from inundations.

Between 580 CE and 600 CE there were several or a succession of wet years. Tree lines by this time were falling and the glaciers were advancing again. The climate was getting much colder. This was a run of disastrously wet years in the 580s. These were isolated events in the grand scheme of things.

"The History of the Franks" written by Biship Gregory of Tours: "in 580 CE in Louraine, one morning before the dawning of the day, a great light was seen crossing the heavens, falling toward the east. A sound like that of a tree crashing down was heard over all of the countryside, but it could surely not have been any tree, since it was heard more than fifty miles away… the city of Bordeaux was badly shaken by an earthquake… a supernatural fire burnt down villages about Bordeaux. It took hold so rapidly that houses and even threshing floors with all their grain were burnt to ashes. Since there was absolutely no other visible cause of the fire, it must have happened by divine will. The City of Orleans also burned with so great a fire that even the rich lost almost everything."

In the Winter of 581-582 CE in Western Europe, the heat during the winter caused the trees to bloom in the month of January. This month also was filled with violent rain, lightning and thunder.

January 582 produced heavy rains accompanied by lightning and thunder in Western Europe. The winter of 582 CE was very soft and gentle in France. Many trees flowered. There were frequent storms.

The Winter of 583-584 CE was of such persistent gentleness; that in the month of January one could see roses.

The winter of 584 CE was very gentle and sweet in France. Roses bloomed in January. Then there was a hailstorm and a drought. This was followed by a white frost, a hurricane and several disastrous incidents of hail that ravaged successive harvests of crops and vineyards. At the same time there was an excessive drought. The year produced almost no grapes. Desperate farmers delivered their vines at the mercy of the herds. But the trees, which had already borne fruit in July, producing a new crop in September, and some even bore again in December, and the vines offered at the same time well-formed clusters.

In 584 CE in Western Europe, an immense drought finally ruined the vineyards and the harvest, which was already compromised by earlier hailstorms and frosts.

The spring and summer of 585 CE Western Europe was so rainy, that it could be confused with winter. The bulk of the rains this year caused rivers to overflow their banks and flood the fields and meadows. These floods seriously compromised the crop yields. In 585 CE the Loire River in France flooded. There were floods and famines. During the autumn of 585 CE and the winter of 585-586 CE in France, the weather produced extreme sweetness. The trees bloomed again in September and again before Christmas and bore fruit.

Lake Punta Laguna in Quintana Roo in Mexico showed evidence of a particularly intense drought in 585 CE. It showed frequent and severe dry events between 165 BCE and 1020 CE. There seems to be a see-saw effect between New World and Old World weather patterns.

Because of the warm weather in Western Europe the trees blossomed in the month of July 585 586?, and bloomed again in September 586 and a large number of these who had already borne fruit produced a second crop of fruit until the Christmas holidays. During the autumn of 586 CE and the winter of 586-587 CE in France, the weather produced a strange new sweetness. The grape vines bloomed twice.

There was extensive flooding in 586 in the north of Western Europe.

In 586 CE there were floods and a great storm in the North Sea.

In 587 the great rainfalls caused the rivers to swell prodigiously in France. This flood was especially severe in Burgundy in east-central France. In France in October, after the harvest, new vines covered with grapes appeared and on the trees, new leaves and new fruit. The trees were blooming in the fall and gave fruit a second time after already being harvested once. Roses appeared in December.

In 588 CE in China a red-colored object fell with a noise like thunder into a furnace, exploded, and burnt several houses.

In 589 CE a storm flood occurred at the city of Durham in England. The sea swept away villages and many people were drowned. Durham is around 18 kilometers from the east coast.

In 589 CE there was a sea floods over Hartlepool and the land around it in northeast England. Hartlepool is in County Durham.

After Easter 589 CE, rain with hail fell in Western Europe. In less than two hours, smaller streams were turned into major rivers. These rivers rose to unprecedented heights and overflowed their banks. The trees bloomed again in the autumn of 589 in Western Europe, and then they produced other fruit. The fruit was pink in November.

In 590 CE El Chichon in Mexico erupted with a VEI of 3.

In 590 CE in Italy, there were great floods from tempest; followed by a plague. Rain fell in the months of September and October incessantly for many days and raised such floods in all rivers and lakes in Italy, as to overflow their banks and drown an infinite number of people and cattle. The rain was accompanied by tremendous tempest of thunder and lightning. The river Tiber swelled so high that all the fields, which were not hilly and mountainous, were overflowed. Many people believed it was a second great flood. In Rome, Italy, the Tiber swelled so high that in some places it reached to, and in other places overflowed the cities high walls. And the water rushed in with such fury that is spoiled and defaced the greatest part of the buildings that were near the river. When the floods ceased, the fields were so soft and covered with slime and mud, that they could not be tilled or sown, hence a general famine. The flood not only demolished many stately buildings and ancient monuments, but also got into the church granaries, and carried away many thousand measures of wheat. After the flood, the river brought down innumerable multitude of serpents, and among them a monstrous great one as big as a great beam. All these serpents were swimming down the river into the sea, where they choked, and their carcasses being cast on the shore. There they rotted and by the stench of the slime and mud and excessive moisture, and the air was so corrupted, that a most desolating plague ensued over all Italy, Spain and France. The plague raged and laid waste to many towns. In many 2/3 of the people died. It was most severe at Rome, followed by Liguria in the coastal region of northwestern Italy and the Venetian territories in northeastern Italy, both by floods, famine and plague.

Following the heavy rains, and disastrous floods of 590, the year 591 produced a drought in Western Europe. The summer of 591 was unusually hot in France. The year 591 in Western Europe was divided as it were between an excessive droughts, which ruined all the meadows, and heavy rainfalls followed by floods, which destroyed much of the hay harvest. The excessive dryness of 591 in Western Europe consumed all the fields.

In 592 CE between January and September there was drought in England. There was a cold winter which implied considerable blocking and periods of high pressure.

There was a drought that lasted from 10 January to September, along with a plague of locusts. This produced a famine. There was a remarkably great drought from January to September, attended with a grievous famine and great swarms of locusts, which for two years ate up every green thing and caused a terrible famine in Italy. But they continued for 5 years in Capitaneo, then shifted to another province. Capitaneo may have been end of the Italian peninsula near Bari in the southern Adriatic.

Winter of 592-593 CE in southern Gaul Western Europe was such a severe winter that no one living ever remembered a similar winter. The winter of 593 was "unprecedented" and extremely harsh in Provence, France.

In 595 CE the Chinese Emperor Yank-Kien was forced to move his court from Xi'an to Henan as there was not enough food for his court due to extreme famine.

In Southern Anatolia in Turkiye at the end of the Sixth Century CE farmers abandoned fields and orchards and returned to pastoralism. Cultivated cereals and walnuts gave way to steppes and forests of pines and cedar.

You are probably asking why are we reading about famines and roses blooming and floods? These are weather and as we progress we shall see connections between events more closely.

600 CE

Around 600 CE Sheveluch in Kamchatka erupted with a VEI of 5 and Ata in Japan erupted with a VEI of 4. Other eruptions were Tungurahua in Ecuador with a VEI of 3, and Izu-Oshima with a VEI of 3. Opala in Kamchatka erupted with a VEI of 5 in 610 CE.

In the 7th century CE there were periods of generally rather colder and more disturbed climate. In the northern parts of the Mediterranean and in northern and western Europe there was an increase in wetness. This wetness was separating periods of extreme dryness.

In Italy and perhaps elsewhere in the northern Mediterranean, the dryest periods were interrupted by the influence of the cold and wet climate that was affecting northern and central Europe. Over wide areas further south the dryness persisted.

In Arabia, places where agriculture had been carried out with elaborate irrigation works, which had survived earlier periods of desiccation or dryness, were abandoned. There was widespread drought over all of these areas. This was after The Marib Dam collapsed. Around 600 CE there was massive drought in Arabia and many Arab tribes left the area.

Radiocarbon dating of the old moraines marking former glacial termini in the valley bottom in the Val de Bagnes in southwestern Switzerland, revealed that the positions reached by the glaciers coming down from the heights on either side around 600 to 700 CE showed that they were as far forward as those registered in the Little Ice Age. These glaciers even cut through an old Roman road across the mountains from Italy which passed down through the valley of Val de Bagnes.

In the sixth century there was an increase in storm and sea floods around the North Sea.

There was widespread drought in the eastern Mediterranean in southern Italy, Greece and Turkiye and also in Syria during the seventh and eighth centuries. There were reports of migration to the coast and to the islands, leaving a depopulated countryside behind. This was the time of the decay of Ephesus and the depopulation of the area around it and of the abandonment of many early Christian churches and communities in Turkiye and Syria. In the Levant, the cities of Ephesus, Antioch and Palmyra decayed due to drought. There and in southern Italy and Greece, people were migrating to the coasts and leaving a depopulated hinterland. This was the same in Anatolia in Turkiye.

Most parts of the Northern Hemisphere south of 35° North, continued to be as warm as or warmer than before, through these centuries.

The yearly floods of the Nile, that were supplied by the summer monsoon rains over Ethiopia, were low: but the winter flow of the Nile, which depended on rains near the equator, was high, as was the level of Lake Rudolph in eastern equatorial Africa. This meant that the equatorial rains had a restricted seasonal migration north and south at the time. They seem to have supplied more water to equatorial Africa, and therefore to the White Nile, than in years of drought in the 1970s in the Sahel and Ethiopia.

After 600 CE there were large, raised field systems around Lake Titicaca on the Altiplano of Bolivia. The raised field system spread across the wetlands covering an area of about 190 square kilometers by 800 CE to 900 CE. This expansion coincided with several centuries of elevated rainfall.

In the 600s CE there was a marked increase in wetness in Europe.

Between 600 CE and 1400 CE the Nile floods in Egypt were generally lower.

Around 600 CE there was heavier flooding of the Rhone River in Southern France due to more rainfall overall which continued.

Around 600 CE there was flourishing agriculture in what is now the Negev Desert in Israel due to wetter conditions.

Palestine which had a moist climate during the Climatic Optimum faced a contraction of farming in the Seventh Century. This was in the Golan Heights and other places.

Sand covered over the irrigation system of the Moche in Northern Peru, and they abandoned their capital. The Moche moved east to higher terrain with more water. There was a shift in climate to a warm period and there was a massive El Nino period.

There was expansion of Alpine Glaciers in Switzerland after this period.

At the end of the Western Roman Empire a period of great climate extremes with more intense cold and heavy rainfall started.

The winter of 602 CE was very severe in France. The sea froze in places that were normally sheltered. Many fish were decimated. As a result, there was a famine.

In 602 CE normal rainfall fell in the Moche Empire in Peru after drought periods.

The winter of 603 CE was unusually cold and destroyed the vineyards in France. The unusual cold of the year 603 in Western Europe killed much of the vineyards.

In the winter of 604 in Scotland there was four months of frost, followed by dearth famine. The frost was also severe in England. In the year 604, there was the severest frost for 4 months, chiefly in Scotland, followed by a dearth. In Europe in 604, there was the most severe rigorous winter. The grapevines mostly died in all places. The Sea was frozen and killed the fishes in it. This produced a great famine.

The winter of 604 CE was very cold in France and then there was a famine.

In 604 CE there was severe frost in England.

Also in 604 CE there was a severe winter in Scotland with four months of frost.

The winter of 605 CE was very severe in the Oise region of France. It destroyed the grape vines.

In 605 CE there was a drought and great heat in England. There was a drought with scorching heat.

Also, in 605 CE There was excessive heat and drought, hence a famine and plague on man and beast in Italy.

In 610 CE Opala in Kamchatka erupted with a VEI of 5. Opala was enough to create a volcanic winter.

By the look of it Kamchatkan eruptions do not have a great influence on weather. Or it appears that way at the time.

Starting in 610 CE and ending in 650 CE there was a wet period around Lake Titicaca in Bolivia.

In 616 CE in China, ten deaths were reported from a meteorite shower and siege towers were destroyed.

In 620 CE the Mono-Inyo Craters in California erupted with a VEI of 4.

In 625 BCE Izu-Oshima in Japan erupted with a VEI of 3.

Greenland ice cores linking layers of ammonium to high-energy atmospheric interactions with objects coming from space were in 539 CE, 626 CE, 1014 CE and 1908 CE, the Tunguska Event. Where and what was the 626 cosmic interaction.

In the year 629, there was a great snowstorm in Scotland. "It lay a Fortnight 5 Foot in Scotland."

In 630 CE Sheveluch erupted with a VEI of 3 in Kamchatka.

In 630 CE there was a Thames flood in London.

In 634 CE there were great snows in the Country of Berg (Germany) that killed many people. The Country of Berg today is within the Nordrhein-Westfalen State in western Germany, east of the Rhine river, south of the Ruhr.

In 634 CE in winter there was heavy snow in Ireland that killed many in Ulster. In southwestern Ireland, there were floods in Munster.

Between 635 and 785 CE Witori/Pago on New Britain Island in Papua New Guinea erupted with a VEI of 6 which would have caused a volcanic winter. Dakataua and Rabaul, also in New Britain, had very large eruptions as well. We only had to find when the next volcanic winter event would occur to confirm when this gigantic eruption would occur.

Between 636 and 645 there was a drought again in the Moche Empire in Peru.

In 638 CE there was an eruption with a VEI of 6 in Rabaul in Papua New Guinea. This eruption might have formed the Rabaul Caldera sea-inlet.

Ice cores with high acidity levels were found in Antarctica and Greenland from 639 CE to 640 CE may be the caused by the cluster of New Britain eruptions in Papua New Guinea.

In August 640 CE San Salvador in El Salvador erupted with a VEI of 3 and Taveuni in Fiji had a VEI of 2.

In 641 CE Harrat Rahat in Saudi Arabia erupted with a VEI of 2.

In the 640s some cold years were noted in Britain.

In 642 CE the winter in Europe was severe. The Black Sea was frozen. There were snowdrifts 90 feet (27 meters) deep.

In 649 CE in Cheshire and Lancashire, England, there was great damage from an inundation of the sea.

Now you are thinking, where is the volcanic winter after the New Guinean eruptions? Due to the lack of climatic recording in this period, this part of the Dark Ages, will have to remain a mystery.

650 CE

Sheveluch in Kamchatka erupted with a VEI of 5 and Arenal in Costa Rica had a VEI of 4 in 650 CE. Harrat Khaybar in Saudi Arabia had a VEI of 2, and Krasheninnikov in Kamchatka had a VEI of 2.

From 650 CE to 1150 CE there was more rain in the area around Lake Titicaca. Other sources state that there was a dry period between 650 CE and 760 CE. Which is right?

Michael Purser mentions frescoes showing that during the siege of Constantinople that year there were falling stars. These falling stars were associated with a disturbed sea event.

From 650 CE to 850 CE China and Japan had a warm period whilst Europe had a cold period. This is another see-saw event. Warm in east Asia and cold in Europe.

In 650 CE the Saxon village at West Stow in East Anglia was abandoned. In its heyday since it was founded in 400 CE it had grown barley, oats, rye and flax until the ground became marshy. The village had been established on a little sandy knoll on a river. The climate had changed.

The Moche State in Peru collapsed in 650 CE in part due to disastrous drought cycles and massive El Ninos.

In 650 CE Sheveluch erupted in Kamchatka with a VEI of 5.

In 653 CE Dakataua in New Britain in Papua New Guinea, erupted with a VEI of 6. This would have guaranteed a volcanic winter.

In 654 CE Izu-Oshima erupted in Japan with a VEI of 3.

In 660 CE Ata in Japan erupted with a VEI of 4.

In 664 CE a pestilence reached Ireland which killed many people.

In the year 669 a great famine struck France. The king even sold his jewels to relieve the poor.

In Ireland there was a great scarcity in 669 and in the following year.

In the Winter of 669-670 CE there was a fatal frost in England.

The winter on the coast of Constantinople was very severe and long and a large number of people and animals perished.

In 670 CE Sete Cidades erupted in the Azores with a VEI of 3.

The winter of 670 CE produced an abundance of snow and ice in France. As a result, wildlife could be picked up by hand. The winter was most severe and long. It killed many people and cattle.

In 675 CE Nevado del Ruiz erupted with a VEI of 3 in Colombia.

In 675 CE there were three months without rain. The drought was extreme. The wells were completely dry at Chalons in Austrie (Austria) until early August.

The summer of 675 CE was unusually hot in France.

In the year 676 in Rome and Italy, there were 4 months of constant rain, thunder, lightning, fatal to people and grains.

In 679 CE the monastery at Coldingham in Berwickshire, Scotland was destroyed by fire coming down from heaven.

In 680 BCE Izu-Oshima off the coast of Honshu in Japan, erupted with a VEI of 4.

In 680 CE in England, there was famine from a drought that lasted for three years. In the days of Ethelwald, King of Saxons, was a great drought for three years. This drought caused such a famine that people pined with hunger and long fasting, went in companies, and climbing some precipice, joining hand in hand, threw themselves either over a rock or into the sea. Æthelwold, a 7th century king of East Anglia, the long-lived Anglo-Saxon kingdom which today includes the English counties of Norfolk and Suffolk.

The drought which ended in 681 and had lasted for three years was broken on the day that Bishop Wilfrid converted the South Saxons to Christianity. The drought was known as St. Wilfrid's Drought. He had actually converted the king at the time who then imposed the religion on his court and people.

In 682 CE the same Irish pestilence returned.

In 683 CE Rabaul volcano erupted on New Britain in Papua New Guinea with a VEI of 6. This eruption formed the Rabaul Caldera sea-inlet. This would also have caused a volcanic winter.

In 684 CE in Ireland, it was so cold that lakes, rivers and seas froze. This appeared to be anticyclonic/blocked with an easterly type resulting. Was this due to the Rabaul volcano erupting the previous year?

In 684 CE the winter in Scotland was very cold. Many lakes, rivers and the sea froze.

In February 685 CE Vesuvius in Italy erupted with a VEI of 4.

On 26 November 684 (Julian calendar) or 29 November 684 (Gregorian calendar), a massive earthquake magnitude 8.4 struck the Hakuhou Nankai region of Japan. In 684 on the 14th day of the 10th month (ancient Japanese calendar) at the hour of the boar (10 p.m.), there was a great earthquake. Throughout the country men and women shrieked aloud and knew not East from West. Mountains fell down and rivers gushed forth; the official buildings of the provinces and districts, the barns and houses of the common people, the temples, pagodas and shrines were destroyed in numbers, which surpass all estimates. In consequence, many of the people and domestic animals were killed or injured. The hot springs of Iyo were dried up at this time and ceased to flow. In the province of Tosa today the Kōchi Prefecture on Shikoku more than 500,000 shiro of cultivated land were swallowed up and became sea. Old men said that never before had there been such an earthquake. On this night a rumbling noise like that of drums was heard in the East. Some said that the island of Idzu had increased of itself on two sides, the north and west, to the extent of more than 300 rods, and that a new island had been formed. The noise like that of drums was thought to be the construction of the island. Then on the 3rd day of the 11th month, the Governor of the province of Tosa reported that owing to a great tide, which rose high, and an overflowing rush of seawater, many of the ships used for conveying tribute had been sunk and lost. The first recorded tsunami in Japan struck on 29 November 684 CE off the coast of the Kii, Shikoku, and Awaji region. The earthquake, estimated at a magnitude of 8.4, was followed by a large tsunami. The earthquakes were the Hakuho and the Nankai earthquake. This was the Hakuho earthquake and up to 1000 people might have died.

In April 685 CE Asamayama erupted in Japan with a VEI of 3.

In 685 CE in Ireland, there was a great inundation of the sea.

In 686 CE Yakedake in Japan erupted with a VEI of 2.

In 690 CE Witori/ Pago in Papua New Guinea erupted with a VEI of 5 and Newberry in Oregon erupted with a VEI of 4. The other eruption was Irazu in Costa Rica with a VEI of 3.

In 690 CE at Venice and Liguria Italy, there were great floods from violent rainstorms. Venice, Italy and other regions had frequent rains, thunder, lightning and great floods. In Venice and Liguria, Italy, happened the greatest tempest of rain, thunder, lightning and inundation, felt or seen since Noah's Flood, with the greatest damage.

In 693 CE there was flooding in Ireland due to heavy and prolonged rainfall. Rivers in Leinster flooded for three days and nights.

In 695 CE the Thames in England was frozen for six weeks, when booths were built, and a market held upon the ice. There was also a great frost in England.

700 CE

Around 700 CE Mount Churchill in eastern Alaska erupted. Mount Churchill was the source of the White River Ash, deposited during two of the largest eruptions in North America during the last 2000 years. The twin-lobed tephra deposit covers more than 130,000 square miles of eastern Alaska and northwestern Canada. The larger eastern lobe stretched over 500 miles, and the total volume of ash exceeds twelve cubic miles or roughly 50 times the volume of Mount St Helens in 1980. The VEI was 6. Other eruptions were Raoul Island in the Kermadec Islands in New Zealand with a VEI of 3, Bezymianny with a VEI of 4, Sheveluch with a VEI of 3 in Kamchatka, and Izu-Oshima in Japan with a VEI of 3.

In 700 CE in England and Ireland, there was a famine and pestilence during three years, "so that men ate each other". In 700, our Saxon ancestors being yet heathens were plagued with such severe famine for three years together, that many died of hunger. And in Sussex, England many were so tormented with it, that sometimes groups of 40 people would get up on the rocks by the seaside and throw themselves down headlong into the sea and were drowned.

In 700, a great drought prevailed in the Auvergne, France. The summer of 700 CE was unusually hot in France.

In the Lake Chad Basin in North Africa there was a maximum occurrence of the plants of the Sudan-Guinean monsoon zone flora between 700 and 1200 CE and these and other water-demanding plants declined rapidly between 1300 CE to 1500 CE.

Between the eighth and eleventh centuries Carioco Basin cores and borings from Yucatan lakes indicate a linkage to the Intertropical Convergence Zone lingering in a more southerly position than normal.

In North America from about 700 CE onwards the climates of the Midwest became moister than before with the prairie giving way to trees, until an abrupt reversal around 1200 CE. Farming people were spreading northwestward on the plains, moving northward into Wisconsin and on up the Mississippi and other valleys into Minnesota as early as the eighth century. They maintained a thriving culture until 1200 CE when drought and vegetation change occurred, and they disappeared. Such a change in the region is explained by increased sway of the westerly winds, intensifying and extending the rain-shadow of the Rocky Mountains, as the thermal gradient increased with the cooling of the Arctic and then setting in.

Every spring the Colorado River to the east of Mojave rose in flood, alternating among channels in its enormous delta. Sometime around 700 CE a prolonged shift of the river caused water to flow into the low-lying Salton Basin which then filled like a bathtub to a depth of about 42 feet. At this depth this inland sea was 115 miles long and up to 35 miles across and 314 feet deep. It became one of the largest lakes in North America. This great lake survived for more than six centuries as an overflow for the Colorado River until rising silt levels blocked the entry channel. The lake then became a closed basin and dried up within half a century. Before this and thanks to the Colorado floods Lake Cahuilla remained relatively stable for hundreds of years with its water level varying by three feet or so. The lake was actually full during most of the warm centuries. The sudden appearance of an enormous lake in the midst of an arid landscape was a Godsend to local hunter gatherer groups though it was too salty to drink. Wildlife flourished around it. In the late Twentieth Century, the Salton Sea was a brackish wasteland.

There was a warmer climate in South Island in New Zealand between 700 CE and 1400 CE than in the centuries before and after it.

In much of North America but not all, a moister period of climate started that ended around 1200 BCE. At this time people were moving northward and westward into Wisconsin with agricultural settlements spreading up the valleys.

In 701 CE in Lincoln, England, a storm (hurricane) threw down above 100 houses.

In 703 CE in Italy, there were three years of famine.

In September 708 Aira in Japan erupted with a VEI of 3.

In 710 CE Pago, east of Kimbe in New Britain in Papua New Guinea erupted. It had a VEI of 6 and ejected 30 cubic kilometers of tephra. Pago is only 70 kilometers southeast of Dakataua. Another name for Pago is Witori. Also, in 710 CE Calbuco in Chile erupted with a VEI of 4.

In 711 CE Aira erupted in Japan with a VEI of 3.

In 713 CE Izu-Oshima erupted with a VEI of 3.

In 716 CE Aira erupted again in Japan with a VEI of 3.

In 717 CE in Rome Italy, the Tiber River greatly overflowed from rain. The Tiber River in Italy overflowed its banks in Rome and in low lying places, the river flowed over the city walls, overturning houses, laying waste to the land and destroying corn grains. The Calvisio says in 717 in the region of Tracia and on the side of Constantinople now Istanbul, Turkiye, the winter was so violent that the horses and camels of the Saracen army perished in great numbers. Tracia is Italian for Thrace. Thrace designates a region bounded by the Balkan Mountains on the north, Rhodope Mountains and the Aegean Sea on the south, and by the Black Sea and the Sea of Marmara on the east. In Constantinople now Istanbul, Turkiye, the winter was so severe that the horses and camels of the Saracen army that was besieging the city perished in large numbers.

In 719 CE in Ireland, there was a rainy summer and a great inundation of the sea. Also, in 719 CE there was firey hail, that burnt ships while the sea boiled up.

In 720 CE Ata in Japan erupted with a VEI of 4 and Fujisan in Japan erupted with a VEI of 2.

In 720 CE Wales had a very hot summer.

Lake Punta Laguna in Quintana Roo in Mexico showed evidence of a prolonged dryer period between 725 CE and 1020 CE.

On July 15th 726 CE Santorini/ Thera erupted with a VEI of 4. This was a major eruption next to Theia, Palea Kameini, produced a third island and ejected ash as far as Macedonia north of Greece and Anatolia in Turkiye. And this was only a VEI 4 eruption.

In 730 CE Karkar Island in the Bismarck Sea, thirty kilometers from the coast of mainland Papua New Guinea erupted with a VEI of 4 and Tungurahua in Ecuador erupted with a VEI of 4.

Between 730 and 1130 CE Ceboruco in Mexico erupted with a VEI of 6. This was a major eruption that would have caused a volcanic winter. Ceboruco released 11 cubic kilometers of tephra.

In 737 CE there was a great drought in London and the south of England.

In 738 CE there was a flood in Scotland due to heavy and intense rainfall that caused 400 families to be drowned in Glasgow.

In 740 CE Cotopaxi in Ecuador erupted with a VEI of 4.

On December 28th 742 CE Kirishamayama erupted in Japan with a VEI of 3.

In the winter of 748 CE in Ireland a great snow destroyed herds and would have been a major disaster.

A devastating earthquake known in scientific literature as the Earthquake of 749 CE struck on January 18, 749, in areas of the Umayyad Caliphate, with the epicenter in Galilee. The most severely affected areas were West and East of the Jordan River. The cities of Tiberias, Beit She'an, Pella, Gadara, and Hippos were largely destroyed while many other cities across the Levant were heavily damaged. The casualties numbered in the tens of thousands. There were also reports of a tsunami in the Mediterranean Sea. Towns were also swallowed up by the earth.

750 CE

Cerro Bravo in Columbia erupted in 750 CE with a VEI of 4. Arenal in Costa Rica also erupted with a VEI of 4 in 750 CE. Other eruptions were Mutnovsky with a VEI of 3, Krasheninnikov with a VEI of 3 and Sheveluch with a VEI of 3 in Kamchatka, and Miyakejima with a VEI of 3 and Kikai with a VEI of 3 in Japan.

From 750 CE to 900 CE there was a shift to drier colder climate in China with a series of three multiyear droughts. These dry cycles coincide with dry cycles recorded in the Cariaco Basin in the Caribbean off Venezuela. These were a result of the Inter Tropical Convergence Zone, ICTZ, moving south.

In 753, at the time of the taking of Clermont in the region of Auvergne, France by Pepin the Short, king of the Franks, there was over all of France a horrible storm. This thunderstorm lasted 22 hours. It spoiled wine cellars. Three thousand people and more than twenty-four thousand animals died of fright during this storm. I suggest these deaths were more likely attributed to lightning strikes. Clermont today is called Clermont-Ferrand and is located in the south of France.

In 759 CE in Ireland, there was a great famine throughout the kingdom, which lasted for several years.

In the winter of 759-760 CE there was a cold winter in England.

There was a dry event in Central America in 760 CE.

A deep-sea core from the Carioco Basin in the southeastern Caribbean shows that the major droughts were in 760 CE, 810 CE, 860 CE and 910 CE. There were roughly fifty-year intervals. These droughts were during a generally dry period. The collapse of the Maya occurred first in the central and southern lowlands where access to groundwater was limited, and farmers relied heavily on rainfall. The northern Yucatan fared better, because here the collapsed sinkholes, known as cenotes, provided groundwater.

In the Winter of 760-761 CE according to the Helgoländer Chronik (of Helgoland and Norddeutschland in northern Germany, the winter was very severe. It began in October when the open sea and large lakes were clogged with ice for many miles. More than 20 Ellen 46 feet, 14 meters of snow fell. In the following February, the ice broke with the most unheard-of incredible bang, that could be conceived on heaven or earth. Some of the ice was like high trees or mountains. The ice was 31 Ellen or 71 feet or 22 meters thick.

Starting in 760 CE there was a multi-year drought in Central America. These droughts recurred at fifty-year intervals. 760, 820, 860, and 910 AD.

Starting in 760 CE and ending in 1040 there was a wet period around Lake Titicaca on the Altiplano of Bolivia.

In 762-763 CE there was a drought in England. The summer was so hot that the springs dried up.

In the winter of 763-764 CE the winter was noted as severe and was followed by a long and terrible drought in the spring/summer of 764 CE. This suggests abnormally persistent blocking and a high-pressure situation with the primary jet perhaps shunted well to the south. Some sources note great snow with an intense frost. In London it was recorded as one of the most severe winters known in history. This probably affected large areas of continental Europe, again suggesting a "Scandinavian High" situation. This winter was recorded as the earliest winter to be documented in many parts of Europe with enormous snowfalls and great losses of olive and fig trees in southern Europe. There was even ice on the Dardanelles between Europe and Asia in Turkey.

Theophanes the Confessor wrote that in 763 CE "it was bitterly cold after the beginning of October, not only in our land, but even more so to the east, west, and north. Because of the cold, the north shore of the Black Sea froze to a depth of 30 cubits (~ 45 feet) a hundred miles out. This was so from Ninkhia to the Danube River, including the Kouphis, Dniester, and Dnieper Rivers, the Nekrophela, and the remaining promontories all the way to Mesembria and Medeia. Since the ice and snow kept on falling, its depth increased another twenty cubits (~ 30 feet), so that the sea became dry land. It was traveled by wild men and tame beasts from Khazaria, and the lands of other adjacent people. By divine command, during February of the same (winter in 764 CE) second indication the ice divided into a great number of mountainous chunks. The force of the wind brought them down to Daphnousia and Hieron, so that they came through the Bosporos to the city (Constantinople) and all the way to Propontis, Abydos, and the islands, filling every shore. We ourselves were an eyewitness and, with thirty companions, went out onto one of them and played on it. The icebergs had many dead animals, both wild and domestic, on them. Anyone who wanted to, could travel unhindered on dry land from Sophianai to the city and from Chrysopolis to St. Mamas or Galata. One of these icebergs was dashed against the harbor of the acropolis and shattered it. Another mammoth one smashed against the wall and badly shook it, so that the houses inside trembled along with it. It broke into three pieces, which girdled the city from Magnaura to the Bosporos, and was taller than the walls. All the city's men, women, and children could not stop staring at the icebergs, then went back home lamenting and in tears, at a loss as to what to say about this phenomenon.

The Danube Delta on the Black Sea is located in eastern Romania and southwestern Ukraine and the Dniester and Dnieper rivers are located in the Ukraine. Mesembria is an ancient name for Nesebŭr, Bulgaria. Medeia is a city on the eastern Black Sea coast of present-day Turkey. Propontis is an ancient name for the Sea of Marmara, between the Bosphorus Strait and the Dardanelles Strait in Turkey. Abydos is the name of an ancient city located in the south of Nara Point (Nara Burnu) in the Dardanelle Strait of Turkey. From the 6th century, the port of Çengelköy in Turkey was called Sophianai because of the palace Justin II built nearby for his consort Sophia. Chrysopolis or Üsküdar is a municipality of Istanbul, Turkey. St. Mamas or Galata is a district of Istanbul, Turkey. The Magnaura was a large building in Constantinople. It was located east of the Augustaion, close to the Hagia Sophia and next to the Chalke gate of the Great Palace.

In another account the Byzantine historian Nicephorus described the winter in Constantinople that same year. In the beginning of autumn, winter has come with abnormal colds; also, saline waters are frozen which affected inhabitants of the city severely. One hundred mile (161 kilometer) stretch of the sea is covered by ice like in the regions north of the Black Sea. Ice invaded most of the rivers; the coasts of Mesembria and Medeia were a solid mass of ice that was

30 coudée thick (13-14 meters). Also, snowfall was so heavy that this ice is enclosed by 20 coudée of snow and all morphological differences between sea and coast disappeared. Now a white cover unified sea and land. All parts of the North Sea (Black Sea) facing north were solidified. Especially the areas of Hazars and round the Scythian's Lands were inaccessible and unsuitable for human and animal life. After a while this significant crystal crust broke into several pieces and these were uplifted in the middle of the sea like pyramids. Most of them, dragged by winds, were smashed and sunk in the opening of the Bosphorus to the Black Sea near Daphnousia, which was a powerful castle. Most of them entered into the Bosphorus. They filled up all the curls of the water way and connected Asia and Europe. They formed a land bridge between two continents and it was easier to pass the strait by walking instead of using boats. Accumulated ice masses in the Bosphorus without any delay were dragged into Propontis (the Marmara Sea) and even reached Abydos. There they accumulated again in a perfect way to form a structure like a monolith and Propontis lost its sea characteristics. One of these huge icebergs was grounded in the bottom of the Constantinopolis Fortress and shook the city walls so that inhabitants were excited. Icebergs accumulated in front of the fortress, then invaded all waterways. They accumulated to the same height as the city walls. As a result, inhabitants of the city were able to go out of the city from the harbor by crossing these icebergs and they can walk to the Galata Castle on the other side from Constantinopolis Fortress. Hazars and around the Scythian's Lands today possibly refers to northwestern Turkey, Georgia and Abkhazia.

During the winter of 763-764, very severe cold reigned in Gaul in Western Europe and in Illyria (the western part of today's Balkan Peninsula) to the shores of the Black Sea. According to the Frankish chronicles, this cold was of exceptional severity and could not be compared with any previous cold winter.

In Gaul Western Europe from 1 October 763 to February 764 there was a very severe frost. The olive and fig trees were damaged because the soil froze to their roots. As a result, over vast regions of the earth a terrible famine broke out in the following years, which killed many people. During the winter of 763-764 in Western Europe, the cold started October 1st and ran until February 764. The winter of 764 CE was exceptionally cold in France; first in October, then from 1 January to late February. There were up to 10 meters 33 feet of snow in places. The olive trees in the south froze. There were frozen seas. The rivers were frozen "to the bottom". Then there was a famine.

On October the 1st, came a most rigorous bitter frost, which lasted until February. It affected not only Europe but also all over the North and the East. The main sea was frozen near the pole and snow laid 20 feet deep upon the ice. It killed most vegetables and many sea animals. The snow destroyed many forests.

In 763 in England, there was a violent frost, which continued about 150 days.

In January 764 CE Aira erupted in Japan with a VEI of 4. This is enough to cause a volcanic winter. Ata also erupted in Japan with a VEI of 4. Double the Volcanic winter in the northern hemisphere.

Between 764 CE and 860 CE there were very severe winters in Europe with the great rivers frozen and in a number of cases thick ice on the Black Sea, the Bosphorus and Dardanelles and the fringe of the Adriatic.

In 764 CE a meteorite struck a house in Nara in Japan.

On July 20th 766 CE Aira erupted again in Japan with a VEI of 3.

Ata in Japan erupted again with a VEI of 4 in 770 AD. Cotopaxi in Ecuador also erupted with a VEI of 4 and Taveuni in Fiji erupted with a VEI of 2.

In 767 CE in Asia there was a great drought. So great a drought in Thracia in Southeast Europe without either rain or dew, that all springs, fountains and rivers at Constantinople were dried up.

In 768 CE in Ireland, there was famine and an earthquake.

Between 770-800 CE there was a period of high frequency cold winters. This suggests that there was blocking of the main Atlantic, westerly flow by often slow-moving, intense anticyclones or an increased frequency of east or northeast flow with higher pressure to the north of the British Isles. It is suggested that this would tie in with the idea that Scandinavian exploration/raids were assisted by a lack of westerly storminess.

Between 770 CE and 800 CE there was a more than usual tendency for cold winters. The other seasons, although more dry than wet, revealed drought years and some years when floods created difficulties in Europe. It appeared that there was a blocking of the westerlies in 45° to 65° North was frequent.

In 772 there was an epidemic at Chichester in Sussex, England, where 34,000 people were killed.

In 774 CE in Scotland, there was a severe famine with a plague.

In 775 CE in England, there was a drought with excessive heat, after a great frost.

The winter of 775 CE was so hard that the Euxine Sea (Black Sea) was quite frozen over. The ice was 30 foot or cubits thick. People could walk 50 or 100 leagues (150 to 300 miles, 240 to 480 kilometers) on the ice from the Danube River to the Euphrates River. On the ice fell 30 cubits deep of snow. When the ice broke, it appeared like great mountains on the sea, which demolished and carried down whole villages standing on the shore. This winter was succeeded by so excessive heat during the summer that all springs dried up. The Danube River probably refers to the Danube Delta in Europe, eastern Romania and southwestern Ukraine. The Euphrates River rises in Turkey, passes through Syria, and joins with the Tigris River in southeastern Iraq to form the Shatt al Arab, which empties into the Persian Gulf.

Because of the famine in Northern Italy in 776 CE, the supply of Lombard slaves sold by the Greeks to the Arabs increased. Some free men boarded the slave ships voluntarily becoming slaves in order to survive the severe effects of starvation.

In 780 CE El Chichon in Mexico erupted with a VEI of 5.

In July 781 CE Fujisan erupted with a VEI of 3.

In 783 CE in Germany, the summer was so burning that many people perished from the heat.

Also in 783 CE in France the summer was unusually hot. The heat during the summer of 783 in southern France was so extreme that many people died from it.

Between 785 and 935 CE Puyehue in the Andes of Chile erupted with a VEI of 5.

On October 15th 787 Vesuvius erupted in Italy with a VEI of 3.

On April 18th 788 CE Kirishimayama in Japan erupted with a VEI of 4.

The archaeologist Richard Gill used intervals of severe cold in Swedish tree rings and the last dates inscribed on stelae at abandoned Maya cities to propose a tripartite collapse of the Maya Civilization. The collapse started in 810 CE and affected Palenque and Yaxchillan amongst other cities. In 860 CE there was another drought that affected the great cities of Caracol and Copan. Finally in 890-910 CE Tikal, Uaxactun and other major cities collapsed. Evidence from Tikal showed that human remains found in a house midden showed signs of burning and chewing that could have only come from survival cannibalism. These cities' populations either perished or abandoned the cities after losing their faith in their leaders who were powerless to bring rain.

In the Winter of 790-791 the vines in Provence in southeastern France suffered very much and the flocks came into the stables. The winter of 791 was very cold. The grapes in Provence, France froze. The livestock was decimated. There was an abundance of snow and ice. Wild animals could be taken up by hand.

On 7 November 792, there was an inundation of the West Sea in Friesland in the Netherlands. Countless people and cattle were devoured.

Around this time, the island of Helgeland, Heligoland, in the North Sea was more than 2 miles long and one mile wide.

In 793 CE there were frightful thunder and lightning storms, especially at Northumberland in England soon after followed by a severe famine. In England, there was a great thunderstorm. In England, there was a famine "after many meteors;" and this famine was spread through other parts of the world. The famine was so great that people made bread from acorns. In England in the year 793 after many meteors, there was a famine under the Pontificate of Sabinianus; a famine and plague under Boniface the 4th; and a famine and plague under Phocas the Emperor.

In 798 CE the snow that fell in Ireland caused men and animals to die.

During the winter of 799-800 CE a huge storm surge damaged Helgoland now Heligoland - a small German archipelago in the North Sea, which then consisted of two mesas, which were of almost identical size.

Prior to 800 BCE there were perhaps eight to ten million Maya living in the lowlands of the Peten-Yucatan Peninsula. This was a staggeringly high population density for a harsh tropical environment with low natural carrying capacity. Lake cores from salty Lake Chichancana in the Yucatan showed climatic periods including droughts over the past nine thousand years with an accuracy of twenty years. The three cores that were eventually sunk showed there had been three major droughts in Yucatan over the past 2000 years. These were from 475 BCE to 250 BCE, when Maya civilization was still forming, the next lasted from 125 BCE to 210 CE which coincided with the peak of El Mirador, one of the greatest of the Maya Cities which was abandoned around 150 CE. The third drought was 750 CE to 1025 CE. This was the period of the Great Maya Collapse.

800 CE

In 800 CE Dakataua on the northern tip of the Willaumez Peninsula in New Britain in Papua New Guinea erupted with a VEI of 6. Other eruptions that year were Kostakan with a VEI of 3 in Kamchatka, and Almolonga in Guatemala with a VEI of 3.

When was the beginning of the Medieval Warm Period? 830 CE, 850 CE? Some sources state 800 CE. This was a period of warmer winters and longer summers though there was very little difference in temperatures, being only a few degrees.

Over the next three or four centuries after 800 CE the climate warmed up in Europe. Cultivation limits were higher on the hills than they have ever been since. Trees were spreading back towards the heights. The upper tree line in parts of central Europe was 100 to 200 meters higher than it became by the seventeenth century.

Around 800 CE the Caspian Sea was at a very low level.

The hordes of barbarians from the east were caused by drought in central and western Asia.

In the Mediterranean, north Africa, and far to the east into Asia, there were severe droughts in 800 CE.

Around 700 CE the Eskimos or Inuit of the Dorset Culture were widespread across the eastern Canadian Arctic, who had returned to high latitudes after 800 CE, had been moving south. This was possibly partly due to another Eskimo culture, developed near Thule in northwest Greenland, was more resourceful in hunting the resources of the Far North. It is also probable that the increasing ice and dwindling seal and walrus populations were making the competition more difficult.

In the ninth century Lake Chad in Africa was a very high 286 meters above sea level. Between the ninth and thirteenth centuries the Sahel was much wetter. Arab travelers wrote of standing lakes and rich pastures in this period but by 1450 CE Lake Chad's water level dropped by five meters then rose by three meters by 1500 to fall again half a century later. During the seventeenth century it was back to five meters higher than today.

In Central America as early summer insolation waned, the region started to get dryer with a peak in aridity around 800 CE to 1000 CE. This coincided with the collapse of the Mayans. The huge populations and sizes of the Mayan states had placed great pressure on available resources such as food, fuel and water. At Tikal and Calakmul builders gave up the wood they had long used for beams, (Manilkara zapot) and turned to substitutes. They also stopped using lime as a material for plaster. The Intertropical Convergence Zone shifted and brought stresses to Central America. The reduction in rainfall was forty per cent and proved difficult for the Mayans.

In Central America a cold period arrived around 800 CE, which was contemporary with Europe and much of North America. Notably the United States east of the Rockies and the central Canadian Arctic. They were probably moister then in North America than in Central America, if the summer rains did not move so far north.

In the ninth century there was a low titanium interval in the Carioco Cores. Higher titanium is a product of high rainfall and low titanium concentrations indicate low rainfall.

Around 6,200 BCE Lake Chichancanab first filled in at a time when Caribbean sea levels and the Yucatan's freshwater aquifer rose sharply. They continued to rise until 4,000 BCE. The core shows relatively wet conditions until 1000 BCE when conditions became dryer and carbonate levels in the lake sediments increased as a result. Drying continued peaking between 800 CE and 1000 CE, at the very time of the Maya collapse. The drought cycle of these two centuries was the driest period of the last eight thousand years. This was indicated by low titanium concentrations.

In Japan the springs (season) were the earliest in the 9th Century. They were their latest in the 12th Century.

This was a cold time generally in and around the wide expanse of the North Pacific Ocean. Part of the explanation of the medieval warmth in Europe and North America, extending into the Arctic in the Atlantic sector and in at least a good deal of the continental sectors either side, must be that there was a persistent tilt of the whole circumpolar vortex, and of the climate zones which it defines, away from the Atlantic and towards the Pacific, which was rather infrequently affected by outbreaks of polar air.

There is evidence of widespread bushfires in this period in Costa Rica during the same two centuries suggest widespread aridity.

Between 800 CE and 1000 CE Lake Patzcuaro, in Michoacan, Central Mexico was at a very low level.

Between 800 CE and 1250 CE there was a remarkably low concentration of El Nino events on the west coast of South America.

Around 800 CE the Ancestral Pueblo, originally known as the Anasazi, the "Ancient Ones", built some of the largest towns in North America. They were subsistence farmers but with the ability to survive in the arid southwest of the United States. They adapted to the dry San Juan Plateau by being expert farmers and selecting soils with the correct moisture retaining properties on North and east facing sites that received little or no direct sunlight. Every farmer planted on river floodplains and arroyo mouths where the soil was naturally irrigated. They diverted water from streams and springs using every drop of runoff that they could find. Everything was done to reduce the risk of crop failure. They dispersed their gardens widely over the landscape to minimize the risk of local drought or flood and also learned how to shorten the growing season from the usual 30 days to 120 days by planting on shaded slopes at different levels and in different soils. They were expert farmers though they lived in huge multi-story houses.

By 800 CE the Anasazi lived in large communities where small hamlets became clusters of rooms and storehouses built up into contiguous blocks that formed much bigger communities. Population densities also rose in places such as Chaco Canyon, New Mexico, and Moctezuma Valley and Mesa Verde in the Four Corners region of southern Colorado. Four Corners consists of the southwestern corner of Colorado, the southeastern corner of Utah, the northeastern corner of Arizona and the northwestern corner of New Mexico.

Extended droughts in the Mojave Desert in North America showed that water sources of all kinds were abridged and there was reduced spring discharge after 800 CE to 1300 CE. Habitation was only near major springs and in perennial oases along the Mojave River.

Between 800 CE to 900 CE water levels in Lake Titicaca on the Altiplano of Bolivia were the same as those of the late twentieth Century.

In England on the 9th of January 800 CE came a most prodigious hurricane from Africa, with irresistible force. It cast down to the ground and destroyed infinite towns, houses, villages and trees. The same year happened a very great inundation of the sea, which carried away much cattle.

On April 11th 800 CE Fujisan in Japan erupted with a VEI of 4.

The winter of 801 CE in France was very cold from 11 November to 12 March. A plague ensued.

On December 24th 800 CE in northern Europe and England there was a gale called the Great Southwest or West Wind. Cities were destroyed. One storm flood reduced the size of the island of Heligoland more than half by the year 1300 CE.

An earthquake originating in the Central Apennines was felt in Rome and Spoleto on 29 April 801. It is reported in two independent contemporary sources, Einhard's *Royal Frankish Annals* and the *Liber Pontificalis*. The information provided by the written sources has been augmented by archaeology. Both the *Annals* and the *Liber* date the event to 30 April according to contemporary Roman practice, whereby the day began at sundown. The *Annals* specify that it happened at the second hour of the night, which corresponds to 20:00 (8:00 p.m.) on 29 April by modern reckoning. The *Annals* record the event from the perspective of the Frankish king Charlemagne, who had been crowned Roman emperor on 25 December 800 and had left Rome on 25 April for Spoleto, where he was staying when the earthquake struck. They do not record the damage at Spoleto. Charlemagne was unharmed, but perhaps spooked. The triumphal arch in San Paolo, was seemingly damaged during the earthquake of 801 and repaired by Leo III.

The quake was severe, with an estimated magnitude at Rome of VII–VIII on the modified Mercalli intensity scale and 5.4 on the energy magnitude scale. The epicenter was probably somewhere between Spoleto and Perugia. The *Annals* refer to its effects throughout Italy, but without specifics. Both sources attest its damage to the roof of the basilica of San Paolo Fuori le Mura in Rome. The archaeologist Rodolfo Lanciani concluded that the damage was even more severe than the sources let on and that Pope Leo III had to rebuild the basilica "from the introit to the presbytery, from the marble floor to the summit of the roof." The *Liber* mentions the earthquake mainly to introduce Leo III's works at San Paolo, but it leaves the impression that the collapse of the roof caused extensive damage to the interior furnishings (including the silverware stored beneath the altar) and the porticoes. The 5th-century mosaic on the triumphal arch appears to have suffered stylistically from repairs, probably associated with the earthquake of 801. The presbytery was completely rebuilt.

During the Winter of 801-802 CE the Pontus Euxine Black Sea was totally blocked by the ice.

On March 17th 804 CE there was a tornado, wind and lightning in Ireland that killed 1010 men. Was this a storm due to a major depression or a small-scale tornado event?

In 806 CE Bandaisan in Japan erupted with a VEI of 3.

In 806 CE the Rhône River in France was frozen over.

In 806 CE a sea floods hit the archipelago of Heligoland in the North Sea off Germany and much of it was lost. Hundreds of people died.

In 806 CE there were floods in the Netherlands all winter.

On November 1st 807 CE Akita-Yakeyama in Japan erupted with a VEI of 3.

The flood of 809 CE surpassed all known floods in Western Europe. The floods drowned the harvest fields and residents and forced the inhabitants of riverbanks to seek refuge on higher ground. The floods were caused by an abundance of rainfall. The floods reached their climax on December 28.

In 809 CE flood records were broken on the Loire River at Burgundy, France. Residents were forced to take refuge in the hills. The floods peaked on 28 December.

It is recorded that in 810 CE a meteor startled Charlemagne's horse, throwing him to the ground. Charlemagne was making his last expedition against Gofrid, King of the Danes, in Saxony and was moving out of camp and beginning his march before sunrise. He suddenly saw a meteor rush across the heavens with a great blaze and pass from right to left through the clear sky. Whilst all were wondering what this meant, suddenly the horse that he was riding fell head foremost and threw him so violently to the ground that the girdle of his cloak was broken, and his sword belt slipped from it. When his attendants ran up to help him, they found him disarmed and disrobed. His javelin, which he was holding in his hand at the time of his fall, fell twenty paces and more away from him. This is from Eginhard "The Life of Charlemagne", translated by A. J. Gent. How close was the meteor?

There was a dry event in Central America around 810 AD.

Winter of 810-811 CE was rough and lasted until the end of March. The winter of 811 seemed very harsh in Western Europe in the north. It lasted until the end of March.

The winter of 811 CE in France was very cold through March.

In 813 CE in England, there was a great overflow of the River Severn in the night; 2000 people and 7,000 cattle drowned.

816 CE using tree ring dating was the warmest year until 1999 which exceeded it and was the warmest day in the whole millennium apparently. This was from studying ancient pines in Mongolia.

On December 25th 817 CE many rivers and lakes were frozen in Ireland until February 22nd 818 CE. This would have affected the British Isles as well. There was abnormal ice and much snow from the Epiphany to Shrovetide. The Boyne and other rivers were crossed dry-footed: lakes likewise. Herds and hunting parties were on Lough Neagh and wild deer were hunted. This was from the annals of Ulster, 818 CE.

Starting in 760 CE there was a multi-year drought in Central America. These droughts recurred at fifty-year intervals. 760, 820, 860, and 910 CE. The 810 CE drought lasted three years after a long period of increasing aridity.

In 818 in England, there was a plague on man and beast from 3 rainy years and moisture. The grain and grapes were all rotten.

In 818 or 820 in France, there was great rains and floods; many and great floods under Boniface the 4th from rain. You cannot plant crops in long rainy periods and already planted crops spoil.

The year 820 CE was a rotten year in France. The Seine River flooded and spoiled the fruit crop. This was followed by famines and plagues.

In 820 excessive rainfall caused rivers to overflow their banks in Western Europe. The rainfall prevented the autumn sowing. The rain and humidity corrupted the grains and vegetables. The lack of warm temperatures, combined with the excessive rains and the humidity, impoverished and deteriorated the crop of wine. There were even countries where the farmers could not sow their seeds during the spring.

In 818 or 820, from long continued rains in France, and moisture in the air for two or three years, came a terrible plague on man and cattle, far and near. All corn and other grains were rotten. Wine was useless. There were great floods and stagnant air. No hard corn was sown in England before the next spring.

In Western Europe, the summer of 820 was strangely cold. There were abundant and persistent rains, which caused inundation of the fields because many rivers overflowed their banks. This was especially true for the Gironde River near Bordeaux, France. The grains and vegetables were spoiled by the wetness and could not be stored without rotting. The grape harvest was very mediocre, because of the lack of heat. The wine produced was quite tasteless. Because of a roughness of the weather, an infectious disease raged among people and the cattle. No part of Gaul in Western Europe was spared from this scourge, and to make this misfortune worse, the flooding also prevented the autumn sowing.

In 820, it rained a lot in Germany and spoiled the grain in the field. This was followed by a great plague, in which people as well as cattle frequently died. The winter was cold and grim.

In the year 821, there was the greatest frost in England following 2 or 3 years of rains.

The abundant rainfall of 821 in the north of Western Europe prevented the sowing of autumn fall crops. Famine and starvation would follow.

During the summer and autumn of 821 CE there were torrential rains in France, which spoiled the crops.

Between 821-822 CE there was a severe winter in England.

In the Winter of 821 CE-822 CE the great rivers in France and Germany were frozen for 30 days. Snow was on the ground in Vienna, Austria from 22 September 821 until 12 April 822.

In 822 CE Izu-Oshima in Japan erupted with a VEI of 3.

In the year 822 in Europe, heavily loaded carts crossed the ice on the frozen Danube, the Rhine, the Elbe and the Seine rivers, for more than a month. The Rhône, the Po, the Adriatic Sea and several Mediterranean ports were frozen.

The Danube River is located in Central and Eastern Europe. The Rhine River runs through Switzerland, Germany and the Netherlands. The Elbe River is located in Central Europe. The Seine River is located in northern France. The Rhône River runs through Switzerland and France. The Po River is located in northern Italy. The Adriatic is the Adriatic Sea. During the winter of 821-822 in Western Europe, the frozen rivers bore the weight of carriages for more than thirty days.

In the year 822, very heavy rains in France spoiled all the fruits of the earth (crops destroyed), which could not be planted until the next spring. The rivers came out of their beds and the water flowed far into the country. "These evils followed a long and very severe winter, so that not only the streams and small rivers, but also the great rivers, the Rhine, Danube, Elbe and Seine were frozen, and wagons drove on the ice." The ice caused great devastation along the banks of the Rhine River to the banks of Meierhöfen, Austria. Maierhofen is located in southern Bavaria near Bregenz, Austria, on the banks of Lake Constance. In the year 821, the winter was so long and frosty that not only small brooks, but streams and rivers including the Rhine, Danube, Albis Elbe, and Seine, and generally all great rivers both in France and Germany were so hard frozen, that for 30 days loaded carriages went over the rivers as if the ice were bridges.

In 822-823 in England and Scotland, thousands of people starved.

The winter of 823 CE was unusually severe. Winter's cold began on 22 September. Frost and snow continued until 12 April. For a month, all the rivers of Europe bore the load of the heaviest wagons almost without interruption.

In 823, there was a famine in England, Wales, and Scotland. Thousands starved.

In 823, there was a famine in Scotland; many thousands starved.

In 823 thunder and lightning this summer did great damage by killing people and cattle. Hail destroyed the corn. In England, there were tempests of hail, thunder, lightning that was fatal to grain, grass, people and cattle which lasted all summer.

In the year 823 or 824, lightning set fire to a multitude of buildings and killed many people and huge hail ravaged the countryside in France. In addition, all historians assure, that we dare not believe without the unanimity of their testimony, that by the summer solstice around 20 or 21 June in Autun in the region of Burgundy, France, was seen falling from the sky, following a sudden storm and amidst a terrible hailstorm, real ice blocks (we are sure of these measures) of 4.6 meters (15 feet) long by 1.8 meters (6 feet) wide and 0.6 meters (2 feet) thick. These facts were confirmed in the Annals of Einhard, the chronicle of Adhemar, the short Chronicle of Reims, the Annals of Fulda, the Chronicle of Hermann, all contemporary sources. Paradin, in Annales de Bourgogne, bk. 1, p. 149, also speaks of a miraculous ice stone that fell in the year 956 in Germany, and another one that fell in April 1562 in the Beaujolais region of central France.

On March 1st 823 CE Popocatepetl in Mexico erupted with a VEI of 4.

The Winter of 823 CE-824 CE in Gaul was severe and lasted longer than usual. Many beasts and even humans were subjected to extreme cold. A disease followed and snatched away many people of both sexes and ages. The winter of 824 in Western Europe in the north was as long as it was rigorous.

In 823, a bitter sharp and long winter ensued. A load of snow fell, which laid 29 weeks, even to Easter. This also was fatal to many people and cattle in England. In the year 822 or 824, there was a great snow that killed many people and cattle in England.

The winter of 824 CE was very long and very cold in France. This was followed by epidemics that affected all age groups.

From 824 to 825 Ireland was afflicted with a great dearth famine.

On December 31st 826 CE Fujisan in Japan erupted with a VEI of 2.

In 827 CE there was such a severe winter that the Thames in England was frozen for nine weeks. The frost in England lasted nine weeks.

In 828 CE "In Italy, raised scorching winds, accompanied by fiery meteors. But this year was very fruitful."

In 829 CE there was ice on the Nile in Egypt. The Patriarch of Antioch, Dionysius of Telmahre, went with the Caliph Al-Ma'mun to Egypt; they found the Nile River frozen.

Europe experienced a very severe winter in 829.

The year 829 in Western Europe produced famines, plagues, and all kinds of evils. In England there was a great hurricane with twinkling fires like stars running to and fro.

In 1830 CE Fujisan in Japan erupted with a VEI of 2.

North America, Europe and Asia exhibited warmer temperatures between 830 CE and 1100 CE. In Australia and South America, the warm period was from 1160 CE to 1370 CE.

After 830 CE there is evidence of cooling in the tropical Pacific Ocean.

In 831 CE the Seine River in France flooded. Tree rings in giant oaks showed this year was a very wet year.

On June 23rd 832 CE Miyakojima in Japan erupted with a VEI of 3.

In 836 in Wales, due to a famine, "the ground was covered with dead bodies of men and beast."

In 837 CE Kirishimayama in Japan erupted with a VEI of 3.

The Winter of 837-838 was completely taken up by rain and wind. Thunder was heard from January to mid-February, just as in March, and the extraordinary heat of the sun dried up the earth.

The year 838 produced diluvial rains, heavy rainfalls that produced floods in Western Europe. There were massive rainfalls in 838 in the north of Western Europe, which ruined the entire crops. Famine as usual followed.

In Germany this year was marked by unusual atmospheric changes. A terrible burning sun scorched the earth during the summer.

On August 2nd 838 CE Kozushima in Japan erupted with a VEI of 4. Izu-Oshima also erupted in Japan with a VEI of 3. These two volcanoes are only forty miles apart. Kozushima is south of Izu-Oshima. Are they both joined to the same undersea magma tunnel or source?

On December 26th 838 CE there were sea floods in Friesland in the Netherlands. This was a major North Sea storm surge that affected Dutch coastal communities and possibly the worst such storm of the 9th century with a high loss of life. There was a severe storm surge which reached the height of the dune ridges and overran the coast. This was probably the first emergence of Leybucht, the second largest bay in East Frisia in northwest Germany after the Dollart. Many of the survivors between the Ems and Weser Ostfriesen emigrated to North Friesland. There were more than 2,400 deaths.

In 840 CE Furness in Iceland erupted with a VEI of 4.

In 840 in Germany, the Rhine River flooded from rains. From excessive rains, the Rhine River overflowed.

Between 840 and 860 CE the rains faltered in the area of the Anasazi in New Mexico and southern Colorado and the inhabitants reverted to their previous survival tactic and lifestyle of being nomadic. They left their large pueblos and dispersed themselves over the landscape. These were mainly communities in the Dolores Vally region to the north which housed dozens of families.

From 750 CE small pueblos had been built by the inhabitants of Chaco Canyon who took advantage of seeps and springs to grow maize in favored spaces. During the ninth and tenth centuries summer rainfall was highly variable. By the tenth century instead of dispersing due to erratic rainfall, the Chaco people built three great houses at the junctions of major drainages.

In 841 in Herbipolis now Würzburgnorth, Bavaria in southern Germany, people cattle and the lands were greatly harassed by hail, whirlwinds and unusual temperatures.

In 841, there was a terrible storm accompanied by lightning, thunder and heavy rains on the territory of Glandfeuil Le Thoureil, Maine-et-Loire in western France.

The cold during the winter of 842 in Western Europe in the north was neither less intense nor less permanent. A lot of snow fell on the night of April 14. The winter of 842 CE was intense and lasting. The Seine and Yonne rivers in France flooded.

The winter of 842/843 was very long and very cold, producing many diseases and in agriculture, the weather injured cattle and bees. The winter of 843 CE was very long and cold in France. The cold weather decimated the livestock and the bees. Then a disease struck.

In 843 CE Kirishimayama in Japan erupted with a VEI of 2.

It was regarded as a cold winter in England in 844-845 AD.

The 847 Damascus earthquake occurred (probably on 24 November) in 847 CE. Recent scholarship suggests that the earthquake was part of a multiple earthquake stretching from Damascus to the south, to Antioch in the north and to Mosul in the east. There were an estimated 20,000 casualties in Antioch according to the 13th-century historian and writer Al-Dhahabi, and 50,000 in Mosul. It is thought to be one of the most powerful earthquakes along the Dead Sea Transform. A number of other towns and cities in the Middle East also suffered major destruction in 847 A.D., probably on the same day (24 November). The earthquake in Antioch may have been the same one which destroyed much of Damascus, Syria on 24 November 847. The Damascus earthquake began around dawn, lasting until at least midday; part of the Umayyad (Great) Mosque was destroyed and its minaret fell down. Bridges and houses collapsed, and huge stones were displaced. Other towns near Damascus were destroyed including Darayya. There was destruction in towns in Homs (Syria), in Lebanon, and also in the region of the Jazira (Upper Mesopotamia). There was also a large earthquake in Mosul (now in Iraq), in which up to 50,000 people were killed.

In Winter 848-849 the Seine River in France was frozen; so, people used the river as a bridge. The winter of 849 CE was very cold in France. The Seine River was passable on the ice as if it was dry land. We crossed the Seine River in France on the ice on 6 January 849. In the year 849, the winter in Gaul Western Europe was very harsh.

850 CE

In 850 CE Miyakeima in Japan erupted with a VEI of 4. Raoul Island in the Kermadec Islands in New Zealand erupted with a VEI of 4 and Kinenin in Kamchatka also erupted with a VEI of 4 and Kizimen with a VEI of 3 in Kamchatka.

In 850 CE geological records of cosmogenic isotopes such as Beryllium-10 and Carbon-14 which originate during solar driven reactions in Earth's upper atmosphere, indicate that solar irradiance increased during the Medieval Climate Anomaly, MCA. This was indicative of an increase in solar radiation.

The increase in solar radiation may have contributed to the El Nino-Southern Oscillation, ENSO, and other ocean-atmosphere interactions like the North Atlantic Oscillation, NAO, that explain the dry periods of the Medieval Climate Anomaly, MCA.

Along with ENSO there may have been a second climate oscillation, NAO, the North Atlantic Oscillation. Regional climate shifts that occurred in the MCA are broadly consistent with ones that develop during a positive phase of the NAO today. The same as ENSO, atmospheric pressures govern the NAO. A greater pressure difference between the subpolar and subtropical Atlantic produces stronger westerlies during the positive mode of the NAO. Northern Europe and the United States Atlantic coast experienced warmer and wetter winters, while the Mediterranean, Greenland and northern Canada tended to be cold and dry.

These are the same hydroclimatic patterns as during modern La Ninas which suggest that a prolonged La Nina period occurred during the Medieval Climate Anomaly.

After 850 CE the North Atlantic became warmer with less ice floes.

There was especial dryness and increasing warming around 850 CE in Equatorial Africa. This area experienced severe megadrought periods.

From 850 CE in China there was a weakening of the monsoon until 940 CE. This appeared to be related to a southward shift of the Intertropical Convergence Zone. Nomads had an advantage during cold periods in war and often gained victory due to this. Nomadic groups expanded south into the central plains.

During the years 850-851 in Italy and Germany there was a drought with famine.

In 850, a famine prevailed in Paris, France.

In 851 there was a famine in Italy and Germany.

In 850 CE a terrible famine struck Europe.

In the year 851 and 852, the sun was glowing extremely hot in Gaul Western Europe Germany and Italy. The drought was so great that food shortage for the cattle occurred. It became clear that a terrible famine was beginning, which continued to the year 855. "You could see parents eating their own children." In 851 and 855, there was so great a drought over all of Italy and Germany as caused such a famine that parents eat their own children and children their parents.

Pollen studies in lake sediments and bogs on either side of the highlands of Guatemala near 17° North shows a quick change of the surrounding vegetation from grassland to deciduous forest about 850 CE to 90 CE.

The Viking colonizers managed to support themselves with resources from Greenland during the Medieval Warm Period of 850 CE to 1100 CE. Then the temperature cooled by 4 degrees Celsius in eighty years. And you thought that it was the other way around?

Eventually there was a population of 80,000 in Iceland during the MCA but overgrazing, deforestation and a very slow rate of soil replacement and volcanic eruptions made it hard to sustain the population.

In Norway around 850 CE Nordic Sagas refer to the harsh conditions affecting high-latitude areas as the Fimbul Winter. This was possibly connected with a downturn in solar/sunspot activity and /or increased cosmic rays. The subsequent reduction from the sun led to cooler and wetter conditions.

After 850 CE warming started in Scandinavia. A longer growing season and shorter winter contributed to population growth in Scandinavia. This aided Viking travel as there were fewer ice floes as before.

There was especial dryness and warming after 850 CE in Mexico. This area experienced severe megadrought periods.

There was especial dryness and warming in the Middle East after this date. This area experienced severe megadrought periods.

After 850 CE there was increasing wetness over Northern Europe.

There was especial dryness and warming in Southern Europe after 850 CE. This area experienced severe megadrought periods.

There was especial dryness and warming after 850 CE in the southwestern United States. This area experienced severe megadrought periods.

In 852-853 CE Mount Churchill in Alaska erupted. This was the White River II Ash eruption. The White River ash covered vast expanses of Alaska and western Canada and has been found as far as Europe. Mount Churchill in eastern Alaska erupted with a VEI of 6. This would have caused a volcanic winter. How exact is this date? Could it be responsible for the dispersal of the Anasazi as a drought-like volcanic winter descended on the region?

On September 14th 854 CE Izu-Oshima off the coast of Honshu in Japan, erupted with a VEI of 4.

In 855-856 CE there was a cold winter. There was great ice and frost until January 7th 856 CE. Rivers and lakes froze. In Ireland there was so much ice and frost that the principal lakes and rivers could be crossed by people on foot and on horse-back from the ninth of the Kalends of December to the seventh of the Ides of January. This was supposed to be tied to a major eruption of a volcano affecting the stratosphere. It was probably Mount Churchill in Alaska. Does this give you an idea of the effects of volcanic eruptions and their long term influences?

On 6 January 856 there was a shocking inundation of the Tiber River in Italy. This was followed by a plague, wherein the throat being obstructed by great defluxions inflammations, the sick died suddenly.

In 856 CE Niijima in Japan erupted with a VEI of 2.

During December 856 CE there was a major earthquake in Corinth in Greece where 45,000 people died.

The 856 Damghan earthquake or the 856 Qumis earthquake occurred on 22 December 856 (242 AH). The earthquake had an estimated magnitude of 7.9, and a maximum intensity of X (*Extreme*) on the Mercalli intensity scale. The meizoseismal area (area of maximum damage) extended for about 350 kilometers (220 mi) along the southern edge of the eastern Alborz mountains of present-day Iran including parts of Tabaristan and Gorgan. The earthquake's epicenter is estimated to be close to the city of Damghan, which was then the capital of the Persian province of Qumis. It caused approximately 200,000 deaths and is listed by the USGS as the sixth deadliest earthquake in recorded history. This death toll has been debated. The area of significant damage extended along the Alborz for about 350 kilometers (220 mi), including the towns of Ahevanu, Astan, Tash, Bastam and Shahrud, with almost all the villages in the area severely damaged. Hecatompylos, now called Šahr-e Qumis, the former capital of the Parthian Empire, was destroyed. Half of Damghan and a third of the town of Bustam were also destroyed. The earthquake badly affected water supplies in the Qumis area, partly due to springs and qanats drying up, but also because of landslides damming streams. The total death toll for the earthquake is reported as 200,000, with 45,096 casualties in the district of Damghan alone.

In 856 CE in England, there were great rains and floods, followed by an epidemic of quinsy. Quinsy is when an abscess which is a collection of pus forms between one of your tonsils and the wall of your throat.

In the year 856 in Scotland, a four-year famine began.

The winter in 856 was very harsh and very dry, a violent epidemic pulled out many people.

The winter of 856 CE was very rigorous and dry in France. It was followed by epidemics.

In 856 CE there was a very great wind, and woods were felled in Ireland.

In the winter of 859 CE to 860 CE there was ice strong enough to bear laden wagons on the edge of the Adriatic near Venice. People went to Venice, Italy on horseback over the frozen water. In 860 in Italy, the Port of Venice was frozen. In the year 859, the ice was so great that carriages were used on the Adriatic Sea.

During the winter most of the rivers in Europe were frozen for two months.

In April 860 CE Ata in Japan erupted with a VEI of 2.

In the year 860, the Rhône River in France froze over its entire length. The last time that this happened was in 1963 and is regarded as unusual. As you can see in this period it was not. The Rhone is in southern France and flows into the Mediterranean.

The winter of 860 CE was very long and very cold in France. It lasted from November to April. There was an overabundance of snow. The winter destroyed the grape vines. Wild animals could be taken up by hand. This was followed by a famine.

The Mediterranean Sea was frozen over, and passable by carts in 860 CE

In 860, frosts and snows were continuously beginning in November until April.

The Ionian Sea froze. The Ionian Sea is an arm of the Mediterranean Sea, south of the Adriatic Sea, bounded by southern Italy including Calabria, Sicily and the Salento peninsula to the west, southern Albania to the north, and a large number of Greek islands. This is a rare event. In 2022 sea ice formed around a small village to a depth of only a few inches. Nothing like the sea being frozen.

The winter was very severe over nearly all of Europe.

The Ionian and Adriatic Seas were frozen during 860 CE The Adriatic Sea separates the Italian Peninsula from the Balkan peninsula. In the year 860, the Adriatic Sea was so frozen that one could travel on foot from the continent and walk to Venice, Italy.

The Rhône River in the south of France also shivered.

In the year 859, there was a very severe and long frost in Britain.

In the winter of 859-860 in Gaul Western Europe and Germany, the winter was very harsh and long. The winter in France lasted from November to April with snow and solid ice.

In Italy, the frost was violent and persistent, and the earth was covered with immense snow. The seeds in the ground and the vines froze and died. The wine froze in the cask, where it has been preserved. The mortality rate among people and animals was large, and then a famine broke out, which was terrible in the next year.

In 860 CE Puyehue-Cordon Caule in Chile erupted with a VEI of 5. Volcanic winter material here.

There was a dry event in Central America in 860 CE. Starting in 760 CE there was a multi-year drought in Central America. These droughts recurred at fifty-year intervals. 760, 820, 860, and 910 AD.

In Spring 860 CE the Rhine-Lek channel to the sea in the Netherlands opened during a North Sea sea floods.

In May 861 CE Chokaisan erupted in Japan with a VEI of 3.

Lake Punta Laguna in Quintana Roo in Mexico showed evidence of a particularly intense drought in 862 CE.

On June 12th 864 CE Fujisan erupted in Japan with a VEI of 3.

On November 9th 864 CE Asosan, also in Japan, erupted with a VEI of 3.

The winter of 864 CE was very cold in France. The Rhône River was covered with ice.

In 864 in England, there was a sharp and long frosty winter. There was a deep snowfall.

In the year 864 in Italy and Germany, the winter was long and harsh.

The Adriatic Sea was frozen around Venice and on its lagoon; riders and wagons laden with goods travelled across the ice.

There was further devastation in the area of the newly formed Rhine-Lek channel in the Netherlands in 864 CE during another sea floods.

In May 866 CE Ata in Japan erupted with a VEI of 2.

On March 4th 867 Yufu-Tsurumi in Japan erupted with a VEI of 3.

On June 20th 867 CE Asosan in Japan erupted with a VEI of 2.

In 868 CE Paris, France suffered again from famine and a great famine afflicted not only Germany, but also all other countries of Europe. In 868 a terrible famine struck Europe.

The heavy rainfalls in 868 caused floods fatal to the grains in Western Europe and there was a flood of many rivers in France.

The winter of 869 CE was cold and produced a famine. As a result, in the province of Quercy in southwestern France, one-third of the population died.

The 869 Jōgan earthquake *Jōgan jishin*) and its associated tsunami struck the area around Sendai in the northern part of Honshu on 9 July 869 (the 26th day of the 5th month in the 11th year of Jōgan; or 13 July 869. The earthquake had an estimated magnitude of at least 8.6 on the moment magnitude scale, but may have been as high as 9.0, similar to the 2011 Tōhoku earthquake and tsunami. The tsunami caused widespread flooding of the Sendai plain. In 2001, researchers identified sand deposits in a trench more than 4.5 kilometers (2.8 mi) from the coast as coming from this tsunami.

In 869 during the summer a terrible famine arose in many provinces of France and Burgundy. A frightening amount of people died. So great was the hunger that people resorted to eating human flesh. Burgundy during this period of time was part of the Frankish Kingdom.

During 870 CE Bardarbunga erupted in Iceland with a VEI of 4. Also, in 870 CE Alamagan in the Marianas Islands in the Pacific erupted with a VEI of 4. Another eruption was Torfajokull in Iceland with a VEI of 3.

In August 870 CE Fujisan in Japan erupted with a VEI of 2.

On May 5th 871 CE Chokaisan in Japan erupted with a VEI of 2.

In Germany and in Gaul Western Europe the summer of 872 CE was characterized by a suffocating heat and almost constant storms. Saint Peter's Church in Worms, Germany was destroyed by lightning. Many people were killed, and the harvests were poor. The summer of 872 in Germany was unusually warm. Because of the heat, the corn grain harvest turned out badly. Then grains became expensive. Due to the extreme dryness during the summer of 872 in the north of Western Europe, almost all the fruit was destroyed. The extreme heat and drought of the summer of 872 destroyed almost all fruit in the north of Western Europe. Lightning consumed many houses with their inhabitants.

In 872 CE England was hit by an all-consuming drought and heat and there was a famine "from ugly locusts."

In the first centuries after the settlement of Iceland, 870-930 CE, as well as of southwest Greenland after 987 CE onwards, sea ice was practically never mentioned in the Viking accounts of voyages: though it had been encountered by earlier Irish monk-explorers in remote waters in the same area in the sixth to eight centuries CE. It was ice filling one of the northwestern fjords, Arnafjord, and all the fjords on the northern side of the island in the cold spring of 865 CE that gave Iceland its name. This was after the first Viking overwintering. From that time until 1200AD there were only two years each century when ice was reported to have caused any difficulty in sailing about Iceland and southern Greenland.

In 873 CE in Paris, France, there was much suffering from a famine whilst a terrible famine struck Europe. The summer of 873 CE was very hot in western Europe. There was a plague of locusts in Germany and Spain. In the year 873, the city of Worms, Germany was burnt down from thunder. Which year was this? 872 or 873?

In the north of Western Europe, the winter of 873-874 was harsh and prolonged, characterized in particular by the prodigious mass of snowfall.

During the winter of 873 to 874 CE it was very cold in Britain, specifically a cold winter. There was a great frost from November to April and the thaw brought floods.

On March 29th 874 CE Ata in Japan erupted with a VEI of 4.

During the reign of Emperor Hi-Tsong 874-889 CE in China, there was a grievous famine, which was caused by the overflowing of the rivers and by vast swarms of locusts, which destroyed most of the corn grain.

In 874 CE in France the heat of summer and long duration caused the pastures to dry up and this resulted in a shortage of grains. As a result of the famine and plague in France, one third of the population was swept away.

In 874, a plague of ugly deformed locusts ate up the fields in France. They had six feet and two teeth harder than stone. So numerous were they, that they darkened the sun. In one day and night they eat up all greens and trees. But strong winds drove them into the sea where they drowned. The waves cast their bodies ashore where the putrification proved fatal to many. So that by famine and plague, a third part of the people died. The long summer of 874 in Western Europe produced a long drought that was so great that it destroyed the hay and grains.

In the Winter of 874-875 CE cold weather brought great frost to Scotland from November to April. The winter thaw produced floods.

In 874 there was a very severe frost in England. The winter was long. There was snow from November to the end of March. Because of the deep snow, the forest was inaccessible for the supply of fuel firewood.

In the year 874, the Rhine and the Meuse rivers in Germany remained frozen for a long time and were accessible to pedestrians.

In the year 874, the winter in Gaul Western Europe, was so long and so strong in frost and snow, that, as the chronicles of St. Denys records, "No man who lived at that time had seen such a severe winter." The winter lasted from September to March. The snow fell in such a large quantity that the forests had become inaccessible and as a result people could procure no wood. The earth was covered with snow for five months, and the effects of this winter were very disastrous. The domestic animals, especially the horses died in great numbers, as did many people from the cold. This was the same as in England.

The famine and the diseases that followed this winter snatched up, according to the chronicles of Fulda, a third of the population of Bavaria in Southern Germany. Italy felt similar effects of the snow and the cold.

In 875 CE Brennisteinsfjoll in Iceland erupted with a VEI of 2.

The winter of 875 was very long and very cold in France. It lasted from early September to the end of March. There was so much snow that many forests were inaccessible. There was up to 15 feet of snow in the mountains of Burgundy. There were 5 months of heavy snow. The Rhine and Meuse rivers were frozen and could be crossed as if on dry land. There was a catastrophic thaw, which caused flooding. Horses were decimated. There were epidemics. One-third of the population of France died from this winter.

The winter of 875 was sharper and longer than ordinary. The Earth was covered with snow and ice from November to the vernal equinox around March 20/21.

In 876 CE in Saxony now part of northwest Germany, there were great rains in June that produced extensive flood damage. There was a sudden tempest and inundation of rain in Saxony, to the ruin of many men, beasts, buildings and trees.

In 877 CE the Vatnaoldur series of craters erupted after a basaltic dyke intrusion from the Bardarbunga volcano. These eruptions were 40-mile, 65-kilometer, long volcanic fissures within the area of a lake in Iceland.

In 878 in England, there was a famine "from ugly locusts."

In 880 CE Pacaya erupted in Guatemala with a VEI of 3 and Taveuni in Fiji erupted with a VEI of 2.

In China dry conditions occurred between 880 CE and 1260 CE. There were major lake level changes throughout the entire country.

In the winter of 880 to 881 CE there was a very cold winter in England.

A new temperature record was made for the period 880-920 CE by studying the proportion of the isotope oxygen$_{18}$ found in Greenland ice which indicated that there was a notably cold event. Was something going on in the North Atlantic?

In the year 880 in Germany, the winter was severe and of extraordinary length. The Rhine and the Main rivers froze for a long time and individuals could travel across these rivers on the dry ice.

Tree rings of oak trees indicate that in Sweden between 880 to 920 CE there was a growth anomaly.

The winter of 880-881 was very long and very cold in Western Europe in the north. The Rhine River and the Main River froze for a long time. We crossed the firm ice on foot. The frosts persevered until the spring. The Main River is located in southwest Germany. The winter of 881 CE was very long and very cold. The Meuse river was also frozen for a long time. There were frosts in the spring. As a result, the pastures did not grow and there were famines.

The winter of 880-881 was very cold and persistent in France, Flanders now Belgium, and Germany. The winter showed itself to be very dangerous for several species of domestic animals because "in the spring, the very severe frost compressed the earth" and yielded no green feed. The cold and famine by the very barrenness of the previous year brought suffering to the utmost. When Flandria (Flanders) appeared in the 8th century, it was a Frankish fief centered in Bruges in northwestern Belgium.

In Norway up until this time wheat was grown in the Trondheim district and hardier grains were even north of Malangen, 69.5 N° in North Norway. This started in 880 CE and ended around 1000 CE. They had given up by the fifteenth century. Wheat and possibly corn were also grown there. In Central Norway the area of settlement, forest clearance and farming which had been more or less static since early Iron Age times, spread 100 to 200 meters further up the valleys and hillsides. This height change indicated a rise of prevailing summer temperatures of 1° over 200 years from 800 CE to 1000 CE. Most of this land was abandoned after 1300 CE as the climate cooled down with advent of the Little Ice Age.

It was a warm period from New Mexico to northern Canada, where forest remnants between 25 and 100 kilometers north of the present limit were dated between 880 CE and 1140 CE.

In November 882 CE Ata erupted in Japan with a VEI of 2.

In 884 CE Zaozan in Japan erupted with a VEI of 3.

On August 29th 885 CE Ata in Japan erupted with a VEI of 4.

On June 24th 886 CE Niijima in Japan erupted with a VEI of 4 and on the same day Izu-Oshima erupted, also with a VEI of 4. The island of Izu-Oshima is around forty kilometers north of Niijima Island off the coast of Honshu in Japan.

On 6 February 886, the Loire River in France flooded.

In March 886, the flooding of the Seine River in France helped the Parisians that were besieged by the Normans.

In 886, it rained day and night almost without interruption during the months of March, June and July in the vicinity of Mainz in west-central Germany, which led to flooding of a frightening part of the Rhine River and other rivers.

Winter of 886-887 was very severe and of unusually long duration. It was accompanied by such a severe disease among the oxen and sheep, that in France few animals of this kind remained. The winter of 887 in France was very long or maybe the term "unusually" long would be appropriate. It caused the near extinction of oxen and sheep.

In 887 CE Niigata-Yakeyama in Japan erupted with a VEI of 4.

On 26 August 887 CE, there was a strong commotion in the Kyoto region, causing great destruction. A tsunami inundated the coastal region and some people died. The coast of Settsu Province (Osaka Prefecture)) suffered especially, and the tsunami was also observed on the coast of the Sea of Hyuga (Miyazaki Prefecture).

In 887 in England, there was a grievous famine that lasted 2 years.

In 889, extraordinary floods desolated northern Italy. The mass of rains in Thuringia in central Germany destroyed in a short time three villages, and three hundred people drowned there.

In 889 CE a terrible famine struck Europe.

In 890 CE Mont Pelee on Martinique erupted with a VEI of 4 and Galeras in Colombia erupted with a VEI of 2.

In 890 CE in Scotland, there was a great dearth famine.

The winter of 891 was very cold in France. The vineyards and herds were decimated. The Meuse River froze. In the year 891 in Flanders now Belgium and Holland now the Netherlands, the winter was severe.

The months of April and May 892 in Western Europe produced an extreme drought. In April and May 892 CE .it was extremely dry in France.

In 892 in Western Europe, excessive drought struck. Beginning on the first of April and continuing through May. Then disastrous frosts struck on May 18 and July 17. These late frosts destroyed the grape vines and wheat.

On 11th November 892 CE a great gale hit Ireland, and many trees and houses fell.

In the Winter of 892-893 the vine and fruit trees in France were killed by the extreme cold. The winter of 893 CE was very cold in France. The Rhône River was frozen. Livestock was decimated. There were 5 days of very heavy snowfall in March. The winter of 893 was at once so hard and so long, we could see in some places a foot of snow for five days in March. This cold led to a great scarcity of wine in the territory of Bayeux in northwestern France. The winter of 893 was severe and of a longer duration than usual.

Several earthquake catalogues and historical sources describe the 893 Ardabil earthquake as a destructive earthquake that struck the city of Ardabil, Iran, on 23 March 893 CE. The magnitude is unknown, but the death toll was reported to be very large. The USGS, in their "List of Earthquakes with 50,000 or More Deaths", give an estimate that 150,000 were killed, which would make it the ninth deadliest earthquake in history. Although the Ardabil area is prone to numerous earthquakes and was struck by a major earthquake in 1997, the 893 event is, in fact, considered to be a "mistaken" earthquake, derived from misreadings of the original Armenian writings about the 893 earthquake in Dvin, Armenia; the Arabic name for Dvin is *Dabil*.

At about midnight on 28 December 893, the night after a lunar eclipse, Dvin, then the capital of Armenia, was devastated by an earthquake. Most buildings were destroyed, and at least 30,000 people died. This event was recorded by contemporary Armenian and Arabic chroniclers, including Ibn al-Jawzi. However, the Arabic name for the city is *Dabil*, and this led the 14th-century writer Ibn Kathir to place the earthquake in Ardabil in Azerbaijan. Ibn Kathir was then quoted by al-Suyuti in the 15th century. Further writers also placed the earthquake in Ardabil, and added some details, such as waters drying up, while changing others, such as making the eclipse preceding the earthquake solar instead of lunar. It is clear, however, that all these reports are descriptions of the 893 Dvin event.

In March 894, a foot of snow fell in 5 days. As a result of this severe winter in Bavaria, there was almost a total lack of wine and many sheep and bees were killed.

In 895 in Ireland, there was a famine from an invasion of locusts.

In 895 in York, England there were hailstones like ducks' eggs.

In the year 895, part of the town of Shaftesbury in Dorset, England was burnt by lightning.

In 895 in the north of Western Europe, the trees bloomed a second time in December.

From 896 to 899, Paris, France, once again suffered from a famine.

In 896-899, Paris, France suffered from a famine.

In 897, there was a great famine in Germany.

In 898, the famine was so great in France that people out of necessity ate one another.

In 898 in France there was a sore famine.

900 CE The Medieval Warm Period

In 900 CE Tolbachik in Kamchatka erupted with a VEI of 4. Other eruptions were Kikhpynch with a VEI of 3 in Kamchatka, Glacier Peak in Washington with a VEI of 3, and Planchon-Peteroa in Chile with a VEI of 3, and Kryusuvik-Trolladyngja in Iceland with a VEI of 2.

There is a meteor crater ten kilometers south of the town of Whitecourt in Woodlands County in Alberta, Canada that fell around 900 CE and is 36 meters across.

During Medieval warmth there was up to 7 per cent higher average yearly rainfall with lower rainfall in summertime.

In the ninth century CE there were periods of generally rather colder and more disturbed climate. In the northern parts of the Mediterranean and in northern and western Europe there was an increase in wetness.

In the 10th Century there were westerly winds in the northwest of England could be consistent with the pattern of westerly winds and anticyclonic conditions producing drought in the southeastern half of England and in Germany.

During the Medieval Warm Period there was reduced El Nino activity and cooler conditions over the eastern Pacific.

In the ninth century there was an increase in storms and sea floods around the North Sea.

Between 900 CE and 1400 CE the level of the Caspian Sea rose 8 meters due to increased rainfall.

Around 900 CE world sea levels for a brief period were high.

By 900 CE the Mediterranean ecotone shifted further north. For the next four centuries, summers passed with good harvests and enough to eat.

The summers were regarded as being notably hot in Europe in the 10th century. This could have been due to a remarkable amount of anticyclonic weather over Britain, Germany and southern Scandinavia. Giving low rainfall, rather warm summers and rather cold winters.

In the Medieval Warm Period which lasted four centuries average summer temperatures in the West (Western Europe) were between 0.7° and 1.0° Celsius above twentieth century averages. The summers were also longer, beginning in June and finishing in August and beyond. Because of this, growing seasons lengthened and for the first time in a millennium, grapes were growing in England, as they are now, with vineyards flourishing in southern and central England. This was a climatic Golden Age. Crop failures were rare in this period so that there was less famine in Europe than normally. By medieval standards it was an age of plenty, though most peasants still lived only one harvest from famine.

In the Medieval Warm Period from 900 CE to 1300 CE Europe basked in warm, settled weather with only occasional bitter winters, cool summers and larger than normal storms. Summer after summer passed with abundant harvests and sunny weather.

The lower precipitation after 900AD in East Africa reduced Nile Floods.

In the Americas there was severe drought, hunger, warfare in the north and the collapse of two major civilizations. This was evidenced by deep-sea cores, pollen samples, tree rings and ice cores from the Andes.

For five centuries sudden aridity descended from the California coast to the Maya lowlands of Central America to Lake Titicaca in the Bolivian Highlands. In the Western Hemisphere savage droughts toppled states. Once again we have the climatic see-saw between North America and Western Europe.

Between 550 CE to 1150 CE there was a long cold drought period in which the population of Cantona, west of Mexico City, dropped from a high of 70,000 people in 900 CE to 5,000 people in 1050 CE.

In 900 CE England was visited by a sore famine.

There was significant warming in Greenland from 900 CE to 1200AD.

In the Viking colonies in west and southwest Greenland the soil was not frozen and the inhabitants were able to bury their dead deep in soil that is now permanently frozen. Corn pollen was found in the turf at Eric the Red's original settlement, Brattahlid, now Qagsiarrsuk, in the sheltered fjord country near Cape Farewell.

The water in the Greenland fjords was at least 4° warmer than the present time in the early 21st Century.

Around this time the southern boundary of the northern pack ice was 120 miles offshore. In the late twentieth century it is 60 miles from the Greenland coast.

The coast was ice-free for most of summer around Greenland. It was warmed by the north-flowing Greenland current that hugs the shore. The West Greenland Current flows into the heart of Nordrseter and Baffin Bay here it gives way to much colder, south-flowing currents. Baffin Island is only 200 miles over the Davis Strait from this part of Greenland which the Vikings could have seen if they just ventured a small distance offshore. The colder waters on the west side of Davis Strait along Baffin Island, Labrador and Newfoundland experience a longer ice season and heavier ice cover that can last into the summer. During this warm period this would be much easier to coast along, and with less ice.

In Greenland at this time milder summers allowed the growth of hay for winter fodder and to plant barley. This was impossible after 1200 AD. For good growth to begin days need to be warm and nights need to have temperatures ranging above 40° F or 4.44° C. Summer temperatures now have highs of 50° F or 10° C which are not warm.

In North Africa between 900 CE and 1100 CE there was an abrupt transition to much more unstable weather, mirrored by increased monsoon variability in the Carioco Basin in Venezuela on the other side of the Atlantic.

After 900 CE the Sahel in Africa grew wetter. The Sahel is the region between the more humid Sudanian Savannas to its south and the drier Sahara Desert to its north. In the Sahel in Africa there was an extreme cooling spike.

Dry cycles in the North American West and Southwest between the tenth and thirteenth centuries appear to be linked to the Pacific Decadal Oscillation and to cool conditions in the eastern Pacific Ocean. The Pacific Decadal Oscillation is a long term ocean fluctuation of the Pacific Ocean which waxes and wanes approximately every 20 to 30 years.

From 900 CE to 1064 CE severe drought affected Mongolia.

There was a long drought between 900 CE and 1350 CE in the highlands of Bolivia and Peru. The Wari or Huari Empire ended around 1100 AD.

There was increased wetness after 900 CE in South Africa.

Newly exposed tree stumps under retreating glaciers in Prince William Sound in Alaska dated from 900 CE to 1300 CE. The ice sheets had advanced during the cycle of wet years. They still do this at the present time during wet winters.

In southern California a 198-meter-deep sea core was taken from the Santa Barbara Channel. Seventeen meters of it represents the Holocene Era which is the most recent era that we live in now. Normally 1.5 meters of foraminifera sediment layers accumulate every thousand years. Climatic conditions as represented by this core showed that they were relatively stable from the end of the Last Ice Age up to about 2000 BCE. At this point the climate became much more unstable. From 450 CE to 1300 CE sea temperatures dropped sharply to about 1.5° cooler than the median sea surface temperature in the Santa Barbara Channel for the Holocene Period as a whole. For the period from 950 CE to 1300 CE, marine upwelling was especially intense, making local fisheries extremely productive. After 1300 CE water temperatures stabilized and became warmer. Two centuries later upwelling subsided and marine productivity stopped.

The colder sea surface temperatures and increased upwelling generally coincided with droughts of varying severity. These were documented in tree rings from the southern California mountains. The cold period from 450 CE to 1300 CE saw frequent climatic shifts, especially between 950 CE and 1500 CE which was a period of extreme drought. The most intense drought cycles were from 500 to 800, 980 to 1250 and 1650 to 1750.

Between 900 CE and 1300 CE, and possibly later, there appears to be no indication of increased spring activity or high lake levels in the Mojave Desert in North America. There was less winter rain than in later times.

In the tenth century the Chaco People in North America built three great houses at the junctions of major drainages. From the 750s small pueblos had been built by the inhabitants of Chaco Canyon who took advantage of seeps and springs to grow maize in favored spaces. During the ninth and tenth centuries summer rainfall was highly variable. By the tenth century instead of dispersing due to erratic rainfall built the three great houses. The largest community was Pueblo Bonito which stood five stories high along its rear wall and remained in use for more than two centuries. At its peak in the eleventh century Pueblo Bonito had at least 600 rooms in use and housed about a thousand people.

In 901 in England, there was a frost of 120 days that began at the end of the year. This would have created a famine.

In 902 the Nile River in Egypt only rose to thirteen cubits during the peak of the annual inundation. Generally, this means that Egypt experienced a strong famine during the harvest in the following year.

In 908 CE there was a severe winter, and most English rivers were frozen for two months. The River Thames in England was frozen for two months.

In 910 CE Brennisteinsfjoll in Iceland erupted with a VEI of 2.

In Central America there was a dry period around 910 CE. Starting in 760 CE there was a multi-year drought in Central America. These droughts recurred at fifty-year intervals. 760, 820, 860, and 910 CE. This drought lasted six years.

Around 910 CE the Maya Civilization in the southern and central Yucatan Lowlands collapsed. The very last Maya inscription dates from 910 CE.

In Germany between 910 CE to 930 CE oaks from the lowlands showed via their very narrow tree rings that there was a major drought.

In 910 CE Owens Lake in California was severely desiccated. This was a period of drought also shown at Lake Walker also in California. Mono Lake, Lake Walker and Owens Lake show the same cycle. This drought cycle started in 910 CE and lasted until about 1100 CE. This was a ninety-year drought.

Between 910 CE and 930 CE there were extended droughts in the British Isles and the summer half years were warm or very warm more often than not. There were some notably hot summers.

In the year 911, there was a great snowfall in Scotland. Snow "lay deep in Scotland, the Death of much Cattle."

In 912 to 913 CE there was a severe winter in England.

On August 17th 915 CE Towada on Honshu in Japan erupted with a VEI of 5. It devastated the surrounding area with pyroclastic flows and lahars and covering most of the Tohoku region of Japan with volcanic ash leading to crop failures, climate change and famines. The area around Lake Towada remained largely wilderness until the end of the Edo Period between 1603 and 1868. More than likely this contributed to a massive volcanic winter.

In 917 CE there was a severe winter that was the called the "Great Snow". Lakes were frozen. This implies a blocked pattern with occasional Atlantic incursions. And here was the Volcanic winter as well. They are usually two years afterwards.

In 918 CE in Scotland, rains extended over five months producing floods. In Ireland, a great flood occurred.

In 920 CE Katla in Iceland erupted with a VEI of 4 and Eyjafjallajokull in Iceland erupted with a VEI of 3.

In 920 CE the Thames River in England was frozen for 13 weeks.

In 925 CE Vulcano in Italy erupted with a VEI of 3.

In 927 CE cold weather struck Byzantium. The frost lasted 120 days. Byzantium at this time included Turkey, Bulgaria, Macedonia, Albania, Greece and the southern and eastern Italian peninsula.

The Winter of 927-928 CE was very severe in northern France and Flanders now Belgium.

The winter of 928 CE in France was very cold in the north. Ten thousand people died.

The winter of 928 CE was an extremely severe winter. The whole River Thames in England was frozen for 3 months.

In 928 the army of the first emperor at the siege of Brandenburg, Germany set up camp on the frozen Havel River that was frozen solid enough to bear the weight of his army. This was Henry the Fowler who established the Ottonian dynasty of kings and emperors and the founder of the Medieval German State.

In 928 CE. in Reims in Marne in northeastern France, the harvest was almost finished before August.

From 928 CE to 961 CE the Pacific Region was much warmer.

In the year 929, the River Thames in England was frozen for 13 weeks. The last modern freezing of the Thames in London was in 1963 and regarded as rare. As you have read, it was very frequent in this period due to the very cold climate over the previous few hundred years.

In 930 CE Ceboruco volcano erupted in Nayarit, Mexico. This was a Plinian eruption releasing eleven cubic kilometers of tephra. This would have created a volcanic winter. The VEI was 6. Another eruption was Guagua Pichincha in Ecuador with a VEI of 5. There were very narrow tree rings in this period. This eruption was named the Jala eruption.

Between 930 CE and 1070 CE there were record low flood levels in the Nile River. This would indicate that there was little monsoonal rain in the Ethiopian Highlands. This seemed to be an ongoing problem with the Ethiopian monsoon rains.

On November 19th 932 CE Fujisan in Japan erupted with a VEI of 2.

In 933 in England, there was a frost of 120 days that began at the end of the year.

The winter of 934 CE in France was very cold from 30 November to March. The frozen Meuse River was passable as if on dry land.

934 CE A terrible whirlwind (tornado) blew down Saint Maximinus's Church at Treves (now Triers) in southwestern Germany.

In 934 CE to 939 CE, Eldgja erupted in Iceland with a VEI of 6. 219 million tons of sulfur dioxide were emitted. This was the largest effusive volcanic eruption in history. It covered 780 square kilometers with lava. There were widespread impacts on the Northern Hemisphere climate as recorded by Chinese, European and Islamic records. It produced a notable cooling of the climate with resulting cold winters and food crises across Eurasia. This was the largest Holocene eruption of the Katla system. 18 cubic kilometers of tephra was ejected. This was listed as a VEI 4 eruption though it appears much larger as it was one of the largest lava eruptions in the past ten thousand years. It is the amount of sulfur dioxide released that is the main creator of volcanic winters and this amount was massive.

It involved a 75-kilometer-long area of the Katla volcano and the Eldgja Fissure. The climate impacts of the event have been recorded as far away as Australia and in tree rings which recorded a cooling of 0.7°- 1.5° in the northern hemisphere during 940 CE. This was most pronounced in Alaska, the Canadian Rocky Mountains, Central Asia, Central Europe and Scandinavia. In Canada and Central Asia, it lasted until 941 CE.

There was increased flooding in Europe after the eruption of 934 CE. Tephra covered an area of 20,000 square kilometers in Iceland.

The first of the four driest drought periods started in the American West in 935 CE. This was all within a four-hundred-year period of overall aridity.

Was this related to the massive volcanic winter caused by Ceboruco volcano in Nayarit, Mexico and Hekla-Eldgja in Iceland. Volcanic winters can last several years. Hekla-Eldgia lasted until 941 CE, and possibly longer.

In 936 in Scotland, after the appearance of a comet, there were four years of continuous famine "till people began to devour one another." These reports of cannibalism during famine are so common in this period and again in the Little Ice Age.

On December 18th 937 CE Fujisan in Japan erupted with a VEI of 2.

In China winter was severe in 939 CE and there were food crises.

In England in 939 CE there was a cold winter.

In Europe winter was severe in 939 CE with the sea and canals freezing while drought occurred during the summer months. The winter was extremely severe in Germany and France. The harvest was destroyed by the bad weather. There was famine and disease, and the mortality among cattle was particularly large.

There was a decrease in human activity in Ireland in 939 CE.

As volcanic aerosols often weaken the monsoons that feed the Nile River in Africa, during 939 the water levels of the river were unusually low.

There was famine in the Maghreb of Arabia, the Levant in the Middle East and Western Europe in 939 CE.

In the winter of 939 to 940 CE it was very cold, and lakes and rivers froze in Ireland.

In 940 CE Merapi in Indonesia erupted with a VEI of 4.

The winter of 940 CE in France was very cold. The cold decimated the herd of cattle. This was followed by epidemics and famines.

In 942 CE., a famine struck Upper India. "In 330 A.H. (941-942 CE .), a comet made its appearance, the tail of which reached from the eastern to the western horizon. It remained in the heavens eighteen days, and its blighting influence caused so severe a famine that wheat, the produce of one jarib a measuring chain of land was sold for 320 miskāls of gold. A miskāl is a unit of weight equivalent to approximately 4.25 grams. When the value of a spike of corn was esteemed as high as the Pleiades star cluster, conceive what must have been the value of wheat!" "The famine in the land was so severe that man was driven to feed on his own species, cannibalism, and a pestilence prevailed with such virulence that it was impossible to bury the dead who fell victims to it."

In 944 CE in England, a great storm raged in and near London, which destroyed 1500 houses. In 944 CE there was a severe gale throughout the whole of England. Different accounts from early chroniclers report that between 1000 and 6,000 people were killed.

In Western Europe in the year 944, the frost froze the grapevines in the beginning of May. It rained constantly during the whole summer. In the north of Western Europe, the frost burnt the vines around May 1, 944. Then the weather turned rainy which lasted all summer. The entire summer of 944 was rainy in the north of Western Europe.

In 945-946, a famine struck France.

In 945, there was a famine over all Italy, which with war reached France in 946.

In 946, there was a shocking famine in Italy.

On November 15th 946 CE Paektu or Baekdu volcano on the border of China and Korea erupted. The eruption was one of the largest and most powerful eruptions on earth in the last 5,000 years. The mountain's caldera was formed by the eruption that is now filled by Heaven Lake. Tephra from the eruption has been found in the southern part of Hokkaido in Japan and as far away as Greenland. Historical records state that white ash rain fell in Nara in Japan, around 680 miles, 1100 kilometers, away. This eruption is regarded as equal to the 230 CE eruption of Lake Taupo in New Zealand and the 1815 eruption of Mount Tambora in Indonesia meaning that this eruption would have created a volcanic winter. The VEI of the eruption was 7. It released 100-120 cubic kilometers of tephra. Paektu is also called Changbaishan. This would have been added to the 934 CE Eldgja volcanic winter. This period of massive volcanic eruptions was amazing, having apparently started in 934 CE.

950 CE

The warmest part of the Medieval Warm Period in North America was between 950 CE to 1200 AD. This was the same in European Russia and Greenland.

Now we get to another amazing year of volcanic eruptions apparently.

In 950 CE Bezymianny and Tolbachik in Kamchatka erupted. They were both VEI 4. Other eruptions were Cotopaxi with a VEI of 3, and Bezymianny with a VEI of 4. Mutnovsky in Kamchatka with a VEI of 2, Mount Adams in Washington with a VEI of 2, Stromboli in Italy with a VEI of 2, Brennisteinsfjoll in Iceland with a VEI of 2, Oddnyjarhnjukur-Langjokull in Iceland with a VEI of 2, and Sete Cidades in the Azores with a VEI of 2.

Remember that I do not show volcanic eruptions with a VEI of 2 and under. Eruptions of VEI of 3 and higher can all create volcanic winters.

In 950 CE a serious drought struck the coast of California.

Desertification increased in China during the cold period in the late Tang Reign after 950 CE.

From 950 CE to 1300 CE the temperatures were warmer in China than the long term mean of the time.

Between 950 CE and 1072 CE Egypt experienced droughts that became ten times more frequent than the previous centuries. Over a period of 125 years there were 27 years with low Nile floods. The Nile flood is primarily controlled by monsoonal precipitation in the Ethiopian Highlands, driven by the seasonal migration of the Intertropical Convergence Zone, ITCZ. Persistent La Nina conditions and a positive mode of the NAO, North Atlantic Oscillation, reduced precipitation in the Nile River Basin, creating low floods and creating famines. This was not helped by massive volcanic eruptions either.

In Iceland grain was being grown since the first settlement until its abandonment in the late sixteenth century. There were also more scrub birch woodlands on Iceland but the settlers were responsible for their destruction. The forest was reduced from around a fifth of the country to one per cent.

From 950 CE to 1250 CE a stronger monsoon prevailed in Southeast Asia.

Southampton, England, was nearly destroyed in a storm by lightning in 951. There was a terrible lightning storm at Southampton, England that lasted 4 days. In 951 in England, there were great thunders.

In 952 CE in Bagdad, Iraq, half of the city was inundated from a great overflow of the Euphrates River.

In 954 in England, Wales and Scotland, there was a great famine which lasted four years. There was pestilence in Scotland in which 40,000 people perished.

In the summer of 955 CE, it was a hot summer in Wales. This would have affected other parts of Britain if the summer was notably hot as far west as Wales.

In 956 a very severe winter followed by a grievous famine, especially in France and Burgundy. Paradin, in Annales de Bourgogne, bk. 1, p. 149, speaks of a miraculous ice stone that fell in the year 956 in Germany.

In 960 CE Ljosufjoll in Iceland erupted with a VEI of 2, as well as Katla in Iceland with a VEI of 3.

960 CE was the beginning of the Northern Song Dynasty in China. The Song dynasty prospered in this period. Strong monsoon rains benefited it, and the Song Dynasty made strong use of rice as the population tripled to 100 million people by 1100 CE.

In 962 CE in England, the frost was so great as to cause a famine. There was a most severe winter, a great famine and a horrible fire.

In 963-964, an intolerable famine visited Ireland, and parents are said to have sold their children in order to get money with which to buy food.

In Winter of 963-964 CE up to the beginning of February, the winter in Western Europe was very hard and rough. The winter of 964 CE in France was cold until February. The excessive harshness of the cold in 964 in Western Europe in the north persisted until 1 February.

The lowest Nile flood peak ever known seems to have been that of 966 CE, when the waters rose only to twelve cubits, seventeen digits. The famine the following year swept away 600,000 people in the vicinity of the city of Al-Fustat, Egypt. Al-Fustat was the first capital of Egypt under Arab rule. Today Al-Fustat is contained in the older part of Cairo. The death toll could have even been greater except for the actions of G'awhar, a lieutenant of the Caliph Mo'izz. G'awhar organized relief measures. The Caliph lent him every assistance by sending many ships laden with grain. But the price of bread still remained high. G'awhar, being a food controller who had no patience with persuasive methods, ordered his soldiers to seize all the millers and grain dealers and flog them in the public marketplace. The administrator then established central grain depots and grain was sold throughout the two years of the famine under the eyes of a government inspector. During the famine, G'awhar allowed the natives to cast their hundreds of unburied dead into the Nile River, thereby contaminating the waters all the way to the sea and ushering in plagues of disease.

In 967 CE due to poor Nile floods, 600,000 people died of starvation and famine-related diseases. This was one quarter of the population of Egypt.

In May 968, there were very tempestuous and strong winds, which corrupted the corn grains, vines and fruit trees; hence arose a great famine.

In 968 CE there was a famine in all of Europe, but chiefly Germany and Scotland. In Europe, there was a tempest of wind and rain, that rotted the corn grain and caused a famine.

In 968 in the Persian Gulf, there were severe irruptions following an earthquake tsunami. Several cities were destroyed, and new islands formed

On December 1st 968 CE Vesuvius erupted in Italy with a VEI of 4.

In May 969, the corn grains burnt by the winds, died; hence a sore famine. In 969 in England, "All grain burnt by the winds." As a result, there was a famine.

In 970 CE Sheveluch in Kamchatka erupted with a VEI of 4.

In 973 CE there was a Thames flood in London.

The winter of 974 to 975 CE was severe until March 11th 975 CE. Europe experienced a cold winter which began on 1 November 974 lasting to 11 March 975. A most rigorous strong frost took place from 1 November to 11 March. Famine affected those that lived in the mountains.

In the year 975 in Gaul Western Europe, the winter was "long, dry and hard." A heavy frost lasted from early November until 22 March. In mid-May heavy snow fell. The winter of 975 CE was very cold for a very long time. It lasted from November to 22 March. As a result, one-third of the population of France was decimated by famine and epidemics. It snowed in May. The winter of 975 in the north of Western Europe was dry, despite its great snow. The winter of 975 in Western Europe in the north was tough, long, dry, and accompanied by deep snow.

In England, the frost was severe in 975 CE. The frost was the severest.

In 975 CE in Paris, France, a great number of inhabitants were carried off by famine.

In England, famine scoured the hills. In 975, there was a terrible famine in the mountains of England.

After a hard winter, on 14 May 975, there was a very heavy snowfall in Germany.

In 979, there was a grievous famine in England.

Between 980 and 1180 CE Lake Mashu in Hokkaido, Japan, erupted with a VEI of 5.

The Winter of 984-985 was a very cold winter in Germany, which lasted from November to May.

Lake Punta Laguna in Quintana Roo in Mexico showed evidence of a particularly intense drought in 986 CE. The drought ended abruptly in 1020 CE. Within a century the lake experienced some of the wettest conditions of the past eight thousand years. Since then, the climate has been marked by repeated wet and dry cycles that have lasted a decade or so.

The Carioco Basin in Venezuela showed evidence of severe drought in this period. This was what Lake Chichancanab in Mexico also showed.

In 987 CE in England, the extreme heat of summer killed many people; and they harvested almost nothing in fruits of the earth. In England, there were tempests all winter.

There was a dearth famine in Albania. A great dearth in Albania; but the unseasonableness of the weather brought barrenness of lands, and a grievous famine on many countries.

In 987, the sudden thaw of unusually high snow brings powerful floods to the river Weser in Germany. In 987 a strong storm struck Germany. The great storm winds and high-water occurred almost everywhere causing a lot of damage.

The extreme summer heat of 987 in Western Europe caused a great reduction in crop yield. The summer of 987 CE produced extreme heat in France. The summer of 987 in Western Europe produced frightening storms with extraordinary lightning and thunder.

In Winter of 987-988 CE the River Thames in England was frozen for 120 days which began on 22 December 987. In 987 there was a frost in London, England. It began on 22 December and lasted for 120 days.

In 988 Byzantium experienced a very cold winter.

In England, there was a great drought with excessive heat, both years (988-989). In England from mid-July to mid-August in the year 988, there was such searing heat that many people were affected. The harvest of the fruits of the earth was much lower than usual. The burning sun and the drought consumed all and then there came a famine. In 988 in England, there was a famine from rains and barren land. In 988, there was an excessive drought and a most scorching heat.

Repeated flooding preceded the great summer of 988 in the north of Western Europe. In 988 CE the summer in Germany was extremely dry. This was soon followed by a famine. A harsh winter was followed by long copious rains. Then there was sudden heat, sustained and passionate. This characterized the weather in 988 in the north of Western Europe. From 15 July to 13 August 988 in the north of Western Europe, a scorching heat burnt the harvest. The heat broke out suddenly after a very cold winter and a great flood.

In the year 988, the winter in Western Europe was rough. The winter crop was destroyed by the cold. There was a drought in the spring and a great famine ensued.

The winter of 988 CE was very severe in France. There were droughts in the spring. Then also came famines.

In 989 CE in England, there were floods all of the winter. There were great and often inundations in winter and violent winds, which threw down many buildings. In spring there was so great a drought, that it hindered sowing. The heat of the summer was past enduring. Hence came a famine. Then there were unseasonable snows and continual rains at harvest time. This prevented both plowing and sowing. In 989 in England, there was a grievous famine from a rainy winter and bad spring. There was neither plowing nor sowing. There was a snowy harvest. In 989 in England, there was a snowy rainy harvest and as a result no sowing during the fall planting followed by a rainy winter.

A great famine occurred in Albania and Saxony in Germany.

In 989 CE there was a large flood in Friesland in the Netherlands. This was a "storm surge" which was an inundation by the sea.

In the year 989 in the north of Western Europe, an excessive drought during the spring prevented the first seeds from being planted. In the north of Western Europe in 989, there was a drought during the spring, which did not allow the planting of first seeds. Abundant snows immediately followed this excessive drought. Then heavy rainfalls completely prevented the planting of the seeds in the fall. The snowfalls and the spring drought of 989 in the north of Western Europe were followed by continuous rains that prevented the sowing of autumn fall crops.

In the 990s there were extended droughts with regularity in Europe. The summer half years were warm or very warm more often than not. There were some notably hot summers.

In 990 CE a terrible famine struck Europe.

In Germany the tree rings of oaks after 990 CE were extremely narrow indicating drought.

In Winter of 990-991 CE the vine and fruit trees in France were killed by the extreme cold. In the year 991 in Western Europe, the vines were suffering much from the severity of the cold, the animal died from lack of food in the stalls, and there was a famine. The winter of 991 CE in France was very long and hard. The grape vines were decimated. The herds of domestic animals were depleted. The wheat crop was destroyed. Then came famine and pestilence.

In 991 CE there was an extremely severe and long-lasting frost in England. Crops failed, and famine and pestilence ended the year.

In 991 CE Vesuvius erupted in Italy with a VEI of 3.

In March 999 CE Fujisan in Japan erupted with a VEI of 2.

In 992 CE there was a storm flood or tempest and high winds that submerged an island fort in Wicklow. This appears to be a "storm surge" event.

The winter of 992 CE was hard in France.

The year 992 was a troubled year producing many dreadful fire signs in the sky in Germany. The winter was long and hard, such that it was still frozen very hard shortly before Pentecost late May. One observed northern lights at the beginning of the year.

Analysis of silver pine tree rings in New Zealand indicate that there was a sharp and sustained cold period in New Zealand between 993 CE and 1091 CE. This was the most extreme over the previous 1100 years.

In 993 CE in Germany from the Feast of St. John 24 June to 9 November, throughout the summer and the fall, the drought and the heat were extraordinary. Many of the fruits of the earth did not come to maturity and were burned by the sun's heat. This was followed by great disease and mortality among humans and domestic animals.

During the summer of 993 CE there was a drought from 24 June to 9 November in France. Then came epidemics. The crops were burned from the great heat and destroyed.

The summer of 993 CE in England was so hot that the corn grain and fruit dried up. Corn grain in this period in England refers to wheat.

In Germany, the winter was very harsh and the freeze lasted almost without interruption from 12 November to mid-May. The spring and summer brought the plagues of every kind and a violent epidemic raged among men, and among cattle, sheep and pigs. The year 994 was a very hard winter in Germany with heavy frost, from the beginning of the winter months, until May. At the end of the month of July, the water in the lakes froze so hard that fish died. The trees, fruits and grazing livestock were destroyed. Then came famine and plague, which emptied the houses. The winter was full of rough weather, pestilence, storms, severe cold and unusually dryness. In 994 there was frost from 14 October until the middle of April. In the following summer, dryness made streams dry up in Germany.

In Italy, the rivers were covered with ice and the plants froze.

During the winter of 994 CE there was a very cold winter of unusual length in France. It lasted from 15 November to 15 May. During the summer of 994 CE there were frosts in July in France. There were droughts and famines. The rigor of the winter of 994 in Western Europe ran from 15 November to 15 May. Then, there were very dangerous cold winds. And, still later, severe frosts lasted until 12 July.

In England in the year 994, the frost was strong and hard. From November 1st to May was a most severe winter. Cold pestiferous winds blew at the same time. About the end of July, from the severity of the frost, ice was frozen so hard on ponds and rivers, that most fish died, and the water was unfit for human use. Trees, corn and pastures were burnt up as though there had been a fire under the earth's surface. Finally, famine and dire pestilence made most terrible havoc of man and beast. So great was the deaths that many houses were left desolate without inhabitants. In 994, there was the severest plague on man and beast from severe frost and famines. A destructive storm struck London, England, blowing down fifteen hundred buildings and killing several hundred persons.

In 995 CE there was summer cold throughout Europe. There was severe frost and ice in July.

In 998 CE there was a possible severe winter as the Thames River in England was frozen for five weeks.

In 999 CE Vesuvius in Italy erupted with a VEI of 3.

The cold weather in Baghdad, Iraq around the year 999 killed palm trees.

In the years 999 and 1000 in Germany, there were two hot and dry summers that caused all the streams and rivers to shrink dry up.

1000 CE

Tolbachik and Ksudach both erupted with a VEI of 4 and Kistakan erupted with a VEI of 3. All three volcanoes being in Kamchatka. Another eruption was Kelud in Indonesia with a VEI of 3.

The eleventh and twelfth centuries were much warmer and wetter than the previous centuries.

There was a slightly higher stand of the sea or highstand as the sea gradually rose globally during the warm time as glaciers and ice masses melted. Particularly in the area around the southern North Sea where the land-sinking due to folding of the earth's crust in that basin was going on. This indicated that the earth was warming up.

Around 1000 CE it was a moister period than now in the Mediterranean region. There were bigger and more permanent rivers and more frequent stream flow in the wadis of the African and Arabian deserts.

There was a period of renewed moisture in North Africa, including the fringes of the Sahara. There were bigger and more permanent rivers and more frequent stream flow. The great Arab geographers wrote that journeys across the desert from North Africa to Ghana, Mali and Kufra took about two months but even in that region a large number of wild cattle often approached the caravan so close that they could be hunted with bows and dogs. The desert had its northern limit around 27° north.

In the Kufra region in Libya the rearing of beef cattle had been given up where huge herds formerly pastured in regions that are now deserts.

Remnants of a cedar forest on the island of Flores in the Azores are buried by a thin layer of peat which formed from about 1000 CE onwards. This indicates that the medieval period in the Azores was moist.

There were great intrusions of the sea in Belgium where Brugge, Bruges, was a major port. The coastal plain of Belgium had a fluctuating population in the 11th and 12th centuries, as the state of flooding varied, leading finally to more general emigration into Germany.

In Labrador and the neighboring Ungava region in Canada there was there no sign of a medieval interruption in the cooling off that began 3,000 to 3,500 years ago and put the forest into retreat before the advancing tundra.

Tree rings point to cold in Central Asia in the eleventh Century which experienced a big chill.

There was a vast loss of population in China in the eleventh century. CO_2 may also have decreased after forests reclaimed agricultural lands after the pandemics occurred and there was sharp population decline.

From 1000 CE temperatures were warmer in Northern China than in the South.

Around 1000 CE there was a cold climatic period in China and the Far East.

In the 11th and 12th centuries the climate of China took a much colder turn. There were frequent references to snow in the winter and snow a month later during the spring than in the late twentieth century. The plum trees were disappearing in northern China and frosts killed the mandarin trees in the coastal province near Shanghai. In parts of the south lychees died from the cold weather.

In Britain a tidal inlet extended as far as Norwich in Norfolk. In 1000 CE there existed an East Anglian inlet like a Danish fjord in which the sea reached inland to Norwich in England. This was around 34 kilometers long from Lowestoft and up the existing river valley of the River Yare.

In England there was a channel called the Minster Fleet that separated the Isle of Thanet from the mainland of England when sea levels were higher. The Minster Fleet still separates the Isle of Sheppey from England. Thanet has only been joined to Kent during the last thousand years at most. Previously it was separated from Kent by the Wantsum Channel.

In East Anglia there was a shallow fjord with several branches that led inland toward Norwich.

The English Fenland south of the Wash provided an extensive watery landscape of shallow brackish channels and low islands, fringed by reeds and brushwood, in which the Island of Ely was so cut off that the Anglo-Danish inhabitants held out for ten years after the Norman Conquest.

In Egypt there was failure of the Nile floods that caused famine.

In the Alps there was concern about droughts and the Oberriederin Canal was built on the upper reaches of Aletsch glacier in Switzerland. This was run over by the Aletsch glacier when the Little Ice Age arrived, as evidenced by larches growing there at the time and now unearthed. The canal head is still covered by the glacier today and this shows that the glacier had retreated much further during medieval times. Similar water supply installations were engineered in the Saastal, also in Switzerland, and in the Dolomites, only to be overwhelmed by glaciers between 1200 CE and 1350 AD.

In Europe there were ancient gold mines in the Hohe Tauern in Austria and other sites in Central Europe, that had been abandoned in the time of Christ, that were being worked. Underground water began to cause difficulties about 1300 CE. At Goslar it was reported in 1360 CE that water had been increasing in the mines in the Harz Mountains for some time. In Bohemia some mines had been closed again since 1321 CE. In the Alps some of the mine entrances were again closed by glaciers.

In Greenland and Iceland warmer conditions started after 1000 CE. In the waters of Greenland and Iceland there were more weeks with ice-free conditions. In the eleventh and twelfth centuries there was little sea ice seen on the Arctic seas near Greenland.

Between 1000 and 1200 CE there were the warmest ocean surface conditions of the southeast Greenland shelf over the late Holocene from 880 BCE to 1910 CE. It was characterized by abrupt, decadal to mutidecadal changes such as an abrupt warming of 2.4° C during a 55-year period around 1000 CE. These temperature changes are rare in the North Atlantic. There appeared to be a lag of about fifty years in ocean surface warming. This was possibly due to increased freshwater discharge from Greenland or intensified sea ice export from Antarctica. Both indicate that the increasing of temperatures to cause this.

Near Thule in northern Greenland there were the ruins of abandoned stone and turf houses roofed with whalebone beams. These were built by the Thule People who had moved across an enormous strip of the Arctic from the Davis Strait to northern Alaska. They had moved eastward from Alaska in pursuit of the Bowhead Whales and in the warmer climate with the advent of the Medieval Warm Period they had established permanent winter camps. The Thule may also have been attracted to the iron on the York Peninsula in Greenland. In the Greenlandic language the name of the settlement close to the cape, Savissivik, means place of meteorite iron which alludes to the numerous meteorites from around 12000 years ago which fell here. The original meteorite was estimated to have weighed 100 tonnes before it exploded. The iron from the meteorite attracted Thule People followed by Inuit. The earliest Thule settlements possessed fragments of iron and other Norse artifacts indicating that trade was ongoing between the two communities.

In Greenland around 1000 CE pastures were richer than they are now. They were necessary for growing cereals both for human and bovine consumption. Hay was grown here for the dairy cattle that were the main animal crop of the Norse.

In the 1000s a ship channel in Limfjord in Denmark that joined the North Sea to the Baltic Sea to avoid the Skagerrak was still in existence due to higher sea levels.

There were sea level rises in this period that were 24 inches to 31 inches in the North Sea. This was sufficient to cause catastrophic flooding when high tides coincided with storm surges and the sea level rise in this period altered the configuration of low-lying coasts.

Between 1000 CE and 1100 CE there was a rather warm and moist climate regime in Mexico with the climatic zones shifted North at the time. The valley of Mexico may have been particularly moist at the time with a lot of lakeside flooding.

The coastal plain of the Netherlands had a fluctuating population in the 11th and 12th centuries, as the state of flooding varied, leading finally to more general emigration into Germany.

This was the same in North America with warm and moist climate. In North America east of the Rocky Mountains the prevailing temperatures followed a sequence very similar to that in Europe.

Around 1000 CE people had been growing corn all across the high plains from the base of the Rockies, through eastern Colorado and western Nebraska. Farther east there were substantial settlements in the river valleys, where oaks and cottonwoods grew.

Between 1000 and 1164 CE Ubinas volcano in the Andes of Peru erupted with a VEI of 5.

In the Sahel in North Africa there was an extreme cooling spike.

In Scotland the Vikings used a sheltered channel on the south side of Moray Firth joining Lossiemouth to Burghead Bay when sea levels were higher.

There was drying and a big chill in Eastern Anatolia in Turkiye in the eleventh century. In western Anatolia there was pronounced cooling but with average humidity.

The year 1003 CE produced excessive rains and overflowing rivers in several places. The Loire River in France rose to a prodigious height, and it ravaged the coast and there was fear of another deluge. The Loire River is located in south-central and west-central France and drains into the Atlantic Ocean in the Bay of Biscay. In 1003 in France, the summer was unfavorable. The winter in France was harder than usual, and disastrous flooding occurred in the aftermath. In 1003, the winter in southern France was longer than usual.

A famine struck England from 1005 to 1016 during the reign of Aethelred the Unready. "Such a famine prevailed as no man can remember." Chroniclers say that half the population of the larger island perished, although many of the dead were caused by the wars between Aethelred and Sweyn the Dane, the latter being forced by the famine to retire from England for a time. In 1005 in England, "This year was the great famine in England." Sweyn the Dane quits in consequence.

In 1006, there was a great famine in England and over all Europe, such as the living never saw before. They scarce sufficed to bury the dead.

In 1006 on December 31st Vesuvius erupted in Italy with a VEI of 3.

In 1008, there was a famine attended with a plague in Wales.

In the year 1009 in Italy, the troops marched over the frozen rivers.

In England, it was very rainy.

In 1009 in France, the weather was unfavorable.

A hurricane in the Irish Sea did incredible damage.

In 1010 CE Medicine Lake in California erupted with a VEI of 3.

In 10,10-1011 CE there was a very cold winter which even gripped the eastern Mediterranean in intense cold. There were few winters after this for 300 years that were as cold.

The year 1010 produced extraordinary rainfall in southern France. Northern France suffered from alternating periods of droughts and harmful overabundant rains.

In 1011 CE there was ice on the Nile River in Egypt as well as on the Bosphorus in Turkey. Was this due to the eruption of Vesuvius in 1006 which is a high sulfur content volcano?

During the winter of 1011, France experienced a cold winter.

In 1012, there was a terrible famine in England. In England and Germany, there was a great inundation from the sea. In England and Germany, endless multitudes died of famine. An inundation from the sea overwhelmed many towns in England, Germany, etc. and much people, endless multitudes, died of famine and plague. There were great rains.

In 1012, a storm surge caused floods. As a result, dikes were built along the river Weser in northeast Germany. "At that time, the Danube River in Bavaria overflowed its banks, the Rhine and its banks as well. An innumerable amount of people and livestock died, and many buildings and forest were destroyed by the forces of the tide."

In 1012 in the north of Western Europe, there were heavy rainfalls that produced floods. The waters of the Danube and Rhine Rivers caused immense damage.

In 1013 CE in England, there was a great earthquake, and whirlwind or hurricane from the west, throwing down houses and tearing up trees by the roots. There was thunder and lightning in May.

The English coast was attacked by a sea floods on St Michael's Mass Eve, 28th and 29th September 1014 CE. The Anglo-Saxon Chronicles stated that the great sea floods ran so far up as it never before had done. This was a possible major flood or storm surge; some have said tsunami. There was great damage to coastal communities along the English south coast and given the impact on the eastern side of the southern North Sea surely had a significant impact on the English side of the North Sea.

Flanders coast in Belgium was also attacked by a sea floods on 28th and 29th September 1014 CE as well.

The coast of the island of Walcheren was also attacked by a sea floods on 28th and 29th September 1014 CE. Walcheren is a region and former island in the Dutch province of Zeeland in the Netherlands. It was an island in this period.

On 3rd October 1014 CE many English seaports were destroyed by the sea. There were great inundations of the English coasts and a number of seaport towns were demolished. The sea overflowed and drowned many villages and an innumerable multitude of people.

In 1014 in France, the winter was mild.

Greenland ice cores indicating layers of ammonium to high-energy atmospheric interactions with objects coming from space were in 539 CE, 626 CE, 1014 CE and 1908, the Tunguska Event.

Ice cores from Greenland GRIP indicate unnaturally high levels of ammonium between 1012 CE and 1015 CE. This was the largest concentration of ammonium recorded between 596 BCE and 1642 CE. This could indicate an impact event. Refer to Tunguska in 1908 CE. The only cosmic impact that I could find was a doubtful impact at Juzjanan in Afghanistan in 1009. It was listed as a doubtful iron meteorite.

Had this aerial intruder crashed into the North Sea and caused the deluge that struck England. The Netherlands and Flanders?

The work of Sekanina and Yeomans states that a comet came close to the earth in 1014 CE. Was a fragment of this comet involved?

There were great sea floods on the North Sea coast between 1015 and 1017.

In 1016 in France, the winter was mild.

In Ireland, there were excessive rains and floods producing cattle mortality. There was an awful famine throughout Europe. "Hail, thunder and lightning." In July, hail and thunder killed many people. Trees and corn(wheat) suffered much and a grievous famine followed.

In 1016 CE there was a terrible tempest of much hail, rain, thunder and lightning in England, which was fatal to people, corn grain, cattle and trees. In 1016, there was a famine from great hailstorms, thunder and lightning in England.

Many people succumbed to the very harsh winter of 1019-1020 in Germany.

In 1020 CE Arenal in Costa Rica erupted with a VEI of 4 and Sheveluch in Kamchatka erupted with a VEI of 3, and Taveuni in Fiji erupted with a VEI of 2.

The year 1020 was remembered as a hard and cold winter with much snow. This year, the sea poured over North Friesland so much so, that many towns and villages were entirely ruined. This entry was listed under the category "storm surge", which might include an inundation of the sea. In the year 1020, the winter was very harsh and persistent. As a result, there was a huge mortality rate spread over the whole continent of Europe.

In 1020 CE in England, the frost was very severe. As well in England, there were great floods followed by plague and many people were killed by the severe cold winter.

In 1020 the Albis (now called Elbe) and Visurgis (now called Weser) Rivers in Germany rose high, drowning many of their coasters or coastal trading vessels - shallow-hulled ships used for trade between locations on the same island or continent. In July 1020, there were 3 days of very heavy floods along the Weser and Elbe rivers in Germany. Also, this year, a strong storm surge struck along the North Sea.

In 1020 CE there was a severe winter in London.

Between 1021 CE and 1228 CE Lake Cardiel, 120 miles north of Lago Argentina, rose and drowned shrubs that were rooted on the shore.

In 1021 or 1022 in England, there was excessive heat, "yet marbles sweat profusely." In 1022 in England, there was a strong, droughty heat wave in the summer. Men and animals died.

The French and Germans place this heat wave in 1022 and say that so great a drought and heat arose that many people and cattle died of it. In this heat, marble pillars sent forth so profuse a sweat. The summer of 1022 in France was superb.

In the year 1022, during the synod and royal assembly at Aachen in west-central Germany, such a strong heat occurred that many people suffocated and died a sudden death. Many animals also died. The plaster and the marble columns of the temple were sweating as if there was considerable moisture.

In 1022 in Ireland, there was a great shower of hail, the hailstones as big as crab apples. There was also great thunder and lightning which killed an infinite number of cattle.

In England, this summer was extremely hot and dry. In a great part of Germany heat that accompanied terrible storms proved fatal for humans and cattle.

In Hindustan (India) in 1022, during the reign of Masaood I, the area experienced a great drought followed by a famine. The whole country was entirely depopulated. Hindustan is the land between the Himalayas and the Indian Ocean. About 1022 CE there was a great famine in India. There are records that indicate whole provinces in India were depopulated by famine. This year was remarkable for drought and famines in many parts of the world.

The English and German coasts were flooded by another sea floods in 1024 AD.

In 1024 CE in Russia, there was a major famine. There were 38 major famines in Russia between 1024 and 1936. Many of these famines were accompanied by such horrors as eating tree bark, grass, and dung, and cannibalism.

In 1024 CE to 1025 CE Fatimid rulers of Egypt, responding to severe drought, had to seize grain transports and open granaries to feed the population.

A famine took place in Egypt in 1025, during the rule of the Caliphate of Zahir. The suffering was widespread. It became necessary to prohibit the slaughter of cattle and there was no meat to be had anywhere, as fowls, the common meat of Egypt, had quickly disappeared. The stronger among the population turned brigand and began to prey upon the weaker members of society. Caravans and pilgrims were attacked by Syrian bands, which began to invade border towns. People flocked to the palace in masses crying piteously for relief at the hands of the Commander of the Faithful. But no help was to be had at that quarter because the palace was very short of provisions. When the banquet for the Feast of the Sacrifice was spread, the slaves of the royal household broke in and swept the food from the tables. Slaves began to rise in revolt in all parts of the country and it became necessary for citizens to organize committees of safety for self-protection, the government granting permits to kill the bondmen. With an ample rise of the Nile River in 1027, however, the period of suffering came to an end.

In 1025 CE in England, there was a famine and plague from the greatest rains.

In Flanders now Belgium, it rained constantly from 15 October to April. This was followed by a plague, which swept away the greatest part of men. Afterwards there was a great famine.

In 1025 CE at Lake Walker in California dead pine tree trunks showed that the level of the lake had dropped by 131 feet below the shoreline of the early twentieth Century. This was followed by a brief wet cycle, then another drought. This was the same at Mono lake in California.

In 1030 the volcano Billy Mitchell in Bougainville erupted with a VEI of 5 and Arenal in Costa Rica had a VEI of 4. Another eruption was Gorely in Kamchatka with a VEI of 3. This is volcanic winter material.

In France from 1030-1032, the whole course of nature seemed to be upset, and there was intense cold in the summer and oppressive heat during the winter. Rains and frost came out of season and for three years there was neither a period for planting seeds nor for harvesting. The miseries of mankind in France at that time were incredible. Also there was a fear of the coming of the end of the world coinciding with the 1000th anniversary of the Crucifixion. Thousands upon thousands died of starvation, and the living were too weak to bury the dead. There were many horrible instances of cannibalism and human flesh was said to have been exposed for sale in the market at Tournus. In their maddened condition, the peasants exhumed human bodies from graves and gnawed the bones.

One of the harrowing incidents of the time, which will give some idea of the insanity which suffering induced, occurred in the wood of Chatânay, near the town of Macon. A traveller and his wife stopped at a hut supposedly occupied by a holy hermit. Scarcely had they entered the adobe, however, when the woman discovered a pile of skulls in the corner. She and her husband fled to the town and when an investigation followed, it was found that the hermit had murdered and partly devoured 48 men, women and children. Chatânay is located in Burgundy in east-central France. Grass, roots, and white clay were the ordinary articles of food for the poorer classes during these terrible years. And as a result, the sufferers almost ceased to resemble human beings. Their stomachs became greatly distended, while almost all the bones of their bodies were visible beneath their leathery skin. Their very voices became thin and piping. Packs of raging wolves came out of the forests and fell upon the defenseless peasants. It seemed as if mankind in France could never recover.

Raoul Glaber (from Medieval France) tells us that in the years 1030-32, the whole earth was so inundated with "continuous rain for three years" that there could not be found a furrow in the field for sowing. It followed that these floods caused an awful famine. Excessive rainfall and humidity were the main cause of the terrible famine of 1030 to 1033 in France. The ground was incessantly drenched by rainfall. Farmers waited in vain for a favorable time for sowing their crops. The soil remained so soaked for three years, that it didn't offer a single furrow to receive grain. These floods offered a sad triumph over the weeds in the fields. A bushel of seed brought only a pint in the best land and the pint itself only a few grains.

In 1031 CE in England, there were extended general floods from rains. In England, there was a famine from great rains and locust. There were terrible tempests and great rains. This caused such inundations in rivers near the sea as overflowed the lands. Famine and plague followed. At the same time, famine and plague grievously oppressed Cappadocia (Turkey), Armenia, Paphlagonia (Turkey), and almost all the East. Many were forced to leave their country.

In winter 1032 to 1033 CE there was a cold winter.

On January 19th 1033 CE Fujisan in Japan erupted with a VEI of 2.

In 1033 CE in Gaul Western Europe the winter was severe.

In Switzerland, the Imperial Army of Emperor Conrad II suffered much from the cold.

On 9 July 1033, a bolt of lightning set the Church of Saint Michael's on fire in Hildesheim, Germany. This would place the event on the year the church construction was completed, and the church was consecrated.

In 1033, there was a severe drought in Southern China during the period of 29 July - 27 August. Because of the severe drought, the locusts, the famine and the plague, 20% to 30% of the people died. In 1033 during the autumn, there was a great drought in China. Typhus fever raged. The drought was accompanied by a plague of locusts. Because of the hardship, soup kitchens were opened. The weather in 1033 was ominous.

An earthquake struck the Jordan Rift Valley on December 5th 1033 CE and caused extreme devastation in the Levant region. It was part of a sequence of four strong earthquakes in the region between 1033 and 1035. Scholars have estimated the moment magnitude to be greater than 7.0 M_w and evaluated the Modified Mercalli intensity to X (*Extreme*). It triggered a tsunami along the Mediterranean coast, causing damage and fatalities. At least 70,000 people were killed in the disaster. Heavy damage was reported in a north–south trend for 190 km (120 mi) from the Dead Sea to the Sea of Galilee. One-third of the city of Ramala was destroyed. Half of Nablus was destroyed and 300 residents died. The landscape around the city was also devastated. Acre experienced great damage and a high death toll. The cities of Banias and Jericho were also among those who suffered the greatest destruction. A landslide buried the village of al-Badan, killing all its residents and livestock. Landslides also destroyed other villages and killed most of their population. Banias was partially destroyed. In Syria, entire villages were "swallowed" by the earth, causing fatalities. In Gaza, a mosque and the surrounding minarets collapsed. A lighthouse in the city sustained heavy damage. Reports of serious damage also came from Ascalon. Damage was reported as far away as Egypt. Sahil A. Alsinawi and others reported a death toll of 70,000.

Parts of the Walls of Jerusalem collapsed and many churches were damaged. A side of the Temple Mount and the so-called mihrab Daud, located near the Jaffa Gate, collapsed. The entire southern section of the city walls which enclosed Mount Zion above the Kidron Valley, which were built by Aelia Eudocia (the fifth-century wife of the Byzantine emperor Theodosius II), were abandoned by Fatimid caliph Al-Zahir li-i'zaz Din Allah who established major restoration projects that lasted from 1034 to 1038 . It is believed to be the largest restoration project in the city's history. The Dome of the Rock was enforced with wooden beams to strengthen the structure. Wooden beams and mosaics were added to the al-Aqsa Mosque. Solomon's Stables and al-Aqsa Mosque were among the structures that underwent restoration.

A tsunami struck the coastal city of Acre. It was reported that the city port became dry for an hour, and a large wave arrived. Waves were also reported along the coast of Lebanon. Greek seismologist Nicholas Ambraseys reported that the tsunami caused no damage or casualties, but this is thought to be a confusion with the 1068 earthquake. Destruction was reported in Acre due to the tsunami. People who scoured the exposed seafloor drowned when the waves arrived. Additional shocks in April or May 1035 CE caused further damage and might be associated with tsunamis.

The temperature in Gaul in Western Europe in 1033 CE was so unfavorable that farmers could not sow or harvest because the fields were constantly flooded. Because of the incessant rains, it was believed that it would take 3 years for the soil to become suitable for sowing furrows. A bushel of grain was sown in the fertile land. When harvested the grain yielded only a sixth of a bushel, hardly a handful.

A plague started in the East. After ravaging Greece, the plague came to Italy and spread to Gaul, Western Europe, and did not even spare England. Individuals were forced to eat grass, and animals that had fallen or dead animals. The people killed themselves in order to consume themselves. Some children were tempted with an egg or an apple in order to lure them away. These children were then killed for food to satisfy their hunger. This madness, the frenzy grew so that the animals were safe to escape death; when the people nourished themselves on human flesh, even though this crime was punishable with burning at the stake. Some people who starved so long that when someone arrived to nurse them back to health, they ate a full meal and fell over dead from refeeding syndrome. It was generally believed; the order of the seasons and the elements had ceased. This is an excellent example of a volcanic winter. Refer back to Billy Mitchell erupting in 1030.

In 1034 Sheveluch in Kamchatka in Siberia erupted with a VEI of 5. This was also enough to cause a volcanic winter. Do volcanic winter affects become cumulative?

In 1034 in France, the weather was warm and produced an abundant harvest equivalent to five years of normal harvest. The harvest was rich in cereals, wine and fruits of all kinds.

The second of the four driest drought periods started in the American West in 1034 CE. This was all within a four-hundred-year period of overall aridity.

In 1035 CE a frost in England on Midsummer Day was so vehement, that the grain and fruits were destroyed. Midsummer Day is 21st June. This produced a dearth or limiting of harvests.

On January 27th 1037 CE Vesuvius erupted in Italy with a VEI of 3.

In December 1037 there was a major earthquake in Taizhou in Jiangsu, China that killed 22,391 people.

In 1038 there was a terrible famine in Constantinople (now Istanbul, Turkey) in the Byzantine Empire.

In 1038 in England, there was a famine with a plague.

In 1040 CE Cayambe in Ecuador erupted with a VEI of 4 and Vulcano Island in Italy had a VEI of 2.

In 1040 CE there were great inundations in Germany.

In 1040, there was a famine in England, more severe than any other.

Between 1040 CE and 1450 CE there was a dry period around Lake Titicaca on the Altiplano of Bolivia. This was during a four-hundred-year period that was marked by extraordinarily low ice accumulation in the Andes. Lake Titicaca lake cores showed that there was a complete stop of organic desedimentation reflecting a decline in the lake level of between 12 and 17 meters.

After 1040 CE snow accumulation in the Upper Andes declined sharply. The drought cycle peaked around 1300 CE and persisted with less intensity until around 1450 CE. Was this initial lessening of snow in the High Andes caused by the eruption of Cayambe in Ecuador?

Also in this period Lake Winaymarka, the smaller basin of Lake Titicaca, seemed to have evaporated completely. The lakes shrinking was between 1030 CE and 1280 CE. This indicated a severe and prolonged drought with a ten to fifteen per cent decline in rainfall from the late twentieth century average. The raised fields around Lake Titicaca were abandoned and unusable. The water table lowered, and the springs and aquifers were not fed.

Around 1040 CE the Turks in central Asia, due to the effects of cold on the one-humped female camel that the Turks bred with two-humped male camels, produced a camel well-suited for Silk Road trade. The one-humped camel did not tolerate cold well, so the cooling climate forced camel breeders to head south.

Also, around 1040 CE in Central Asia the new cold climate propelled people westwards. These included the Pechenegs, Oghuz and Seljuk Turks.

There was a massive cold spell in Central Asia. In Iran cotton cultivation had created much prosperity but the cold shift of weather struck as cotton production declined in northern areas and as seminomadic peoples moved into Iran.

Also, around 1040 CE a chronicle described that the city of Nishapur in Persia, Iran, now lay in ruins and a great number of people had died from hunger. The weather was bitterly cold, and life was becoming hard to bear. Disorder in Iran set off a diaspora that increased the influence of Persian culture in South Asia.

On 11 January 1041 there was a strong storm surge along the North Sea coast. In Flanders now Belgium, the sea broke down its banks, and carried off all, far and near, with it into the ocean.

The whole year, 1041 was frightful in England, in both temperature and great excessive rains. This damaged corn grains and caused a great death of cattle. This destruction was far greater than anyone living ever remembered. Then began a famine that lasted seven years. On 3rd November, there was a fearful tempest and great rain

In 1042 CE Hakusan erupted in Japan with a VEI of 3.

On August 12th 1042 an earthquake struck Palmyra in Syria, Baalbek in Lebanon, Syria and Lebanon. 50,000 people died.

In England in 1042, about this time came such a famine that a sextarius of wheat, which usually is a load for one horse, sold for 5 solidi and more. This shortage lasted for seven years.

The summer of 1042 in northern France was very wet.

In November 1042 CE some sort of major flooding occurred on the eastern side of the southern North Sea in Dutch coastal communities. This could have possibly affected the English side of the North Sea as well.

In 1043, tempest and profound summer rains, harvest snow, scarcity of wine and corn grain prevailed in France and Germany. The year 1043 in northern France produced great rains. The summer was rainy and wintery. There were tempests all summer. The rains and storms in the summer of 1043 in France made the summer similar to the winter. There was very little fruit and poor harvests.

In 1043, there was a grievous famine over all England. Corn (grain) was the dearest ever known by anyone living.

The winter of 1043 to 1044 CE was regarded as cold in Britain. In 1043, a great snow fell in harvest.

There was frost from 1 December to 1 March in Normandy, France. In 1044, the vine and fruit trees in France were killed by the extreme cold. In 1043 in northern France, heavy frosts lasted from early December until early March. During the winter of 1043-44, the winter was very harsh in Germany and France, and accompanied by frequent snow. The vines were so damaged that the wine was extremely rare. The loss of the harvest produced a famine so great that many people were forced to eat unclean animals. The mortality was considerable.

In Western Europe in 1044, the year was remarkable for the great abundance of rain showers and the unusual lack of fruits of the earth. The winter of 1044 in France was cold.

In 1044, there was "hunger all over England with corn grain dearer than ever known."

In 1045 CE in Flanders now Belgium, there was an inundation from the sea.

In England in 1045, it rained all November. There were summer rains, harvest snows and tempests with scarcity afterwards.

There was such a large amount of snow in Western Europe, that the forests were inaccessible.

In 1047 in Ireland, there was a great famine and snow.

On 1 January 1047, snow fell to a great depth in the west of England. In England, there was a famine from snow and frost. In 1047, there was a famine in England that was caused by great snow and frost. All summer there were tempests with rain, thunder and lightning, dearth and death. On 1 January, there fell in the west of England, a very great and deep snow, which broke down most woods. The snow laid on the ground until 1 March. The summer after had such tempests of thunder and lightning, that the growing of corn (grain) was burnt and blasted. Several towns were struck by lightning and reduced to ashes. There followed a great dearth (famine), and the death of people and cattle. There was a terrible thunder and lightning storm in England.

In 1047-1048, famine in Scotland extended over two years. The winter of 1047 CE was reputed to be the worst winter in living memory. There was severe frost and heavy snow.

1050 CE

In 1050 CE Cerro Bravo erupted in Colombia with a VEI of 4 and Tolbachik erupted with a VEI of 3 in Kamchatka.

In the area of Rio Catari, Katari River, which flows into Lake Titicaca, in Bolivia, after 1050 CE the large highly organized communities with raised beds and irrigation were abandoned and were replaced by small villages covering only one hectare. This happened suddenly and lasted for the length of the three-hundred-year drought and rainfall patterns did not revert to pre 1050 CE times until the 1450s CE.

Around 1050 CE colder drier climate returned to the southern plains of Canada and the southern plains suffered from drought. The moist climate gradually retreated southwards and eastwards and the people went with it.

In Colorado and New Mexico from 1050 CE to 1100 CE the rains were again plentiful and Chaco Canyon and its outliers flourished. Chaco Canyon was the center of at least seventy communities that were dispersed over 65,000 square kilometers of northwestern New Mexico and parts of southern Colorado. It was an intensively important and sacred place with ceremonial trackways leading to it. The population continued rising steadily and all was fine while the rain fell.

In 1151 CE Lago Argentina in Patagonia drowned southern birch trees that had grown up in the previous half a century to a century. This was a sudden rise in water levels.

In 1151 CE there was a great barrenness of the land in England, and dearth, famine, want of bread and great mortality. In 1151, there was a dreadful famine in Lincolnshire, England.

The entire year of 1151 in northern France was very wet. The summer of 1051 in France was very poor.

In 1151 there was a famine in Mexico that caused the Toltecs to migrate.

Between 1052 CE and 1160 CE German oak chronologies show ring widths that were 35 to 80 per cent wider than in the tenth Century. This indicated more rain. There was more moisture than in the 900s and more general warmth of the growing season.

Between 1052 CE to 1060 CE in India and Afghanistan there was a famine. In Hindustan, from 1052- 1060, there was seven years of drought in Ghor province in central Afghanistan so that the earth was burned up, and thousands of men and animals perished with heat and famine. There are records that indicate whole provinces in India were depopulated by famine in 1052 CE. The famine of such severity swept over Hindustan that the Mongol emperor himself was unable to obtain the necessary food for his household. Hindustan is the land between the Himalayas and the Indian Ocean.

In 1052 CE on St. Thomas's Eve, December 20, was such a hurricane in England as demolished many churches, blew down innumerable houses and broke down and rooted up trees.

1053 CE was a year of heat and drought, which extended to the north of France. The summer of 1053 in France was a great excessively hot summer.

In 1053, a persistent drought burned lands in the country of Caux in Switzerland.

In 1053 there was a famine in England after a comet. The famine lasted two years.

In 1053 during the period between 13 November and 12 December, a drought engulfed China. The drought was accompanied by a plague of locusts.

In 1055 CE in London, England, nearly 400 houses were blown down by a storm. A hurricane at Coventry, England blew down 400 houses. In 1055, there was also a great famine or a continuation of the previous one.

In 1055 CE there was a famine in Ghor, now a ruined city, in the Maldah district of West Bengal, India.

In 1059, there was a great comet seen in Poland, followed by a severe famine.

In 1059 in England, there was a severe winter with a great frost. This produced a severe plague and famine in 1060. The winter in England in 1059 was cold and long, very injurious to corn grain, hence followed a famine and plague in 1060. The winter of 1060 was unusually hard and strong. It caused a very significant loss in the wheat and grape harvest. A terrible famine affected many people there.

In 1060 CE eruptions were Nejapa-Miraflores in Nicaragua with a VEI of 3 and Medicine Lake in California with a VEI of 3.

In 1061 CE the Thames River was frozen in London for seven weeks.

In 1062 in England, there was a terrible thunder and lightning storm with subterraneous motions.

In 1063 CE there was a severe winter in London and the south of England. The River Thames in England was frozen for 14 weeks this time and there was a great frost in England.

In 1063 CE the winter in Europe was long and intensely cold and many people perished by cold and hunger. All the rivers of the continent were frozen, and even south of the Alps, the Po River in northern Italy and many other streams were blocked by ice.

In April of 1063 in Western Europe, there was a tempest for four days together of cold, wind and deep snow which killed all fowls and cattle and damaged trees and vines.

In the middle of April 1063 in France, there came four days of winter so harsh with winds and snow, as most trees and grape vines were killed, and the birds and livestock died from exposure to the cold.

In China in 1064 CE at Chang-Chou, a daytime fireball was seen, and a meteorite fell. Fences were burned.

A terrible 7-year famine struck Egypt beginning in 1064. To the hardships of starvation were added the miseries of civil warfare. Nasir-ed-dawla, commander-in-chief of the Fatimid army, upon being deposed by the Caliph Mustansir, quickly gained the support of bands of Arabs and Berbers. Black regiments were soon in control of all Upper Egypt. Forty thousand horsemen of the Lewata Berbers descended upon the delta of the Nile River and swept all before them, cutting dikes and destroying canals with the malign purpose of spreading starvation. Both Al-Fustat and Cairo in northern Egypt were cut off from supplies, and to add to these tribulations, the Nile failed to come to a flood in 1065.

The peasantry, not daring to venture into the fields for fear of the armed bands of brigands, were unable to carry on any agricultural pursuits; so that the dearth of one year's harvest was prolonged into seven. Prices soared to heights never before reached in the near East. A single cake of bread sold for 15 dinars. Five bushels of grain sold for 100 dinars. Eggs were scarce at a dinar each. Cats and dogs brought fabulous prices, and women, unable to purchase food with their pearls and emeralds, flung their useless jewels into the streets. One woman according to a historian gave a necklace worth 1000 dinars for a mere handful of flour. The caliph's stable, which had numbered 10,000 horses and mules, was reduced down to three scrawny nags.

Rich and poor suffered on equal terms. Finally, the desperate people resorted to revolting cannibalism. Human flesh, which was sold in the open market, was obtained in the most horrible manner. Butchers concealed themselves behind latticed windows in the upper stories of houses, which looked out upon busy thoroughfares. Letting down ropes to which were attached great meat hooks, these anglers for human flesh snared the unwary pedestrians, drew their shrieking victims through the air, and then prepared and cooked the food before presenting it for sale in the stalls on the street level.

Beginning in 1064 in Egypt and lasting for seven successive years, the overflow of the Nile River failed and with it almost the entire subsistence of the country while the rebels interrupted supplies of grain from the north. Two provinces were entirely depopulated. In another province, half of the inhabitants perished. While in Cairo, the people were reduced to the direst straits. Bread sold for 14 dirhems to the loaf. All provisions were exhausted. The worst horrors of famine followed. The wretched resorted to cannibalism. Organized bands kidnapped the unwary passenger in the desolate streets, principally by means of ropes furnished with hooks and let down from the latticed windows. In the year 1072, the famine reached its height. It was followed by a pestilence, and this again was succeeded by an invading army.

In 1064 CE severe cooling and severe drought ended in Mongolia.

In the late summer and autumn of 1066 CE after a long, dry summer, west to west-northwest winds prevailed in the English Channel all through September. It was only the breaking of this anticyclonic northwesterly spell that gave William the Conqueror his chance to cross the English Channel on the 7th October 1066 CE.

In 1066 to 1067 CE the watery state of the English Fenland, among the rush and willow girt islands, were evidence of high sea levels.

In 1066, there was a tremendous storm surge that destroyed the castle and village of "Mellum", an east Frisian island in the Netherlands. Mellum is now uninhabited. There are still small sand dunes on Mellum as well. Mellum is now in Germany and surrounded by mudflats off the end of the Butjadingen peninsula.

Around this time the town of Beccles in Suffolk was a thriving herring port. It is now seven miles inland from the North Sea. Before the Conquest of William of Normandy, local fishers had supplied the Abbey of St Edmund with thirty thousand herring annually. William the Conqueror doubled the assessment. Beccles was once a flourishing Anglian fishing port.

Sea floods hit Oldenburg in Holstein, Germany, on the Baltic Sea in 1066 CE.

The winter of 1066 to 1067 was regarded as cold in Britain.

1066 CE is regarded as the end of the Viking Age. Around this time there were about one thousand farms in Halogaland in north Norway, and they grew barley, oats and rye. By the 1430s these farms were deserted as these crops could not grow there anymore as the climate changed.

The winter in Europe in the year 1067 was long and intensely cold and many people perished by cold and hunger.

The winter of 1067 in France was cold and there were 6 consecutive weeks of frost. In 1067, the vine and fruit trees in France were killed by the extreme cold. In France, a terrible winter began on 13 November 1067 and lasted until 12 March 1068.

During 1067-1068, in France, the winter between St. Brice to St. Gregory (from 13 November 1067 until 12 March 1068) was extremely severe. The vineyards and forest trees bore no fruit. The mishap brought forth by this and the previous year's infertility produced in England such a famine, that the unfortunates were forced to eat dog and horse meat, yes, even to eat human flesh.

Two major earthquakes occurred in the Near East on 18 March and 29 May, 1068. The two earthquakes are often amalgamated by contemporary sources. The first earthquake had its epicenter somewhere in the northwestern part of the Arabian Peninsula around Tabuk, while the second was most damaging in the city of Ramala in Palestine, some 500 km to the northwest. The March earthquake affected the southern portion of the Dead Sea Transform (DST) fault system. The combined events were responsible for an estimated 20,000 deaths, of which some 15,000 occurred in Ramala alone, and caused damage throughout Greater Syria, including Palestine, where a tsunami devastated the Mediterranean coast, in Egypt and the Arabian Peninsula, and in areas to the east along the Euphrates such as al-Rahba and Kufa. Other strong earthquakes have occurred in the southern portion of the DST throughout history, impacting the wider region. The earthquake's effects were seen from as far north as Banias at the southern foot of Mount Hermon, to the Hejaz region of modern-day Saudi Arabia. The ancient city of Ayla, located at the northern end of the Gulf of Aqaba where modern Aqaba stands today, was destroyed. Palaeoseismic investigations have revealed more than 12 kilometers (7.5 mi) of fault rupture, beginning just north of Aqaba/Eilat, that were dated between 900 and 1000 years before present. A magnitude of at least 7.0 was presented based on the reported damage and the extent of the observed fault breaks.

In 1069 CE the rivers froze in the north of Germany. In the year 1069 in Germany, the winter was harsh and long. There was a shortage of wine and fruit because of the extreme cold. The rivers were frozen over and King Henry IV came to the countries of the Saxons and caused such carnage that the area was depopulated.

In England in 1069, there was a great dearth. The peasants of the north, unable any longer to secure dogs and horses to appease their hunger, sold themselves into slavery in order to be fed by their masters. All the land between Durham and York in northeastern England was laid waste, without inhabitants or people to till the soil for nine years. Some of the destitute resorted to cannibalism. A factor that contributed to this hardship was the taxes exacted by the conquerors. Peasants became discouraged, realizing that the fruits of their labor were taken from them as fast as they were earned.

In 1069, the Normans desolated England, and in the following year famine spread all over England, "so that man, driven by hunger, ate human, dog and horse flesh;' some to sustain a miserable life sold themselves for slaves. In 1069, there were plenty of good grapes in England, but all wild fruit trees were barren.

The winter of 1069 in France was cold.

In early summer and probably June 1070 CE there was a violent storm in the North Sea as the Danish king Sven II, was returning to Denmark with looted treasures from the East Midlands and Fenlands after being defeated and then reconciled with William the Conqueror in early June.

In 1070 CE there were high flood levels in the Nile River due to increased monsoonal rainfall in the Ethiopian Highlands. This lasted from 1070 CE to 1180 CE.

In 1073 CE Vesuvius erupted in Italy with a VEI of 3.

In 1073 CE in England, famine was followed by more death, so fierce that the living could take no care of the sick, nor bury the dead. In 1073 to 1074 CE there was a severe winter in Britain.

In 1074 CE the cold was very lively in France and most of the rivers froze.

The winter was so severe, that all the rivers in Flanders now Belgium, and Germany were completely frozen.

In France in 1074, there were great frosts from 1st November to mid-April. The winter of 1074 in France produced a frost that began in November and lasted until April. There was a cold dry violent north wind. The mills were paralyzed by the cold and the army of Henry IV sorely lacked bread. The little grain that was available could not be ground into flour because the extreme cold caused a shutdown of the mills.

In the year 1074, a very severe frost in Western Europe lasted from early November until mid-April. The cold, dry and cutting sharp wind was so intense that the rivers not only froze on the surface; but the rivers turned into solid blocks of ice.

In 1074 in the vicinity of Shanghai, China, it rained continuously from the 1st to the 6th month. The lakes overflowed. The land could not be cultivated. Houses were destroyed and the inhabitants discarded their lands and went away to beg.

In 1075 CE the San Francisco Volcanic Field in Arizona erupted with a VEI of 4.

From 1075 CE to 1375 CE in Northern China there was an extended dry period which was followed by a four century wetter period. This coincided with the dates from the Quelccaya ice Core from the Southern Andes, 12,400 miles away.

The year 1076 was another very cold winter in France, which destroyed many trees and vines.

The Rhine River in Germany by the Feast of St. Martins was frozen from 11 November 1076 until early April 1077. People crossed the river on the ice.

The winter of 1076-1077 in France, England and Germany was so severe, that the oldest people could not remember experiencing a similar cold winter; nor had anyone heard speak of it. The snow lasted from 1 November to 26 March. One could travel on the ice on the Rhine River from the Feast of St. Martins (11 November) until the end of March. The frost lasted 4½ months in the interior of France. The ground was frozen down to the roots of the grape vines in several areas. The shortage of grain was so great that few people had wheat from this year's harvest.

The Great Winter in 1076-1077 in northern France was accompanied by snow. The snow began to fall at the end of October 1076 and did not stop until 27 March 1077.

In England, there was frost from 1st November to 15th April. "In the tenth year of William the Conqueror 's reign which began on Christmas 1066 CE, the cold of winter was exceedingly memorable, both for sharpness and for continuance; for the earth remained hard frozen from the beginning of November until the midst of April then ensuing." A frost in England from November to April (Some of these accounts show the frost occurred in the winter of 1075/1076, but the tenth year of William the Conqueror's reign would begin in Christmas 1076).

In 1076, there was dreadful frost in England from November to April. The River Thames in England was again frozen over.

The winter of 1077 was an extremely severe winter (the Canossian winter) throughout Europe that lasted from late October to 15 April. The rivers were frozen from 26 November until mid-March. This winter was one of the coldest and longest winters within the memory of man – The Rhine River was frozen into a solid mass from November till April.

The winter of 1077 was extremely cold in Europe. Lake Constance situated in Germany, Switzerland and Austria near the Alps froze. The trees and vines were destroyed. The winter was so cold that the land remained barren for several years afterwards. The Rhone, Danube, Po, Tiber, Elbe, Vistula and Loire rivers froze. The Rhine River froze from 17 November to 7 April.

The first frost struck Augsburg, Germany on 1 November; Lagny, France on 17 November; and Saint-Amand, France on 19 November. The frost ended on 18 March at Saint-Amand; 1 April at Augsburg; and 22 April at Lagny. Europe experienced cold weather from November to April.

France experienced 4½ months of frost over the winter.

In France in 1078, the year was marked by drought and heat. As a result, the grass withered. But nevertheless, the year produced a good harvest with fruit in June and the wine was very abundant. The summer of 1078 in France was very hot and very dry. The farmers harvested in August. The wine was plentiful and very good. The summer of 1078 in France was very hot and the harvest was a month early.

A famine occurred in Constantinople in 1078 which was caused not by weather but by mass migration. Many Asians for fear of the barbarians laying waste in the East fled to Constantinople but were pursued by famine and a grievous plague. So, the living were too few to bury the dead. In 1078 in Constantinople, there was a famine "from the multitudes of strangers" that fled to Constantinople.

In 1079 CE in southwestern Italy, the Calore Beneventano River (also called the Calore Irpino River) was so frozen that you could cross it safely on foot.

In 1080 CE Mashu in Japan erupted with a VEI of 5. This would cause a volcanic winter.

In 1080 CE there was a great famine in Denmark.

The winter of 1080 in France was cold.

In 1080 during the period between 5 February and 6 May, a drought engulfed the "north and west marches" in China. Then during the period between 22 May and 19 June, a drought engulfed China. The north and west marches may refer to the boundary provinces of western China of Kansu, Sauchuan and Yunnan. Between these three provinces lie the vast area of Southern Mongolia and Tibet.

In 1082 Ubinas erupted in Peru with a VEI of 5. This could also cause a volcanic winter as well. Were we accumulating the results of volcanic winters as we were still in the previous volcanic winter?

The winter in the year 1082 was severe in Italy. In the month of December, King Henry IV marched with his soldiers and a great number of citizens on the completely frozen Po River.

On April 17th 1083 Fujisan in Japan erupted with a VEI of 2.

The weather was erratic. In Germany during the summer of 1083, the heat from the sun's glow was so strong that not only did men die, but also the heat brought about the demise of the fish in the ponds.

In 1085 CE Miyakejima in Japan erupted with a VEI of 2.

In 1085 or 1086 CE there was a severe winter in London and the south of England. This was a sorrowful year in England, full of miseries for the great death of cattle, late ripening of corn grains, and all fruits, abnormal temperatures, terrible thunder and lightning which was fatal to many. In 1085, there was an earthquake in England, followed by great cold.

In the winter of 1085-1086 CE in England "The weather was so inclement that the unusual efforts made to warm the houses caused many accidental fires, nearly all the chief cities in the Kingdom were destroyed, including a great part of London and St. Paul's."

From the 11th of November 1085 to the 1st of April 1086, there was "so great a frost, the frozen Rhine River was passable on foot. There were excessive rains and great water floods in Italy, Flanders now Belgium, and England. These floods softened the hills and overwhelming villages, carrying along with them much people. There was a great death of cattle this year and a sore distemperature of the air. Hence a great death of people both from fevers and famine. In many places, but chiefly in Italy, so prodigious were the inundations, that rocks by their fall demolished many towns with landslides.

The same year in England, peacocks and other tame fowl, left the houses and fled to the woods. Fishes were dead in the waters. There was terrible thunder and lightning, fatal to many people and much cattle; thence the scarcity of corn grain and death of cattle".

In 1086 CE in England, there were heavy floods from rain.

"In the twentieth year of William the Conqueror, there fell such abundance of rain that the rivers did greatly overflow in all parts of the Realm. The springs also rising plentifully in divers numerous hills, so softened and decayed the foundations of them, that they fell down, whereby some villages were overthrown by landslides. By this distemperature of weather many cattle perished, much grain upon the ground was either destroyed, or greatly impaired. Thereupon ensued first a famine, and afterwards a miserable mortality of men." In 1086 in England, there was a great murrain of animals, and such intemperate weather that many people died of fever and famine. The famine was caused by excessive rains. A murrain is a highly infectious disease of cattle and sheep. It literally means "death" and was used in medieval times to represent just that.

Sea floods from the North Sea hit the county of Norfolk in 1086 or 1087.

In 1086 CE there was a wet year and much famine. There was want and pestilence. 1086 CE was noted as a very thundery year with much flooding and many people killed by lightning. In England there was great thunder and lightning. One half of all the people of England were seized with a violent burning fever, which began in 1086 and proved very fatal to multitudes. There was a general famine. There was a great thunder and lightning storm in England.

In 1087 in England, pestilence was followed by famine, which caused great suffering in the 21st year of William I.

In 1087 CE a sea floods created the Winchelsea Channel in East Sussex in England near Rye.

In Norfolk a sea floods hit St Benets Abbey, now about 8 miles or 12 kilometers inland from the North Sea. St Benets Abbey at Hollme lies in the Broads, close to the meeting place of the Bure and Ant rivers. This sea floods would have crossed over several villages and communities.

In 1087 in Denmark, King Olaf I inherited the surname the 'Hungry" in consequence of the famine in his reign.

In 1088, there was a famine in England from bad air, thunder and lightning. In 1088 in England, there was a great scarcity of corn grain. Some crops did not ripen until the end of November.

In 1089 CE according to studies of the tree rings of bristlecone pines from the White Mountains of eastern California, the period of 1089 CE to 1129 CE was the wettest cycle in California in the past one thousand years.

In 1089, there was a dearth in England. In the summer there was a great scarcity of fruits, and the harvest of grains was not complete until 30 November.

Before 1090 CE at a thousand-year-old farmstead in Iceland, the Kvisker Farm, the forest surrounding the farm contained birch stumps of a good size, that have never attained this size since. These were found buried under the volcanic ash layer of 1091 CE. The volcano Grimsvotn had erupted the previous year. These large stumps indicated a warmer climate in this period. Grimsvotn erupted in 1090 but no VEI is known.

In England on the 5th of October 1091 CE, a great hurricane came from the southwest and struck several parts of the country. A mighty storm struck on 5 October 1091 in several parts of England. The sky went very dark; the winds came from the southwest. Many churches were destroyed; and in London 500 houses fell. A storm struck in several parts of England on October 5th, especially at Winchelscomb (Winchcomb), in Gloucestershire, when the steeple of the church was thrown down. Many churches were destroyed. It took off the roof of St. Mary-le-bow Church in London and carried it a good ways. There were 4 beams in the church that were 26 feet or 8 meters long. These beams fell with such great force that they were driven into the streets (which were not then paved, but of Moorish ground) that they sunk down 20 feet (6 meters) into the street. As they could not be pulled up again, people were forced to saw them even with the level ground.

On 17th October 1091 CE there was a violent whirlwind, probably a tornado, in London. More than 600 houses were destroyed and there was much damage to churches. There was also damage to the not-long built Tower of London. This might have been a gale as it spread over a wide area. A storm in London, England on October 17th, unroofed Bow Church and at Old Sarum (Salisbury), threw down the steeple along with many houses. It caused the death of 200 to 400 persons. The roof and tower of Salisbury Church in southwestern England was broken down by thunder. This account places this event in 1092. On 1 October in 1090 or 1091, there was a tempest of thunder and lightning in England; there were hurricanes and floods constantly. On 1 October, a terrible tempest of thunder and lightning struck several parts of England, but especially at Winchcomb in southwestern England, where it did great damage to a church, and left a most intolerable stench behind. On 17 October there was a most dreadful hurricane, which rent, blew down and scattered many thousands of houses in London, Salisbury, etc.

In 1091 at Constantinople (Istanbul, Turkey) there were great clouds, which demolished houses, filled valleys with water like a sea, drowned many people and cattle. It was a great hurricane.

On 5 November 1091 at Coutances in northwestern France, a violent storm accompanied by lightning, thunder and earthquake occurred.

On 6 November 1091, on the Feast of St. Edmund, there was a great flood in England. London Bridge was swept away by the force of the water.

In 1091 CE "In this summer Vsevolod, who was trapping animals near Vyshegorod (Ukraine) …. saw a serpent falling from the clouds; all the people were frightened. All of this time the earth was rattling." Was this a comet or a bolide?

In 1092 CE there was a very wet year overall in London and southeast England. The winter in England was severe. The great rivers were frozen. "The great streams of England were congealed in such a manner that they could draw two hundred horsemen and carriages over them; whilst at their thawing, many bridges, both of wood and stone, were borne down, and divers numerous watermills were broken up, and carried away. In 1092 in England, there was a terrible flood, followed by a great frost, followed by a second flood "as the like was remembered by none." Many bridges were destroyed.

In the winter of 1092 to 1093 CE there was a severe first in winter. English rivers were frozen so hard that horsemen and wagons could travel on them. When the thaw came, drifting ice destroyed bridges. In England, there were great floods, and afterwards severe frost. In England in 1092 or 1093, there were great rains, then a sudden frost and a ruinous flood after.

In 1093, there was a great frost in England. The River Thames and all the English rivers were so heavily locked in with ice that when the thaw came, bridges and mills were carried away. In England fell excessive rains, which raised such floods in 1093 as had not been known long before. All low grounds were flooded. After that came a sudden frost. The ice of the thaw carried down most of the stone and wooden bridges and water mills. In England, there was great famine and mortality. Was this a volcanic winter created by Grimsvotn in Iceland in 1090?

Also, in 1093 plagues and famine prevailed in France and Germany wherewith the poor being afflicted, vexed the rich with thefts and fires.

In Ireland in 1093 there were "Great rains and inundations in summer and autumn."

In 1094 CE there were great inundations throughout Ireland.

The year 1094 was called the rainy year. From October to April, the rain never ceased. This caused a plague and famine over England, France and Germany. In England, the famine was made significantly worse by King William Rufus's strangling taxes. As a result, there was so great a mortality that scarce did the living suffice to bury the dead. In 1094, there was the heaviest rains, famines and a desolating plague.

In 1094, there was a sea floods in the North Frisian countries in Schleswig-Holstein in Germany. This entry was listed under the category "storm surge", which might include an inundation of the sea.

The drought in 1094 in France was extraordinary. The summer of 1094 in France produced an extraordinary drought. The winter of 1094 in northern France was harsher than usual. The cruel winter raged for eight straight weeks and the severity of the cold froze animals and men.

Pope urban II instigated the First Crusade in November 1095 CE at the Council of Clermont in France following the appearance of massive meteor storms over Europe. These meteor showers which were observed right through Europe generated widespread eschatological expectations concerning a popular view of the Final Judgement that would eventually occur in the same way at the End of Times. These responses arose as a result of "great earthquakes in divers places" and of "stars in the sky that were seen throughout the whole world to fall towards the earth, crowded together and dense, like hail or snowflakes". "A short while later, a fiery way appeared in the heavens; and after another short period half the sky turned the color of blood". The earth was obviously going through a meteor storm.

In 1095 CE a great snowstorm struck Ireland on January 3. Multitudes of people were killed.

In England, the winter of 1095 was very severe. In England, there were terrible tempests in 1095 and excessive summer rains. Therefore, the corn grain and fruits in many places were not good. It was a late harvest. Most of the corn grain was not harvested before November the 10th. After the rains, there was a great intemperate of the air and a most severe winter. All the rivers were so frozen that horses and loaded wagons went over them. There was a terrible thunder and lightning storm in England. In England, there was a famine from summer rains, tempests and bad air. The rains took up in harvest, then came most pernicious frosts which caused dearth and famine in England. In England, there was a famine from rains in the summer. "Heavy-timed hunger that severely oppressed the earth." The famine was caused by "summer rains, tempests and bad air.

In 1097 CE in England, the winter was still very severe.

In Europe, the winter was very mild and produced many diseases. The large amount of rain that fell caused the rivers to overflow their banks.

The winter of 1097 in France was very mild, the sweetest in many years. The great flood of 1097 in northern France did not permit the fall planting.

In 1098 CE it was a very wet year across England. Contemporary reports suggest heavy rain events throughout the year rather than just concentrated in one season or a cluster of months. In England it was another oppressive year for endless taxes and gelds, and great rains, which scarce ever ceased; all low marshy grounds perished with floods and water. In England, there was a famine from rains. In England in 1098, there were tempestuous seasons, rains, corn rotten, low grounds perished. This was a repeat of 1095.

On the 3rd of November in 1099, "as well in Scotland as in England, the sea broke in over the banks of many rivers, drowning divers towns, and much people, with an innumerable number of oxen and sheep, when Earl Godwin's lands in Kent were covered with sands." In 1098 the land of Goodwyn Sands, Godwin's Sands, was swallowed up by the sea. This is also recorded as happening in 1100 CE. Godwin was Earl of Wessex.

In 1099 CE in England, there were rains and sea floodss, "fatal to much people and cattle." The Thames was much flooded on the festival of St. Martin (November 11). In England, the winter was very severe. There were great inundations both by sea and rivers, drowning many cattle, people and towns in England. There was a severe winter and a great dearth of grain. In England, there was famine from excessive, heavy, long rain and floods.

On 11 November 1099, there was a great inundation of the sea in England. A sea floods hit Kent and the Thames Estuary. A tidal flood affected the River Thames estuary and adjacent areas of North Kent. According to legend this inundation was responsible for the formation of the Goodwin Sands. "On the third day of the nones of November, the sea came out upon the shore, and buried towns and men very many, and oxen and sheep innumerable." "On St. Martin's mass-day, the 11th of Novembre, sprung up so much of the sea flood, and so myckle harm did, as no man minded that it ever afore did." In London on the Festival of St. Martin, the sea floods sprung up to such a height and did so much harm as no man remembered that it ever did before.

Across the English Channel the coast of the Netherlands was also hit by the same sea floods. In all 100,000 people drowned.

1100 CE

In 1100 CE Lolobau in Papua New Guinea erupted with a VEI of 4 and Akita-Komagatake in Japan had a VEI of 2.

From the Twelfth Century to the fourteenth century there was very heavy rainfall in the North Sea area.

Between 1100 CE and 1300 CE Europe had its warmest temperatures when dry summers and mild winters were the norm. This was more than likely Western Europe. There were benign climatic conditions in Europe and villages were developed on previously unworked land.

The MCA warm period helped create a climate suitable for farming. The warm period allowed farmers to grow crops at higher elevations and at higher latitudes. The MCA was the Medieval Climate Anomaly or Medieval Warm Period.

In the Balkans in Eastern Europe, there was pronounced cooling with average humidity.

Around Lake Titicaca on the Altiplano of Bolivia lake cores indicate that the fine organic layers produced by good rainfall disappeared abruptly around this time. On land, the raised field systems of the Catari vanished almost entirely within half a century. Only fields where there was locally high groundwater flourished.

In Canada Norse iron artifacts as trading goods have been found in Inuit dwellings on Baffin Island. Strands of Norse woven wool were also found that are the same as those found in the Western Settlement which was the northernmost Norse community in Greenland.

Viking trade goods such as non-native copper and iron, fragments of chain mail and carpenters' tools as well as boat nail rivets and fragments of woolen cloth were also found on Ellesmere Island in the Arctic.

The Merewether Impact Crater north of the treeline in Labrador was created in 1100 CE. and is 200 meters across. There is a string of three small craters near it. The Merewether Impact Crater is in the Torngat Mountains in Newfoundland. Canada. Had the Norse in western Greenland seen it falling? The Merewether Crater is around 700 miles west of Garda, the Norse capital of Greenland on the west coast.

In the 12th Century CE, the Song Dynasty lost control of its lands in the north to a Jurchen invasion. The dynasty did not fall but moved to the south and established a new capital at Hangzhou.

Also, in the twelfth century CE China lost large amounts of its population. CO^2 may also have decreased after forests reclaimed agricultural lands after the pandemics occurred and there was sharp population decline.

There was clearly some difference between the sectors of the northern hemisphere and the situation over east Asia where the climatic zones appear to have been pushed south over a long period of which the twelfth century was the climax. The swing to the southeast of the isotherms and of the flow lines of the circumpolar vortex from a northwards displacement, or ridge, over the Indian sector to a southwards displacement, or trough, over east Asia is a pattern which seemed liable to have introduced an anticyclonic tendency over Thailand and northern Indochina and therefore reducing rainfall there. There were dryer conditions over the Khmer empire in Cambodia and it was swallowed up by the jungle by 1300 CE.

On Dartmoor in England there were small communities of farmers where none were after the Medieval Warm Period or even in the twentieth century. This was the same in Northumberland on the Scottish border as well.

English vineyards prospered between 1100 CE and 1300 CE. Classic winemaking grapes fare badly at extremely cold temperatures. England was not extremely cold at this time.

Average summer temperatures in England were between 0.7° and 1.0° C warmer on average than the late twentieth century average when grapes are being grown again in England.

This was the peak period of English grape growing and wine making due to less liability to frost in May between 1100 CE and 1300 CE.

In York, England remains of *Heterogaster urticae* were found from this period. *Heterogaster urticae* or the stinging nettle bug, indicate that temperatures here indicate prevailing temperatures higher than the late twentieth century. Also abundant in York was the *Aglenus brunneus* beetle whose habit is to live in high temperatures generated by decaying vegetable refuse.

In 1100 CE there was much loss of life from a storm in the North Sea area.

In 1100 in England, there was a surprising great tide. In the spring, the River Thames rose up with such high tides that many towns were drowned. It did great damage to London and other places. In England in the year 1100, there was a long and severe winter frost.

In 1100 CE the sea overflowed 4,000 acres of Earl Godwin's land, in Kent, England since called Goodwin Sands. In the English Channel, Earl Godwin's lands, exceeding 4,000 acres, were overflowed by the sea, and an immense sandbank formed on the coast of Kent, now known as the Goodwin Sands. There are various dates and explanations for this event. You have already met it once before in the previous year.

In 1100 CE a terrible famine struck Europe.

In Central Europe average temperatures were 1.0° C to 1.4° C warmer than current temperatures in Central Europe in the late twentieth century.

Between 1100 and 1300 CE the May frosts that had plagued growing crops for centuries were virtually unknown in Europe. Warm summers and mild winters allowed people to take risks with planting marginal lands and at higher altitudes where previous cold temperatures would have precluded any cultivation. A growing farming population spread northward and uphill. The growing season for cereals was three weeks longer. And summer unsettled weather would begin in June and extend through July and August in the harvest days.

The winter in 1100 in northern France was excessive.

In Africa in this period there was more flow of streams in the wadies of North Africa and Arabia.

In India there was more rain in the dry northwest that was adequate for the population.

These climatic features were possibly explained partly by a displacement of the anticyclone belt of the desert zone during the warm epoch north of its present usual positions to an axis from the Azores to Germany or Scandinavia as in some of the fine summers in the late twentieth Century. Such partly meridional wind circulation patterns, with a cold trough deformation of the circumpolar vortex, commonly thrust cold surface air south over Eastern Europe and western or even central Asia, and from there it would be deflected by the mountains westward and southward toward the Mediterranean. This is an eastern position for such a development in the circumpolar vortex, requiring a longer wavelength, or spacing of the troughs and ridges, than commonly prevails in the upper wind flow from the more or less fixed disturbances over North America caused by the Rocky Mountains. Such a longer wavelength would be likely to occur at a time when the main flow of wind was displaced towards higher latitudes and particularly when, as in the thirteenth century, Arctic cooling strengthened the thermal gradient and the winds.

In the 1100s there was an early cooling period which was a general feature of the high Arctic and was reflected in a sharp increase of Arctic Sea ice between Iceland and Greenland.

In the Twelfth Century CE Norse settlements in Greenland were enjoying considerable prosperity when the summer pack ice frontier lay far north of Iceland and voyaging conditions in the North Atlantic were relatively easy during the summer.

The Norse in Greenland relied on a dairy economy which involved growing hay for winter fodder. This required a cool but equitable climate. No ice-covered wasteland here.

The waters west of the Norse settlements in Greenland were at least as warm as the warmest periods of the late twentieth century. This is evidenced by the presence of the abundance of cod in this period. The bones were found in the middens of the inhabitants of Greenland.

Further to the south off Cape Farewell in Greenland there was possibly an even warmer water anomaly. It is recorded in Norse sagas that at least one person swam in the fjordwaters and survived. This would need a minimum temperature of 10° C for a person to survive. This is 4° warmer than mean temperatures in the late twentieth century.

Old Norse burial grounds have been found in deep ground, whereas in the late twentieth century the ground is permanently frozen.

There were dry periods as well as a big chill in Iran in the Twelfth Century CE.

In Italy in the 12th century, the rivers Erminio and San Leonardo were navigable, which is now impossible to navigate even with the vessels of that time. Very long bridges were built in

Sicily. The one at Oreto in Sicily at Palermo is much longer than necessary for the rivers in the late twentieth century.

In Greece there was more general flow of the streams then the late twentieth century.

In the eastern Mediterranean there was growing dryness from the twelfth century on.

After 1100 CE climatic conditions in the southern lowlands of Mexico became more humid but Mayan civilization never recovered from the last great drought. It only survived in Yucatan thanks to natural sinkholes called cenotes in the limestone that enabled people to reach the subterranean water.

Between 1100 CE to 1400 CE in Norway spruce trees spread across Trondelag to Trondheimsfjord in Norway.

Farmers grew wheat as far north as Trondheim in Norway and settlements reached farther north in Norway during the MCA.

In Sweden settlements extended into areas previously inhabited by Sami who were semi-nomadic reindeer herders of Northern Scandinavia. They are called Lapps in English.

By the end of the 1100s farms were already being abandoned in northern Norway, accompanied by an expansion of areas used by Lapp hunters and a drift to the south and towards the coastal fisheries.

In Scotland Kelso Abbey had well over 250 acres under cultivation at an altitude of over 980 feet. This was well above late twentieth century limits. Fourteen hundred sheep and sixteen shepherd's households thrived on the Abbey's land. It is impossible to do this now.

In the Swiss Alps farmers planted crops deep in mountain river valleys that had been covered with glaciers two hundred years earlier.

So, are we in a cooler period now than in the MCA?

Around 1100 CE a strong El Nino struck the area along the Lambayeque River in northern Peru. Deep floodwater cascaded through the great city of Batan Grande and others carrying everything before them. It was total devastation. Instead of rebuilding the inhabitants piled up wood and brush and burned the pyramids and urban complexes to the ground. Legend tells us that the last ruler, Fempellec had moved the sacred green idol from its ancient resting place and the offended gods retaliated with heavy rains which fell for thirty days and nights. Enormous floods swept away the people's fields. In savage anger, his subjects bound Fempellec's hand and foot and cast him into the Pacific Ocean.

In northwestern Iowa the Indians of the Mill Creek Culture grew corn (maize) in an area which is somewhat marginal as regards enough rainfall for the crop. Elk and deer, both woodland animals, which they hunted accounted for most of the flesh in their diet before 1100 CE. After the 12th Century the proportion of these among the bones in the middens rapidly declined and was overtaken by bison, an animal of the open plains. The abundance of bison bones increased towards the west where climates are drier, in the rain-shadow of the Rocky Mountains.

Tree rings tell us that in 1100 CE persistent and increasing droughts hit Chaco Canyon. Within half a century the Great Houses of Chaco Canyon were deserted.

On the heights in California the tree ring record indicates that there was a sharp maximum of warmth, much as in Europe, between 1100 and 1300 CE.

Around 1100 CE Mono Lake in California rose abruptly by 62 feet. This was during a very brief wet period when the rainfall was higher than in any year in modern times, the fourth highest of the past four thousand years.

There were extended periods of high rainfall in Southeast Asia in the twelfth century.

There was a one-hundred-year long drought in Uganda There were other one-hundred-year droughts starting in 1550 CE and 1750. Sediments from Lake Kitagata and Lake Kibengo in Uganda confirmed this.

In 1102 CE there was drought in Britain accompanied by excessive heat. In England, there was a drought with excessive heat. It was an excessively hot summer.

In 1102 and 1116, high flood waters swept over a good portion of the land of the island of Helgeland which fell down and was thrown over washed away. Helgoland is Heligoland which was an archipelago north of Germany in Schleswig-Holstein.

In August 1103 CE on St. Lawrence's Day, 10th August, a great storm occurred in England which did much harm. The storm occurred in the morning and did more damage than anyone could remember. There was considerable stress to agriculture that year which suggests both cold and wet conditions consistent with a significant storm turning in the summer months.

In 1104 CE Hekla in Iceland erupted with a VEI of 5. After this eruption Cistercian monks wrote that Hekla was the Gateway to Hell.

There was a very great snow in February 1104 and great floods on the land. There was a mighty scarcity of corn grain and dearth from endless taxes and wars in England.

In autumn 1104 CE crops perished in England. This happened again in 1105 and 1109. It was very thundery across England which implies cooler than normal conditions aloft. This would be associated with a dust veil cooling of the atmosphere. This was probably the result of the eruption of Hekla in Iceland that same year which affected the climate in the immediate and downwind or European area for several years.

And we believe that we can control the weather?

What is it about the gradual change of environment? Seems rather fast to me.

In 1106 CE in England, there was an inundation from the sea. On the first day of the Ides of June in England, there was a great earthquake and several inundations of the sea. This event may have been a tsunami caused by a massive earthquake/landslide. Men, cattle, grain, lands and buildings suffered much from thunder, lightning, rain, hail, high winds and tempest. Grains and fruits were beat down and broken. Barrenness of land from inundations, dearth from scarcity, plague from famine, all prevailed. In England, there was famine from barren lands and then plague.

On August 29th 1108 CE Mount Asama, or Asamayama, on Honshu Island in Japan erupted with a VEI of 5. This was called the Tennin eruption. This was a Plinian eruption and contributed to extreme weather that caused severe famine, torrential rain and consecutive cold winters in Europe. Rice paddies and fields could not be farmed due to being covered in thick ash.

In 1108 CE a great part of Flanders, now Belgium, was overflowed by the sea. Most of Flanders was drowned by the sea. The town and harbor of Ostend were totally immersed. In Flanders, a terrible inundation forced many of the inhabitants to leave the country. Some settled in England. Nearly the whole of this country is believed to have been covered by the sea in early times. Ostend is now located in northern Belgium.

In England, the year 1109 was remarkable for thunder and lightning.

In 1109 in France, particularly in Orleans and the province of Chartres, fever and sickness struck causing some deaths. Excessive rainfall drowned the crops. The grape harvest was almost a total failure. In areas where both the grain and grape harvest were affected, terrible famines decimated the population everywhere. For three consecutive years from 1109 to 1111, there was a terrible famine in France.

In Russia, the water was high in the Dnieper, the Desna and the Pripet Rivers. The Dnieper River runs through Belarus, southern Russia and Ukraine. The Desna River is a tributary of the Dnieper River. The Pripet River runs through the Ukraine and Belarus and is a tributary of the Dnieper River.

In 1110 CE Irazu in Costa Rica erupted with a VEI of 3.

In England on 5 May 1110 CE, there was a great frost, which killed the blossoms of the trees. The River Trent was dry at Nottingham for 24 hours. There was tempest pernicious to corn grain and destruction of all fruits. The people over all England were afflicted with sore diseases, especially an epidemic of Erysipelas a type of skin infection wherein many died, the parts being black and shriveled up. Nottingham is located in Nottinghamshire in the east Midlands of England. In 1110, the River Trent at Nottingham, England was dry from morning until 3 p.m., for a mile (1.6 kilometers) in length, so that it could be passed with dry feet. In England, there were many tempests in the year 1110.

Around 1110 CE there was the start of a colder, drier period in Greenland.

In 1110 to 1111 CE there was a long, hard winter and bad weather with both planted crops and trees severely affected in London and the south of England.

In England in 1111, there was a long and severe frosty winter, very harmful to corn grain. There was barrenness of land. There was great dearth, mortality of people, a grievous murrain death caused by highly infectious disease of cattle, and the death of fowls. All tame fowls fled to the woods and fishes died in the water. In England, there was a famine from frost and barren land.

On March 9th 1112 CE Kirishimayama in Japan erupted with a VEI of 2.

On February 27th 1113 CE Kirishimayama in Japan erupted with a VEI of 2.

In England, there was a drought, and it was "so hot that grain, and some forests of wood, took fire. In England, the heat was in June 1113, so strong that the crops and even forests fell into fire.

In 1113 CE there was a sea floods that caused another mass emigration of the population from Flanders in Belgium. It is a wonder that there was anybody left there. By the breaking in of the sea, a great part of Flanders now Belgium was drowned; whereupon a great number of Flemings fled to and became subjects of King Henry the 1st of England for some place to inhabit. And he gave them Pembrokeshire in southwestern Wales, where their posterity remains to this day.

At Parma in northwestern Italy and Ravenna in northeastern Italy, it rained blood both in the towns and fields. It was the same in Emylia Emilia in northern Italy: so excessive was the heat of this month that corn and some woods took fire and burnt. After this the people were afflicted with grievous and long diseases, especially dysentery, and a most destructive plague.

In England on 11 November 1113 or 1115, there was a tempest and hurricane.

In 1114 in England, the tides went out instead of coming; hence the rivers dried up. In 1114 in England, the drought that began in 1113 continued and there was a great want of water.

In 1114 Irish texts state that there was very severe weather with frost and snow from the fifteenth of the kalends of January to the fifteenth of the Kalends of March, or a little longer. This made great havoc of birds and cattle and people; from which rose great scarcity and want through all Ireland and in Leinster especially.

On 4 April 1114, the River Thames in London, England was so dry that children waded over the river between the bridges and the town. Under London Bridge, the water was only knee deep. The river was dry for two days.

In 1114 in Germany, the Hever Strom rose so high that the built up "Capell" by the grandstand was washed away by the water. The Hever Strom is a channel located in Schleswig-Holstein in northern Germany.

On 23 April 1114, there was so great a snowfall in Flanders that it broke down the trees in many places.

In May of 1114 in England, there was a great drought and want of water.

On 6 October 1114, the River Medway, in Kent, England was almost dry. The Saxon Annals give these two dates April 4 and October 6 as both occurring on October 10.

1114 CE was considered to be one of the driest years on record. On October 10th the Thames at London was so low that men and boys were able to wade across the river. This was the combination of a notably low tide and the drought. On 10 October 1114 in England, the rivers Thames and Medway dried up so that men could wade across on foot.

In October, a terrible hurricane struck destroying houses, villages and woods. The sea shrunk in from its old boundaries, seamarks, and ordinary heights that a man might have walked on foot on the dry sands a whole day. Great rivers which used to ebb and flow twice in 24 hours became shallow, that in many places people might safely walk over. The River Thames was so low, that horses, men and children passed over it between London Bridge and the Tower. And under the bridge, the water scarcely reached the knee, the whole day and night of October 15. The water level in the River Medway at Kent was so low that the smallest vessels could not pass in the midst of the channel.

In October and November 1114 CE there were great winds in October and a terrible storm upon the night of November 18th which damaged buildings and trees in Britain.

On 11 November 1115 there was a most destructive hurricane. There were many storms and a great death of cattle this year. The winter of 1115 in England was very cold. The English Channel froze. Even stones were split in two from the cold.

During the winter of 1114 and 1115 CE there was an outstandingly severe winter. The frost lasted for 9 to 11 weeks and nearly all of the bridges in England were damaged by ice.

The winter of 1115 to 1116 CE was also notably severe in the British Isles. In England there was a great frost; timber bridges broken down by weight of ice. This year was the winter so severe with snow and frost, "that no man who was living ever remembered one more severe; in consequence of which there was great destruction of cattle" The winter was most severely cold, with great frost and snows. At the thaw, most of the bridges in England were broken and carried down.

From 1116 CE to 1117 CE it was a year of excessive rains. Heavy rains from August to Candlemas ruined crops in Britain. In England, this was a sad rainy year. The summer began with terrible thunder and lightning, which did great damage. The rains began on 1 August and continued until the Feast of Candlemas on 2 February. There was great destruction of corn grain and all fruits.

In 1116 in Ireland, there was a great famine, "during which the people even ate each other."

In 1116, high flood waters swept over a good portion of the land of the island of Helgeland which fell down and was thrown over washed away. Helgeland is Heligoland.

In 1117, Hopei (now Hebei province) in northern China at Ho-chien and Ts'ang experienced extreme flooding with over a million people drowned.

In Jerusalem in 1117, there was famine. In May was a great plague of locusts at Jerusalem, which ate up the herbs, trees, vines and sown corns. Plagues of locust can be triggered by weather conditions.

On 1 November 1117 CE there was a great tempest of thunder, lightning, clouds and hail in England.

In 1117 CE there were heavy rains all year which was disastrous for corn. On December 1st there were violent storms with hail in England. In England there was a famine from tempest, hail, and a year's incessant rains. There was a scarcity of corn grain from great hail and tempest and incessant rains which lasted almost the entire year. Most bridges in England were broken down by floods and rains.

The town of Wells in Somerset, England was partly burned down by lightning in 1117 CE.

There were several other terrible and fatal tempests in December 1117 CE in many places at different times, as at Leodium, now Liège, in Belgium.

In 1117 CE torrential flooding along the River Arno in Italy swept away the original Roman-era wooden Ponte Vecchio, which straddled the river in Florence.

On May 14th 1117 CE there was a sign by thunder at ten o'clock during evening service in St. Sophia in Novgorod Russia; one of the chanters, a clerk, was struck by the thunder, and the whole choir with the people fell prone and knocked unconscious, but the people remaining were alive.

In December 1117 CE there apparently was a very great wind on St. Thomas' Day, 21st December. It did damage to houses and trees. On the Feast of Saint Thomas December 21, there was a great hurricane in England.

During Epiphany Week, Epiphany celebrated on January 6 1118, there was great thunder and lightning, which killed many.

In February, there was tempest, thunder, lightning with great hail and rain in England.

On May 7, 1118, a severe frost destroyed the grape vines in most areas but especially in Auxerre located in the Bourgogne region of northern France.

On 21 December 1118, one of the strongest gales in living memory, struck in Normandy, France causing terrible havoc.

Between 1118 CE and 1167 CE there was a savage drought in the Southern Sierra. This was one of the four warmest periods that occurred in a major drought in Southern California.

In 1119 CE there were constant inundations for so long that corn grain could neither be sown, nor reaped, not only in Poland but also in its neighboring countries.

In England, there was a violent tempest the whole day of Christmas 1119 CE.

In 1119 during the previous winter, torrential rains fell in France. High floodwaters swept through homes in Rouen and Paris. Crops were damaged by the raging torrent of the overflowing Seine River. The winter of 1119 was extremely wet in France. The Seine River overflowed in Paris. The floodwaters swallowed up homes.

From 1119 CE to 1139 severe drought affected Mongolia.

Then during Lent 1120, a very strong wind blew over the Seine River and dried it up. One could cross from bank to bank on foot if he had the courage.

In 1120 CE the monk and historian William of Malmesbury described travelling through the Vale of Gloucester in the West Country in England. He described seeing vineyards that were the best in England. The grapes were planted out in the open, trained up on poles and not protected against cold winds by strategically placed walls. There were numerous vineyards in England at the time. These were much further north than the northernmost vineyards in France and Germany in the 1960s. During the 12th and 13th centuries England had such a temperate climate that her merchants exported large quantities of wine to France whose growers complained loudly.

In 1120 CE there was a famine in Jerusalem, Israel not caused by weather. In Jerusalem in 1120, a famine was caused by a "plague of mice and locusts."

In July 1120, there was a horrible tempest of hail at Treves, now Trier, in the Rhineland-Palatinate in Germany. It overthrew many buildings. It did much damage at Halberstadt in Saxony-Anhalt in Germany, so that the ground in a nine miles radius bore no corn. It killed most small birds and oxen. In Germany, the wolves tore apart and destroyed many people. Halberstadt is around 240 miles northeast of Trier.

In 1121 CE in England, all three spring months were dry with excessive heat.

In England in 1121 or 1122, there was a drought, and all three spring months were dry and had excessive heat. Also, in 1121 or 1122, there was a "great famine from long and cruel frost."

In England, after the Nones of April the rains ceased and a dearth ensued, the corn being parched in the ground from the excessive heat and drought of three spring months.

On 18th October 1121 a violent northeast gale did much damage in London.

During the winter of 1121-1122 in England, the frost killed the grain crops, "and much people and cattle;" and famine followed.

In England, on December 25th 1121 there was a terrible and general hurricane. Soon after there was a severe winter which not only killed the sown corn grain but people and cattle; hence a famine.

In 1122, England experienced the greatest dearth. This famine was made worst by "taxes, and Danegelt, endless." Danegelt or Danegeld was a land tax originally levied in Anglo-Saxon England during the reign of King Ethelred to raise funds for protection against Danish Invaders.

Around the year 1122 there was a drought in the former province of Quercy, France. As a result, there was a religious procession to pray for rain.

In 1123 or 1124 in England, France and Germany, there were famines from terrible weather and the greatest plague.

In 1123 or 1124, there was a plague from great snows and frost, intemperate air, to March. Then changes to hailstorms, snow, rain and frosts.

In the Winter of 1123 /1124 CE in England there was a very great and destructive snowfall.

In 1124, a terrible plague and so great a famine afflicted Germany, that the third part of the people died; and scarce were there survivors to bury the dead.

In 1124 the vine and fruit trees in France were killed by the cold.

In 1124, a very hard winter occurred in Germany and many people and especially poor people froze to death. The birds in the air and the fish in the ponds were cold. Even the winter wheat in the field was completely frozen out.

In 1124 CE the many tempests in England were pernicious to corn grain and all fruits, so that at Candlemas on 2 February, they were sold at a great price. The famine was made worsebecause of the "scandalous adulteration of money and grievous taxes". In 1124 in England, "Such a famine prevailed that everywhere in cities, villages and crossroads lifeless bodies lie unburied." "By means of changing the coine all things became very deere, whereof an extreme famine did arise and afflict the multitudes of people, even to death."

After Whitsuntide, White Sunday - Pentecost, a sharp frost killed the trees. At Pentecost was a hard frost, which did harm to fruit trees and grape vines. Pentecost is fifty days after Easter Day.

In 1123 or 1125, terrible was the famine in England so as in towns, villages and highways, dead bodies lay unburied, dissolving into stinking slime. In May trees scarce budded, the ground was so chilled. In 1124, there was so great a dearth in England that a horse load of wheat sold for six shillings.

In 1124 there were great inundations at Rome in Italy and a famine so great that multitudes of both sexes died of hunger.

In 1124, the frozen Rhine River in Germany was crossed by pedestrians. This winter of 1124-25 was harsher than usual, because of the accumulation of snow that fell incessantly. A significant number of children and even women died from the extreme cold. In ponds, the fish were trapped under the ice. The ice was so thick and firm that loaded wagons and the horses traveled on the Rhine River as on the mainland in Germany. A strange incident occurred in Brabant: Countless numbers of eels were driven by the cold from the swamps and found refuge in barns, where they sought to hide; but the cold was so great that they died from lack of food and rotted. The cattle died in many areas. The bad weather was prolonged so that only in May 1125 did trees begin to bud and the grain and other cultivated plants begin to grow.

Winter in 1124 to 1125 CE was exceptionally severe in France and the Netherlands. This would have probably affected Britain, especially in the south and southeast.

The year 1125 was a very hard winter and snow in Germany. Many poor people traveling, the birds in the air, the grape vines, and the trees froze. The fish in the ponds died because of the thick ice. The choughs bird - member of the crow family crawled into the haystacks and froze. In June, the Wednesday after Pentecost, there was a big snowstorm especially in the Bohemian Mountains. Then the following Sunday came a big frost that froze all streams so hard that they could be crossed on foot. Then throughout the whole country, this was followed by a cruel famine, and death.

In the famine of 1125 in Germany, one half of the population died of hunger. During the Middle Ages it was the custom of the city authorities to drive the poor of the cities outside of the city walls in time of famine, where they were "left to die and devour one another".

The winter of 1125 was cold and very long. There was extraordinary cold in Germany, France and Italy. Frozen rivers were passable on horseback. The leaves appeared on trees in May. The grape vine and fruit trees in France were killed by the extreme cold.

In 1125 in France, the winter had cold more severe than usual and was accompanied by a large amount of snow. Alternating snows, rains and frozen juice freezing rain/sleet continued until March. Then continual rains destroyed all the seeds. During the winter, far greater and more frequent snows than ordinary fell whereby many poor people's children were killed, as were the fishes in ponds, even eels themselves. After this followed a great plague on man and beast, and great intemperature of air, even till March. From the variety of weather, snow, rain, hail, frost, etc. came great damage. The spring came on slowly from cold nights, and daily heavy stormy showers. All seeds were drowned. Hence a plague in France.

During the winter of 1124-1125 in France, the thick ice on the rivers could carry loaded wagons. Many children and women died from the cold. Alternating thaws, rains and snows gave way to very severe cold that lasted until the middle of March. The trees did not begin blooming and the earth was not covered with greenery until the month of May.

During the winter of 1125 in England, the frost was "so intense that the eels were forced to leave the water and were frozen to death in the meadows." It was the dearest, scarcest year known for wheat. In 1125 in England, there was a famine from excessive rains that were incessant all summer.7

In 1125 CE in the Novgorod Republic, now part of Russia, there was a great storm with thunder and hail; it damaged houses and tore tiles off shrines; it drowned droves of cattle in the Volkhov River, while others hardly saved their lives.

In 1125, there were excessive constant daily rains for the whole summer in England. Hence the most terrible famine through the whole nation on man and beast. On St. Lawrence's Day was such a flood, as drowned many towns and much people. It carried down bridges, destroyed corn grain and meadows. A plague accompanied the famine, and the weather was so bad that it destroyed the corn and all fruits, as none living ever saw before. In England a sextarius of wheat sold for 20s shillings." St. Lawrence's Day is on 10th August, and many towns, bridges and lowland crops were ruined in England. This suggests the culmination of excessive rainfall for at least a couple of months beforehand. Many villages were destroyed in southeast England by a sea floods as well.

In Germany and Italy raged famine and pestilence.

The summer of 1125 was rainy and damp in France. This led to poor harvests and cruel distress famine. In 1125 in France, continuous rain took away almost all the seeds after the month of May.

In the Winter of 1125-1126 CE the winter was again severe and the spring unhealthy. In France there was a great famine. The harsh winter of 1126 in northern France lasted six weeks. The grape vine and fruit trees in France were killed by the cold.

In Bohemia, now western Czechia, the trees burst from the extreme cold and the rivers were frozen over in 1126.

In 1127 CE in Novgorod a blizzard fell thick over land, water and houses during two nights and four days. The water was high in the Volkhov River, and snow lay on the ground until St. James Day on May 1. In the autumn the frost killed all the corn and the winter crop; and there was famine throughout the winter; an osminka about 11½ pecks of rye cost half a grivna a circular ingot of silver.

In 1128 CE the famine that began in 1127 carried over and intensified. Becoming very cruel in the Novgorod Republic now part of Russia. One osminka about 11½ pecks of rye cost a grivna which is a circular ingot of silver. The people ate lime tree leaves, birch bark, pounded wood pulp mixed with husks and straw; some ate buttercups, moss, horseflesh. Many people dropped dead from hunger, their corpses lay in the streets, in the marketplace, and on the roads, and everywhere. They hired hirelings to carry the dead out of the town; the serfs could not go out; woe and misery on all! Fathers and mothers would put their children into boats in gift to merchants, as slaves to the overseas merchants, who came up the river Volkhov from the Baltic to Novgorod in boats or else put their children to death. Others were dispersed over foreign lands. Thus did our country perish. This year, the water was high in the Volkhov River and carried away many houses.

In 1128 CE there was a severe winter with heavy snow on Easter Day, 22nd April that year. It would be remarkable if heavy snow fell in late April in the early 21st century.

Between 1128 CE and 1437 CE wine was produced in eastern Prussia at 55 degrees north, as well as in southern Norway. The Black Forest in Germany had vineyards up to 780 meters above sea level.

At this time summer temperatures were 1.8° to 2.6° Celsius higher than those in Central Europe in the 1950s.

In 1130 CE Cotopaxi erupted in Ecuador with a VEI of 5.

In 1130 CE in England, there was the greatest drought and hottest years.

In 1130 there was a great famine in Rome, Italy.

In the famine in France during 1130-1132, one man butchered and ate 48 people.

The largest community in Chaco Canyon was Pueblo Bonito which stood five stories high along its rear wall and remained in use for more than two centuries. At its peak in the eleventh century Pueblo Bonito had at least 600 rooms in use and housed about a thousand people. In 1130 CE fifty years of drought hit Chaco Canyon. Maize yields plummeted and within only a few years Pueblo Bonito emptied. At the end of the fifty-year drought Chaco Canyon was virtually deserted. The Anasazi or Ancestral Pueblo had moved away and possibly settled with relatives living in places with better water supplies.

Between 1130 CE and 1180 CE there was major drought in New Mexico. Chaco Canyon was the center of at least seventy communities that were dispersed over 65,000 square kilometers of northwestern New Mexico and parts of southern Colorado. It was an intensively important and sacred place with ceremonial trackways leading to it. In 1130 CE fifty years of drought settled onto the region and the canyon and the Chaco people had only one avenue of survival. That was to move. Within a few generations the great houses were empty and well over half of Chaco's population had dispersed into villages, hamlets and pueblos far from the canyon. Almost everyone was gone. To survive you had to stay mobile.

1131 CE in England was the greatest drought and hottest years.

In 1131, there was a great famine in Rome, Italy.

In 1131, there was so great a drought in France that all the lakes, rivers, springs and wells dried up. This year and some after, was so great a dearth of domestic animals, as few survived. Oxen died so fast that out of 10 yoke, not one was left; and of every 200 or 300 swine, scarce one remained alive. Fowls also died, hence a great dearth of flesh animal & fowl, butter, cheese and eggs.

During the winter of 1132/1133 CE the cold was so intense in Italy, that the Po River was frozen from Cremona (in northern Italy) to the Adriatic Sea. The wine froze and burst the casks, and the trees split with a great noise.

During the winter of 1133 the Rhône River in southern France froze as well as the wine cellars froze. Wine froze in the cellars. Since the water in tree sap acquires greater volume when it freezes in extreme cold, trees burst apart with a loud noise. In Strasbourg in northeastern France more fruit trees burst when the cold reaches -16° Reaumur (-20° C, -4° F). A great number of trees in France burst in the winter of 1133. Heavily loaded carts passed along the frozen Rhône River in France on the 5th of January as they would on the mainland. The Rhône River in France was frozen, and people crossed it on the ice.

In the year 1133 in Italy, there was a very severe winter. A tremendous amount of snow covered the roads, rivers and streams were all frozen, everything, even the wine was frozen, and the oak and walnut trees were split, with a crash and were torn, and the olive trees and vines withered. This produced a very terrible shortage of food, which forced, the following year, the inhabitants of the area of Padua, to feed on grass.

In 1133 CE in France, there were great floods from rain. In 1133, there were severe rains in France and a great intemperature of air.

On 2 August 1133 in Germany, there was an eclipse darkness of the sun of the type that the sun was completely covered. This was a solar eclipse. Following this for an entire month was unusual and unstable weather. The effects only began over a half a year and lasted two whole years. Without a doubt a part of the effects was the powerful hot summer of 1135. Due to great heat much water, lakes and ponds almost dried up and many forests caught fire because of the great heat and there was much loss of timber.

Winter of 1133-1134 CE On 28 December 1133, a blizzard struck Normandy France. "On Innocents' Day very heavy snow fell, which covered all the face of the earth and blocked the doors of houses with its drifts so that next day men and beasts could scarcely leave their shelters or find any means of providing for their needs. Many of the faithful never entered a church on that feast day, and the priests themselves in many places were totally unable to cross the thresholds of the churches because their way was blocked with snowdrifts. After six days the wind veered to the west and the snow melted, suddenly causing great floods. The rivers, swollen by the snow water, burst their banks and caused widespread loss to some people and gain to others. In villages and towns nearby the floods rose to the roofs and drove people from their homes. Large stacks of hay were swept from the meadows, and barrels full of wine and other container vessels were carried away with all kinds of precious belongings."

Was all of this rain and cold the consequence of Cotopaxi erupting in Ecuador with a VEI of 5in 1130? Remember that VEI 8 is the maximum for eruptions.

In 1134 at Cashel in south-central Ireland, there was a great hail shower.

The extreme drought of 1134 in northern France caused the failure of oats, barley and vegetables. In June 1134 in Normandy France, blazing heat scorched the earth for fifteen days. Streams and pools of water dried up and flocks and herds suffered terribly from thirst. "Then one Saturday a great number of thirsty people plunged into the waters to cool themselves, and many in different places were drowned in the space of a single hour. In the region around us of which we were well informed thirty-seven men were drowned in the waters of pools and rivers."

In the first week of September 1134, many cities and villages in Normandy France were burnt to the ground including Le Mans, Chartres, Alençon, Nogent in Perche and Verneuil. The cause of the fires was undetermined and attributed to the flames of God's wrath.

In September or October 1134 in Flanders now Belgium, "one night the sea poured over the land and, spreading rapidly for seven miles, overwhelmed churches and castles and cottages alike, and involved countless thousands of men and women of every order and rank in a common catastrophe." "The flood swept away all, fair and ugly, men and women, alike and, choking them with water, quickly dragged them down to death." "One poor woman, who had recently given birth to a child, heard the roar of the raging water and leapt from her bed terrified, but she kept her head, snatched up the baby and a hen with its chickens, and quickly climbed on top of a haystack that was outside her cottage. The force of the rushing water, which was submerging everything, lifted up the hay and carried the stack for a long distance, eddying to and fro. So, by the pity of merciful God, the woman was saved from imminent death and snatched to safety by heaven, with the few poor possessions she had with her." "A twelve-year-old boy told me that on that occasion he climbed quickly on to the gable of the roof and there escaped death; his father and mother, however, who were lower down, perished." The sea broke in on the land and overflowed a great part of Flanders and the neighboring countries, killing many people and cattle. This was a rainy year.

In October 1134 CE a significant storm affected the Dutch coastal community and possibly also the English side of the North Sea.

On December 31 1134 CE in the Novgorod Republic, which is now part of Russia, bad weather set in; frosts and blizzards were very terrible!

In France, the winter was rough and long. For the fruits of the soil, this year was very unfavorable. As a result of these unfortunate events, a terrible famine gripped the land.

On the 9th of January and 16th February 1135 CE, sea floods hit northwest Germany. One hundred thousand people died.

In the area around Prague, Czechia, during the weekend of Pentecost (around the 20th of May) in 1135 CE, a thick snow fell in some wooded areas. The next day there was a very lively cold. The frost damaged the crops of every kind. The weather especially damaged the autumn planted crops and the grape vines. The cold also destroyed a large number of trees, as well as bushes, which were frozen down to the roots. The cold froze gently running water streams.

1135 CE in France and England, there was a great drought.

In 1135 in Europe, the heat and drought were extremely high. The pastures and the crops were scorched, and it was followed by a great dearth and famine. The rivers and springs dried up. The heaths of the mountains small shrub with tiny evergreen leaves and pink or purple flowers and the dry forests caught fire, allegedly from the glow of the sun's rays.

The Rhine River in Germany was almost completely dry and could be crossed on foot in several places. In 1135, so great a drought and heat, that all grass and corn grain were burnt up. Dearth and a great famine followed. Rivers and springs were dried up. Mountains and woods were burnt up. And many places were said to be set on fire by the sun. The Rhine River was so dried up, that one might safely ford it in any place. In the year 1135, there was a mighty fierce summer in Germany. Due to great heat much water, lakes and ponds almost dried up and many forests caught fire because of the great heat and there was much loss of timber in the powerful hot summer of 1135. Due to great heat much water, lakes and ponds almost dried up and many forests caught fire because of the great heat and there was much loss of timber. The water in large lakes, streams and ponds dried up. Many forests were burned up with great heat. The summer of 1135 was hot and dry.

"Burnt earth" was everywhere in France.

On 28 October 1135, a violent wind sprang up during the night in Normandy, France. The howling was dreadful. It carried away the roofs from countless homes along with churches and high towers. It cleared woods by tearing down a great number of trees.

On 28th October 1135 CE sea floods hit Holland.

On 28th October 1135 CE sea floods hit North Germany.

On 1 December, there was such dreadful thunder and lightning, which was very uncommon in England in the winter. Then came a tempest or hurricane.

From 1135-37 in England, there was a great drought and famine.

Between 1135 to 1137 CE the weather in Britain was relatively dry with one exceptionally dry year in 1136 CE.

In Britain there was a particularly fine, hot and dry summer in 1136 CE.

At the summer solstice in 1136 in France, the weather was unusual due to the intense heat. This heat had a disastrous effect on people, the flocks and the fruits of the soil.

On 27 October 1136 in France, there was a wind so violent that it knocked down many buildings.

The waters of the English Channel overflowed and swallowed part of Flanders now Belgium with its inhabitants.

Analysis of Silver pine tree rings in New Zealand indicate that there were above average temperatures in New Zealand between 1137 CE and 1177 CE. Temperatures were 0.6 to 0.9 degrees Fahrenheit above the twentieth century average for the region in the 1950s.

1137 CE In England, there was a general drought, with great heat: hence famine.

In Europe, the summer of 1137 was very hot and dry. The navigable rivers were so dry that they were crossed on foot in some places.

In France, the springs and wells gave no more water, and many villagers were dying of thirst.

In the midst of this consuming heat, several towns were burned on the same day, among others the cities of Mainz (Mayence) and Speyer in Germany. From excessive heat, many towns took fire and were totally burned down as Moguntia (Mainz, Germany), Spira (Speyer in southeastern Germany), on the same day.

Underground fire appeared in Italy for three years and it was this year that Vesuvius erupted. The ordinary state of the waters was finally restored in the year

The drought of 1137 in France broke out in March and lasted until September. The drought caused wells, springs and rivers to dry up. In 1137 in northern France, the summer heat was stifling. It burned and overwhelmed. The summer 37 was dry from March to September in France.

Tree ring data shows that there also was a drought in Algeria during this time.

The summer of 1137 in England was exceedingly hot and droughty. Navigable rivers were so dried up in many places that they might be walked over on foot.

In 1137, the whole world, likely Europe, suffered a severe drought. Brooks ran dry. Pools and water tanks and cisterns dried up, and some rivers ceased to flow. Men and beasts suffered terribly from thirst. In some districts, men traveled seven leagues or 21 miles, 34 kilometers, in search of water. Some died of excessive heat while carrying water on their backs for themselves and their households.

In July and August 1137 in France, the burning summer heat scorched mortal men. This heat wave ended on 13 September. Many kinds of pestilence struck people down.

From this drought, and the inexpressible cruelties and barbarities of King Stephen's reign, arose a great dearth and famine in England in 1137.

The 1138 Aleppo earthquake was among the deadliest earthquakes in history. Its name was taken from the city of Aleppo, in northern Syria, where the most casualties were sustained. The earthquake also caused damage and chaos to many other places in the area around Aleppo. The quake occurred on 11 October 1138 and was preceded by a smaller quake on the 10th. It is frequently listed as the third deadliest earthquake in history, following on from the Shensi and Tangshan earthquakes in China. However, the figure of 230,000 deaths reported by Ibn Taghribirdi in the fifteenth century is most likely based on a historical conflation of this earthquake with earthquakes in November 1137 on the Jazira plain and the large seismic event of 30 September 1139 in the Transcaucasian city of Ganja.

The worst hit area was Harem, where Crusaders had built a large citadel. Sources indicate that the castle was destroyed, and the church fell in on itself. The fort of Athareb, then occupied by Muslims, was destroyed. The citadel also collapsed, killing 600 of the castle guard, though the governor and some servants survived, and fled to Mosul. The town of Zardana, already sacked by the warring forces, was utterly obliterated, as was the small fort at Shih.

The residents of Aleppo, a large city of several tens of thousands during this period, had been warned by the foreshocks and fled to the countryside before the main earthquake. However, many people did not take the warnings of the foreshocks seriously and decided to stay. This mistake cost many people their lives because the next day (October 11) the main shock occurred which caused the collapse of many buildings, killing thousands of people.

The walls of the citadel collapsed, as did the walls east and west of the citadel. Numerous houses were destroyed, with the stones used in their construction falling in the streets. The cracks and holes in the foundations of the walls and buildings also caused further problems for the people of Aleppo. The holes allowed Crusaders and people from Muslim factions to invade the city, and another citadel in Aleppo was breached. Contemporary accounts of the damage simply state that Aleppo was destroyed, though comparison of reports indicate that it did not bear the worst of the earthquake.

Other reports claim that Azrab, which is north of Aleppo, experienced the worst of the damage. Reports claim that the ground split in the middle, swallowing the village. This was most likely the result of a landslide from the earthquake. Reports also state that the main earthquake and its aftershocks were felt in Damascus, but not in Jerusalem. Accounts of men being swallowed by holes opening in the ground at Raqqa were erroneously attributed to the Aleppo earthquake, and based on the confused late twelfth-century account of Michael the Syrian.

On June 1st 1139 CE Vesuvius erupted in Italy with a VEI of 3.

The 1139 Ganja earthquake was one of the worst seismic events in history. It affected the Seljuk Empire and Kingdom of Georgia; modern-day Azerbaijan and Georgia. The earthquake had an estimated magnitude of 7.7 M_{LH}, 7.5 M_s and 7.0–7.3 M_w. A controversial death toll of 230,000–300,000 came as a consequence of this event. Mkhitar Gosh, an Armenian scholar and writer, quoted Job 9:6 and Psalm 103:32 from the Holy Bible to describe the earthquake. He wrote of tremendous damage in the P'ar'isos and Xach'e'n districts of Syunik. The city of Ganzak was also devastated, leaving many of its residents buried under ruins. Many structures including monasteries and churches castles and villages in the mountainous region were totally destroyed. Strong shaking triggered massive landslides off the sides of mountains and canyons in the Caucasus Mountains region. Parts of Kapaz Mount collapsed, and the resulting landslide blocking the Kürəkçay River, forming Lake Göygöl. Another six lakes formed, including Maralgol and Lake Ağgöl. The number of people who died in the mountains is not known, described as "incalculable". Estimates of the death toll range between 230,000 and 300,000 making it one of the deadliest earthquakes in history. Among the dead were two sons of then ruler of the Seljuk Empire, Qara Sonqor. The death toll figure remains controversial with some authors stating it is an exaggeration considering the population of the area at the time of the disaster. Others argued that this was a conflation of information about the 1138 Aleppo and 1137 Jazira earthquakes. King Demetrius I of Georgia took advantage of the earthquake and looted the city. Troops stole many artifacts and prized items from the city, including the Ancient Gates of Ganja, which was utilized as a trophy. The city was reconstructed by Qara Sonqor, where it began to flourish.

The extreme cold period and severe drought ended in Mongolia in 1139.

In May 1141 CE at Welsburn, now Wellesbourne, in Warwickshire, there was a very violent whirlwind, tornado, that sprang up. It rose from the earth to the sky, a tempest of whirlwind and thick darkness tornado. A hideous darkness extended from the earth to the sky and the house of a priest was violently shaken and all of his outbuildings were thrown down and broken to pieces. Some forty houses were severely damaged and large hailstones as large as pigeon's eggs fell. One of these hit a woman and killed her, and possibly one other.

In 1141 a famine began in England, which lasted twelve years. It was a most dreadful and desolating famine and there was a long, rigorous, tempestuous frosty and snowy winter.

In the winter of 1141 CE, there was very cold weather with snow in December in England and also Europe.

During the winter of 1142-43 in England, after the rains was a very hard winter. The River Thames and other rivers were frozen; so, as men, horses and burdens loaded wagons might safely pass and repasts on the ice. The earth was covered with thick deep snow. In 1142 there was a frost in England.

In France in 1142 the ground remained buried under a deep layer of snow from December 6 1142 to February 2nd 1143. In France, violent winds in January 1143 CE overturned many buildings and tore up the trees.

In the year 1143, in France and Germany there was a severe winter, and very thick snow covered the earth from the beginning of December to February. A terrible storm tore down houses and churches. When the thaw came, the melting snow brought floods. In Germany, the trees burst from the extreme cold and the grape vines froze. The famine continued to decimate the people.

In 1143 CE there were catastrophic autumn rains that destroyed the harvest and caused hunger. This was from the "Chronicle of Novgorod" in Russia. In Novgorod all the autumn was rainy, from Our Lady's Birthday September 25 to Korochun the Winter Solstice – 21 or 22 December & warm and wet; the water was very high in the Volkhov River and everywhere, flooding carried away hay and wood. Lake Ilmen at the north end lies Novgorod froze in the night, and the wind broke up the ice and carried it into the Volkhov River, where it broke the bridge. It carried away four piles or bridge supports, never heard of again.

In 1144 CE in England, there was a drought that lasted during all the harvest months and long after. There was a terrible famine.

On 14 February 1144, there was a dreadful hurricane. England was almost consumed by a general sore famine and civil wars. There was a most droughty harvest. There was neither rain nor dew until the Feast of Saint John the Baptist June 24 and then no more for a long time afterwards.

In 1145 CE catastrophic Autumn rains fell again around Novgorod in Russia and the same results as 1143 reoccurred. In Novgorod there were two whole weeks of great heat, like burning sparks, before harvest; then came rain, so that we saw not a clear day until winter; and a great quantity of corn and hay were unable to be harvested; and that autumn the water was higher than three years before; and in the winter there was not much snow, and no clear day, not until March. Corn in old English is derived from Old Norse korn and means grain.

In 1146 CE in France there was a famine. The year 1146 was wet. This caused a calamity in wheat production, which led to famines. There were poor harvests in Reims, France and Aachen, formerly Aix-le-Chapelle when it was in France.

In Germany the Rhine River flooded.

In 1148 back in Novgorod there was rain with hail on June 27, a Sunday; and thunder set fire to the Church of the Holy Mother of God in the monastery of Zverinets.

The Winter of 1148/1149 CE was dry and warmest to 1st April, then coldest to 15th May in England.

From 1149 CE to 1150 CE there was a severe winter. The Thames being frozen solid. The frost lasted from December to March and the frozen river was crossed on foot and on horseback. The very intense cold began on December 10th and continued until at least February 19th. The Thames was frozen over at London Bridge and supported loaded wagons. In England, the year was full of tempests of thunder and lightning, hail, rain, etc., which did inestimable, harm. The summer and harvest were excessively rainy. These rains did great damage to standing corn grain so that a dearth famine followed. There was a terrible whirlwind, which broke down many houses and tore up trees by the roots (tornado). The earth was very barren.

In 1149, the sea was frozen off the coast of the Netherlands.

In the year 1149, the winter was more severe in Flanders, now Belgium, than usual and lasted from early December to March. The sea was completely frozen and passable from a distance of more than three miles from the coast. The frozen waves appeared in the distance like towers. In Tournai, Belgium, there was great shortage of food.

From 1149 CE to 1220 CE the Pacific Regions was much warmer.

1150 CE

The warmest part of the Little Ice Age in Europe was between 1150 to 1300 AD.

In 1150 CE Vesuvius in Italy had a VEI of 3, Sheveluch in Kamchatka had a VEI of 3, Karymsky in Kamchatka erupted with a VEI of 2, and Cofre de Perote in Mexico had a VEI of 2.

During the winter of 1150 off France, the sea was frozen 3 miles from shore from December to February. The winter of 1150 in northern France was no less rigorous and continued for three months. Several people had their limbs frozen. This winter did not allow the spring farm work.

The onset of a drought occurred at Tiwanaku, Tiahuanaco, on the shore of Lake Titicaca in Bolivia in 1150 CE. The lake level dropped between 39 and 56 feet over five decades, which indicated a drop in rainfall of ten and twelve per cent. Tiwanaku, dependent on raised-field agriculture nourished by streams and high groundwater levels, collapsed with its agriculture unsustainable. The population dispersed into smaller communities and turned to the herding of native, drought-adapted llamas and alpacas and opportunistic farming.

Precipitation increased in Nazca in Peru between 1150 CE to 1450 CE.

Around 1150 CE at Mill Creek in O'Brien County, Iowa, pollen spectra from the middens of settlements of the Milll Creek Culture on the northern plains of Iowa that were abandoned from 1150-1200 onwards indicate a rapid decline of oaks and the rise of grass pollens. From bones found the people went over abruptly to bison hunting and predominance of meat in their diet.

The third of the four driest drought periods started in the American West in 1150 CE. This was all within a four-hundred-year period of overall aridity.

Around 1150 CE the former homelands of Pueblo Bonito, Mesa Verde, Chaco Canyon and others were abandoned in favor of positions along bigger stream courses and more water control channels, roads and signaling stations.

Between 1150 CE and 1450 CE profound changes affected Hohokam Culture in Arizona and the Southwest. Major droughts descended onto the Southwest. This caused the abandonment of Chaco Canyon and its great houses, and later large pueblos at Mesa Verde and in the nearby Moctezuma Valley. Snaketown lost its population. By 1450 CE the cultures collapsed.

In the Panhandle region of Texas and Oklahoma, conditions for agriculture improved and it became moister.

From 1150 CE to 1200 CE volcanic activity was reduced globally and sunspot activity was high. Both of these contributed to a cooler La Nina-like state across the eastern Pacific.

In 1150 land was being opened up for farming on the heights of the Lammermuir Hills in Scotland. This continued until 1250 and recorded in the annals of the Scottish border abbeys. These farms were abandoned by the mid sixteenth century. Eventually 4,800 hectares of farmland were abandoned. The upper limit of the farms was 400 meters above sea level

In the area around Belgium-Northern France, the year 1151 promised abundant harvests, but the rain which fell beginning on the Feast of St. John, June 24 and continuing without interruption until mid-August, destroyed the goods of the earth. Very little fruit came to maturity. The wine was missing, because of the small quantity of grapes collected. And the wine produced turned into vinegar.

The summer of 1151 in France was rotten. There were continual rains from 24 June to mid-August. There were frequent storms, gales and fogs. A promising crop was destroyed. Excessive rainfall on the Feast of Saint John to mid-August in the year 1151 in northern France stemmed the crops from maturing and caused enormous devastation. In 1151 or 1152, there were great and excessive rains, which fell this summer, hindered the growth of corn grain, hence a famine, together with a great mortality of people.

In 1151-52, there was a famine in Europe.

There was a famine in Palestine. Palestine is the region between the Mediterranean Sea and the Jordan River.

In 1152 CE in England there was a drought from "13th March to harvest, neither rain nor dew. First, cold nights: frost, northerly winds; then greatest heat and dry weather with flies and gnats."

In Germany, there were great floods on the Rhine River from excessive rains.

In 1152 in France there was a grain famine.

In 1153 CE in Ireland, a great famine raged in Munster, and spread all over Ireland.

In November 1154 CE Miyakejima erupted in Japan with a VEI of 3.

In 1154 CE there was a great frost in England with thunder, lightning, rains, and a horrible tempest. In England, there was a famine from rains, frost, tempest, thunder and lightning.

There was a general famine over all of Europe

In 1154 CE there were famines in Scotland along with the plague.

In 1155 CE there was a cold spring and summer with late snow and periods of strong winds and gales in England.

In 1156 CE in England, there were rain and floods, lasting all the harvest.

In 1156 in England, thunder and tempests were very frequent in July. An abundance of rain followed which began on 11 August. The rain hindered the reaping and sowing of corn grains; hence many great and long floods, which carried down houses, churches, etc. Then came the frost.

In 1156 in southern France, excessive drought brought about winter.

In 1157 Mount Etna in Italy erupted with a VEI of 2 and Pico de Orizaba in Mexico had a VEI of 2.

In 1157 CE there was a great frost in Italy. During the winter of 1157 there was an enormous amount of snow and the violence of the frost destroyed a large portion of the grape vines. In summer afterwards was excessive heat and drought, followed by the plague. In 1157, the summer was extremely hot and dry in Italy.

In 1157 there was a terrible thunder and lightning storm in Normandy that was fatal to people and cattle. In June there was a great tempest, which did much damage to corn, trees, and buildings.

In 1157, the summer was abnormally very hot in Germany.

After a year of foreshocks, an earthquake occurred on 12 August 1157 near the city of Hama, in west-central Syria (then under the Seljuk rule), where the most casualties were sustained. In eastern Syria, near the Euphrates, the quake destroyed the predecessor of the citadel al-Rahba, subsequently rebuilt on the same strategic site. The earthquake also affected Christian monasteries and churches in the vicinity of Jerusalem.

In Novgorod in Russia, in autumn there was very terrible thunder and lightning, and hail the size of which was larger than apples. This hailstorm occurred on 7 November at 5 o'clock at night.

In 1157 or 1158 in Italy, there was a great overflow of the Tiber River in Italy.

In Normandy, there were great floods.

In 1158 a large flood tore down many churches and houses and drowned many people and livestock in Germany. This entry was listed under the category "storm surge", which might include an inundation of the sea so this would have been on the North Sea coast.

In July 1157 or 1158, there were several lightnings and tempests in Normandy. In several places, many people were killed from lightning. A great inundation followed which hindered the reaping, fetching home and sowing of corn grain.

There was a great inundation of the Tiber River in Italy.

The River Thames in England dried up. In 1157 or 1158 from tempests, floods, corn grain was neither ripened, reaped, got, nor new sown for fall planting.

In Novgorod there was great mortality in the people and a large number of horses also died; so that it was not possible to walk to the marketplace through the town, nor along the dike, nor out to the fields, because of the stench. Horned cattle also died.

On January 19th 1158 CE Hekla erupted in Iceland with a VEI of 4. Would this lead to a volcanic winter? Icelandic eruptions are quite sulfurous.

In 1158 CE because of an earthquake the Thames at London was waterless, and it was crossed dry-shod. It was much more likely that it was a dry year which coincided with a minor earth tremor.

In 1160 CE Pacaya erupted in Guatemala with a VEI of 3, Taveuni in Fiji had a VEI of 2, and Mount Etna in Italy had a VEI of 2.

In Australia and South America there was a warm period from 1160 CE to 1170 CE.

In 1161 CE in Sicily Italy, there was an inundation of the sea that drowned 5,000 persons; there were "floods in many rivers and multitudes of people lost."

In May 1161 in Italy, there was a tempest with hail the size of goose eggs. There were also several tempests, inundations of rivers, and loss of much people.

There was a great thunder and lightning storm in England that was fatal in 1161 CE.

At the village Landaaren, now Landavran, in Brittany in northwestern France at noon rose out of the earth a terrible whirlwind tornado and floods on high. The noise of spears and lances were heard in it, but no hand was seen. On the top were seen fowls flying in and about debris. Soon after a grievous plague raged both there and in several places in Normandy and the neighborhood.

In 1161 CE there was a great famine and earthquake in several places including Antioch (Turkey), Tripoli (Libya), Damascus (Syria), etc. wherein 20,000 men were killed. There were also several tempests, inundations of rivers, and loss of much people.

In Novgorod the sky stood clear all summer and all the corn was scorched, and in the autumn, frost killed all the spring corn. During the winter the whole season stood with heat and rain, and there was thunder. We bought a little barrel 11½ pecks for seven kunas with marten skins used as money. Oh, there was great distress in the people and want! The marten is a weasel-like mammal and is valued for its fur.

In 1162 CE there was much loss of life in the North Sea area due to a massive storm.

In 1162, there was said to have been a great famine all over the world. This was probably referring only to Europe. In 1162, there was a general famine in Europe that was still terrible. But in Poland, it was a great famine.

In 1162, there were several large snowfalls in Germany. It covered tall houses and trees. One could not travel by horseback or on foot.

In 1162 CE in the Netherlands, there was an inundation from the sea; many people and cattle were lost. There was a tempest of hail, and the sea overflowed.

The sea overflowed farther in Friesland in the Netherlands than ever was known. Even Hadelen, and all the low country of Albia and Wirra were flooded, and many thousands of people and cattle drowned. At the same time hail made fearful havoc of men, beasts, trees, and horses.

Much of Friesland is currently under the Zuider Zee. The remaining Friesland (Frisian) territories consisted of: West Friesland, which remained a part of Holland and became a part of North Holland around 1800. The current region of West Friesland is smaller than historical West Friesland Friesland is now a Dutch province. East Frisia became a part of the Kingdom of Prussia and was formerly a district of the federal state of Lower Saxony in the Federal Republic of Germany. North Frisia was a part of the Danish duchy of Schleswig and is now part of the German state of Schleswig-Holstein. The Frisian Islands off the coast of the Netherlands and Germany are the leftover dunes of flooded lands.

On 10 February 1162, there was a great flood, and many thousands of people drowned in the Elbe and Weser rivers, along with countless cattles. One called this flood "the human trough" because people drowned one by one. In the district of Dithmarschen in northern Germany, a collapse of the water pond released so much water that in the whole parish of Brunsbüttel only 30 people survived. In Lübeck Germany, there was a tempest of wind, thunder and lightning.

On 14 March 1162, there was a great tempest of wind, thunder and lightning at Lübeck in northern Germany, which burnt and overthrew many houses.

In Mediolana Mediolanum or Milan, Italy fell twelve great snows, which greatly afflicted both animals and vegetables. A great famine still reigned over most parts of the world. Famine, plague and war sorely afflicted the people of Mediolana.

In 1163 CE there were famine and plague in Aquitania. Aquitania or the Duchy of Aquitaine, currently is a region of southwestern France between Bordeaux and the Pyrenees.

On 17 February 1164, a storm surge caused a flood. The southern North Sea coast was probably more severely affected than the west coast of Schleswig-Holstein. It affected mainly East Friesland and the Weser-Elbe region, said there were initial dips of the Jade Bay in northwestern Germany. The flood affected 20,000 people between the Rhine and the Elbe rivers. The first break of the Jade Bay was directed to the southwest. Along the Weser River, salt water covered the country up to 12 miles inland. This was the Saint Juliana Flood.

In 1165 CE in Sicily, there was an irruption of the sea. On Sexagesina Sunday (second Sunday before Lent), the sea swelled and rose three days together, and in Sicily it drowned 12000 people.

In Italy during the summer, red-hot winds dried up all the plants. The summer of 1168 was hot and dry.

In June 1168 the Sarthe River in France dried up.

The 1169 CE Sicily earthquake occurred on 4 February at 08:00 local time on the eve of the feast of St. Agatha of Sicily (in southern Italy). It had an estimated magnitude of between 6.4 and 7.3 and an estimated maximum perceived intensity of X (*Extreme*) on the Mercalli intensity scale. The cities of Catania, Lentini and Modica were severely damaged, and the earthquake also triggered a paleotsunami. Overall, the earthquake is estimated to have caused the deaths of at least 15,000 people. The tsunami affected most of the Ionian coast of Sicily and caused inundation from Messina in the north to the mouth of the Simeto River in the south.

On 4 February 1169, a tsunami affected most of the Ionian coast of Sicily.

In 1170 CE Furnas erupted with a VEI of 4 in Iceland.

On November 1st 1170 CE in Holland, Friesland, and Utrecht, there was a terrible flood. In the latter province the water rose to so great a height that the people were able to catch fish with nets within the walls of the town. Utrecht is now located in the northern Netherlands. In 1170, sea water was driven up very high by the great storm winds in Germany. This was called the All Saints Flood. During this storm the islands of Texel and Wieringen were separated from North Holland. Wieringen was rejoined to the mainland by draining, dikes and landfill from 1924 to 1932.

In 1170 CE London's population exceeded 30,000. This was a vast city by medieval standards in Europe. Most people were subsistence farmers who lived from harvest to harvest. There was never a great chance of a harvest surplus as much of the grain was kept as seed. These farmers were living on subsistence levels even in good years.

The 1170 Syria earthquake was one of the largest earthquakes to hit Syria. It occurred early in the morning of 29 June 1170. It formed part of a sequence of large earthquakes that propagated southwards along the Dead Sea Transform, starting with the 1138 Aleppo earthquake, continuing with the 1157 Hama, 1170 and 1202 Syria events. The estimated magnitude is 7.7 on the moment magnitude scale, with the maximum intensity of X (*Extreme*) on the Mercalli intensity scale. Severe damage was widespread from Antioch in the north
to Tripoli and Baalbek in the south. Several places that had been badly damaged by the 1157 Hama earthquake, suffered again, such as Hama, Aleppo and Antioch. The fortress of Krak des Chevaliers (Hisn al-Akrad) was badly damaged by the earthquake. Although the extent of the damage is unclear (one source says that no traces of the walls were left), the castle underwent major rebuilding in subsequent years. 80,000 people died in Aleppo and 25,000 in Hama.

On 3rd November 1170 CE a sea floods hit Borkum Island in Lower Saxony in northwest Germany. Much of Borkum was lost. Borkum is in the Westerems Strait which forms the border with the Netherlands and is the largest and westernmost of the East Frisian Islands in the North Sea.

The 1170s were regarded as very dry summers in Britain and Western Europe.

In 1170 CE the practice of sending an annual trading ship from Greenland to Norway ended due to ice buildup in the ocean. The last bishop of Gondar, the capital of Greenland, died in 1170 CE and he was not replaced.

In 1171 CE in England, there was an inundation of the sea; harvest destroyed in many places. On Quadragesima Sunday occurring after Ash Wednesday, there was a great inundation of the sea. The harvest in many places was lost and carried off by the waves. A plague on man and beasts.

On 25 December 1171, there was terrible thunder and hail in England, which killed birds, beasts and people. The storm struck England, Ireland, France and Scotland. At night fell a most terrible tempest. The lightning did great damage.

During the winter of 1171-1172, a storm with the intensity of a major hurricane slammed into the shores of South Wales and other western seaports. "At the time when Henry II, King of the English, was wintering in Ireland, a curious phenomenon occurred here at Newgale Sands, Wales. The wind blew with such unprecedented violence that the shores of South Wales were completely denuded of sand, and the subsoil, which had been buried deep for so many centuries, was once more revealed. Tree-trunks became visible, standing in the sea, with the tops lopped off, and with the cuts made by the axes as clear as if they had been felled only yesterday. The soil was pitch-black, and the wood of the tree-trunks shone like ebony. By this strange convulsion of nature, the element through which ships were wont to move so freely became impassible to them, and the sea-shore took on the appearance of a forest grove, cut down at the time of the Great Flood, or perhaps a little later, but certainly very long ago, and then by slow degrees engulfed and swallowed up by the waves, which encroach relentlessly upon the land and never cease to wash it away. The tempest raged so fiercely that conger-eels and many other sea-fish were driven up on the high rocks and into the bushes by the force of the wind, and there men came to gather them."

In 1172 CE in Ireland, "great floods destroyed numbers of men."

There was a terrible tempest of thunder and lightning with a hurricane in Britain and France.

After Christmas in the year 1172, there was a great mortality in Ireland. As a result, King Henry II was forced to quit the country.

In Germany, there were great floods on the Rhine River.

In Europe, the winter of 1172-1173 CE was so mild that the trees remained covered with green foliage. Towards the end of January, the birds were nesting and in February, they were having their young. There were also large storms and much rain. In January, there was often thunder, and the fire from heaven damaged many houses and churches. What is the "fire from heaven" that appears every now and again?

The winter of 1172 was an extremely mild winter in France.

In late January 1173 in Belgium, there were leaves on the trees. By mid-February, the birds had built their nest and their eggs hatched.

In 1173 CE in the Netherlands, a great flood considerably extended the limits of the Zuyder-zee. The Zuyder-zee or Zuiderzee at the time was a shallow bay of the North Sea in the northwest region of the Netherlands. Today the Zuyder/Zuiderzee as such no longer exists. It is now two lakes. After centuries of North Sea storm surges that breached barrier dunes and dikes, it is now enclosed by a 32-kilometer dam wall, the Afsluitdijk. This was constructed in the aftermath of the flood of January 1916. At its completion in 1932, the Zuiderzee became the Ijsselmeer Lake, enabling large areas of water to be reclaimed for farming and housing. Then in 1975, the IJsselmeer Lake was split in two by the completion of the Houtribdijk, now called Markerwaarddijk, 28 kilometers long creating the Markermeer Lake and making it hydrologically separate from the IJsselmeer Lake. The province of Flevoland was created in 1986 from the polders reclaimed from the IJsselmeer. Polders are low-lying tracts of land enclosed by embankments (barriers) known as dikes.

In France in the year 1174, the rain lasted from the Feast of St. John's, June 24, to the end of the year. There was a lack of wine and all fruits. The area around Metz in the Lorraine region of northeastern France experienced a flood. The year 1174 was a year of universal rains.

In 1174 the whole world was afflicted with cloudy corrupt air, which occasioned a most universal cough and catarrh: a disorder of inflammation of the mucous membranes, which was fatal to many.

In 1175 CE Pico de Orizaba erupted in Mexico with a VEI of 3 and Kirishimayama erupted in Japan in January 1175 with a VEI of 2.

As a result of the previous year's corrupt air and its consequences, in 1175, both England and the neighboring countries groaned under a grievous mortality of people, soon followed by a great dearth and famine. In 1175 in England, pestilence, was followed by a great dearth.

In 1175 in France, there was a very great flood in November. The flood destroyed the farms and the seed. In 1175, the Seine River in France flooded several times. The summer rains in 1175 in northern France prevented the grain harvest of August and the grape harvest in autumn. There was disastrous overflow of several rivers, particularly the Seine River, around Christmas.

In the severe winter of 1175 to 1176 CE there was frost and snow from 25th December to 2nd February in Normandy in Northern France and possibly England.

In 1176 CE in the marshes of Lincolnshire, England, there was an inundation of the sea which swallowed up much cattle and people. It took two days for the waters to return to their normal boundaries.

There was also a great inundation of the sea that struck the Netherlands.

In England, there was constant hard frost from Christmas to Candlemas. In England there were many and great hurricanes during the year and a most temperate winter.

In Wales there was a great famine and mortality.

In 1176 in northern France, a large frost stretched from December 13 to March 15.

In the Novgorod Republic now part of Russia in the springtime, the Volkhov River flooded for five days.

In 1177 CE Katla erupted in Iceland with a VEI of 3 and Hakusan in Japan erupted on May 18th 1177 with a VEI of 3.

In 1177 in Western Europe, the summer was very dry and very hot, and the drought was so strong that the seeds were lost, and there were no grain or hay. The harvest began in August, and the wine was excellent. In 1177, the summer and harvest was so great a drought that the seed sown was lost. During the harvest, there were great rains, floods and shipwrecks. On the Feast of Saint Mary Magdelen, July 22, there was thunder and a storm, which laid flattened corn and killed birds. No corn grain or hay was harvested. On 3 December at night rose a most violent tempest. Their came a terrible hurricane with southwest winds which overthrew churches, houses, trees, etc. It was a most tempestuous stormy winter.

In 1178 CE there was a shower of great hail that killed men, sheep and goats. Later there was a tempest of thunder and lightning at York, England. On the Ides of January, the sea broke in on the marshes and drowned people, villages and cattle innumerable.

In 1178, there was a drought in Quercy, France and a famine.

In 1179 CE Reykjanes in Iceland erupted with a VEI of 2.

On January 7th-8th 1178 CE major storms affected large areas either side of the southern North Sea.

In 1179 in Europe in the second week of January, the snow fell in abundance and a very strong and unpleasant frost ensued and lasted until mid-February. During the remaining part of this month and also in March and April a continuously blowing cold east wind was still palpable. Very large mortality among the cattle and sheep resulted from this cold.

The year 1179 produced in Germany a cold winter and much snow that began on New Years Day and lasted until Candlemas, 2 February. As the winter progressed, there were great floods that filled all the cellars in Wroclaw in Silesia, now a city in Poland, such that one could go to many places in the city by boat. The flooding did great damage to bridges and mills. Due to this cold winter and large floods over 2000 people perished.

On 5 January 1179, there was terrible thunder and lightning with a hurricane and hail in Kent, England.

In 1179 in England there were "Many floods from a most severe winter."

The French historians cite great mortality of animals due to cold during 1179. There was a severe winter. There were several great inundations carrying down bridges, houses and people.

From 1180 CE to 1260 CE there were low Nile floods in Egypt due to low Monsoon rainfall in the Ethiopian Highlands. This is primarily controlled by monsoonal precipitation in the Ethiopian Highlands, driven by the seasonal migration of the Intertropical Convergence Zone,

From 1180 CE to 1190 CE severe drought affected Mongolia.

On 3 June 1181 in Novgorod Russia, the Varangian church in the marketplace was set on fire by thunder at 10 o'clock in the evening, and the church of St. Ioan in Ishkovo was burnt. The same year a fire broke out in Slavno, from Kosnyatin's, and two churches were burnt: that of St. Mikhail and that of the Holy Fathers, and many houses along the bank, even as far as the Stream.

There was a general and great famine over England and Wales with terrible thunder and lightning on 16 August 1181.

On 19th December 1181 a sea floods devastated Friesland in the Netherlands and a town was largely decimated.

On May 28th 1183 CE Zaozan erupted in Japan with a VEI of 4.

In 1183 in France, an unusual drought was accompanied by great heat. It withered away in many places the rivers, springs and wells. During the summer of 1183 in France, the weather was extraordinarily hot and dry.

In 1183, there was a great famine severely afflicting both England and Wales.

On February 7th 1184 CE Kirishimayama in Japan erupted with a VEI of 2.

On 7 June 1184 in northern France, frost burnt the grape vines and grain harvest.

In 1184-1185 in Germany, there was a very severe winter. But the following year 1186 in January the weather was like the month of April. In February, there were beautiful spring days. At the end of May began the grain harvest. In August there was the grape harvest.

The year 1186 CE in Germany the winter was warmer than had been known for a long time. The vegetation was very advanced. The harvest took place in May and the grape harvest in August. The year 1186 was "a strange year," in Germany. First there was a very mild winter in 1185-1186. Then in January 1186, the trees began to bloom. The chicken and forest birds laid eggs and hatched off in February. In May, the grains were harvested. In early August, the grapes in the vineyards were ripe, so they began to harvest grapes. There is much goods and food grown.

In France in 1186 the trees were blooming in the middle of winter.

In 1186 in France, the harvest took place in May and the grape harvest for the wine in August.

In 1186, there was an eclipse of the sun in Poland and Russia and the hottest winter that ever was felt in these parts. The harvest was in May and the vintage in August the earliest harvest. Then came a sweeping plague. In Corinthia, great swarms of locusts, with prodigious large bodies eat up all sorts of green vegetables, hence a barrenness of land, dearth, famine and pestilence. Corinthia may refer to the Duchy of Carinthia, which is today part of southern Austria, north Slovenia and northeast Italy. There is also a Corinthia in Greece.

In 1187 CE in England, there were great floods. There was a grievous and pestilent mortality of men and cattle in England.

In 1187, a terrible famine struck Europe.

The summer of 1187 was excessively hot causing a drought in France. The city of Chartres in northern France caught fire.

In 1187 in Germany, the winter lasted so long that spring planting was delayed until the end of May.

In Novgorod in Russia, there was very terrible thunder and lightning. The people having come with crosses from St. Sophia to St. Michael's and singing nine hymns, the thunder and lightning struck, and all the people fell, and the church caught fire, but by the mercy of God and by the prayers of St. Michael, there was no harm done to the church; but two men were dead.

In 1188 CE in England, there were inundations of the sea that "killed very much people and cattle." On Sunday 6 July 1188 CE rose a tempest of wind, rain, thunder, lightning, and hail the size of pigeon's eggs. The sea overflowed its banks a great height and killed many people and cattle in England. The summer of 1188 produced extraordinary heat and droughts. In many places, rivers, springs and wells dried up.

France also suffered another misfortune because of the multiple fires that were spawned. The heart and drought of 1188 in France produced similar effects as the weather of 1183. It withered away in many places the rivers, springs and wells. The dryness also led to a large number of fires at Tours, Chartres, Beauvais, Auxerre, Troyes, etc. Tours is located in central France. Chartres, Beauvais, Auxerre, Troyes are located in northern France. The summer of 1188 in France was extremely hot and dry.

In 1188, there was a great scarcity of food in the north of Ireland.

In 1190 CE El Chichon in Mexico erupted with a VEI of 4.

In 1190 CE the cold period and severe drought ended in Mongolia.

The year 1191 was a damp year. There were heavy rains in Savoy, France near the western Alps. A landslide blocked the Romanche River in southeastern France.

In 1191 there was a famine in East Riding of Yorkshire in England.

In 1193 CE weather in England was regarded as unreasonable with thunder and lightning often noted throughout the year. It was regarded as a wet year which indicates that it should have been dryer.

In England and France there was a famine that led to pestilential fever. This lasted from 1193 to 1195.

From 1193 to 1196 due to incessant rains in England, most of the corn grain perished and was lost. "The common people perished everywhere for lack of food; and on the footsteps of famine, the fiercest pestilence followed, in the form of an acute fever."

In 1194 CE Mount Etna in Italy erupted with a VEI of 2.

In 1194, there was a famine in France. In 1195 and 1196, there was a famine for 4 years. This was from rain and floods; years together. This year there was a terrible dearth in France, Flanders and England from excessive and unseasonable rains from some years past. Hence an epidemic and acute fever. Most of the vulgar died of the famine; then came the plague. This dearth began some years before and continued four years together. Quickly after the Octaves of Pentecost, began this great mortality, which was ushered in by long wars and famine. This fever of burning ague raged six months and vanished this winter. There was so great a mortality, that there not being living healthy persons enough to bury the dead. Funerals were neglected. The dead were thrown on heaps into pits made on purpose.

In 1194 CE a violent storm almost desolated a great part of Denmark and Norway.

There was great thunder, lightning, hail and rain at Beluata in Macedonia which broke down all fruit-trees, vines and corn grain. Many villages were burnt down from the lightning.

Another tempest struck at Laudun in southern France.

In Germany there was so great a heat and drought at Thuringia, that in many places of their river, people walked over dry footed. The year 1194 was remembered as a very warm and dry summer in Germany. The mills would not work because of a lack of water.

There was a famine in France. The year 1194 was the driest year "without history" in France.

The heavy rains in the month of February 1195 in France made the rivers overflow and occasioned much damage. The summer of 1195 was very wet in France. This was the first year of three rotten years 1195-1197.

During the fall of 1195, a plague of locusts struck Europe, from France to Hungary.

In June, there was a hurricane in England as well as excessive rains. Most corn grain was lost.

In 1196 CE in England, there were great floods in March from rains. In 1196 and from some years past in England, there was constant rain, dearth and death.

On the 3rd of the Nones of November in France there was a hurricane. There was almost incessant rain for two months in 1196. This produced a terrible flood of the Rhône and Saône rivers in France. The waters of the Rhône River in France overflowed its banks and spread into the coastal plains, flooding everything in their paths. Several towns and villages sitting on the banks of the river were partially submerged and destroyed. In March 1196, the Seine River in Paris, France flooded for 16 days. The summer was very wet and this caused a scarcity famine.

In March 1196, there was a sudden and great inundation, which carried away many places, towns, villages and inhabitants in France, England and Belgium.

In 1197, the famine in England and France was still calamitous. The summer of 1197 in France was very wet and this caused a scarcity famine.

The situation of the Old Norse in Greenland in 1197 to 1203 CE had become critical with the increase of ice encroaching on the seas that were used for their links with Iceland and Europe clearly had to do with the cooling of the Arctic. We are now heading for the Little Ice Age.

In 1198 CE about the Feast of John the Baptist, June 24, dew fell in France, as sweet as honey.

In July 1198 in France there was a grievous tempest and great hailstones, which broke down houses, woods, grape vines and corn grain.

In England on 13 August 1198 there was a tempest.

In 1198 a cruel storm with high winds and water caused great damage to the marshlands in Germany. At the beginning of October 1198, Philip of Swabia, from Bavarian Germany, in his disputes with Otho, arrived on the banks of the Moselle River, and found the waters lower than they had been for centuries.

In 1198 CE seventeen years of famine hit the Novgorod region in Russia.

From 1199 CE –1202 CE there was a famine of great severity for deficient rise of the Nile River. There was a great dearth in Egypt from 1199-1202. The flooding of the Nile River was the life-giving inundation which yearly fertilized the crops in Egypt. This annual flood generally peaked in September near Cairo. During the growing season (after the inundation had receded) the Egyptians planted their crops - around October and November - and tended to the fields. The Egyptians watered their crops using an irrigation system of canals or by bringing water to the fields in basins or by using the shaduf, to raise water from the river to the bank of the Nile. By the time the Nile reached its lowest level, sometime around March or April, the crops would be ready for the harvest. The highest point reached by the annual inundation, and very rarely reached, is a little above nineteen cubits. In this case, much cultivable land remains so long submerged that the sowing cannot take place; and it is as barren as a desert for that year, while in some spots which are ordinarily dry, yielded a rare harvest. But at this level, the inundation is accompanied by a great destruction of dwellings and of livestock. When the rise reaches eighteen cubits, there is great rejoicing, for the produce is then sufficient for two years' consumption, after the government dues are paid. When it reaches sixteen cubits, there is enough produce for the wants of the year; and this was called, " the Sultan's flood," because then the Sultan claimed his taxes. Below sixteen cubits, there is more or less scarcity. In these cases, the south wind has prevailed, whereas during the good years, the north winds prevailed. The cubit at the Nilometer was equivalent to 19½ inches. There were 28 digits in a cubit. A Nilometer was a ancient structure used to measure the level of the Nile river during floods at Elephantine Island. The Nile River inundation peaked on 9 September 1199. It stood no higher than twelve cubits, twenty-one digits; and it then began to decline.

The lowest Nile peak ever known seems to have been that of 966 when the waters rose only to twelve cubits, seventeen digits: and the next lowest was in 1199, when it rose only four digits higher. At this point it became obvious that Egypt was at the verge of a great famine; a wrath that could annihilate all the resources of life and all the means of subsistence. The price of provisions began to rise. The inhabitants of the villages and country estates began to relocate to the great provincial towns. Many Egyptians in large numbers began to flee to Syria, to the Maghreb in North Africa, to Hedjaz coastal region of the western Arabian Peninsula bordering on the Red Sea, to Yemen, to Mosul and Baghdad in Iraq, to the countries of Greater Khorasan Iran, Afghanistan, Turkmenistan, Uzbekistan and Tajikistan, to the Greek empire, and to other parts of Africa.

In 1199 CE in England, there were serious floods from rain. In England, the rains were often, great and heavy and produced great floods.

In 1199 at the rise of the Teutonic Order, strong north winds blew in Prussia for 12 years together, which was the cause of very great tempests. Old Prussia during this time period consisted of Poland, east of the Vistula and southwest Lithuania, and the historical ethnographic region of Lithuania Minor.

There were several heavy rains and great floods in many parts of England, which carried down Berwick Bridge, etc. with many houses and much people. On October last, there was frightful thunder. On 4 November, there was terrible thunder. The Berwick Bridge spans over the River Tweed in Northumberland, northeast England.

1200 CE

The largest eruption in 1200 CE was San Salvador in El Salvador with a VEI of 4. This was enough to cause a Volcanic winter. Brennisteinsfjoll in Iceland had a VEI of 2, Vulcano Island in Italy had a VEI of 2 and Kostakan in Kamchatka had a VEI of 2.

Concentrations of CO^2 dropped from 284 ppm around 1200 CE to 272 ppm by 1631. Concentrations of CO^2 dropped after Northern Hemisphere cooling. Greater solubility of gases in the ocean as temperatures cool can in principle reduce atmospheric CO^2 concentrations. CO^2 may also have decreased after forests reclaimed agricultural lands after the pandemics occurred and there was sharp population decline.

From 1200 CE to 1400 CE Lake Naivasha in central Kenya experienced a long period of intense aridification. This also happened to Lake Victoria, Lake Tanganyika and Lake Malawi.

There was severe and prolonged drought in the middle of the thirteenth century in Europe.

There was an exceptional series of great North Sea floods, mainly on the Dutch and German coasts. There were great losses of land and people in the marshlands between Hamburg and Jutland. Some of these storm floods are believed to have affected the English coast. H. H. Lamb stated that there was a marked increase in North Sea storminess during the 13th century, relative to what had gone before.

From 1200 CE to 1300 CE world sea levels were high.

Around 1200 CE the Caspian Sea rose sharply over the next two to three centuries. By the 1400s the water level was eight meters more than the late twentieth century.

In 1200 CE in England the winter was excessively cold.

A colder and more disturbed climate set in during the 1200s in Europe.

There was lowering of the treeline in the Swiss Alps and heights in Central Europe.

There was increasing wetness of the ground and spread of lakes and marshes in many places in northern, western and central Europe and northern Russia and Siberia, plus swollen rivers and an increase in landslides and similar difficulties. There was an increasing frequency of freezing of rivers and lakes.

Around this time rainfall fell by ten per cent on average in Europe and temperatures rose between 0.5° to 1° Celsius. Rose from which base temperatures though, as they had already risen?

In Europe there was evidence of increasing severity of windstorms and resulting sea floods and disaster by shifting sands in 50-60° North, particularly in the thirteenth century. These were from Brittany in France to the Hebrides and Denmark.

In Brittany in France there was the overwhelming of a number of coastal places by blown sand that continued to 1800 CE.

In what is now Belgium there was the overwhelming of a number of coastal places by blown sand that continued to 1800 CE.

In Denmark there was the overwhelming of a number of coastal places by blown sand that continued to 1800 CE. This was the same as Belgium.

In Germany there was the overwhelming of a number of coastal places by blown sand that continued to 1800 CE.

In the Hebrides there was the overwhelming of a number of coastal places by blown sand that continued to 1800 CE.

There was an increased incidence and severity of windstorms and sea floods in the thirteenth century in Europe.

In the 13th Century there were four sea floods of the Dutch and German coast. The death toll was estimated at around 100,000 people or more. The worst case estimate was 306,000.

In Europe, the forests and swamps had covered around four fifths of the temperate and central regions in 500 CE. By 1200, with the need for more food for a quickly expanding population, there was less than half. Most of it had been done in the Medieval Warm Period.

In 1200 CE there was fierce storm surge in the English Channel and the North Sea.

In the North Atlantic, Arctic sea ice spread into the northernmost Atlantic and around Greenland. This was the earliest advance of Arctic Sea Ice off East Greenland was in 1200 CE. This was the first increase of ice in the Greenland Current. Previously oats and barley could be grown in Greenland but after this, oats were given up and barley declined by half.

There were advances of the inland ice and permafrost in Greenland and of glaciers in Iceland, Norway and the Alps.

Between 1200 CE and 1250 CE Inuit and Norse first met in Greenland. At first some trading took place between them.

In the thirteenth and fourteenth centuries grain growing was given up in Iceland there was a shift of Iceland's population to the coast and export of cod and whale and fish oil increased to pay for grain. The growing of oats was given up about 1200 CE and reduced the amount of barley grown by half. This was at Kvisker in Iceland and possibly the rest of the island as well. The crop paddocks within one hundred years were covered by river gravels and part of a glacier.

There were advances of the inland ice and permafrost in Greenland and of glaciers in Iceland, Norway and the Alps.

A severe decline set in in old Norse society in Iceland in 1200 CE that continued over six centuries. The population of the country fell from about 77,500 as indicated by tax records in 1095, to around 72,000 in 1311 CE. By 1703 CE nearly down to 50,000 and after some severe years of ice and volcanic eruptions in the 1780s it was only about 38,000 people. The average stature also declined from 173 cm, 5 ft 8 in., to 167 cm, 5 ft 6 in., from the 10th century to the 18th century. It is clear from surviving records that years when the Arctic sea ice was close to the Iceland coast for long months, usually between January or March and any time from June to August, played a big part of it. In such years the spring and summer were so cold that there was little hay and thousands of sheep died, especially in the northern and eastern part of the country. The shellfish by the seashore were also destroyed by the ice. Gradually all attempts at grain growing were given up and the glaciers were advancing.

There was as much sea ice from the late 1200s through the 14th Century, and then some improvement before the drastic increase of ice in the late 1500s and after.

In 1200 Ireland was struck by a famine. In Ireland it was "a cold foodless year".

In the twelfth century in Japan there was a particularly marked cold period. The mean blossoming of the date of the flowering of cherry trees in Kyoto was ten days later than the average of the last one thousand years. This suggests an eccentric position of the circumpolar vortex. This was the opposite of the warming period in Europe and the North Atlantic.

From 1200 CE there was an end to the long cold period and warming peaked in the early 1200s in Mongolia.

In 1200 CE a sea floods devastated Friesland in the Netherlands and 100,000 people were estimated drowned.

In the Netherlands there was the overwhelming of a number of coastal places by blown sand that continued to 1800 CE.

The peak of the medieval warm period in New Zealand was from 1200 to 1400 CE.

There were advances of the inland ice and permafrost in Greenland and of glaciers in Iceland, Norway and the Alps.

The population of Norway started declining during the thirteenth century.

In the European Alps some trouble was caused by advancing glaciers.

Between 1200 CE and 1400 CE there was a culmination of El Nino events in the eastern Pacific.

Between 1200 and 1350 CE the walls of a medieval course, the Oberriederin, that led off from a subglacial stream emerging from the Aletsch glacier, high up in the Swiss Alps, indicate that the structure was destroyed by the advancing glacier. The site is at present under the glacier (1984). Ancient documents state that the Oberriederin was out of use by 1385 CE. In the thirteenth century there were glacier advances in the Alps.

After the long warm period of the Medieval Climate Optimum the peoples of the Mississippian Culture had left Cahokia in Illinois as well as other sites in the Ohio River and Mississippi River Valleys. Shifts in hydroclimate may have placed Moundbuilders under stress, either through greater aridity or greater flooding. Long droughts between the mid-twelfth and the early thirteenth century weakened the Mound Builder societies. The cycle of drought undercut intensive agriculture in the Cahokia region. The water table fell and reduced precipitation, rain, threatened the intensive Maize cultivation necessary to sustain the comparatively dense population. Drying out would have also lowered the stock of fish.

By 1200 CE in North America the scene changed rapidly. There was change in the valleys close to the watercourses, but the oak trees disappeared in most places where they had grown, the overall number of trees had declined in favor of prairie plants. The shorter grasses gained at the expense of the bigger, more moisture-demanding ones. At one site in northwestern Iowa the increase of grass pollen went from a negligible proportion to about seventy per cent of the not-tree pollens took only forty-five years or less. The forest animals such as deer gave way to bison in the peoples' diet.

In North America there are signs of a significant decrease of rainfall which accords of increased dominance of the west winds, generated by the increased north to south gradient of temperature at a time of cooling of the Arctic. This would extend the rain-shadow effect of the Rocky Mountains further east than before and intensify the dryness within it. Smaller villages in the driest areas were deserted and people tended to congregate in the larger places in the river valleys. Ultimately, even the biggest one, Cahokia in Illinois with a population of 40,000 people was also abandoned.

The 200-year period of extreme dryness coincided with the strong development of the circumpolar vortex which carried the westerlies mostly far to the north in the European sector, accounting for the warm periods of the 13th and 14th Centuries there but also for the vigorous development of cyclonic rains in Europe around 1315 CE when the westerlies came further south. Much further south, over the southern plains in northwestern Texas and adjacent parts of Oklahoma, there would be an increase of rainfall that may have been substantial. It was presumed that it was towards these regions that the former populations of the northern plains went. Then numbers of people inhabiting the Panhandle region rapidly increased around 1200 AD.

Damage from greater flooding at the end of the arid period in the early thirteenth century caused much damage to Cahokia in Illinois which is on the east bank of the Mississippi River opposite St. Louis in Missouri.

There was a lowering of the tree line in the Rocky Mountains in the United States.

With the advent of the Little Ice Age the Vernagt glacier in the Tyrol in Austria made some initial advances. This was the first recorded glacier advance of the Little Ice Age.

The settlement at the mounds near Kincaid in Christian County in southern Illinois reached its peak in the early 1200s, but mound building ended around 1300 and the settlement was abandoned by around 1450 CE.

Around 1200 CE the civilization of Chaco Canyon in New Mexico collapsed. The surviving population departed, and the great buildings were abandoned. Chaco Canyon had been built between 800 CE and 1150 CE and comprised large multi-story houses connected by roads. Chaco Canyon appeared to trade in turquoise. They used dams and canals, where small dams intercepted runoff of rain and steered water into canals that led to larger dams. There were several dams including a large masonry dam that was 130 feet wide. All the New Mexico sites were abandoned.

The 1202 Syria earthquake struck at about dawn on 20 May 1202 (598 AH) with an epicenter in southwestern Syria. The earthquake is estimated to have killed around 30,000 people. It was felt over an extensive area, from Sicily to Mesopotamia and Anatolia to upper Egypt, mostly affecting the Ayyubid Sultanate and the Kingdom of Jerusalem. The cities of Tyre, Acre and Nablus were heavily damaged. A magnitude of M_s 7.6 has been estimated with damage up to XI on the Mercalli intensity scale. The earthquake was felt from Sicily in the west to northwestern Iran in the east, and from Constantinople in the north to Aswan in the south. The affected areas, listed by decreasing order of the intensity, were, in today's terms, Lebanon, central Palestine, western Syria, Cyprus, northern Israel, Jerusalem, Jordan, southern Turkey (Antioch, Lesser Armenia, eastern Anatolia), Sicily, Iraq and Iran, Egypt (as far south as Aswan), Constantinople and Ceuta. The greatest damage was reported from Mount Lebanon, Tyre, Acre, Baalbek, Beit Jann, Samaria, Nablus, Banias, Damascus, Hauran, Tripoli and Hama (VIII–IX on the Mercalli intensity scale). The tsunami probably associated with this event was observed in eastern Cyprus and along the Syrian and Lebanese coasts.

During the flood season in Egypt, on 4 September 1200, the peak reached fifteen cubits, sixteen digits. But the floodwaters began to drop almost immediately. This flood was referred to as a phantom inundation, a ghost that would appear as if in a dream and then immediately vanish. Only the level lands profited by this inundation. Only the lower provinces were sufficiently watered. But the famine had already taken a vicious toll and the villages by this time were entirely emptied of cultivators and laborers. In many cases these watered lands remained untilled because the proprietors could neither provide the seed nor pay the expenses of cultivation. Of the fields, which were sown, many were laid waste by vermin, which devoured the seeds.

1201 CE and 1202 CE were recorded as consecutive wet years. In 1201 the summer experienced severe thunderstorms with notable hail in the London area in mid to late June. Other sources state that 1201 experienced a notable heat and drought episode and the harvest was over by June 25th. The drought continued through July and August. In England the spring had glutting and continual rains and very great floods. On 25 June and 10 July, there were great tempests of thunder, lightning, hail as big as eggs, and prodigious rains. This destroyed the corn (grain), cattle, people, meadows etc. and burning towns (from lightning strikes). The rains continued from the Feast of Pentecost to the Feast of the Nativity of the Blessed Virgin, September 8, which not only hindered corn and fruits from ripening, but also rendered them mostly useless and unprofitable. A great dearth of animals followed, but chiefly of sheep.

In 1201 and 1202, there was a great famine in Egypt. During January 1201, the waters of the Nile River sank considerably and continued to fall until men and horses could ford the river in several places. On 20 May 1201, a powerful earthquake struck Egypt adding to the devastation. The epicenter of the earthquake was in Syria and caused approximately 30,000 deaths throughout very wide area, from Sicily to Iraq and Anatolia to Upper Egypt. It is believed this earthquake actually took place on 20 May 1202. Thus, this entire stream of famine events in this account may be offset by a year. During the next flood season, the peak occurred on 1 September 1201 reaching a height of one digit under sixteen cubits. After two days at this height, the waters began to decline slowly, and to flow away very gradually. As the famine first took hold in 1199, the infinite number of people who fled to Cairo and Al-Fustat, Egypt experienced a frightful famine and mortality. They ate carrion, corpses, dogs and the dung of animals. As the famine grew very severe, they went even further; devouring little children. The commandant of the city tried to halt the practice by sentencing all who committed this crime cannibalism as well as anyone who ate the meat to be burned alive. But the hunger drove the people. In the space of a few days, as many as thirty women were burned alive, every one of whom had confessed that they had eaten several children. But shortly after their execution their bodies disappeared, because the bodies of these "already cooked" criminals were in turn stolen and eaten. Therefore, the authorities found it very difficult to stamp out the practice. Instead, cannibalism extended over all Egypt. The horror that was first associated with this crime ceased to be felt and people became indifferent and viewed it as an ordinary thing. There was not a single inhabited spot where the practice of eating human flesh did not become extremely common. Syene Aswan, Kous, the Faioum Faiium, Mahalleh Alexandria, Damietta, and all other parts of Egypt, witnessed these scenes of horror. At Cairo and Al-Fustat, and in the neighboring places; wherever one went, there was not a spot in which one's feet or one's eyes did not encounter a corpse, or a man in the agonies of death. Day by day, from one hundred to five hundred dead bodies were taken from Cairo, to be carried to the place where they might have funeral rites. At Al-Fustat the number of dead was incalculable. They were not buried, but merely cast out of the town. At last, there were not enough living left to carry away the dead, and they remained in the open air, among the houses and shops, or even in the interior of dwellings. You might see a corpse falling to pieces in the very place where a cook or a baker, or other tradesman, was carrying on his business. A traveler often passed through a large village without seeing a single living inhabitant. He saw the houses standing open, and the corpses of those who had lived there stretched out opposite one another — some decayed, and some recently dead. Very often, there was a house full of furniture, without any one to take possession of it. One could travel for several days together, and in all directions, without meeting a single living creature, nothing but corpse. A great mortality and pestilence happened in Fayum, Faioum, Faiium, in the province of Garbiyyeh, Gharbiyah, and at Damietta and Alexandria. Nor in many cases, did those that fled Egypt fare better. According to the testimony of a great number of witnesses, the road between Egypt and Syria was like a vast field sown with human bodies. It had become as a banquet-hall for the birds and wild beasts, which gorged themselves on their flesh; and the very dogs that these refugees had taken with them, to share their exile, were the first to devour their bodies. The inhabitants of the Hauf, (a district to the east of the Nile, below Cairo,) when they fled to Syria to find pasturage, were the first who perished upon this road; long as it is, it was strewn with their corpses, like locusts which have been broiled.

The death toll in the end was horrendous. At Al-Fustat of the nine hundred machines for weaving mats; now only fifteen remained A loss of 98%. We have only to apply the same proportion to the other trades, which are carried on in that town, to the shopkeepers, bakers, grocers, shoemakers, tailors and other artisans. The numbers employed in each of these were reduced in the same proportion or greater than the mat weavers. At Maks Izab al Maksa? there was a hill on which human remains had accumulated in great quantity. Abdallatif or Abd-ul-Latif was a celebrated ancient physician and traveler estimated that there were twenty thousand corpses. When from the height he looked down, upon the place called the Basin, and which is a considerable hollow, we saw skulls, some white, some black, and others of a deep brown: they were in layers, and heaped up in such a quantity that they covered up the other bones: one would have said that there were only heads without bodies: and one might suppose that one saw melons which had been gathered, and which were thrown into a pile, as we heap sheaves upon a granary floor. Days later he returned to the same spot. The sun had dried the flesh: the skulls had become white, and they appeared like ostriches' eggs piled together

The Winter of 1201-1202 CE in England was severe beyond any in the memory of man for extreme cold, and long continuance. Frozen ale was sold by weight. It snowed for many days and was very deep.

In 1202 CE in England, after the frosts followed the like of tempests of thunder, lightning, rain, and hail as big as hen's eggs. This destroyed corn (grain), fruits, young cattle and horses, etc. As a result of the rains of 1201, a bad crop, and the corn for seed marred, there came a dearth. In 1202 in London, England, there was a great storm of hail and rain.

In 1202 there "fallen grete reynes, and hailstones as gret as an eg, medlyd with reyn; where thorough trees, vines, cornes, all manner frutes, were moche destroid: and the people were sore abaysshed, for there were seyn foules fleynge in the eyre berynge in their bills brennyng coles, which brenden many houses." (Translation: In 1202 there fell great rains, and hailstones the size of an egg, mixed with rain; where all the trees, vines, grains and all manner of fruits, were much destroyed: and the people were mortified, for there were seen fowls flying in the air bearing in their bills burning coals, which burned down many houses.) Another account gave this event as occurring in the year 1203.

In London "there fel (fell) great raines (rains), thundrings (thunder), and hales (hail) (stones as big as eggs), whereby many trees and corne (grain) were destroyed; and birds were seen flying in the ayre (air) with fyre (fire) in their mouthes (mouths), and to set fyre (fire) in houses and burn them. in England during the years 1201-1204, the long rains caused famine, dearth and great mortality.

All summer 1202 was troubled by storms, which delayed many fleets leaving from the Flemish ports of the Netherlands bound for the Third Crusade.

On 20 May 1202, a tsunami probably associated with this event was observed in eastern Cyprus and along the Syrian and Lebanese coasts.

In 1203 there were heavy rains in London and there was a very sore famine. Multitudes of poor died. There were bad seasons. In England in 1203, there was a great mort (death) and famine from long rains. In England during the years 1201-1204, the long rains caused famine, dearth and great mortality.

In Ireland in 1203, there was a great famine "so that the priests ate flesh meat in Lent". Christians during this time era abstained from eating meat during the Lenten period as a sacred obligation.

In the Winter of 1203-1204 CE from late January to May in 1204, there was a continuous drought and a burning heat in summer. This season was very destructive to the fruits of the earth, and as a result a very large famine and mortality occurred in England, France, Spain and Italy. In April 1204, a famine still prevailed in the North and the East.

In 1204 there was mortality (death). In Ireland, a prodigious number (of people) perished. There was a general plague throughout Europe.

In London, England alone, 200 were buried daily in the Charterhouse yard. In England during the years 1201-1204, the long rains caused famine, dearth and great mortality.

In 1204 in Italy, the summer was extremely dry and hot.

From late January to May 1204 in France, the heat and drought were unusual. In 1204 from February to April (in France), it was very dry. It did not rain or rained very little during the months of February, March and April. The hot weather followed three months of drought.

In 1204, Auge and the neighborhood of Caen in northern France were almost submerged by floodwaters.

In 1204, there was a terrible great flood of water in Germany. The greatest since the birth of Christ. All marshlands were inundated, and many people and livestock drowned.

In England the winter of 1204 and 1205 CE was one of the most severe winters of history and most rivers, including the Thames, were frozen completely. The frost prevented plowing, and all agricultural work was suspended from 14th January to 22nd March. The winter seed was destroyed and there was widespread famine. There was frost until March 22 with deep snows. "Wheate was sold for a marke (mark was currency equivalent to 160 pence) the quarter (quarter ton), which before was at 12 pence." "Frozen ale and wine sold by weight." "In the seventh year of King John began a great frost, which continued till the 22nd March, so that the ground could not be tilled, whereof it came to passe, that in the summer following a quarter of (a ton of) wheat was sold in many places in England for a mark, which for the more part of the days of Henry II, was sold for 12d., and a quarter of (a ton of) beans and peas for a noble, and a quarter of (a ton of) oats for 3s. 4d., which were wont to be sold for 4d." (d. = pennies; s. = a shilling worth 12d.; a mark worth 20s.; a noble was a gold coin worth 6s. 8d.)

In 1204 or 1205 on the Nones of December began a most violent rigorous frost, which continued to 12 April. So, the ground could neither be plowed nor sown. Hence there came a dearth. But there was the fertility and plenty from the little corn that was sown with difficulty. The frost killed much sheep and cattle with their young. In England, the cold of Christmas lasted until the vernal equinox (around March 20 or 21).

The winter of 1204-05 was very harsh in France, Flanders (now Belgium) and England.

On the Continent (Europe) a great mortality of the animals, especially sheep and birds occurred. A famine followed this severe weather.

The cold of 1204 in northern France surpassed everything we had seen in living memory. The winter of 1205 was very cold in Brittany and Normandy, France during mid-January to mid-March.

Between 1205 CE and 1225 CE Genghis Khan and the Mongol Horde erupted deep into European Russia, to the Indus and to the gates of Beijing in China. What is known of cooling in high latitudes from the isotope record in northern Greenland and the great advance of Arctic Sea ice towards Iceland, brings up the possibility that colder Arctic air than previously invaded the heart of Asia. The effects of this would be particularly noticeable if it occurred in summer. Some scientists speculate that China had long been experiencing a cold regime and that this anomaly gradually spread westward until it enveloped Europe in the Little Ice Age of later centuries.

On December 4th 1206 CE Hekla erupted in Iceland with a VEI of 3.

In 1206 there was a great flood of water in Germany. The Main River which flows into the Rhine rose to 32 Ellen (74 feet, 22 meters). The Rhine River destroyed certain protective dams and drowned several thousand men, women and children.

During Winter 1206-1207 CE a terrible flood succeeded by lightning and thunder in December 1206, brought a tremendous disaster. We do not recall ever having seen a similar flood. The Seine River in Paris, France broke three arches of the bridge "Petit Pont" and all the streets of Paris were so flooded that they could only be accessed by boat.

On 5 December 1206 in France, lightning and thunder accompanied abundant rainfall. These rains brought excessive floods. In December 1206, the greatest flood ever seen took place in France. (Various accounts of this greatest ever seen St. Nicolas flood in France are also described in years 1207, 1208 and the Winter of 1209/1210.) In France in December 1206 on the eve of the Feast of St. Nicolas, December 6, there were violent burst of lightning and thunder. Winds and a raging storm accompanied the rains of spring and summer and the cold winter. People travelled throughout the city by boat.

A blizzard struck England on 27 January 1207. Many houses fell in. The storm left deep snowdrifts.

In Germany, many travelers froze to death on the roads. On January 17th 1206 or 1207, about the middle of the night, there suddenly rose such a tempest of wind, as blew down many houses. The area was buried in snow and drifts. Many flocks of sheep and cattle were destroyed.

In 1207, the River Thames in England was frozen for eleven weeks. In 1207, the frost in Britain lasted fifteen weeks.

The summer of 1208 in France was hot. The grape vines flowered in May.

In December 1208 in France, there were terrible rains and great floods, destroying bridges, houses etc. "Greatest ever seen in France." In 1208 there were such terrible rains, thunder and hail, which killed men, destroyed grape vines, trees and corn (grain). In December was the greatest inundation in France that the oldest of that age had seen, overthrowing bridges and buildings.

Were these consecutive floods or separate floods?

In 1209 CE Old London Bridge was built. Because of its construction it allowed buildup of water upriver, particularly when debris clogged the gaps. Even without such problems high-water levels could lead to a significant difference between up-river and down-river sides. Differences of several feet are mentioned. Conversely tidal rise and fall was dampened by the bridge, decreasing the chance of tidal flooding above the bridge. However, the bridge increased the chance of fluvial flooding upstream. Bermondsey in London experienced flooding that year and is only one mile east of the bridge.

The winter of 1209 to 1210 CE was regarded as a severe winter in Europe and England. There was severe frost in January and early February in London and southern England.

In the Winter of 1209-1210 CE in England there were great floods on St. Nicholas Eve (December 5), "after a tempest of thunder and lightning." Thunder and lightning causing many houses to be burnt, followed by very high floods, which caused great damage. Wind blew down houses and trees. In England, in the year 1209, the winter was long and severe and followed by a dearth. In the year 1209 or 1211, there was a severe long winter in England followed by a dearth. In England in 1209, there was a famine from a rainy summer and severe winter. In 1209 or 1211, there were terrible thunders this summer, severe heavy rains, a stormy and cold winter, hence a scarcity and famine.

In 1210, the frost was the hardest in England from 1 January to 1 March and a dearth followed. There were great deposits of snow.

In the year 1210 in the beginning of January, a very severe frost began in France, which lasted about two months and the winter crop was spoiled for the most part, and the little that they reaped in some places in wheat, barely gave back as much as they had sown. This winter was very disastrous for the cattle. In 1210 in northern France, we experienced a very sharp frost at the beginning of January, which continued nearly two months. It prevented the sowing of the winter crops and destroyed many seeds sown. The French historians cite great mortality of animals due to cold during 1210.

In 1210 CE Katla erupted in Iceland with a VEI of 4 and Reykjanes, also in Iceland, had a VEI of 3. There should be a volcanic winter as a result of this.

In 1210 or 1212, there were great floods in the Rivers Tay and Anan in Scotland. The city of Perth was overflowed (flooded) and most houses broke down and many people were drowned. The King lost his youngest son and nurse in it and twelve more of the court ladies. The King and his brother with great difficulty escaped in a boat. There was a strong frost from January to March whereby the grain sown was killed and it yielded not as much a crop as sown. People were afflicted with sundry diseases, and many died. This was a sickly time.

At Perth, about the time of the feast of St. Michael, the flood carried off much of the harvest crops from the field. The water of the River Tay and the River Almond so swelled that the large bridge of St. John was overthrown. "William the King, David Earl of Huntingdon, the King's brother, Alexander, the King's son, with some of the principal nobility, went into a boat, and sailed quickly out of the town, otherwise possibly they might have perished."

In 1210 or 2012 in England, there was a famine caused by last summer's rain and winter's frost.

In 1210 CE Owens Lake in California was severely desiccated. This was a period of drought also shown at Lake Walker also in California. Mono Lake, Lake Walker and Owens Lake show the same cycle. This drought cycle started in 1210 CE and lasted until about 1350 CE. Inflow to Owens Lake appeared to have dropped between 45 to 50 per cent of late twentieth century levels.

Analysis of Silver pine tree rings in New Zealand indicate that there were above average temperatures in New Zealand between 1210 CE and 1260 CE. The temperatures were comparable to warming since the 1950s in the present period.

On August 31st 1211 Eldey erupted in Iceland with a VEI of 4. Can three eruptions in Iceland within one year create a volcanic winter?

Between 1211 CE and 1225 CE there was overall wetter and warmer weather in Mongolia.

In the summer of 1212 CE it was a dry summer and a great fire in London.

In 1212 CE early frost destroyed the harvest in Novgorod, Russia in 1212 CE and children were sold as slaves for bread.

In Sicily Italy, there was an inundation from the sea, "thousands of people swept away by it." In 1212 in Cathinna in Sicily, some thousands of people were swept away by an inundation of the sea and in Italy fell a shower of hail, each stone as large as a goose egg.

The year 1212 was very dry in France. The summer of 1212 in France was very dry. Throughout the rest of the year, the weather was dry.

In 1212 CE sea floods attacked the northern Netherlands. There was an enormous loss of life with 306,000 people drowning.

Between 1212 CE and 1215 CE there were many extraordinarily very severe North Sea sea floodss.

The winter of 1213 was so long and hard before and after Christmas, that in Vienna, Austria, the Danube River froze three times and could be crossed on the ice.

In 1214 the North Sea flooded seven kilometers inland at Brugge in Belgium and many people were drowned.

In summer 1214 CE there was another dry summer in which the Thames was so low in London that women and children could wade across it.

In 1214 CE in Novgorod, Russia on 1st February, on Quinquagesima Sunday, there was thunder after morning service, and all heard it; and then at the same time they saw a flying snake or dragon. In China the term dragons were often used for meteorites or comets. This terminology could have been used in Novgorod as well.

In 1214 the River Thames in England was so low between the tower and the bridges that men, women and children waded over it. The water was only four inches (10 centimeters) deep owing to so great an ebb in the ocean that laid the sands bare several miles from the shore, which continued a whole day. In London, England, the level of the water was so low in the River Thames that individuals waded through it at the Tower after the sea withdrew for several miles. What would cause the sea to recede but with no report of it returning?

In 1215 CE in the English Channel there was a great hurricane off the coast of Calais, France. A number of the Norman nobility on their way to assist King John against the barons were wrecked. Hugh de Beauvais and several thousand foreigners, on their voyage to assist King John against the barons perished.

In Novgorod in Russia during autumn, much harm was done; frost killed the corn (grain) crops throughout the district, but at Torzhok all remained whole. The Knyaz (prince) seized all the corn in Torzhok and would not let one cartload into the city (Novgorod). And in Novgorod it was very bad. They bought one barrel (11½ pecks) of rye for ten grivnas (grivna is a circular ingot of silver), one of oats for three grivnas, a load of turnips for two grivnas. People ate pine bark and lime tree leaves and moss. Oh brothers, then was the trouble; they gave their children into slavery. They dug a public grave and filled it full. Oh, there was trouble! Corpses in the marketplace, corpses in the street, corpses in the fields; the dogs could not eat up the men! Early frost destroyed the harvest throughout the district about Novgorod. People ate pine bark and sold their children into slavery to get money for bread. Many common graves were filled with corpses, but they could not bury them all. Those who remained alive hastened to the sea. Were the three Icelandic eruptions responsible for this cold event in Novgorod?

Andrew Douglass, an astronomer, interpreted ring growth in pines from Anasazi ruins and other sites in New Mexico and found patterns were emerging where every now and then tree rings would indicate periods of minimal growth due to changes in climate such as lack of water. 1215 -1217 CE showed four narrow rings followed by two wide rings with a noticeable narrow ring seven years later. This was the infant science of dendrochronology or the study of tree rings to determine dating.

In 1216 to 1217 CE there was a cold winter in Western Europe. This implied that parts of Britain may have been affected as well.

In 1216 North Sea sea floodss attacked Bremen in Germany and 20,000 lives were lost. Bremen is around 100 kilometers from the North Sea coast.

In 1216 much of the archipelago of Heligoland in the North Sea off Germany was lost in a sea floods. In 1216, there was a great flood that collapsed countries, swallowed up 10,000 people along with livestock and buildings.

Prior to the flood of 1030, Helgeland, Norway had 9 parishes, but now only 2 remained. The great flood of February 1216 began the separation of the islands that exists today from the mainland.

Even as early as the year 1000, there were no islands on the North Frisian coast in the Netherlands. Only with recurrent massive storm surges the islands were created. In 1216 Sylt and Fohr were separated.

In 1216 CE the winter in Italy was similar to the year 1133 CE. The River Po in Italy was frozen to the depth of 15 Ellen (~ 34.5 feet or 10.5 meters). The wine in barrels in the cellars burst from freezing. The Elle is an old unit of German measurement. It is based on the length of the arm bone, which is generally 60-80 centimeters long. The length varied. In the north, the Elle was generally around 2 feet. In Prussia, it was generally 2 1/8 feet. In the South, it was 2 ½ feet.

In 1216, the Rhône River in southern France froze to a great depth.

The winter of 1218 to 1219 CE there was a cold winter in Western Europe.

In the year 1218, a very severe frost began on 27 September 1218, which was very destructive. Seven days later another frost took the grapes, which had been harvested for the most part.

On 29 September 1218 in France, a heavy frost accompanied by snow, reigned for seven days and destroyed, at the time of harvest, most of the grapes. New snow and cruel freezing temperatures struck after the 30th of October and persisted relentlessly until 6 December. People crossed the ice on our larger lakes and the largest rivers, including the Loire and the Seine rivers. The cold dampened somewhat with the arrival of winds from the south, but this did not last long with the sudden return of northerly winds.

On 27 October 1218, it began to freeze, and the cold lasted, with intermittent snowfalls up to St. Nicholas (December 6).

On 17th November 1218 sea floods attacked the Netherlands and Germany. About 100,000 lives were lost. In Germany the Jadebusen basin formed near Wilhelmshaven in Lower Saxony. The extension of Jadebusen, or Jade Bight, and its branches fragmented the free Frisian territory of Rustringen in Bant in the northwest, most of which had disappeared in the waves. On 17 November 1218 there was a terrible abnormal flood that inundated North Friesland houses and villages. One surmises that at that time the north shore at Lundenberger Harder (Lundenbergsand) was separated.

The weather continued so violent, that everything was frozen, the earth, the lakes, the rivers, and especially the Seine and Loire rivers in France. After a decline caused by the west wind brought us the cold north wind then suddenly there was very rough weather including snow and frost, which lasted until mid-March 2019.

In France the frost and snow were continuous until the middle of March. Unbearable cold winds finally followed the killing frost, so that in the middle of May, the bare fields had barely a few ears, and the vines a few buds. In many places, the frost was so fatal that it forced them to plow and sow the fields twice.

In 1218, large rivers in France, particularly the Seine and Loire Rivers were frozen and were crossed on the ice.

During the winter of 1219, the Loire, Seine and Vienne rivers in France froze three times. Then the Seine River in Paris flooded. "The water was up to the second floor."

In 1218 at the siege of Damiata (now Dumyat) in Egypt in winter, the east wind blowing, the Nile River swelled and did great damage to the besiegers.

In 1219 there was a huge storm surge in the English Channel and the North Sea.

In France icy winds blew even after the thaw. Therefore, the fields in May 2019, one could only see an isolated grain on the stalks or weak shoots on the vines. In many places the land was worked over again and replanted.

In 1219 CE in Friesland and Groningen in the Netherlands. On the 16th of January 1219 36,000 people perished when a sea floods came in. This was called the First Saint Marcellus's Flood.

Also, on the 16th of January 1219 land was lost in Nordstrand in the North Sea in Schleswig in Germany during a sea floods. At the time Nordstrand was on the island of Strand.

In October 1219 in Nordland, Germany "The St. Lawrence Lake broke out and drowned 36,000 people besides cattle." In 1219 in Nordland, 36,000 men perished by a sudden flood.

A storm on 16 January 1219 produced a large drowning primarily to the Dutch coast (Friesland), but also on the west coast of Schleswig-Holstein where some 10,000 people were drowned. The Saxon Chronicle speaks of a total of 36,000 deaths from this storm. The storm increased the depth of Jade Bay or Jadebusen. Hail and storm surge accompanied the full moon, when there was virtually no tide. There was enormous damage to the existing dikes (at the time, there was still not a closed dam line of the coast). These seem disjointed as they are different reports of the same event.

On St. Luke's, October 18 1219 there was a tempest and hurricane out of the northwest that struck England. The winter was terrible with thunder and continual rains and hurricanes. In England all winter there were frequent thunders, continual rains, and violent hurricanes.

In 1219 in France, an impetuous west wind blew hard during the first months of March and April. In April, although it did not rain, the (river) waters swelled beyond measure, and ravaged for a month and half the surrounding countryside. In Paris, the Seine River invaded a great number of houses where it was impossible to enter without boats. The unexpected flood even covered the Petit-Pont Bridge. It was still impassable during the first fortnight of May. Soon the rainy season began. The rains began on June 24 and continued without interruption until the following August. This unusual weather therefore delayed the grain harvest and the grape harvest. Then early frosts accompanied by large snowstorms brought about the ruin of the harvest of wine. Then the rains resumed again and continued, still tirelessly, along with terrible floods, until February 1220. The floods submerged almost the entire city of Grenoble in southeastern France.

Also, in 1219 in France, westerly winds blew incessantly during the months of March and April. These were succeeded by long rains that lasted until the Feast of Saint John's. In mid-August, the weather produced extraordinary bursts of lightning and thunder. During the last Monday of August, a hard frost withered grape vines. At the end of September, the weather produced cruel frosts, which lasted three weeks along with copious amounts of snow that stayed on the ground for several days. Sustained rainfall ended this year.

In the year 1219, the wine in France had to endure the harshest adversity. When the grapes were flowering, it rained constantly. In the last days of August to the end of September, the usual harvest time came a frost that destroyed the grapevines. It was very cold for three weeks and the grapes could not ripen. A thick snow covered the ground for several days. As a result, almost all the wine was lost in the Kingdom of France.

From 1220 to 1221 CE there was a famine in Egypt in which 100 to 500 people a day died in Cairo of starvation due to low Nile flood levels. Incidentally it is now known that famine in Egypt almost always coincided with famine in India and vice versa. This is due to the Southern Oscillation.

In Friesland in the Netherlands, there was a considerable inundation in October.

In 1220 or 1221 there were continual great rains all the summer in Poland. Hence there were so great floods. Many villages were swept away. The winter corn (grain) was lost and there was no sowing in the spring. A sharp horrid cold winter followed. Then came three years of famine and plague wherein myriads of people and cattle died. There was so great a mortality in foreign countries. The number of individuals that were well were not enough to take care of the sick and bury the dead. In cities, towns and villages the mortality was so great that sometime only three or four survived, yet these had multitudes of dead to bury.

On 18th October 1220 CE there was a violent northeasterly gale that did much damage in London.

In England in April 1221, there was a prodigious snow, which broke down many trees. The frost that followed killed far more, so that in many places no leaves appeared on them that summer. There were no apples in most places. After this there was so great a drought that most late sown seeds died. On Holy Rood Day (September 14), there was a terrible and destructive thunder and lightning with profound rains, long and deluging floods. On November 30th there was a tempest of thunder and lightning producing great damage to England. At the same time a great hurricane overthrew houses and trees. This storminess continued until Candlemas, February 2nd 1022.

In 1222 CE Hekla in Iceland erupted with a VEI of 2.

The 1222 Cyprus earthquake occurred at about 06:15 UTC on 11 May. It had an estimated magnitude of 7.0–7.5 and triggered a paleotsunami that was recorded in Libya and Alexandria. The strongest shaking was felt in Nicosia, Limassol and Paphos. Many people died, although there are no estimates for the total number of casualties. Much damage was caused at Limassol and Nicosia and other parts of the island, but the greatest damage was done at Paphos, where there was great loss of life. Paphos Castle, a Byzantine fort, was destroyed and had to be rebuilt by the Lusignans. A modern excavation at Paphos Castle found the remains of a man who apparently climbed into the castle's main drain to escape the earthquake but was trapped there by falling masonry. The sea retreated from the harbour but returned and flooded the town. A church is said to have fallen, burying the bishop and his congregation. Monks of the Franciscan order abandoned their church in Paphos after the earthquake. The castle of Saranta Kolones, built only 30 years earlier overlooking the harbour, was destroyed by the earthquake. It was never rebuilt, as it was no longer needed to protect the port, which had dried up.nd The earthquake permanently changed Paphos, rendering the harbour unusable, and moving the shoreline seawards; it no longer had a protected anchorage.

In 1222 CE there was a dry and hot summer in London and southern England.

The summer of 1222 was very hot in France.

In the winter of 1021/ 1222 in England and Wales, the land was so inundated with continuous rains that scarcely an article of food was raised. Violent rain occurred in London, England. The storm threw down several churches. From 14 September 1021 to 10 February 1022, there were terrible tempests of thunder, lightning, rain and dearth. Lightning and thunder, so dreadful, as to throw down several churches in February 1022. In England in the year 1022, high tides and continuous rainstorms did great damage. In the "seventh yeare of Henry III, on Holy Rood Day (September 14), was a great thunder and lightning tempest throughout all England, and such great floods of water followed with great winds and tempests, which continued till Candlemas (February 2), that the year following wheat was sold for 12 shillings, the quarter."

On 8 February at Grantham in Lincolnshire there was such thunder and lightning, as filled the church with a most offensive smell, that the people fled out of it. On the Day of the Exaultation of the Cross (September 14), there was thunder throughout all of England. There was a most shocking winter for thunder, lightning and hurricanes, which demolished many buildings as houses, churches, steeples, etc. These misfortunes caused a dearth of corn (grain) in 1223. On Saint Lucia Eve (12 December), there was a most destructive tempest of wind. The sea also rose with higher tides and springs than ordinary. Thunder killed many people chiefly in Warwickshire.

In 1223 CE there was a very wet year with much flooding.

In 1223 CE Reykjanes erupted in Iceland with a VEI of 3.

The rains that fell in April, May, June and July 1224, completely destroyed the nuts and grains in France. In 1224 in France, there was so much rain mixed with winds and clouds, from April to August, that the wheat and nuts died penniless. The harvest in turn was reduced to almost nothing by the frosts of autumn. Then came a winter with so rough and violent a wind that it threw down the towers of churches in many locations.

In Western Europe in 1224, the heat of the summer was so strong that the grain dried up. Violent winds prevailed during the entire month of August, completed the devastation of the fields.

In England in 1224, there was a very dry winter and bad time for seeds; whence followed a great famine. There were several great rains and thunders, hailstones as big as eggs. These destroyed trees, grape vines, corn (grain) etc. In England, there were terrible hurricanes. There were great tempests, destruction of corns, trees and buildings and shipping. Yet there was so great a drought in winter, as hindered sowing of corn; hence a scarcity.

The winter of 1224 to 1225 CE was a severe winter in London and southern England. "The winter stretched from St. Denis (9 October) until the feast of St. Mark's (25 April) and was very severe. A violent wind struck down the harvest and also tore down church steeples in several places in France and Normandy. A very strong famine prevailed in the whole continent (Europe) but particularly in Flanders (now Belgium), but we have, thank God, not heard that any man had died because of it."

On 18 October 1224 in London, England, there was a great windstorm (hurricane/tornado) that caused significant damage. In "1224, in this yere, upon Seynt Luke's day (October 18) there blew a gret wynd out of the north, whiche caste downe manye houses, steples, and torrettes of chirches, and turned up so downe trees in wodes and in orchardes; at which tyme fyry dragons, and wykkes spirytes, grete noumbre, were seyn openly fleying in the eyre." (In 1224 in this year, upon Saint Luke's day (October 18), there blew a great wind out of the north, which cast down many houses, steeples, and turrets of churches, and turned upside down trees in the woods and in orchards; at which time fiery dragons, and (weak?) spirits, in great number, were seen openly flying in the air.)

In early times, when events happened that defied logical explanation, they were sometimes attributed to mythical creatures. It was common in China to describe tornadoes as dragons. The roar of a tornado became the roar of a great dragon. The debris picked up through the tornado's funnel cloud and thrown miles away became spirits flying through the air. Only a great dragon could swoop down from the top of a cloud and instantly cause great devastation by uprooting or breaking massive oak trees in two or utterly destroying homes and buildings and then as quickly as they appeared, they disappeared. Or this could be referring to a meteor shower. If it was lightning, it would have been written as such but it was not.

At the extreme of coldness of 1224 in northern France, joined a violent wind, which uprooted the crops in several places and overthrew the towers of churches. The winter of 1224/1225 was very cold and unusually long (in France). The frost lasted from 9 October until 25 April. The storms in Normandy were so fierce that church towers fell.

There was a long and severe winter in 1225 in France, followed by an unparalleled famine, fatal to many.

In 1225 there was a great death of sheep in England. In 1223 or 1225 in England, there was a great and deep snow. It snowed many days and began on 14 January.

On July 15th 1226 CE Reykjanes erupted in Iceland with a VEI of 4. This can still cause stratospheric disturbances that can affect the weather. Especially with the higher sulfur levels.

In 1226 CE the Rhône River in France overflowed its banks and combined with the effects of war caused great destruction to the fortified city and the people of Avignon in southeastern France. This unfortunate city had, like the Count of Toulouse embraced the cause of the Albigenses. Louis VIII came with a considerable army to besiege Avignon in the beginning of the year 1226. This siege lasted three months until the city capitulated. Louis VIII forced the Avignonais to raze the walls that protected their city against foreign enemies. The flooding began a few months after this siege. The Rhône River overflowed its banks on September 17th. The floodwaters found no obstacle (city walls) to block it from entering the city, spread with strength to the lower parts of the city. It caused great damage and added to the misery of the people.

The extreme drought in France in the year 1226 brought about the ruin of almost all the summer crops. The summer and fall of 1226 also proved hot and dry. Across France, this dry heat produced a prodigious quantity of wine. The summer of 1226 in France caused a drought. The wines were wonderful.

In 1226 snow fell in Syria.

In England 1226 there was a terrible hurricane with a north wind.

In January 1227, thick sea ice on the Baltic Sea allowed a German Army of Monks (the Livonian Brothers of the Sword) to march from the mainland of Estonia to the islands of Muhu and Saaremaa and to capture these islands.

In France, a very heavy frost and dry clear weather, preceded by a warm, dry autumn reigned continuously from 1 November 1226 to 5 February 1227 and killed the olive trees.

The winter of 1227 was very dry and cold and lasted till February in France. The drought lasted until February 1227. The frost lasted from 1226 in northern France, in clear weather and dry, since the first days of November until 5 or 6 February.

In 1227 CE there was famine throughout Ireland and much sickness and death among men from various causes including cold, famine and every kind of disease.

In 1228 CE in Friesland (the Netherlands), an irruption of the sea caused 100,000 people to drown. In 1228 in Friesland the sea overflowed its banks, demolished towns, churches, castles innumerable, and drowned over 100,000 people. Why would people return to Friesland when it was being frequently drowned by sea floodss with great loss of life.

In England there were terrible thunder and lightning all summer, ruining houses, killing man and beast. The summer was so hot that the harvest was fully ended by midsummer. During the harvest there were excessive rains. There was a terrible thunder and lightning storms along with meteors all summer in England. In 1228 all summer was terrible with thunder and lightning, killing man and beast; in harvest there was incessant rains.

In Europe, the summer of 1228 was very hot. The harvest was all finished by the Feast of Saint John (24 June).

In England, lightning killed many people and animals.

The summer of 1228 was hot in France.

In 1228, there was a very mild winter and spring in Germany. In April the grape vines bloomed. The grape harvest was over before the feast of John the Baptist. In late July, the grapes were ripe.

In Novgorod in Russia during autumn, great rains came down day and night. From our Lady's Day, the Assumption, 15 August, until St. Nicholas Day,19 December, we saw not the light of day. People could not get the hay nor do the fields. Also, there was great flood in the Volkhov. Around the lake and along the Volkhov River, the flood waters carried away the hay. Then the lake froze, and the ice stood for three days, a south wind drove it up and having broken the ice carried it into the Volkhov River, tore away nine stays of the great bridge, and carried down eight by night to the Pitba stream on St. Nicholas Day, and the ninth stay was carried away on 8 December, St. Potapi Day.

In 1228 in England after a rainy harvest followed by a frost, the heavens had fiery flames all winter.; and in winter a hard frost. There was great and deep snowfalls that lay on the ground for a long time.

In October 1229 CE Zoazan erupted in Japan with a VEI of 3.

On December 31st 1229 CE Asosan in Japan erupted with a VEI of 2.

There was a two-year famine between 1229 to 1230 CE in Russia. By the second year there were many incidents of cannibalism over the whole district of Russia with the exception of Kiev.

In 1230 (the 2nd year of Kwanki) there were universally poor crops in the country of Japan. Starved corpses lay uncared for along the roads. Poor people, hard pressed for a living, sold their wives and children. In extreme instances, not a few men sold themselves as slaves. Whereupon, with the idea of meeting the extraordinary situation with an extraordinary method, the sale of the human body was publicly permitted. That it was permitted to those who saved starving men and women to make them their slaves was noted in the order of the Shogunate Government, dated April 17th, 1228 (1st year of Sho-o) as recorded in a compilation of Government orders. Was this due to the Zaozan eruption the previous year?

In Italy in 1230, there was a great overflow of the Tiber River, which overflowed the low city. This was followed by a famine. The Tiber River overflowed, so that it reached to the stairs of Saint Peter's Church. The lower city was drowned. Then followed such a famine that scarce one in sixteen persons survived. In July and August, it was so burning hot that men roasted eggs in the sand.

There were floods in France.

In 1230 in Rome, Italy, there was a famine after a deluge of the Tiber River.

In 1230 CE a great famine consumed Russia. "For what is there to say, or what to speak of the punishment that came to us from God? Of the bitter and sad memory of that spring! How that some of the common people killed the living and ate them; others cutting up dead flesh and corpses and ate them; others ate horseflesh, dogs and cats. Those found who committed such acts were punished: some they burned with fire, others they cut to pieces, and others they hanged. Some fed on moss, snails, pine-bark, lime-bark, lime and elm-tree leaves, and whatever each could think of. And again, other wicked men began to burn the good people's houses, where they suspected that there was rye; and so, they plundered their property. Instead of repentance for our wickedness, we became more prone to wickedness than before, though seeing before our eyes the wrath of God: the dead in the streets and in the marketplace, and on the great bridge, being devoured by dogs, so that they could not bury them. They put another pit outside at the end of Chudinets Street, and that became full, and there is no counting the number of bodies in it. And they put a third at Koleno beyond the Church of the Holy Nativity, and that likewise became full, there was no counting the bodies. And seeing all this before our eyes we should have become better; but we became worse. Brother had no sympathy with brother, nor father with son, nor mother with daughter, nor would neighbour break bread with neighbour. There was no kindness among us, but misery and unhappiness; in the streets unkindness one to another, at home anguish, seeing children crying for bread and others dying. And we were buying a loaf for a grivna, a circular ingot of silver, and more, and a fourth of a barrel of rye for one silver grivna. Fathers and mothers gave away their children into servitude to merchants for bread. This distress was not in our land alone, but over the whole Russia province except Kiev alone. And so has God rewarded us according to our deeds".

"In Novgorod in Russia, on the Day of the Exaltation of the Honourable Cross (September 26), a frost killed the crops throughout our district and from that there arose great misery. We began to buy bread at eight kunas, a barrel of rye at twenty grivnas, a grivna is a circular ingot of silver, or at twenty-five in the courts, wheat at forty grivnas, millet at eight, and oats at thirteen grivnas; our town and our country went asunder, and other towns and countries became full of our own brothers and sisters; and the rest began to die. And who would not weep at this, seeing the dead lying in the streets, and the little ones devoured by dogs? Vladyka Spiridon put a common grave by the Church of the Holy Apostles in Prussian Street and engaged a good and gentle man by name Stanila to carry away the dead on horses wherever he went about the town and so continuously he dragged them every day; and filled it up to the top; there were 3,030 corpses in it."

On November 29th 1230 CE Zaozan in Japan erupted with a VEI of 2.

In 1230 CE a significant El Nino event took place when one descended on the west coast of South America with significant flooding and damage. The El Nino event devastated Peru's North coast.

In 1231 CE Reykjanes erupted in Iceland with a VEI of 3.

In England in 1232 CE in November, it was thundery. This is a month that is not normally noted for thunderstorms inland. Also, in 1232 CE London experienced fifteen days of thunderstorms.

The summer of 1232 was hot. In the east of France during July and August, one could "cook eggs in the sand".

In Austria, there was a general overflow of the Danube River. In 1232, the Danube overflowed its banks and did much damage. It drowned people, cattle, towns, corn and woods. Hence there was scarcity and famine.

In November 1233 CE the thundery weather occurred again in England. There was a wet summer in England. Heavy rains led to severe and widespread flooding over most of England. Severe thunderstorms occurred on the 10th of February and again in November. This sort of weather would have had major effects upon a mostly rural population, with possible famine and disease. In England and Wales, the land was so inundated with continuous rains that scarcely an article of food was raised. The rain was violent in London, England. In 1233, it thundered for fifteen days together in England. The next year began with terrible tempest of thunder, rain and floods, which spoiled the fruits of the earth. In 1233 in England, there was a great tempest of wind, with rain and it thundered for fifteen consecutive days. (Some accounts place this event in 1231 or 1232.) It thundered 15 days together, with rain and floods that destroyed the fruits of the earth. It began the day after the feast of St. Martins. "Which lasted 15 daies, beginning th' morrow after St. Martin's Day", 11th November. In England, there was a rainy spring with many floods. But on the morrow of Saint Martin's Day in 1232 was great thunder and lightning, which continued 19 days together. In November 1233, there was a tempest with terrible thunder and lightning in England. In England in March 1233, there was thunder, lightning and rain for 15 days together.

In 1233, in England, the frost "lasted till Candlemas."

During the 18th year of Henry III reign (1234) in England, "was a great frost at Christmasse, which destroyed the corne in the ground, and the roots and hearbs in the gardens, continuing till Candlemasse without any snow, so that no man could plough the ground; and all the yeare after was unseasonable weather, so that barrenesse of all things ensued, and many poor folkes died for the want of victualls, the rich being so bewitched with avarice that they could yield them no reliefe."

There was so great a frost in 1233 in Gallia Cisalpine, Italy that the Venetians walked on the ice of the Po River and travelled with coaches and wagons over it as in a land journey. Wine was frozen in bottles and was thawed to melt it. Vines and other trees died. Many people froze to death in bed.

In the winter of 1233 and 1234 CE there was a long and severe frost from Christmas to 2nd February 1234.

In the Winter of 1233-1234 CE the sea between Norway and Denmark, and from Sweden to the island of Gothland, and the Rhine River and Baltic Sea were all frozen and snow fell to a frightful depth.

The winter in Italy in 1234 was similar to the year 1133 CE. It was so cold in Venice, Italy, that the Adriatic Sea froze in 1234. The ice was so thick that it bore the weight of wagons. In 1234, loaded carts and wagons crossed the ice on the frozen Adriatic Sea in front of Venice. In 1234, the Po River in Italy froze.

In England, the frost continued till Candlemas. There was no snow. Corn (grain) was lost. Herbs and roots of trees died. The frost was the hardest and strongest but there was no snow.

In the year 1234, a remarkable winter raged throughout France, England and Italy. On the night of the Feast of the Circumcision of Christ (1 January) joined a very severe and persistent frost. The extreme cold froze the seeds for the most part with the root. During this sad time for the unfortunate people, except the pain of the cold, weighed the torments of hunger.

In 1234, the Rhône River in France froze.

During the winter of 1234, Lake Zurich, the Rhône River and the Bay of Venice were made of ice.

The winter of 1233-34 in France was as long as rigorous. The cold froze the Rhône River and all plants of the south to the roots. A whole pine forest was killed by the cold.

In Germany, the ice of the rivers destroyed bridges, houses, walls and trees.

In 1234 CE famine and plague followed. So great was the famine in France, that men ate grass like oxen, especially in Aquitania.

The plague was so terrible in Pictavia, Scotland, that Saint Maxentius's Church was filled with dead corpse.

In 1234 in England, there was the greatest famine when people ate horseflesh, bark of trees, and grass. A plague followed.

On 25 December, there was great thunder and lightning in England. The rest of 1234 had exceedingly bad weather that was wholly unseasonable.

Hence in 1235 came barrenness, scarcity, dearth and pestilence. Many people died. The famine was so great that people were forced to eat grass, horseflesh, and bark of trees in France and England. In London alone, 20,000 people starved from famine and plague. While backward weather seasons were contributing factors of the great famine of 1235 in England during Henry III's reign, much of the responsibility is laid at the door of the government itself.

On January 25th 1235 CE Kirishimayama erupted in Japan with a VEI of 4.

In 1235 CE the water rose so high in the River Thames in England as to extend round Westminster Hall, to such a depth, that the judges and lawyers were taken from the Hall in boats. (Another account places this event in 1236.) In England in 1236, there were great tempest of rain, which soaked the earth with water and caused monstrous floods. This rain continued all January, February, and part of March. On 10 February, the River Thames rose with such a high tide, as filled Westminster Hall.

The summer of 1235 was hot in France. There was flooding in Paris, France.

In the United States the fourth of the four driest drought periods started in the American West in 1235 CE. This was all within a four-hundred-year period of overall aridity.

The winter of 1235 to 1236 CE was a severe winter in Western Europe.

In 1236 CE there was very heavy rain from January to March. There were two floods. The first of which flooded Westminster Palace early in the year was due to prolonged heavy rain. The second was produced by a storm surge tide in November which drowned many people and a great number of cattle in the Woolwich area. An inundation by the sea in Norfolk destroyed flocks of sheep and herds of cattle, tore up trees and demolished houses. In one village alone about one hundred people died. This must have been a huge wind-driven event caused by a violent depression, very low pressure and high winds. The interesting thing is that the summer of 1236 was noted as dry and hot in London and the southeast.

The English Fenlands in Cambridgeshire were flooded by a sea flood, or storm surge on November 12th 1236. This storm surge inundated the village of Wisbech, and hundreds of people were drowned. Entire flocks of sheep and herds of cattle were destroyed, trees felled, and ships lost. The castle was utterly destroyed. Wisbech is around thirteen miles from the coast. The storm surge most likely came from the wash as land surface levels were only a couple of meters above sea level from the Wash to Wisbech.

In 1236 CE the Danube River was frozen in its full depth for a considerable time. The Danube was frozen over across its entire width and remained frozen for a considerable time.

The Loire River in France was frozen solid. During the winter of 1236 in northern France, the rivers froze. The ice collapsed and destroyed bridges in Tours and Saumur, France. In the year 1236, in France and to the banks of the Danube River, the winter was very hard. The Loire River was first struck with disastrous flooding, and the severe cold and frost came later. The Bridges at Saumur and Tours, France were destroyed by the ice conditions. As a result of these scourges, a famine spread throughout Europe. The summer of 1236 was very hot. Crops failed in Normandy, France because of the drought.

The winter of 1236 was cold in Winchester, England. Rivers froze. There was a terrible thunder and lightning storm in England.

In 1237 CE in February there were heavy rains. The Thames flooded great stretches of the country roughly downstream from Oxford. It is assumed that the winter as while was a wet one. In England in December 1236 and January, February and March of 1237, it was all rainy and great floods. The summer was an excessive drought for five months. However, the earlier rains brought on an epidemic of Ague. This was a rainy, stormy, troublesome and sickly year. Agues were epidemic beyond compare. Wines this year were 16 times as dear as last year.

The summer of 1237 was very hot in France.

In 1238 a sea floods attacked South Jutland in Schleswig in Denmark and destroyed an island, Bollertsand, and its villages which were lost. Bollertsand is now a sandbank off Romo on the west coast of Jutland.

The year was hot and dry in England in 1238 AD.

In 1239 CE Hakusan in Japan erupted with a VEI of 3.

In 1239 in England, there was the greatest famine. It was written that "people eat their children." A plague and such a famine that delicate mothers ate their tender children.

On February 8th 1239 CE Asosan in Japan had a VEI of 2.

In 1240 CE Hualalai in Hawaii erupted with a VEI of 2, and Asosan in Japan had a VEI of 2.

In 1240 CE it was dry from January to March in London and the south of England. In England, the River Thames was greatly flooded from rain and extended above 6 miles at Lambeth.

In France the summer of 1240 was dry and burning hot. The wines produced this year were so strong that they could not be drunk without drinking water. In 1240, there was flooding in Paris, France.

In England in the year 1240 or 1241 for about four months together, it scarcely ever ceased raining. But about Easter, it began to turn clear and fair. Then there was three months of drought. Great famine followed. Wheat rose to 40 shillings. On 7 May, there was a dreadful hurricane.

In 1240 CE the island of Nordstrand in Schleswig in Denmark was separated from the mainland. 62 villages and many parishes were lost. Over half of the agricultural income, at that time, of the Danish dioceses of Slesvig, Schleswig, had been swallowed by the sea salt. Storm floods on the low-lying coasts of the North Sea when the sea level may have been raised after long periods of warm climate and glacier melting and when a cooling Arctic has produced a strengthened thermal gradient in latitudes between about 50° and 65° North, leading to increased storm frequency and severity over this zone.

In 1241 CE from March to October there was a prolonged drought in England. It was dry and hot from 25th March to 28th October in London and the south of England. There was deep snow with great frost after. On St. Mark's night (April 25) there was frost and snow fatal to fruit trees. In England on Saint Lucius's Day fell a prodigious snow with great winds, deep drifts. Many people and cattle were lost. A long and severe frost followed. In 1241 in England, there was a deep snow followed by a great frost.

The summer of 1241 was very hot in France. There was a drought from 6 January to 20 September. The wine was famous but there was a poor harvest of grain crops due to the drought.

In England there was a drought from 1242 to 1245 CE.

On 12th November 1242 there was a sea floods in the Thames which inundated all lower parts of London up to ten kilometers from the river.

Also, on 12th November 1242 sea floods struck Oldenburg in Schleswig in Germany. Netherlands.

On November 19th 1242 CE there was heavy rain and thunderstorms and on many days thereafter. The Thames flooded at Westminster and Lambeth.

On 27th November 1242 sea floods struck the Netherlands.

A winter storm on 6 December 1245 was so cold in Ireland that many people lost toes due to frostbite. The snow was referred to as "poisonous snow".

In England in 1243 and 1244, there were great droughts followed by a most fatal plague. There was a dry autumn in London and the south of England.

In 1245 CE Katla erupted in Iceland with a VEI of 4.

Around 1245 CE the average life expectancy of a Winchester farmworker was around 24 years if he survived childhood diseases.

A major drought hit Peru and Bolivia between 1245 CE and 1310 CE.

On 18th October 1246 Belgium was attacked by a sea floods.

Thousands of people perished in a sea floods in Friesland in the Netherlands in November 1246.

Also, in November 1246 the area where the Weser River flows into the North Sea was inundated by a sea floods. Thousands of people drowned. The Weser River is in Germany and passes through the Hanseatic city of Bremen and the port of Bremerhaven.

In England, from 1247 to 1250, there were several inundations of the sea and great losses.

In 1247 in England, there were no tides for three months together.

In England from St. Valentine's Day to St. Bennet's, not one fair day, then came the plague. There was such rainy weather that scarce there was one day without rain until the Feast of Saint Bennet. Over the past few years, the great drought brought great and fatal epidemic diseases on all of England. But this year in September 1247, the plague raged sore. Thunder and lightning killed several people and broke down trees. The sea overflowed its banks.

On 20th November 1248 a North Sea sea floods hit Flanders in Belgium.

Also, on 20th November 1248 a storm surge from the sea inundated Friesland and other parts of Holland. There were three storm tides in the Netherlands with major inundations that year.

In England, from 1247 to 1250, there were several inundations of the sea and great losses.

On 28th December 1248 a sea floods hit Hamburg and its surrounds in Northern Germany. There was a great famine in Germany. There was an eclipse of the sun and an inundation of the sea.

In the winter of 1248 to 1249 CE there were three major storms that affected the Dutch coastal communities which may have had an impact on the English coast. These were on November 20th and December 28th in 1248 and February 4th in 1249.

On 28th October 1249 CE there was a gale in London and the south of England.

In 1249 CE the winter in England was so pleasant, sweet and warm, that people fancied the season was changed. There was no frost or snow the whole winter. Folks threw off their cloaks and went in the thinnest lightest summer dress. But from the end of March 1250 to the middle of May came as great a cold. In June fell an abundance of rain about Abington, England that the willow trees, mills and houses near the waterside were borne down and overturned. Corn in the fields was beaten down, and bread made from it when ripe was like bran.

In July, Posson, today Poznań, located in west-central Poland) was burnt by lightning and 300 men who came to the horse races were killed on the course.

In Friffingen (Frisingen or Frisange is a commune and town in southern Luxembourg) was such a plague of mice, which ate up corn, hay and all greens. The year 1249 was a rainy year.

In England, from 1247 to 1250, there were several inundations of the sea: great losses.

In 1249, a sea floods struck with such voracity that it caused great damage in Germany.

1250 CE

In 1250 CE the eruptions were Fuss Peak on Paramushir Island in the Kuril Islands with a VEI of 3, Tokachidake with a VEI of 3 in Japan and Mount Etna in Italy with a VEI of 2.

From the mid-1200s onwards analysis of agricultural records of the time suggests that after the mid-1200s, harvests were increasingly subject to failures for various reason such as drought, cold and wet.

After 1250 CE there was an abrupt change in weather in the Pacific Region to cool, dry times with increased storminess that spanned the Little Ice Age from 1350 CE to 1850 CE.

On 1 October 1250, a most dreadful inundation of the sea did great damage to the marsh-ground in Flanders, now Belgium.

In Friesland in the Netherlands there was a storm surge or sea floods on 1st October 1250. On 1 October 1250, a most dreadful inundation of the sea did great damage to Holland beyond sea and the Zuyder Zee started forming in the Netherlands. It eventually became a large inland lake.

Around this same time thousands of acres of coastal Denmark vanished under the North Sea. This is probably now part of the Wadden Sea which is an expansive area of mud flats that stretches from Den Helder in the northwest of the Netherlands, past the great river estuaries of Germany to its northern boundary at Skallingen in Denmark. This is a total coastline of 500 kilometers, 310 miles, and a total area of around 10,000 square kilometers or 3,900 square miles. This is the largest tidal flats system in the world.

Also, on 1st October 1250 in Humber in England there was a sea floods. In England, from 1247 to 1250, there were several inundations of the sea and great losses. In England on 1 October 1250 there was so great and mighty a hurricane both by sea and land that the likes of it had not been known nor heard of. The sea, contrary to its natural course flowed twice without ebbing, sending into the midland to a great distance a frightful hideous noise. In the night it seemed all in a flame; and the waves to fight one against another. Mariners could not save their vessels. Around Winchelsea in east Sussex on the coast, 300 houses and some churches were carried down by the flood. Besides damage done to churches, steeples, mills, etc. In other parts, inestimable damage was done in Holland, the Lincolnshire Fens, and other low places. There was a most rigorous and long winter, very great snows. At the thaw was a prodigious flood, which did much harm. The Lincolnshire Fens are a vast, extremely flat plain between Cambridge and Lincoln in east England.

Also, on this date a major North Sea gale and sea floods caused great damage to adjacent parts of England, the Netherlands and Flanders in Belgium.

AS well around this same time thousands of acres of coastal Germany permanently vanished under the North Sea.

In 1250 CE the *King's Mirror* (Konungs Skuggsja) tells that the east Greenland ice was formidable. Ice Ages, even small ones, seem to be relatively fast occurring.

In 1250 there was a sharp decline in corn growing north of Trondheim, in Namdalen, in central Norway. Rye by this time had been completely given up on. This was due to decreasing temperatures. Corn growing stopped for two hundred years but by then spruce forest had taken over and some of the farms in the neighborhood now remained unoccupied.

Prior to 1250 CE the upper limit on the Lammermuir Hills southeast of Edinburgh, which had been as high as 425 meters, nearly 1,400 feet, above sea level, fell in stages until around 1600 CE until it was 200 meters lower.

Plentiful rainfall gave way in about 1250 CE in California to an intense drought period which lasted for over a century. The lake levels fell precipitously as evidenced by trees growing in the newly exposed bed.

Between 1250 and 1450 CE the southeast of North America enjoyed almost stable summer rainfall while the Colorado Plateau in the northwest suffered unpredictable rainfall and droughts.

On May 18th 1251 CE Akagisan erupted in Japan with a VEI of 3.

On 19th May 1251 CE several houses in Windsor, including one occupied by the Royal Family (Henry III) were struck by lightning during a severe thunderstorm. In England, a storm caused the chimney of the chamber where the queen of King Henry III, and her children lay, to be blown down, and their whole apartments at Windsor Castle shaken. Many oaks in the park were rent asunder, and turned up by the roots, accompanied with such thunder and lightning as had not been known in the memory of man. Windsor Castle is located in Berkshire in southern England. Norwich is located in Norfolk in eastern England.

On the Feast of Saint Dunstan, 19 May, 1251 the air being darkened from all corners, happened such a terrible tempest of thunder and lightning, as none living had ever seen. It began first at a great distance, but soon burst out in most terrible shocking claps, shaking and demolishing houses, rendering oaks, etc. At the same time, the sea on the coasts of England rose with higher tides than ordinary, by 6 feet. The summer was excessive and intolerably hot. So great mortality followed that in many parishes, a hundred died in a month. The harvest was very early and good. In 1251 a sea floods inundated Kent and Lincolnshire with water two meters higher than ever seen before.

In Germany the summer of 1251 was exceptionally and intolerably hot. As a result, there was so great a mortality that they buried in some parishes a hundred people in one month. There was a wine shortage in France.

In Novgorod in Russia heavy rains came and took away all the ploughed fields and crops and hay; and the flood carried away the large bridge over the Volkhov River, and in the autumn a frost struck the crops, but a remnant was preserved.

In 1252 and 1253 CE there were two dry years, considered by some to be the driest pair of consecutive years known in history. The summer and spring in London and the south of 1252 was outstandingly hot and dry with the ensuing drought ruining crops and many people died from the excessive heat. Oddly enough significant flooding also occurred in October 1252 due to heavy rain. In England in 1252, no rain from Whitsuntide, the week following Pentecost, to Autumn. No grass; hence arose a severe famine. There was great mortality of man and cattle; dearness of grain and scarcity of fruit. In England in 1252 there was a long drought from Easter to the harvest. There was no rain or dew. This condition combined with the morning frosts and northerly winds did great harm both to the fruits and corn, grain. As the season wore on and the heat and drought increased, the remainder of the fruit withered away so that only a tenth part was scarce left. The grass was so burnt up that one might rub it to powder between their hands. Cattle were ready to starve. The exceedingly hot nights brought a vermin of fleas and gnats that were very troublesome. Many diseases followed, such as agues, sweats, etc.

At harvest time there was a great death of cattle, especially in the Fens, Norfolk and the south. The infection was such that dogs and ravens feeding on the carrion, swelled and died, so that people did not dare eat the dead cattle. Heifers and bullocks followed the milk cows and sucked on them as if they had been calves. All apple trees and pear trees after they yielded their first ripe fruits blossomed again as when they did in April. The death of cattle came about in this way. After so great a drought that lasted until the end of July, there came a period of rainfall which produced an abundance of greenery. The starved cattle feed so greedily on this new grass, that they quickly became bloated. This condition also led to their death.

At Michaelmas, 29 September 1252, the plague began in London and spread over the whole nation and reigned till August 1253.

In 1252, there was an inundation of the sea in England. "An inundation of the (River) Humber at Cottyngham, Cottingham, destroyed both man, around 35 people perished, and beast, especially at Owythfleet, Saltage Myrton, Tharlesthorpe, Sutton, and Drypool, where nearly all the buildings were lost. After which Owythfleet, Tharlesthorpe, and Saltage were gradually but totally swallowed up by the Humber."

In 1253 CE Harra Es-Sawad erupted in Yemen with a VEI of 3.

In 1253 CE in spring and summer it was noted as dry and hot in London and the south of England. There was flooding in October from a tidal storm surge.

There was a sea floods in Lincolnshire and the Thames Estuary.

In 1253 CE in England a great drought occurred during the spring and summer. At harvest there fell such great rains, which caused deluging floods. The rivers broke down and overflowed their banks, drowning an abundance of land, destroying many people, many villages and houses in sundry places, such as Holderness and other low countries. After Michaelmas, 29 September, returned such a drought that people could have no corn ground who lived within a day's journey of a grain mill. On Saint Lucius' Day, there fell a great snow. There was significant thunder during winter and a great hurricane. Holderness is located in Yorkshire, on the east coast of England.

In 1254 CE there was severe frost from January to March in London and the south. This is possibly the winter of 1254-1255. In England, the frost was severe between the 1st of January to the 14th of March. In England, there was a severe cold winter until the Feast of Saint Gregory in March. There was so great a murrain and death of sheep, that in many places above half had died. The winds came from the north for about three months continuous. They did great harm to the flowers and fruits. On 1 July, there fell such a storm of hail and rain as had not been known in England. The force of the rain and hailstones broke tiles covering the houses, and the boughs of the trees. This storm was incessant downpour for an hour. In England and France, there was a great plague on horses called "the evil of the tongue"

In 1255 CE there was drought in spring and summer in London and the south.

On June 5th 1256 CE Harrat Rahat erupted in Saudi Arabia with a VEI of 3. A lava flow erupted from six aligned scoria cones and travelled 23 kilometers, 14 miles, to within 4 kilometers, 2.5 miles of the Islamic holy city of Medina. It is the biggest lava field in Saudi Arabia.

In 1256 CE in England on 7 November and 17 December, there were terrible tempests of wind, rain, hail and thunder, which did great damage to water-mill wheels, arches of bridges, stacks of hay and corn, houses, children in cradles. These were borne down in torrents of water.

From 1256 to 1258 CE there were three wet years and extensive flooding which led to harvest failures and high grain prices. There was food shortage and starvation as well as distress for poor people.

On July 1st 1257 CE a catastrophic eruption occurred at Samalas/ Mount Rinjani Volcano on the Indonesian island of Lombok. Aerosols injected into the atmosphere reduced solar radiation reaching the earth's surface, causing a volcanic winter and cooling the atmosphere for several years. This led to famines and crop failures in Europe and elsewhere. This eruption may have triggered the Little Ice Age. This eruption had a VEI of 7 and was one of the largest volcanic eruptions during the Holocene epoch. Pyroclastic flows buried much of Lombok and crossed the sea to reach the neighboring island of Sumbawa. The city of Pamatan, which was the capital of a kingdom on Lombok, was buried and has never been found. Ash fell 340 kilometers away in Java. Samalas is also known as Rinjani. Ice cores from the Arctic and Antarctic link the sulfate spike to Samalas.

Then we had a volcanic winter. In these days people lived close to famine most of the time so it did not take much to cause a period of starvation after crops could not be planted, grown or sown.

The eruption of a volcano in the tropics led to a bad harvest in England in 1257 CE to 1258 CE. A monk chronicled "The North Wind prevailed for several months. Scarcely a small flower or shooting germ appeared, whence the hope of harvest was uncertain… innumerable multitudes of poor people died, and their bodies were found lying all about swollen from want…Nor did those who had homes dare to harbor the sick and dying, for fear of infection…the pestilence was immense-insufferable; it attacked the poor particularly. In London alone 15,000 perished; in England and elsewhere thousands died." Another report states that the impact on the climate would have been significant. 1258 was notably cold and wet overall. There was a combination of a cold and backward spring and heavy autumn rains gave rise to a very poor harvest.

1257 CE-1259 CE England. While backward weather seasons were contributing factors of the great famine of 1257-59 in England during Henry III's reign, much of the responsibility is laid at the door of the government itself. The whole kingdom had been drained of its coinage by the taxes, which the king had levied to pay German troops and to buy electoral votes for his brother, the Earl of Cornwall, who was a candidate for the imperial crown of the Holy Roman Empire.

It was during this famine that England for the first time imported from Germany and Holland grain to alleviate the suffering of her poorer classes. The Earl of Cornwall himself sent sixty shiploads of food, which was sold for his account to the starving.

In England in July 1257, there were great floods from rains. In July 1257, there were excessive rains and floods. Lowlands drowned. All the marshes were like a flooded desert. There was great scarcity of horses and cattle in England. In England in 1257, the inundations of autumn destroyed the grain and fruits and pestilence followed.

In the following year, 1258, in England there was a bountiful harvest but destructive rains caused the heavy crops to rot in the fields, and even the grain, which was gathered, became mouldy.

In 1258 in England, north winds in spring destroyed vegetation and food failed. The preceding harvest was small and innumerable multitudes of poor people died. Fifty shiploads of wheat, barley, and bread were procured from Germany; but citizens of London were forbidden by proclamation against dealing in same. "A great dearth followed this wet year pest, for a quarter of a ton of wheat was sold for 15 to 20 shillings. But the worst was in the end; there could be none found for money when – though many poor people were constrained to eat barks of trees and horseflesh; but many starved for want of food – 20,000 in London." In 1258, all summer and harvest there were the greatest floods. Corn (grain) all rotten. Famine. The previous year's excessive and long rains caused a dearth in 1258 over all of England because of a scarcity of corn (grain). A Quarter of wheat (1/4 ton), which previously sold for 2 shillings now sold for 24 shillings. Wheat had become very scarce. Great stores of grain were shipped in from Alamain.

The crop also failed in France and Normandy as well as England. The King of Alamain procured 50 great ships leaden from Dutch lands with wheat, barley, meal and bread, which greatly relieved the poor. But the Londoners bought it up, either to hoard it, or to sell it for a marked-up price, or to send it off to other ports. Many lived on herbs and roots and not a few of the poor were starved to death. Because the winds were keeping north several months; the fruits, flowers and produce from the earth were so hindered that they served no purpose until June was nearly over. There was all summer and during the harvest excessive rains and inundations. Yet a double crop of corn, grain, and grass was on the ground but unfortunately, it was all rotten. Thus were the expectations of the farmers lost. Famine and death went hand-in-hand triumphantly together. People died so fast, they dug great pits in churchyards and filled them with heaps of dead carcasses. But towards the end of the harvest, the weather picked up and so much of this rotten crop was harvested very late. This did much good and lowered the price of corn, grain, half and half. On 1 December 1258 at night a terrible tempest of thunder, lightning, wind and rain occurred.

In 1258 horseflesh was a delicate dish but there was still great mortality.

In 1258 Mediaeval texts describe atrocious weather that summer in 1258. It was cold and the rain unrelenting, leading to flooding. Thousands of people who died in this period were buried in mass graves in London. This was a volcanic winter.

Also, on June 28th 1258 CE there was a storm flood on the Severn River in England. Many people were drowned.

In 1259 CE it was a dry autumn in London and the south.

In 1260 CE Pico de Orizaba in Mexico erupted with a VEI of 3.

In 1260 in England, there was "no rain all the year to August; then moderate showers only; oats and barley lost. In England, there was drought during the summer that was so long, great and severe; that oats and barley sown in due time came not up till near harvest. Then moderate rains fell, they sprang up, grew and shot up. But now it was Michaelmas,29 September, and without any sun to ripen them, they were cut down and dried for cattle fodder.

There was a shocking inundation on the Rhine River in Germany, fatal to many people and cattle.

Between 1260-1285 CE tree ring dating showed that there was a great drought in the American Southwest.

In summer 1260 CE there were frequent and heavy thunderstorms during the summer that produced hail with a diameter of two inches or 5 cms. There was also a drought in the summer in London and the south.

In 1261 CE there was frost and snow during February in London and the south.

In 1262 CE Mount Katla erupted in Iceland with a VEI of 5. This would cause a volcanic winter. Add this to the 1257 CE Samala/Rinjani volcanic winter residue in the stratosphere and you would get a continuation of the previous Volcanic Winter.

In Ireland, there was a great destruction of people from plague and hunger.

There was a great scarcity and famine in Scotland and England from last year's rainy harvest.

On 28th January 1262 CE a major storm/flood affected the Dutch coastal communities and may also had an impact on the English coastline of the North Sea. This was called the Grote Mandrenke or "Great Drowning of Men". An intense extra-tropical cyclone swept across the British Isles, the Netherlands, Northern Germany and Denmark including Schleswig and Southern Jutland. The immense storm tide swept far inland from England and the Netherlands to Denmark and the German Coast. Entire towns and districts as well as islands were wiped out. These included Rungholt that was said to have been located on the island of Strand in North Frisia, Ravensor Odd in East Yorkshire and the harbor of Dunwich in Suffolk.

In 1262 Friesland in the Netherlands was attacked by a sea floods.

Sometime between 1262 and 1460 CE El Chichon in Chiapas, Mexico erupted. It had a VEI of 5. Chichon is an actively stratified volcano and are considered the most lethal in activity, since they do not spew lava, but rather the emissions are mainly gases and ash, essentially aerosols of sulfuric acid that rise vertically very high in the atmosphere. Sulfuric aerosols are what primarily cause volcanic winters.

So here we have a sulfate spewing volcano during the already existing Volcanic Winters caused by Samalas/ Mount Rinjani Volcano on the Indonesian island of Lombok and Katla in Iceland.

In England, "on St. Nicholas we began a month's hard frost." In England, on St. Nicholas' Eve 1263 began a very severe frost, which lasted over a month. Horses and people went over the River Thames on ice. On 1 December and 13 December, there were terrible thunder and lightning storms in England.

The summer of 1263 produced a drought in the Netherlands

On May 13th 1264 there was a gale that affected London and the south of England.

Between May and October 1264 CE Eleanor of Provence (Queen-Consort to Henry III) was frustrated by bad weather in her attempt to bring troops to hear husband's cause. The Queen's fleet was trapped by frequent spells of high wind at Sluis in Flanders before it could cross to the Kent coast. Sluis is now in the Netherlands near the Belgian border. According to H.H. Lamb the 13th century experienced the highest number of severe sea floods along North Sea and English Channel coasts. Although the climate of northwest Europe was still generally being the peak of warmth of the Mediaeval Age, from the middle of the 13th Century, there was an increase in unsettled weather which is believed to be the descent into the Little Ice Age.

A severe famine struck Egypt in 1264. But much of the hardship was averted by the strong leadership of Bibars. Bibars was a native of Kipchak (Gypjak), located between the Ural Mountains and the Caspian Sea. Years before he was sold into slavery and fetched very little on the auction block because of a cataract in one of his eyes. Later he founded the Mameluke Empire. He met the famine promptly and vigorously by regulating the sale of grain wisely and compelling his officers and emirs to support the destitute for three months.

On the night after St. Marcellus feast, one of which was December 29th 1266, a storm struck Friesland in the Netherlands and flooded far and wide and did great damage.

On 14 February 1267, a flood took Friesland.

The summer of 1267 in France was hot.

On 6th January 1268 CE a major storm and flood affected the Dutch coastal communities and possibly southern England. 1268 CE. Friesland.

In 1268 Friesland in the Netherlands was attacked by a sea floods.

A severe winter struck Europe beginning on 30 November 1268. The winter lasted until 25 May 1269.

In England in February of 1269, there were great floods from the winter thaw.

An earthquake occurred northeast of the city of Adana in the Armenian Kingdom of Cilicia (modern day Turkey) on 14 May 1269 at "the first hour of the night". Most sources give a death toll of 8,000 in the Armenian Kingdom of Cilicia in southern Asia Minor, but a figure of 60,000 dead was reported by Robert Mallet in 1853 and repeated in many later catalogues. Several contemporary sources report that the fortress of Sarvandikar was destroyed, killing those inside. Several castles were also ruined, including those at Delnk'ar, Hamus, Harunye and Hagar Suglan. The monastery of Ark'akalin (thought to be located near Sis) was severely damaged, leading to the deaths of priests and monks. Many villages were destroyed across Cilicia, particularly those at the foot of the Nur Mountains. About 8,000 people were said to have died.

In the year 1269, the winter was severe in Northern Europe. During the winter of 1269-1270 the rivers in northern France froze.

In 1269, the Kattegat was frozen between Sweden and Jutland, Denmark. The narrowest part of the Kattegat is around 32 miles but is much wider in places.

During the winter of 1269, the frost was so intense in Scotland that the ground bound up.

In England, the frost lasted from 30th of November to 2nd of February. They were the hardest frosts. In England, there was a continuous frost from the Feast of Saint Andrew,30 November, to Candlemas, 2 February. The River Thames was frozen over. Horses, draughts and people passed over the river on the ice. Merchant goods came to London by land. Ships could not come up the river. On 6 February, there fell such a profound rain, as raised the greatest flood in the memory of man. The River Thames filled the cellars and vaults in London with water, to a great loss of merchandise.7

In the winter of 1269 to 1270 CE a bitter frost persisted for about ten weeks during this severe winter. The Thames froze solid and thick enough for men and beasts to cross over and was closed to shipping so that merchandise had to be transported overland between the Channel Ports and London. Accounts of the winter refer to glazed frosts. A glazed frost or verglas is a smooth, transparent and homogenous ice coating occurring when freezing rain or drizzle hits a surface. The thaw, when it arrived, was accompanied by heavy rain and flooding. There was a flood on the Thames in February that was a combination of heavy rain and inland snowmelt.

In August 1269 CE Asosan in Japan erupted again, still with a VEI of 2.

In 1270 CE Yasur erupted in Vanuatu with a VEI of 3.

During the 1270s there were frequent dry or very dry summers in England.

In 1270 it rained every day for so hard during harvest time that the grain rotted in the fields in Germany. As a result, a famine emerged. The conditions lasted 5 years.

The summer of 1270 was very poor and very wet in Austria, Switzerland, Bohemia (now western Czechia, Netherlands, Germany (Lower Saxony), and France (Carcassonne). This produced poor harvests.

During 1271 CE the church tower of S. Mary-le-Bow in London was blown down and killed several people. Norwich ecclesiastical records show that a great flood occurred in this year and lightning also damaged the cathedral steeple.

On January 5th 1271 CE Asosan in Japan erupted with a VEI of 2.

In England, on the 4th of the Nones of July, 1271, there was a terrible wind and rain rotting and breaking trees, overthrowing houses etc. A great famine over all England followed. On 14 October, there was a great inundation of rain at Canterbury with thunder, lightning and tempests, as their forefathers never saw nor heard. During the whole day and night, thunder never ceased, but roared continually, like one single clap. A very great flood followed, which overthrew trees, vines etc. Men could neither go nor ride. Many were in eminent danger from the force of the water in the streets and houses of the city. The flood carried down many people and buildings.

On April 16th 1272 CE Asosan in Japan erupted with a VEI of 2.

There was famine in England in 1272 AD.

In August 1273 CE Asosan in Japan erupted with a VEI of 2.

In 1273 the sea engulfed Friesland in the Netherlands and 20,000 people drowned.

In 1274 CE Asosan in Japan erupted with a VEI of 2.

The winter of 1275-1276 in Italy was very long and harsh. Heavy snow fell and covered the earth near Parma in northern Italy on 29 November and the snow cover remained until early April. You could sow this year no vegetables, and cereal grains that were planted almost entirely failed. The herds in the Diocese of Parma died out almost completely. During the winter of 1275-1276 in Parma in northern Italy, there was snow on the ground from December to April.

There was a great drought in New Mexico between 1275 CE to 1299 AD.

In 1276 CE it was dry from April to July in London and the south.

In 1276 CE in England there were great floods from the sea and from the rains. In England from long and excessive rains came desolating inundations in many places, so that corn (grains) and grass came not to maturity.

There was a great inundation from the sea at Venice, Italy followed by a great earthquake. Was this a tsunami caused by the earthquake?

In 1276 and 1277, there was a drought in England. It was so hot and dry in the summer that scarcely any fodder, dried hay or feed for cattle, was available.

In 1276 CE Bagdad in Iraq was inundated after the appearance of red flame in the sky.

The great drought of 1276 to 1299 occurred around Mesa Verde in Colorado. Over the course of the next decade very dry conditions expanded over the entire Southwest and lasted until 1299. There was dramatically reduced rainfall but with marked differences from North to south. More than sixty per cent of the shortfall occurred in the northwest, in southern Utah and Colorado, versus ten per cent in the southeast of New Mexico. Tree ring data from Arroyo Hondo confirmed this.

In 1276 CE Mesa Verde in Colorado was abandoned. As were Aztec above the Animas River and Great Houses in the Moctezuma Valley.

In Western Europe the summer of 1277 was hot. There was an exceptional drought. The largest rivers, the fountains, the sources of water were completely dry. As a consequence, there was a large loss of life. The lightning struck during the months of August and September, in many places.

In 1276 and 1277, there was a drought and a famine in England. It was so hot and dry in the summer that scarcely any fodder (was available.

The summer of 1277 in France was hot.

On Christmas Day 1277 the sea engulfed Reiderland in the Netherlands and over fifty villages were lost. This was when the Dollart Inlet was formed. The town of Torum was also drowned. In Holland, now the Netherlands, there were great inundations at Friesland, forming the Dollart Sea. The Zuyder-zee is a great gulf, which penetrates far inland between North Holland, and Friesland, Overyssel and Gelderland on the east. The southern portion was originally a large lake, the barrier between which and the sea was broken through by an inundation in 1225. It is much encumbered with sand banks, and subject to violent storms. The Dollart, a similar inlet between Groningen and Hanover, was formed likewise by an irruption of the sea in 1277. The Lauerzee was also created in this event when the mouth of the Lauwers River disappeared and its tributaries the Reitdiep, Dokkumerdiep and the Ee flowed directly into the new bay.

In 1278 CE in Italy, there was a great overflowing of the Tiber River. There was a great inundation of the Tiber River that went four feet above the altar of Maria Rotunda. The Pantheon in Rome has been used as a Roman Catholic church dedicated to "St. Mary and the Martyrs" but informally known as "Santa Maria Rotonda."

In May 1279 CE there were severe thunderstorms in England. Trees were uprooted, buildings were destroyed, and lakes were flattened. Had the water been blown out of them by the wind? Trees were plucked up by the roots in many places by a tempest, and removed to others, men were wrapped up in the air, and lakes were dried.

In 1280 CE Quilotoa erupted in the Andes in Ecuador with a VEI of 6. Pyroclastic flows and lahars reached the Pacific Ocean and spread an airborne deposit of volcanic ash throughout the northern Andes. This was a Plinian or explosive eruption. It produced 21 cubic kilometers of tephra. It would also cause a volcanic winter. Quilotoa is 186 kilometers from the Pacific coast of Ecuador. This was one very long lava flow. This eruption produced three times the bulk volume of Pinatubo.

We had a very rare VEI of 7 in 1257 CE and now a VEI of 6 only a few years later. Would this continue increasing the possibility of extending a Volcanic winter to become a mini-Ice Age? How much would the intervening VEIs of 5 be affecting this cooling?

In 1280 CE in England, there were great floods all the summer; especially on 2 August. On 2 August, there was a prodigious inundation, which carried off many people, cattle, mills, bridges, houses, trees, hay, grass, etc. In 1280 CE there was considerable damage across East Anglia in England due to floods and storms. This was from Norwich Cathedral records.

In 1280, more than 300 houses were overwhelmed at Old Winchelsea in England by an inundation of the sea. This forced the moving of the town to its present location on the western edge of Romney Marsh and Camber Sands. 300 houses were also recorded to have been lost in 1250 as well. Is this a type setting error or an amazing coincidence?

On 11 November, there was a terrible thunderstorm, which broke down houses and trees. So great a flood was there in the Sequan that it broke down the bridges at Lyons in east-central France.

During the famine in Bohemia, now western Czechia, in 1280-82, individuals resorted to cannibalism.

By 1280 CE tillage reached so high on the Pennines and the Northumbrian moors in England that sheep farmers were complaining that were was too little land left for sheep grazing. Within a short period of time nothing would be grown in these areas as the Little Ice Age literally erupted.

Is it possible that the eruption of Quilotoa in 1280 CE and a very rare but destructive VEI 7 eruption of Samalas Volcano on the Indonesian island of Lombok in 1257 CE as well as the 1262 CE Mount Katla eruption in Iceland with a VEI of 5, to combine to produce a volcanic winter. Add this to the ash being spewed into the atmosphere from the other 28 volcanoes that also erupted in this period.

Such a peak of warmth in the last stages before Europe itself was affected by a downturn of temperatures in the Arctic would be meteorologically consistent with the development of a strong thrust forward of the Arctic regime in the longitude of Greenland and Iceland, distorting the pattern of the circumpolar vortex with a sharp trough there and a recurrent warm ridge over western Europe. Something like this pattern recurred at times in the middle and late parts of the fourteenth century from 1301 CE onwards, bringing notable droughts in Europe after an extremely wet phase which had marked the first break in the early part of the century.

On the 9th October 1280 CE heavy snow fell in London. This would be remarkable even in the 21st Century.

There was a glacial advance in Grindelwald in Switzerland in 1280 CE. This was in evidence with the stumps of larch trees that had been overrun by the ice on the slopes of the Eiger and Fisherhorn and a documentary record of the church at Burbiel was moved out of danger of the glacier and the flood. In this period the local inhabitants wanted the glaciers to go away. It would be years before they started doing this again.

In 1280 CE the Unterergletscher glacier or Lower Grindelwald Glacier at Grindelwald in Switzerland buried a wood. Previously this same glacier had buried trees in a moraine and had attacked parts of the same wood in 200-year cycles until it crushed all of the wood completely. This glacier was the lowest in the Swiss Alps reaching down to less than one thousand meters. It no longer exists in the twenty-first century.

On July 2nd 1281 CE Asosan in Japan erupted with a VEI of 2.

On July 3rd 1281 CE Asamayama in Japan erupted with a VEI of 3.

In 1281 CE there was a grievous famine in Poland and great multitudes moved to Russia and Hungary.

In 1281 in Bavaria in Germany there was a snowfall in mid-June. Crop failure caused great famines.

In 1281, there were very great floods in Paris, France.

In the Winter of 1281/ 1282 CE many houses in Austria were completely buried in snow, and many people perished due to the cold and hunger. In 1281 the snow fell in great abundance in Austria and many houses were entirely buried in the open countryside. A great snowstorm struck Austria in December 1281.

Europe experienced a cold winter.

London Bridge in England was wrecked by ice.

The spring thaw flooded Paris, France. The melting snow and ice in the Seine River in Paris, France also produced a very severe inundation.

In Bohemia, now western Czechia, the freeze lasted until 25 March 1282 and then the thaw and melting of the snow produced a terrible inundation and great need.

In the winter of 1281 to 1282 CE in England there was a notable winter. There was a Great Frost recorded in contemporary records in January 1282. Snow persisted from Christmas to March. The Thames was frozen so hard that people could walk across the river between Lambeth and Westminster. The force of the ice damaged five of the arches of London Bridge though other sources state that the arches collapsed. After the snow fall there were reports of a destructive thaw in early to mid-spring. There were severe floods in 1282 when a great gale brought much destruction and loss of life to Lincolnshire and East Anglia. Rochester and other bridges were wholly destroyed.

In 1282 there was a North Sea floods in the Netherlands that separated the island of Texel in the West Frisian Islands from the mainland.

In 1282 CE during the famine in Bohemia in 1280-82, individuals resorted to cannibalism.

In 1283 CE there was a wet summer and autumn in London and the south.

This year began one of the really notable periods in the Middle Ages of mostly warm, dry summers from 1284 CE to 1311 CE.

In 1285 CE in England, there was a sudden great darkness, and then such drought and heat as killed most grain. Almost all greens died. Then came great and long rains; hence began a famine in England, which continued twenty-three years.

In 1282, there was a great flood and gale at Boston, England. "The Monasterie of Spalding and many churches destroyed. At Yarmouth (Great Yarmouth), Donwich (the coastal town of Dunwich), and Ipchwich (Ipswich), an intolerable multitude of men, women, and children were overwhelmed by the water and drowned, especially at Bostone (Boston).

On January 1st 1286 CE a storm surge reached the east edge of Dunwich in East Anglia and destroyed buildings in it. Other sources say March. At Dunwich the combination of high tides and strong easterly winds caused great devastation. In particular a strip down the eastern side of Dunwich in places one hundred meters wide was eroded by the sea. Residential areas, churches, and a small monastery were carried away. The Priory of Grey Friars was almost totally demolished, only the graveyard and the west wall of its chapel remaining in existence. Massive waves swept across Kings Holme and flooded the lower town. The harbor mouth seems to have stayed open, but further north the River Blyth forced a way through Kings Holme at or near the point where the previous opening had been.

On 9th May 1286 CE a thunderstorm with hail as big as stones fell in England. Crops were levelled, houses damaged, and branches of trees broken. There were squally winds and possibly a tornado.

In 1286 in England, during the night of the Feast of Saint Margaret, Saint Margaret of Scotland – 10 June, fell a great tempest of rain, thunder, and lightning, so great that it drowned all the sown corn (grains). All grains had been cheap. Wheat was 18 shillings a Quarter (quarter ton), but now began a dearth, which continued more or less for 40 years. In England the rains which came too late in 1285 lasted too long in 1286, hence a dearth. There was a 23-year long famine in England that began in 1286.

On 6 July 1286, there was a dismal tempest of hail, thunder and lightning at Magdeburg in Saxony-Anhalt in central Germany.

On August 30th 1286 CE Asosan erupted, again, with a VEI of 2. Asosan is one of the most active volcanoes in Japan and is on the island of Kyushu.

From the excessive rains that fell this winter in England; there were very great floods. On 1 January 1287 the sea from the Humber to Yarmouth broke into the land, overflowing for three or four leagues (9-12 miles, 14.5-19.3 kilometers) in breath, overthrowing buildings, drowning people and cattle. It came so suddenly that there was no avoiding it. It laid the whole Fens of Lincolnshire, England under water.

In Winchelsea, England in February 1287, there was a great inundation of the sea; more than 300 houses swept away. There was a "Charter granted for erection of a new port" but the town was rebuilt on the cliff top behind. The storm was such that that whole areas of coastline were redrawn. Silting and cliff collapses led to towns that had stood by the sea were finding themselves landlocked, whilst others that had been inland found themselves with access to the sea. The town of Broomhill near Winchelsea was destroyed. The course of the River Romney was diverted away from New Romney which was almost destroyed and left a mile for the coast which ended its life as a port. The cliff collapsed at Hastings, taking part of the castle with it which blocked the harbor and ended its role as a trade center. Here are those 300 houses again. Winchelsea was regarded as a very large town but I am not sure how many lots of 300 houses it could afford to lose.

Old Dunwich on the Suffolk coast was a prosperous town with a fine harbor, housing a large fleet of merchant ships. On 1287 CE a violent electrical storm damaged several churches beyond repair, inundated the land and swept away many buildings. At the time Dunwich was an international port similar in size to London at the time. This was during the South England flood of February 1287 CE.

In 1287 CE it was dry from April to July in England.

In England, the winter was excessively rainy producing great floods. On 1st June the sea broke in from the Humber to Yarmouth, forced by the winds. The Humber is a large tidal estuary on the east coast of northern England. Yarmouth or Great Yarmouth is a coastal town in Norfolk in eastern England. In December 1287 CE there was a great storm surge in the English Channel and the North Sea. This was caused by a low-pressure system mixed with a high tide that caused the North Sea to rise over seawalls and dikes, causing a large part of the Netherlands and Northern Germany to be flooded.

In December there was St. Lucia's flood. There was a terrible inundation in East Anglia, particularly Norfolk, coastal areas in 1287. Houses were destroyed and in the village of Hickling the water was so deep that it overflowed the high altar of the priory by a foot or more. Some 500 people perished in this most fatal of all English floods. H. H. Lamb stated this was one of many storm floods along the East Anglian, Kent and Sussex coasts and adjacent continental coastlines. There was also major impact on the English Cinque Ports on the English Channel.

In December there were storm surges on the Suffolk and Norfolk coasts and plague all the year.

In the Netherlands on 31/14 December 1287 there was the Saint Lucia Flood which formed the Waddenzee and the Zuiderzee. This event created direct sea access for the then village of Amsterdam thus allowing for its development into a major port city. In Holland (now the Netherlands), there was a dreadful storm, laid the whole country on both sides of the Zuiderzee under water. To such a height did the water rise that Count Florence took advantage of the circumstance to subdue the inland towns by using armed vessels called "cogs". In Selandia, fifteen islands were submerged by the sea and 15,000 people were drowned. Selandia was a name for Zeeland up until 1600.

As storms and hurricanes continued a Dutch seawall collapsed in December 1287 CE and the Zuyder Zee took some 50,000 lives to 80,000 Lives. In December 1287 the Zuyder Zee enlarged and over 80,000 lives were lost. Two different reports, two different totals.

Also on 14 December 1287, a great storm struck. The chronicle of the monastery Rastede speaks of flooded dikes in Friesland and Stedingen, now a tidal marsh of Lower Saxony, Germany, and thousands of deaths. The water rose "five feet higher" than ever before causing whole villages to be destroyed. People fled before the advancing North Sea, and thereby established new villages as Osteel, Marienhafe or UpgantSchott.

In 1287 Sea floods also hit Denmark, Friesland in the Netherlands and Flanders in Belgium. England was also flooded from the Humber to Kent. East Anglia and the Norfolk Broads were also inundated by sea floodss.

On 4th February 1288 the Isle of Thanet, Romney and Winchelsea were inundated by sea floods. All the dykes were demolished. The Thames Estuary was also flooded as well as villages far from the Thames. This was called the St. Agatha Flood.

Also, on 4th February 1288 all of the Netherlands except for Walcheren was inundated by sea floods of enormous size. There were thousands of deaths.

In the month of March 1288, the Rhine River froze below Basel (a city situated at the border between Switzerland, France and Germany).

The frosts of 1288 in northern France killed the buds of the vines, all woods and orchards. The summer of 1288 in France was hot.

In England during the summer of 1288, there was heat and drought so intense as killed many. There were great deaths. There was plenty. In England during the summer of 1288, it was so exceedingly hot, that in some places men died of the heat. This year and last brought such a plentiful increase that wheat sold for 16d. to 20d. per Quarter (quarter ton). All provisions were very good and cheap. This drought was followed by a great mortality of people because of a severe cold frosty winter and much snow. In England in 1288, it was a good year for great wine, hay and acorns. But in August, there was such great heat that the birds died in the fields. In some places people died of suffocation from the heat. There was famine in England in 1288 CE. The summer was hot and dry in London and southern England.

The winter of 1288 to 1289 CE was severe in London and the south.

On 9 July 1289, there fell the greatest tempest of hail in England than could ever be remembered. This hailstorm was followed by continual rains. So that all corn (grain) turned very dear (scarce). This dearth continued and increased even to the death of King Richard II. In England in 1289, a tempest destroyed the seed, and corn (grain) rose to a great price.

In 1290 CE Cayambe in Ecuador erupted with a VEI of 4.

In England in 1290 CE there was a wet summer and autumn in London.

On July 3rd 1290 CE there was a well-attested meteoric impact event at Veliki Ustyug, Veliki Usting, at Kotova village in Vologda Oblast in Russia with great clouds, ceaseless lightning, the ground swaying, clouds of fire and great heat. The fall was witnessed by local priests.

The 1290 Zhili earthquake occurred on 27 September with an epicenter near Ningcheng, Zhongshu Sheng (Zhili), Yuan China. This region is today administered as part of Inner Mongolia, China. The earthquake had an estimated surface wave magnitude of 6.8 and a maximum felt intensity of IX (*Violent*) on the Mercalli intensity scale. One estimate places the death toll at 7,270, while another has it at 100,000. The earthquake destroyed 480 storehouses and countless houses in Ningcheng. Changping, Hejian, Renqui, Xiongxian, Baoding, Yixian, and Baixiang County were also affected. It severely damaged the Fengguo Temple in Yixian.

In 1291 CE there was a dry summer in London and the south.

In the spring of 1291 CE, the Volkhov River flooded. The horses all died in Novgorod in Russia, and but few were left. The same year a frost attacked the crops throughout the whole of the Novgorod district.

In England, there was a drought all summer followed by a great famine. In England in 1291, there was a most droughty summer, an excessively rainy harvest and a frosty winter. This resulted in an extraordinary scarcity of hay, grass and corn (grains). In the Winter of 1291-1292 CE in England, the frost was severe all winter.

In India, there was a great drought in 1291. No rain fell in the provinces about Delhi and there was in consequence a most terrible famine. In 1291 CE during the reign of Firok Shah, there was a famine in Delhi and its neighborhood in India. "The Hindus of that country, came into Delhi with their families, twenty or thirty of them together, and in the extremity of hunger drowned themselves in the Jumna River."

In 1291 in Damascus, Syria, there was an inundation caused by the overflowing of streams.

In 1292 "the Rhine was frozen over," in Germany and the snow was represented as being of an "enormous depth". In February, one would walk across the frozen Rhine River in Western Europe with dry feet. In 1292, loaded carts crossed the Rhine River at Breisach in southwestern Germany, on the ice.

In 1292, the winter in Germany and Northern Europe was very severe. The Kattegat was covered with ice seven feet thick, and batteries of artillery were moved to and fro on the strait. The Kattegat is between Denmark, Norway and Sweden.

The winter of 1292 in England was very harsh and there was famine.

In 1293 and 1294, there was a drought in England and the summers were excessively hot.

In England, there was a very great drought. In England, there was a grievous famine. Wheat sold from 16s. to 20s. per Quarter (quarter ton). As a result, thousands of poor died. There was so great a drought that springs and rivers were dry. Grass was burnt up. Cattle were kept alive on straw. Corn (grain) was harvested before Saint John's Mass, 23 June, and grapes at the Nativity of the Virgin in September. In England in 1294 there was a very sore famine with a desolating mortality.

The 1293 Kamakura earthquake in Japan occurred at about 06:00 local time on 27 May 1293. It had an estimated magnitude of 7.1–7.5 and triggered a tsunami. The estimated death toll was 23,024. It occurred during the Kamakura period, and the city of Kamakura was seriously damaged. In the confusion following the quake, Hōjō Sadatoki, the Shikken of the Kamakura shogunate, carried out a purge against his subordinate Taira no Yoritsuna. In what is referred to as the Heizen Gate Incident, Yoritsuna and 90 of his followers were killed. A tsunami deposit has been found that is consistent with this age.

On 27 May 1293, a magnitude 7.1 earthquake and tsunami hit Kamakura, then the *de facto* capital of Japan, killing 23,000 in the resulting fires.

In May 1294 CE on the 14th heavy snow fell in London.

The drought of 1294 dried up all the wells and all sources (of water – springs, creeks, small rivers and lakes) in Provence, France. The Huveaune River dried up completely. The water on the Rhône River declined to such an extent that it was no longer navigable, even at its mouth. It was impossible to grind wheat with windmills.

On 18th October 1294 CE the Thames flooded Rotherhithe, Bermondsey, Tothill and Westminster. Not sure if this was rainfall related or a tidal surge.

In 1294 CE the Irish annals record the following "Lightening and meteors destroyed the blades of corn". The Irish knew the difference between lightning or lightening and meteors.

In the winter of 1294 CE, the sea between Norway and Denmark, and from Sweden to Gothland, and the Rhine River and Baltic Sea were all frozen and snow fell to a frightful depth it was that cold.

In 1294, the Kattegat, or sea between Norway and Denmark, was frozen, and that from Oslo in Norway, traders travelled on the ice to Jutland. The Caltégat Sea (Kattegat Sea) was frozen over completely with ice. The Kattegat Sea is the strait between north Denmark, Norway and Sweden.

On 19 and 20 January 1295 CE day and night, a hurricane with violent showers and storms consumed the winter seeds in marshy places. There were great floods in England. There was a great intemperature of the elements this year. On the 3rd of the Nones of April, there was a deep snow. Hail spoiled the corn (grain). In 1295 in England, there were no grain or fruits, "so the poor died of hunger."

Famine oppressed those of Bourbon France as well.

In England on 19 and 20 January, there was a hurricane followed by rain, storm, floods all winter and as a result seeds lost; hence dearth. There was a famine from hailstorms and a great concussion of the elements.

In Ireland, there was a great dearth from 1294-1296.

In 1296 CE in the winter, the Skagerrak, an eastward extension of the North Sea, between Norway and Denmark, and from Sweden to Gothland, and the Rhine River and Baltic Sea were all frozen and snow fell to a frightful depth. The sea between Norway and the promontory of Scagernit froze over and from Sweden to Gothland. Scagernit or the Skagerrak is a strait running between Norway and the southwestern coast of Sweden and the Jutland peninsula of Denmark, connecting the North Sea and the Kattegat Sea area, which leads to the Baltic Sea. Gothland is Gotland Island, Sweden in the Baltic Sea. Similar to this had happened two years before.

The summer of 1296 in France was hot.

Never in living memory, has anyone seen a winter so cruel in France as that of 1296. The Seine River in Paris flooded from 20 December 1296 to 1 January 1297. There was water in all the streets of Paris.

In Scotland in 1297 there was calamitous famine and pestilence.

A time of the most ice and the coldest climate periods in Iceland suddenly occurred in 1197-1198 CE and 1203 CE.

In England in November 1299 CE, there was an inundation from the sea in the River Thames. "In December, great calm, heat, and clearness." In December in England, there was a hurricane. Then great calm, clear and hot. After that there were great floods.

In Persia in 1299 they were ravaged by famine and pestilence.

1300 CE. Little Ice Age

1300 CE is the accepted start of the Little Ice Age, LIA, that lasted until 1850 AD.

The Little Ice Age was a six-century period of constant climatic shifts and decreasing temperatures.

There was a world-wide cooling of the earth. There was a surge in the frequency of El Nino events after 1300 CE. There were more frequent ENSOs or El Nino Southern Oscillation events in the eastern Pacific region during the Little Ice Age which was warmer there.

El Nino and a negative North Atlantic Oscillation prevailed during the Little Ice Age.

In 1300 CE Krafla In Iceland had a VEI of 2, Iwatesan in Japan had a VEI of 3, and Mutnovsky in Kamchatka had a VEI of 2 when they erupted.

There were more storm surges in the first half of the fourteenth Century than before or since in Britain and Western Europe. The total was nine. This was coupled to the changes as the European climate cooled.

Between 1300 to 1600 CE there was a rapidity of decline in temperatures. There was a long period of colder climate culminating in the seventeenth century. Rainfall changes indicated that there was a long period when the totals of rainfall were ten per cent lower than the late twentieth century. This may be explained by colder seas and therefore less water vapor taken up into the atmosphere during the time of colder climate. The warm summers were attributable to the anticyclone belt moving. H. H. Lamb even noticed that there was the appearance of an intriguing oscillation, whereby the summers of the second half of most centuries were wetter than those of the first half.

The frequency of 'severe' winters across Britain during the first three decades of this century was unusually high. Also, analysis of agricultural records, estate reports, tax returns etc., also points to frequent wet-cool summers with failures of harvests and impact on survival of livestock. This probably by extension applies at least to continental NW Europe.

The warm peak of the Little Ice Age ended around 1300 CE when the incidence of severe storms in the English Channel and the North Sea increased and there was low-lying flooding of coasts. From the fourteenth century and through the fifteenth century it was an unhealthy time for humans. There were many troubles with the diseases of mankind, animals and crops. In England the average life expectancy dropped ten years from 48 years on average to 38 years. One of the most horrifying diseases was ergotism or St Anthony's Fire produced by the ergot blight, *Claviceps purpurea*, which blackened the kernels of the rye in damp harvests. Even a minute portion of the poisoned grains, baked in bread, would cause the disease. The course of the epidemics was such that the whole population of a village would suffer convulsions, hallucinations, gangrene rotting the extremities of the body, and death. In the chronic stages of the disease the extremities developed first an icy feeling, then a burning sensation; the limbs went dark as if burnt, shriveled, and finally dropped off. Even domestic animals caught it, and pregnant women miscarried.

How much did Pieter Breughel the Elder observe and paint in this period?

How much did this help stimulate the visions of witches, Satan and demons reported in this period?

The island of Heligoland which was fifty kilometers out in the German Bight and measured over sixty kilometers across, had been reduced to 25 kilometers by 1300 CE, perhaps half of the island disappearing that year.

The frequency of severe winters across Britain during the first three decades of the 14th Century was unusually high. There were frequent wet and cool summers with failures of harvests and impact on the survival of livestock. This was probably the case in northern Europe as well.

Starting in 1300 CE there were abrupt changes of 2.16° Celsius in the sea surface temperatures of the eastern North Atlantic. This meant that the temperature decreased by 2.16° C. At the same time, salinity levels changed which can affect the workings of the ocean conveyer belt that is a fundamental driver of global climate. This transfers heat from the tropics to northern latitudes. The sea surface temperature in the eastern Atlantic has a strong effect on the dry winds that blow across the Sahara. If sea surface temperatures are lower in the eastern Atlantic between ten degrees North and 25 degrees North and higher in the Gulf of Guinea, the monsoon winds are displaced southwards, causing drought in the Sahel and Sahara.

At the beginning of the fourteenth century there was a 25-year period of rainfall in the American Southwest that remained above the long-term average. In Arroyo Hondo the settlement that was established in 1300 CE reached its greatest size, comprising 24 room-blocks constructed around ten enclosed plazas.

Around 1300 CE the former inhabitants of the Anasazi Culture who had left Mesa Verde, Pueblo Bonito, Chaco Canyon and others moved again south and southwest, along the Rio Grande River to the Hopi Mesas area in central Arizona.

After 1300 CE in the American West there was a six-hundred-year long period of persistently wetter conditions which ended in the present period with the return of drought conditions.

Around 1300 CE there was a pronounced cold episode in California.

Mesa Verde in southern Colorado was abandoned around 1300 CE. The Anasazi abandoned their recently constructed cliff dwellings and moved to the south and east.

The great pueblos of the Four Corners region were now silent and empty after the population dispersed itself and joined distant communities elsewhere.

It is believed that there were several large volcanic eruptions in the 1300s. These possibly caused the Little Ice Age. The theory being that several large eruptions in quick succession could have initiated sufficient cooling to trigger sea ice growth. The increased reflection of sunlight by ice and snow, in turn, would create additional cooling. This is the ice-albedo feedback and is one of several feedback that amplify climate change. Then again, we have also seen at least three thirteenth century contenders as well.

Due to decreasing temperatures farmlands and villages were being abandoned in the fourteenth and fifteenth centuries in England, Scotland and Germany, parts of France and Hungary, besides all the Scandinavian countries.

The Arctic ice cap extended and changed the cyclonic pattern leading to a series of disastrous harvests. These in turn led to widespread famine, death and social disruption.

In the 1300s CE in Asia there were no cold temperatures as in the Northern Hemisphere. In Asia it was shifts between droughts and periods of abundant precipitation. It was a switch from persistent La Nina conditions to a climate pattern resembling El Nino. During this time the Intertropical Convergence Zone shifted several hundred miles to the south resulting in a weaker summer monsoon in Southeast Asia.

During the Little Ice Age there were memorably cold winters when the Baltic Sea froze.

In Eastern Europe there was a series of winters of unparalleled severity and depth of snow. Chronicles from Poland and Russia tell of cannibalism, common graves overfilled with corpses, and migrations to the west in search of food. This was before the Black Death came.

In Russia in the fourteenth century the greater proportion of the increasing climatic troubles seemed to have been due to summer droughts than further west. This was confirmed by the general decline of ring widths shown by timbers used in the successive surfacing of streets in medieval Novgorod. There was also an increased incidence of severe winters which were of a severity in terms of famine and loss of life, unmatched in western Europe except in the 1310s, 1430s and 1690s.

Around 1300 local shepherds on the Pennine Moors in northern England complained about encroaching farmland but this would soon reverse as the climate got colder.

In England villages were being deserted due to the cold weather and inability to reliably grow crops. Thousands of villages were abandoned. The price of wheat soared, and the population fell.

There were periods in the Little Ice Age when Winter Markets stood for months on the frozen River Thames in England.

In the 1300s CE there was a turn of the climate of middle and northern Europe toward greater wetness. Underground water began to cause great difficulties in the silver mines in central Europe.

Later in the fourteenth century there were collapses of massive buildings such as cathedrals etc., built in the previous 200 years, as the previously wet ground dried out in the droughts that came during the mid to late fourteenth century.

Over the period from 1300 CE to 1500 CE on the hills of continental Europe, from the Vosges in France in the west to through middle and southern Germany to Czechia, the upper tree line fell by 100 to 200 meters.

After 1300 CE there was changing climate with enhanced short-term fluctuations, including some runs of 3 to 5 years, or even more, of wet, flood-ridden seasons, of drought and either severe or very mild winters that made itself felt further south in Europe.

By 1300 CE France's forests had been reduced from 74 million acres to 32 million acres since 800 CE. This was all for land clearing for crops and timber. Between 1100 CE and 1150 CE more than half of Europe's forests had been cleared.

At Goslar in Lower Saxony, Germany, water had been increasing in the Harz mountains mines for more than fifty years and all attempts to get rid of it now failed. Attempts to reopen mines in Rammelsberg, also in Lower Saxony, were unsuccessful.

In Bohemia, now Czechia, similar troubles had begun in 1315 CE and led to some mines being abandoned by 1321.

In the 1300s CE in Germany villages were being deserted due to the cold weather and inability to reliably grow crops. This multitude of abandoned villages were so common that they were called Wustungen or deserted sites.

From 1300 CE Greenland became drier and the grass cover became thinner.

In Iceland the culminating phase of moist sea ice and coldest climate occurred around 1300 CE.

In the 1300s CE the small Furtwangler Glacier may have formed during the Little Ice Age on Mount Kilimanjaro in Tanzania.

In Norway in the 1300s many farms were abandoned as farmers could no longer reliably grow grain above 1000 feet. This made dependent communities so vulnerable that they left for opportunities elsewhere. Summer temperatures dropped and the previous cultivation up the hill and valley sides ended abruptly.

In Scotland similar was happening as the warm period was coming to its end. At Kelso Abbey in the south of Scotland at an altitude of 300 meters, around 1000 feet, there were over one hundred acres of tillage, 1400 sheep, and sixteen cottages for shepherds and their families. This was no longer possible as the climate got colder.

Between 1300 CE and 1900 CE cooling caused dry conditions in the Sahel, the region bordering the southern Sahara Desert.

In South Africa around 1300 CE the capital of the state of Mapungubwe near the meeting point of the Shashi and Limpopo Rivers was abandoned. Severe drought associated with the onset of the Little Ice Age appears to have done this by contributing to the demise of the city. Climate proxy data from baobab trees shows an early fourteenth century drought.

The monsoon weakened in the fourteenth century in Southeast Asia.

In Switzerland in the 1300s CE glaciers rapidly advanced to bury villages in the Alps. Near mountains, cooling posed a more direct threat to people and communities. Glaciers were expanding and moving downslope towards high Alpine villages. The glaciers in some cases choked off valley floors creating ice dams and when the ice dams broke flooding ensued. The glaciers also destroyed roads and buildings as well as grazing land.

On 6 January 1300, the lakes in Europe collected such an unusual amount of water from streams so that they were 4 Ellen (9 feet, 3 meters) above normal on the highest lakes. This flood devoured towns and villages with people and livestock.

On 15th-16th January at Rungholt, a city in Nordfriesland, northern Germany, the water devoured 7 parish churches along with 7,600 people and innumerable amounts of livestock. This was a storm surge and Rungholt sank beneath the waves. Rungholt was possibly on the island of Strand. There is dispute about the year with some sources stating that it occurred in 1362 though the month and days conflict with the Grote Mandrenke that year which was in October. The remains of Rungholt, which was believed to have five hundred houses and a harbour, were found near the tiny islet of Hallig Südfall in the Wadden Sea.

Also, on 16th of January 1300 half of the former Heligoland archipelago was lost as well as many islands off the German North Sea coast. By the late 20th Century Heligoland was 2.15 kilometers on its longest axis.

On July 11th 1300 CE Hekla in Iceland erupted with a VEI of 4. A VEI of 4 definitely effects the stratosphere and even VEIs of 3 can do this if they generate a lot of sulfur dioxide.

The winter of 1301 CE was warm in Italy.

In early December 1301, a hurricane destroyed homes and other buildings in Germany; it calmed down and the air cleared, and there was such unusual warm that in January 1302 young branches sprouted on the trees. Later the rivers overflowed their banks.

1301 was a pitiful time, a time when nothing but hunger and grief could be found. Hunger was so great that in Erfurt, Germany alone 8,000 people perished.

The winter of 1302 to 1303 CE was a cold one in Western Europe and possibly the same in Britain.

During the winter of 1302-1303, the Rhône River in the south of France was frozen over completely. The winter of 1302 in France was bitterly cold and the olive trees died. In 1302 in Provence, the year produced a severe winter.

On August 8th 1303 in the Eastern Mediterranean a large tsunami hit Crete, Rhodes, Alexandria and Acre in Israel. The tsunami was probably caused by the earthquake in Crete.

The 1303 Crete earthquake occurred at about dawn on 8 August. It had an estimated magnitude of about 8, a maximum intensity of IX (*Violent*) on the Mercalli intensity scale, and triggered a major tsunami that caused severe damage and loss of life on Crete and at Alexandria. It badly damaged the Lighthouse of Alexandria. The earthquake and the tsunami are recorded as having a devastating impact on Heraklion, Crete. Detailed information is available from reports made by representatives from Heraklion (then Candia) to the controlling Venetian administration, written on the day of the earthquake and twenty days later. They describe the extent of damage to the main public buildings of Candia and castles over the whole island. The reports mention that most of the victims were women and children, without giving numbers. There was massive flooding at Alexandria. Many ships were destroyed, some of them carried up to 2 miles (3.2 km) inland. The port city of Acre, on the Levantine coast, was also affected. Buildings were destroyed and people swept to their deaths. In Egypt the earthquake caused severe damage in Cairo, dislodging much of the Great Pyramid's white limestone casing) and toppling minarets on many mosques. In Alexandria the city walls were mostly destroyed. Most notably, the Lighthouse of Alexandria, one of the seven Wonders of the World, was badly damaged. Homes in Alexandria collapsed, killing many people. According to Alexis Perrey, the earthquake was felt on the entire Adriatic coast, up to Venice (about 1600 km or 1000 mi from Heraklion). Modelling of the tsunami predicts a maximum 9-meter run-up at Alexandria, with about a 40-minute delay from the time of the earthquake to the arrival of the first wave in Egypt.

The 1303 Hongdong earthquake occurred in the Yuan dynasty of the Mongol Empire, on September 25. The shock was estimated to have a moment magnitude of 7.6 and it had a maximum Mercalli intensity of XI (*Extreme*). This was one of the most deadly earthquakes in China, in turn making it one of the top disasters in China by death toll. In the nearby towns of Zhaocheng and Hongdong, every major temple and school building collapsed and over half the towns' populations perished. Every building in Huo county, Shanxi was destroyed. In Taiyuan and Pingyang, nearly 100,000 houses collapsed and over 200,000 people died from collapsing buildings and loess caves in a similar manner to the situation that would be experienced 253 years later in the 1556 Shaanxi earthquake. Cracks in the ground turned into miniature rivers, and many canals in Shanxi Province were destroyed, along with city walls. Some reports stated that the earthquake even levelled mountains and hills, altering the topographic make-up of the region. Landslides and soil subsidence and liquefaction triggered by the shaking were a likely root cause of these large-scale environmental changes. Rebuilding was generally slow, owing to the destroyed infrastructure of Shanxi and was interrupted by several other earthquakes in the following years. The 1303 Hongdong earthquake, though currently the last to have occurred on its fault system, marked the start of a centuries-long episode of heightened earthquake activity throughout China, the first of several to occur up to the end of the twentieth century. It was also the first of many examples of earthquakes that demonstrated the tendency of earthquakes in China to strike near loess plateaus.

On 25th November 1304 the Netherlands was affected by a sea floods. This was especially so in Walcheren in Zeeland which had missed out in the 1288 sea floods. Walcheren was originally an island at the mouth of the Scheldt Estuary but is now joined to the mainland.

In 1304 a mighty storm raged in Germany from the northeast along the shores from Rügen and Pomerania. Rügen is Germany's largest island. It is located off the Pomeranian coast in the Baltic Sea. The storm threw down many houses and church steeples. At that time the land that stretched from Mönchgut Peninsula, south almost to the island of Ruden, between the two was such a small stream, that a man was able to leap across it. Now it is around 8 kilometers or 5 miles. The terrible power of the northeast sea caused the creation of a new sea route across the entire strip from Mönchgut to the promontory Thiessow. The newly formed route was called the "New Low." Larger ships that only came from the north in their harbor were now also able to take the newly created passage from the southeast, right from Stralsund.

In 1305 CE there was a hot, dry summer in London and the south of England.

A severe winter over much of western Europe suggests a high frequency of blocked-anticyclonic episodes.

On May 2nd 1305 CE Asosan in Japan erupted with a VEI of 2.

In the summers of 1305 and 1306 in France, the weather was very dry and hot. As a result, the fruits of the earth suffered much. An extreme drought ruled the spring and summer. The cold froze strongly the waters before they had diminished. An extraordinary drought fills the spring and summer of 1306 in France, and major flooding in the following winter. Intense cold froze the rivers quickly before they were able to decrease in river height. This ensured that the spring thaw produced many disasters.

In 1306, there was a great flood at Würzburg and Frankfurt, Germany. The Rhine, Main, Werra, Weser and Saale rivers flooded. The Stone Bridge at Würzburg was damaged. There were 500 people on the bridge when it collapsed, of which ten people died.

The winter of 1305 CE to 1306 was severe in London and the south of England.

There was also a severe winter for much of Western Europe. This suggests a high frequency of blocked and anticyclonic episodes.

In 1305 the Rhône River and all the rivers of France froze. This was two years after the previous freeze.

In 1306, the Baltic Sea was passable by foot passengers and horsemen for six weeks. The Baltic Sea was covered with ice for 14 weeks, between the Danish and Swedish islands.

1306 CE to 1307 CE was a prolonged severe winter in England, the Baltic and Italy.

In 1306 -1307 CE there was reduced growth shown in tree rings in England and Ireland.

Many rivers in Flanders (now Belgium) were also frozen solid enough that carriages could ride across them.

In the Winter of 1306-1307 CE the rivers of France, among others the Seine River, were frozen. The rivers of France froze before the waters, which had risen due to great flooding, could substantially recede. As a result of these the ice conditions, the force of the ice was so great that the bridge, the mills along the rivers and standing houses collapsed. In Paris, at the Port of Grève, a large number of loaded barges sank on the Seine River with people and cargo onboard.

In 1307 CE Izu-Oshima erupted in Japan with a VEI of 3.

On 2nd February 1307 all the English coast was inundated with sea floods.

The first Frost Fair held on the frozen Thames River in London was held in 1309 CE. There would be five Frost Fairs until the last one in 1814. The Thames River had frozen over at least 23 times since 1309 AD. There were periods in the Little Ice Age when Winter Markets stood for months on the frozen River Thames in England well before this date.

During the winter of 1309 to 1310 CE London bridge arches were damaged by ice due to a severe winter. There was a frost fair in London and people walked across the Thames. Contemporary reports state that dancing took place around a fire built on the ice and a hare was coursed (chased) on the frozen waterway.

In 1310 CE Mount Tarawera/Okataina on the North Island of New Zealand erupted with a VEI of 5 which would have caused a volcanic winter.

From 1310 to 1319, the weather conditions were very unfavorable in Germany. Crops failed causing famines and a third of the population starved to death.

Around 1310 CE there was a sharp increase in raininess in Europe.

After 1310 CE the wheatlands and vineyards of northern France shared in heavy harvest failures and the resulting famine and deaths by the millions of people in that decade. There were reports of cannibalism due to starvation.

Was this a volcanic winter as a result of the Mount Tarawera/Okataina eruption in 1310?

In 1311 Kelud in Indonesia erupted with VEI of 3.

There was famine in England in 1311 AD.

In 1312 CE a three-year famine struck Bohemia, now western Czechia, and Poland. This famine was so great and severe that children devoured their parents and parents ate their children. Some fed on the dead bodies of malefactors hung up on gibbets, gallows-type structures from which the dead bodies of executed criminals were hung on public display. Wolves also were so famished, that they devoured all they met and fed on them.

An old chronicle from Würzburg reported: "the year 1312 brought an unusual amount of rain and storm in Germany. Everywhere the streams and creeks came out from their banks and flooded the fields. As a consequence of this, there was a great famine and the most terrible plague, which raged for a long time. In some places there was no one who could bury the dead. Inflation of the following year, 1313, was even greater and those spared by pestilence were now wiped out by hunger. This period of great misery lasted for a long time." Muellner wrote in his Nuremberg Chronicle: " the year 1312 caused the greatest mortality everywhere. One third of the population died. Parents were so hungry that they slaughtered and ate their children. The corpses of criminals were pulled down from the gallows and eaten." Afterwards there were deserted villages. More westungen. This was not a great period to be trying to stay alive.

In England there were wet, cold summers from 1313 to 1317 in an unbroken series. This reduced the harvest considerably. Between 1314 to 1316 CE several famines occurred in England. All three years were regarded as very wet. The famine of 1316 was probably the last really severe famine in England and over this period around half a million people died of causes related to the famine. This was ten per cent of the population. It is suggested that there was an increase in climate variability and an increase in extreme events including windstorms. As well as excessively damp conditions, temperatures were depressed as well.

In 1313 a sea floods attacked Friesland in the Netherlands and 500 people drowned.

In the year 1313 in March, April, and half of May there was a summer-like warmth. On Pentecost Eve, a sudden sharp cold struck Pohlen, Germany. Snow two elbows deep (2 Ellen, 4 ½ feet or 1.5 meters) fell. It was feared that it would destroy all the fruits of the field. But on the 6th day, this snow was melted by a warm rain. The ground was watered like by a sweet dew and made fruitful. There was no harm done to the seeds. On 1 May 1313, there was a high flood in Germany, which did great damage.

Between 1313 CE and 1314 CE and 1317 CE in Czechia, Germany and Poland, there was an extraordinary run of wet summers and mostly wet springs and autumns which continued to the early part of 1321 CE. This followed closely upon one of the really notable periods in the Middle Ages of mostly warm, dry summers from 1284 CE to 1311 CE.

In one, two or three years between 1314 CE to 1319 CE almost every country in Europe lost almost the whole harvest. The poor were reduced to eating dogs, cats and even children.

In 1314 in England, it rained almost ten months continually, but during July and August, the rains were incessant. The husbandmen, farmers, could not get in the small crop they had on the ground, and what they got in, the yield from it was very small. Hence there was a grievous famine in 1315 that lasted two years and from it most mortal dysentery so that it was drudgery on the surviving to bury the dead. Cattle and beasts were being corrupted by the grass whereon they fed and then died; hence people dreaded eating their flesh. Only horseflesh was a delicate dish. The poor stole fat cats to eat. Criminals in gaols, jails, quickly pulled to pieces fresh malefactors and ate them. Malefactors are evil doers or criminals. Or the last imprisoned tore in pieces and devoured the old Goal Birds, jailbirds. Hunger compelled some to eat their own children, and some stole other people to eat.

Few English kings have lived through a greater period of distress than Edward II, who was scarcely able to secure food for his own immediate household when the heavy rains of 1314 spoiled the harvests.

Misery in England was widespread and intense: the dead lined the roadsides; everything imaginable was eaten – dogs, horses, cats, even babies. In England in 1314, grains were spoiled by the rains. Famine so dreadful that the people devoured the flesh of horses, dogs, cats and vermin. Parliament passed a measure limiting the price of provisions.

In Ireland in 1314, there was famine and various distempers.

And you thought that things were bad in 1314?

In 1315 CE to 1317 CE the Great Famine occurred. Other famines in 1321, 1351 and 1369 CE also occurred. The famines were caused by extraordinary cold winters being followed up by wet and cold summers and that the unusual weather patterns were caused by a volcanic event in New Zealand. During these periods of famine life expectancy was reduced to 29 years. The only New Zealand volcano to erupt in this period was Okataina in 1310 CE with a VEI of 5 which is very high.

In England, there were great rains and floods during harvest; much grain spoiled. In 1315 in England, the grain was spoiled by the rain. Famine "so dreadful that people devoured the flesh of horses, dogs, cats, and vermin." Parliament attempted to fix prices; and failed. In 1315, there was a dreadful famine in England. The poor ate horses, dogs, and cats. In King Henry III's reign, in a dreadful famine, the people ate the bark of trees, and 20,000 persons starved to death in London alone.

In 1315 CE there was a cold winter in Western Europe and possibly parts of Britain.

Famine struck Europe between 1315 CE and 1322 CE. Heavy rains had caused erosion and prevented planting. The rains of 1315 were extraordinary. Summer and Fall were cool. The rains persisted daily for five months and the rain and damp ruined crops and swept away soil. This was especially so in the newly settled areas created by expansion during the prior Medieval Warm Period. Swollen streams and rivers carried away mills, bridges and whole villages. Over the following three years the effects of famine were widespread and dire. Crops could not ripen that had not been washed away. It is estimated that 1.5 million Europeans perished as a result of famine and famine related illnesses. "The Great Rains" is regarded as what ushered in the Little Ice Age. There was cannibalism in Europe due to lack of food. Great numbers of sheep and cattle also died in the "murrains" or epidemics of disease which swept the sodden and often flooded landscape. Thereafter the 14th Century brought wild and long-lasting variations of weather in western and central Europe in the later 1320s and 1330s and also the 1380s with mostly warm, dry, droughty summers.

Europe was hit by incessant rains. These were followed by a famine so severe that Polish poor ate hanged bodies. So great was it in Poland and Silesia, that parents abstained not from devouring their own children and the filthiest creatures. Silesia is a historical region of Central Europe located mostly in Poland, with smaller parts also in the Czech Republic, and Germany

In 1315 in France, from the middle of April until late July, there was almost continual rainfall combined with an especially cold summer. The grains and the grapes did not ripen. So terrible was this famine in Thuringia in Germany.

In the walls of Exford, Oxford, in England several people were starved and died.

The famine continued for three years in Lithuania. The Kingdom of Lithuania during this period of time extended west into the Kievan Rus territory of Russia.

In 1315 there was a famine that was worst in England, Thuringia, Poland, and Silesia. This famine lasted years in Lithuania.

There were crop failures and starvation in Northern Estonia.

In 1315 in Ireland, there was famine and various distempers.

In Europe from mid-April to late July 1315, it rained incessantly, and there was unusually cold weather. The cereals and the grapes did not come to fruition. Seven weeks after Easter 1315 sheets of rain spread across a sodden Europe, turning freshly ploughed fields into lakes and quagmires. The deluge continued to September. In that time hay lay flat in the field and wheat and barley rotted in the fields. Grain production over northern Europe had fallen by a third at least. Herds and flocks were reduced by as much as ninety per cent owing to diseases such as rinderpest and liver fluke brought on by wet weather . Fruit rotted on saturated trees. Coastal fisheries and fishponds were devastated. Many of the fishponds were large undertakings built by Monastic and Royal houses to feed many of the occupants. There were at least 1.5 million deaths from starvation in Europe, mainly from the poorest class.

In 1315, the Po River in Italy froze.

In Germany and England, the problem of the abandonment of former settlements all over northern and central Europe was reaching its peak. In Germany alone there were thousands of now-abandoned villages which became prominent in connection with the famines of the decade about 1315 CE. These were the Wustungen.

Of 80 deserted village sites in England for which population records can be deduced from tax records, in two counties in central England only about ten per cent were attributable to the Black Death, but all had suffered severe losses of population in the famine times between 1311 CE and the 1320s.

In 1315 there was harvest failure in Denmark. By 1334 many farms had been deserted and multiple families shared the same farmhouse.

In 1316 England was struck by 2 years of rains and floods. Famine killed thousands. There was a universal dearth and a great mortality, death of large number of people, particularly among the poor. As a result, the living could scarcely bury the dead. There was a royal proclamation – no more beer to be made. As of March 1316, there was a great famine in England. At Nottingham, the season for the past 3 years was so adverse that almost all the grains were destroyed and the people were driven to eat horses, dogs, cats, and vermin, etc. Even children were stolen and eaten. And prisoners were eaten by other prisoners. This was one of the most grievous famines that ever visited Great Britain. The famine was followed by a pestilence scarcely less destructive to life, so that the living scarce sufficed to bury the dead. This famine resulted from two or three years of continued rain, which destroyed the corn (grain) and caused a frightful mortality amongst the sheep and cattle. Corn (grain) was four to five times its ordinary price, (i.e. 60 to 90 shillings); oxen 48 to 70 shillings; fat hog 10 shillings; fat wedder 5 shillings; goose 7½ pence; fat hen 3 pence; two chickens 3 pence; 2 dozen eggs 3 pence. These articles were 6 to 8 times their average price. The famine lasted several years.

In England in 1316, wheat was sold at 40s. and 44s. per Quarter (quarter ton). By reason of the murrain among cattle, beef and mutton were exceeding dear (scarce). After this, both famine and mortality increased greatly, together with a general failure of all fruits of the Earth. This was due to excessive rains and unseasonable weather. Provisions could not be obtained for the King's household; nor other great men to keep up their tables; as a result they were obligated to discharge their servants in great numbers. These servants having lived so delicately and not able to perform other work felt scorn to take up begging. As a result, these servants fell to stealing and robbing, which caused fresh misery to the Nation. So terrible was the famine two years before in 1314 that not only horses and dogs; but also, men and children were stolen for food. All malting throughout the Kingdom was forbidden, even for the King's family. Malting is the process of converting barley into malt, for use in brewing or distilling. When wheat was sold at 10d a bushel, it was so very cheap; but at 10s., it was monstrously dear.

In Ireland in 1316, there was a great dearth. Eight captured Scots were eaten at the siege of Carrickfergus in Northern Ireland.

In the year 1316 in France, at St. Andrew's Day (November 30) began a very hard winter, and this continued until Easter 1317. The pretty rough winter of 1316 in northern France lasted without interruption from late November to Easter. There were bread shortages in France due to the famine.

In Germany, the harvests failed entirely because the cold had destroyed all of the seeds entrusted to the earth. A famine took hold, caused by a lack of food; poor nutrition produced many deadly diseases. People from the German countryside made their way to beg at towns along the Baltic Sea. Some communities disappeared as their inhabitants left. Disorder and lawlessness were occurring. Destruction of seeds slowed down recovery. Crop yields and and mortality from disease rose in weakened populations.

The more settled climate of the Medieval Warm Period was over. Much more unpredictable weather, greater storminess and cycles of very cold winters, shorter growing seasons, or warm summers began, marking the beginning of the Little Ice Age.

Heavy spring rains in spring 1316 meant that crops could not be sown. Flocks and herds withered as there was little or no food.

In Flanders in Belgium there were bread shortages due to the famine.

In 1316 intense gales battered the North Sea and the English Channel.

From 1317 CE to 1464 CE the Pacific Region was much warmer.

In 1317 CE in England, there was a very good summer, and early and plentiful harvest. Wheat, which sold for 10s. per bushel, now sold for 10d. On Saturday, it was 44s. per Quarter (quarter ton); next Wednesday it was sold at 10s. in Leicester Market. At the same time, many who had been rich and had an abundance of all good things, came to want and were forced to beg. In the south (southern England), there was a murrain of cattle. This year and also in 1319, were both very fatal to people and several other animals, over the whole Kingdom. So that the survivors were not sufficient to plow and sow on the ground. Besides, many were still buried daily in every churchyard. This plague was two years in its perambulation over England. Hence there was great desolation from bad food in the famine.

China experienced massive floods in which multitudes drowned.

In England in 1318, there was a murrain of kine (cattle), that dogs or ravens, which ate their flesh were poisoned swelled and died. Therefore, people dare not touch them.

In 1318, the winter was severe in France, Germany and Italy. Wagons crossed on the ice on the Po River in Italy.

In 1318 CE there was a great earthquake at Limburg an der Lahn, Germany. This was followed by a severe famine. Farmers killed travelers and ate them.

On the 23rd of March 1318 CE a storm was noted as having affected both the Netherlands and England.

In 1318 CE there was a great earthquake that was felt throughout England on 14th November.

By 1319 CE the murrain which last year was in the south of England, now reached the north and overspread the whole realm. The carrion was still poisonous.

In 1320 CE numerous sources stated that two decades before the advent of the Black Death in England, a great earthquake was felt throughout the country. Later writers referred to droughts, floods, numerous earthquakes, locusts, subterranean thunder, unheard of tempests, lightening, sheets of fire, hailstones of marvelous size, fire from heaven, stinking smoke, corrupted atmosphere, a vast rain of fire, masses of smoke. There were also reports of a black comet seen before the arrival of the epidemic. There were heavy mists and clouds, falling stars, blasts of hot winds, a column of fire, a ball of fire, and a violent earth tremor in Italy. All of this before the Black Death arrived. There was a wet and cool summer and a disastrous harvest. This was a period of low temperatures and above-average rainfall. Continuing cold weather and parasites killed off many sheep, damaging the important woolen industry in England.

From 1320 to 1330 there was a sudden cooling of the North Atlantic surface water. This was simultaneous with an abrupt downturn in European oak growth. The sea-surface temperature from 1300 to 1318 CE oscillates on a high frequency in a two-to-three-year cycle. After 1318 there was a smoother long-term cycle until 1333 CE.

In 1321 CE there was a hot, dry summer in London and southern England. In England, there was the greatest drought, with heat. The summer of 1321 in England was extremely hot and dry. The springs and the rivers were dried up. The domestic animals and the cattle suffered greatly. Many unfortunates died from lack of water to quench their thirst. In England, there was famine again.

In 1321 CE the rains that had started in Europe in 1315 CE subsided. Since 1315 over a million and a half people had perished by starvation and famine related epidemics.

On 11th November 1321 a sea floods hit Belgium.

It was recorded that in China between 1321 and 1368 an iron rain fell that killed people, animals and damaged a house. Most meteorites are made of iron, so this is pretty self-explanatory.

In 1322 in the Netherlands and Belgium, there was a massive sea floods. Flanders lost all of its coastal islands and there were many deaths, especially in Holland, Zeeland and Flanders.

During winter 1323 to 1324 CE, December and January, there were damaging floods that started on New Years Eve. At Stepney in London, a series of damaging floods with a mighty flood proceeding from the tempestuousness of the sea, which overflowed all banks: as the waters ebbed, they tore a great breach in the wall, allowing subsequent tides to flow across the land.

The winter in 1323 was intensely cold and the Baltic Sea was so firmly covered with ice, from Mecklenburg, Germany to Denmark, that merchandise was conveyed over the Baltic Sea with horses and wagons for six weeks. Mecklenburg was a historical region in northern Germany. The largest port cities of the region at this time were Rostock and Schwerin. In 1323, the Baltic Sea was frozen over, and during three months travelers passed from the continent to Sweden on the ice. Heavy wagon trains were substituted for the traveling vessels.

In 1323, a severe winter struck Denmark. The sea between Denmark and Rügen, Germany was covered with thick ice. Huts were erected on the ice, where one could buy food and drink. This frozen passage was used for 10 weeks.

The winter of 1323 was a very cold winter. At the end of February, the Elbe River in Germany stood hard like a rock. People traveled all over the ice from Teutschland, Germany, and the frozen Baltic Sea.

In 1323, the Rhône River in France froze.

In 1323, the Po River in Italy froze.

In the Baltic Sea, travelers pass on foot and on horseback on the ice between Denmark and Lübeck in northern Germany and Danzig (now Gdansk, Poland).

On September 7th 1324 CE Asosan in Japan erupted with a VEI of 2.

In the summer of 1324 CE there was a drought in London and the south. This was probably the start of ten or so years of warm, often dry summers.

In summer 1325 to 1326 there was severe drought in Britain. Rivers and springs dried up and in both years the Thames was so low that sea water penetrated much further upriver than usual and was regarded as salty for most of the year.

In 1325 CE the Seine River in France froze twice at short intervals. The river was crossed with sleds with heavy loads. The ice on the river was thick enough to support the weight of men and full barrels. During this winter of 1325, the cold was very great. It was even mentioned in the minutes of the Parliament of Dijon, France. In Paris, the ice conditions of the Seine River were so violent that the two wooden bridges were carried away. "We crossed the frozen river at Paris, France with a burden (carrying weight). The strength at its surface was strong enough to support rolling barrels full of wine over the ice. Large snow accompanied the frost. The ice melted completely at Easter.

In Western Europe the summer of 1325 was extremely hot. The drought was so great in the year 1325 in France that there was barely two days' worth of rain in the course of four moons or 4 months. In 1325 in France, there was excessive heat with a severe drought, but no lightning or thunderstorms. The year produced little fruit. Only the wines were better than usual.

In England, the earth was very fruitful with air temperature and sea calm.

On 30 January 1326, there was a great earthquake in southern Germany. There were very cold temperatures and the Lake Constance in Switzerland froze over. This was followed by a hot summer and a poor harvest.

By 1327 the population in England had declined by a third. Only ten per cent were attributable to the Black Death. Out of fifty deserted village sites in Oxfordshire and 34 in Northamptonshire, suffered serious decline between 1311 and 1319. The population decline by 1327 averaged 67 per cent, twice as great as the Black Death. In 1327, there was a famine in Great Britain. Owing to a succession of cold rainy harvests, the whole kingdom experienced a most grievous famine.

In 1327 CE there was a famine in Delhi and its neighborhood in India. During this famine, women ate the skin of horses that had been dead for several months. The skins were cooked and sold in the marketplace. Crowds fought over the blood at the slaughterhouse.

In 1328 CE a great storm filled the harbor of Dunwich on the Suffolk coast. That year the harbor was filled with sand and silt. Little by little as the sea pushed into the town, whole streets and houses vanished. 400 houses were swept into the sea. Dunwich had a population of around 3,000 people at this time. This was similar in size to that of London. After this it was little more than a village and around 16th January 1362 in the Grote Mandrenke Storm much of the remainder of the town was destroyed. Dunwich originally had eight churches, Greyfriars Franciscan Priory, the Leper Hospital of St. James, and a Perceptory of the Knights Templar.

At the beginning of October 1328, the eve of Saint-Denis and during the octave of the feast, strong winds toppled many buildings in France.

In 1328, it was so cold that the Danube River in Europe was frozen over for 17 weeks. It was a fruitful year, but again and again earthquakes.

On June 28th 1329 Mount Etna erupted in Italy with a VEI of 3.

In 1330 CE Cerro Bravo in Colombia erupted with a VEI of 4 and Gorely in Kamchatka had a VEI of 2.

Between 1330 CE and 1350 CE there appeared to be a massive ocean turnover event resulting in mass deaths of fish and other aquatic species. There was also a massive stench from the dead aquatic life. What had affected the ocean?

In England and Wales in 1330, the land was so inundated with continuous rains that scarcely an article of food was raised. The rain was so violent; the harvest did not begin till Michaelmas (29 September). In England there were heavy rains; grain did not ripen; harvest not commenced till Michaelmas. In England, exceedingly great rains fell from May to October. The corn (grains) could not ripen and in most places, harvest did not begin until 29 September. Wheat was not harvested before 21 November. Nor pease (pea) before 30 November

On 23rd December 1330 CE roofs were blown off at Croxden Abbey in Staffordshire, England. From the Croxden Chronicle…. "On the night preceding Christmas Eve at twilight, a very strong wind blew up from the West, and took the roofs off the Abbey buildings and farm buildings throughout the country in a terrifying way. It tore many of them from their foundations and uprooted oaks in the woods., and countless apple and pear trees in the gardens in a remarkable way".

From 1330-1346 CE there was anomalous river flow in Arkansas. The years 1330 CE, 1334 CE and 1338 CE were all abnormally high with the river flow for 1341 being the highest in the entire 950-year tree ring record. In contrast the years 1342 CE and 1346 CE were both anomalously low.

In 1331 CE Azumayama in Japan erupted with a VEI of 2 and Zaozan in Japan also had a Vei of 2.

In April 1331 Asosan in Japan had a VEI of 2 and in December 1331 Asosan again had a VEI of 2.

In 1330 CE and 1340 CE there was a reduction of taxes in Norway on account of lowered farm yields and losses caused by disasters such as rock falls. Owing to the nature of the country there were big variations from district to district and from farm to farm. The decline was on the whole sharpest in Trondelag, the district about Trondheim, which had been richest earlier in the Middle Ages. Wheat had been grown there. The abandonments were when there were periods of climatic stress. Full scale farming was not restarted there until the 1930s.

The upland village of Hoset was marginally situated near the Swedish border at an altitude of 1350 meters, 1150 feet above sea level and east of Trondheim. Originally cleared around the 4th Century CE and then abandoned, repopulated and abandoned around 1445 CE and finally abandoned after resettlement in 1690s and reconquered by the forests and never lived in since.

In Spring 1331 there was a drought which lasted 15 weeks, but a few days before the 17th of June, when a tournament was due to commence at Stepney, the drought was broken, and all the ground was thoroughly watered.

From 1331-1332 CE there was a sudden period of cold years worldwide.

In December 1332 Grimsvotn in Iceland erupted with a VEI of 2 and there was a disastrous crop failure in Europe that year as well.

In 1333 Mount Etna in Sicily erupted with a VEI of 2.

In England in the summer of 1332 CE there was a major drought.

In 1332 CE the Black Death was supposed to have originated in China or Central Asia where it is endemic, during or after exceptional rains and flooding. This flood was one of the greatest weather disasters ever known and is alleged to have taken seven million human lives in the great river valleys of China. The flooding destroyed not only human settlements and their sewage arrangements but also the habitats of wildlife.

In China, Toghon Temur who ruled the Yuan Dynasty, 1333 to 1368, saw China face repeated climatic crises. There were alternately severe droughts and major floods. Epidemics and famines also struck China.

In Florence in Tuscany in central Italy in November 1333 CE, there was a great overflow of the Arno River which swept away the stone bridge Ponte Vecchio, which straddled the river in Florence. This had happened before when the Roman era wooden Ponte Vecchio was swept away.

In 1333-34 the winter was very severe in Italy and Provence. There was snow in Padua, Italy from November 1333 to March 1334. In the year 1334, all the rivers in Provence and Italy were frozen. The frost of 1333-1334 stopped all the rivers of Italy and Provence, France.

At Paris, France, the frost lasted two months and for twenty days the winter of 1333-1334 in France was very wet. In December 1333 and January 1334 in Paris, France and the surrounding country, there were great thunder and lightning storms with wind and hail. These were storms that were normally found in the month of July.

After 1333 CE it got progressively cooler and dryer in the American Southwest.

In 1334 CE Kelud in Indonesia erupted with a VEI of 3.

In early 1334, the Po River in Italy and the Rhône River in France froze.

On the 22nd November 1334 there was a tidal flood on the Thames. In England on 23 November 1334, there was a prodigious inundation of the sea along the coast, especially about the River Thames. The violence of the water broke down the banks and drowned infinite numbers of beast and cattle and turned the pasture ground into salt marshes.

On 23rd November 1334 sea floods hit England, Belgium and Zealand and Friesland in the Netherlands. There were thousands of deaths.

In 1335 CE it rained so heavy in England that the grain was spoiled. In England there were continued rainstorms and there was famine occasioned by long rains. Also, in England there were floods, storms, tempests and meteors. Were these meteors or lightning? Mind you, the educated people in this period actually knew the difference between a meteor and lightning. In England after the abundance of rain came a murrain of cattle and dearth of corn (grain). Wheat sold for 40s. a Quarter (quarter ton). There was so great a number of deaths in England that scarce could the living bury the dead.

In Novgorod, Russia in autumn, ice and snow drifted into the Volkhov River, carrying away fifteen stays of the great bridge.

On February 7th 1335 CE Asosan in Japan erupted with a VEI of 2.

In 1335 CE the pattern of precipitation in the American Southwest shifted toward high annual variability with severe droughts separated by brief wet intervals. Soon after 1335 Arroyo Hondo's population began to decline even more dramatically than it had increased. By about 1345 CE the pueblo was virtually abandoned for the next 30 years. Then, sometime in the 1170s a second phase of resettlement began.

In 1337 CE in England, there was a severe frost without snow. Wheat was very dear.

In 1337, there was a major earthquake in Wuerttemberg, Germany. Then came a severe drought, poor grain and grape harvests, followed by a famine.

In 1337 the whole of Moscow, Russia, was burnt down and then came heavy rains, which flooded everything. The rains flooded both in the cellars and in the squares wherever anything had been carried out (from the fires). In the same year Toropets in east-central Russia was burnt down and flooded.

In 1337 in China, there was a famine, which occasioned a pestilential epidemic. Four million people died in the neighborhood of Kiang.

In 1337 CE a great comet appeared in the heavens. It had a far-extending tail.

In 1337 the town of Rungholt and fourteen smaller places in Schleswig, in what was Denmark, were lost in a sea floods. Other sources say this was in the Grote Mandrenke, the Great Drowning of men in 1362.

A massive sea floods hit the Netherlands in 1337. Was this the same sea floods that sunk Rungholt?

Between 1337-1347 CE was the second driest period in the last 800 years in the American Southwest.

On September 15th 1338 CE Izu-Oshima in Japan erupted with a VEI of 3.

In 1338 CE in England and Wales, the land was so inundated with continuous rains that scarcely an article of food was raised. In England, it rained from the beginning of October to December. In England, there was a severe frost for twelve weeks, after the rain. In England, there was a very rainy harvest, which hindered sowing of winter corn (grain). From 1 December 1338 to 1 March 1339 there was a most rigorous frost, which killed the little sown seed. Yet such a scarcity of money, that grain was not dear. Wheat was 2s. a fat ox 6s., and a sheep 6d. In the winter of 1338 to 1339 CE a hard frost started in December and lasted for twelve weeks. This was in London and the south of England.

In Western Europe, the Meuse River in Germany froze.

In 1338, by a great flood of water, strong currents cut through Eydersted, now Eiderststedt, in Schleswig-Holstein violently and completely demolished Dithmarschen, also in Schleswig-Holstein in Germany. Great quantities of rain fell which caused a dearth and then a great famine in which many people died of hunger.

In Ireland so great a frost was this year, 1338 CE, from the 2nd December to the 10th of February, that the river Liffey was frozen over so hard as to bear dancing, running, playing football and making fires to broil herrings on. The depth of the snow that fell during this frost was almost incredible. It was agreed that such a season was never before known in Ireland.

In the Novgorod in Russia, the water was big in the Volkhov River as it never had been before (a great flood), three weeks after Easter Day on Wednesday, and it carried away ten stays of the great bridge; at the same time, it carried away the bridge over the stream named Zhilotug, and much harm was done.

From 1338-1342 CE great floods occurred in the vicinity of the Rhine River and in France, which could not be attributed to rain alone, for everywhere, even on mountaintops, springs were seen to burst forth and dry tracts were laid under water in an inexplicable manner.

In 1339 CE in northeastern England on 22nd March, there was a great flood of the River Tyne; many lives were lost. In England, on 22 March 1339, in the night there was a great flood in the River Tyne, which broke and carried down six perches of the wall of Newcastle wherein 120 men, several priests, and many women were drowned. This year a Quarter of wheat cost 40d. and sometimes less; barley 10d.; peas and beans 12d.; oats 10d.

In Ireland in 1339, there was a general famine.

In 1339, there was also a famine in Scotland. The crops failed and such a famine ensued that the poor were reduced to feeding on grass. Yet at the same time, wheat in England was only 3 shillings 4 pence per quarter (quarter ton).

In 1340 CE Mont Pelee in Martinique erupted with a VEI of 4 as well as, the island of Eldey with a VEI of 3 in Iceland.

Asosan in Japan erupted on February 3rd 1340 CE with a VEI of 2.

There were major floods in China in the 1340s.

In Indochina rainfall did not shrink all at once but between 1340 to 1380 there was markedly low rainfall.

On May 19th 1341 Hekla erupted with a VEI of 3.

The *Nonarium Inquisitiones*, a valuation of agricultural production in the year 1341 CE, a few years before the arrival of the Black Death, showed that there were large numbers of villages with uncultivated land in every part of England. This was mostly said to be the result of the shrinkage of the population since the famine years earlier. This was also due to soil exhaustion and shortages of seed corn and ploughing teams.

In 1341 CE the cold of this winter in Livonia (currently comprising present day Estonia and parts of Latvia) was so great that many soldiers of the army of the Crusades froze to death or sustained frozen noses, fingers and limbs.

In England and Scotland in 1341, there was a great dearth in this year and the next. People ate horses, dogs, cats, etc. to sustain life.

In 1341, there was a great flood in Germany that drowned many people. This was probably a storm surge from the North Sea.

On 16th January 1342 CE a gale destroyed the tower of the Church of Friars Minor in London, The fall occurred at night and was associated with a violent thunderstorm.

Also, in 1342 CE there was a great drought in summer in southern Britain.

On 9 July 1342, a great flood struck Hannoversch Münden, in Lower Saxony, Germany. It lies at the confluence of the Fulda and Werra Rivers which join together to form the Weser River. The floodwater went through the upper gate and stood for several days in the city. The flood swept away numerous bridges and caused many houses to collapse. The water was 8.91 meters (29.2 feet) high.

The flood of 21st July 1342 was one of Central Europe's largest floods during the millennium. There was a long drought followed by two days of sustained extraordinary downpours. At Würzburg, Germany, the River Main was very close to the cathedral. In the Rhine region, at the Mainz Cathedral, the water was up to a man's waist. In Cologne, boats could travel over the city walls on the floodwaters.

In the chronicles of Regensburg in Bavaria, Germany, Passau in Bavaria, Germany and Vienna in Austria, the St. Magdalen day flood was described as a catastrophic Danube River flood. It was the same with the Moselle, Vltava, Elbe, Werra, Unstrut and Weser rivers.

Even Carinthia in Austria and Lombardy in Italy were ravaged by floods. "This summer (July 1342), there was such a great flood of waters by the whole world, that there was a flood not created by rain but it seemed as though water was gushing from everywhere, even from the tops of mountains. It flowed over the walls of the city of Cologne and boats went over the walls. The Danube, Rhine and Main rivers destroyed towers, very strong city walls, bridges, houses and the bulwarks of cities. It was as if the floodgates of heaven were opened and the rain fell to the Earth as it did in the 600th year of Noah's life. At Würzburg, Germany the River Main smashed the bridge and forced many people to leave their homes." This was on the Feast of Mary Magdalene on 22nd July 1342, poured a great flood and put many places underwater in Central Europe.

In 1342, there was a severe drought in several regions of China. The resulting famine was so severe that cannibalism was practiced throughout these regions. China and Europe seem to play vice versa with weather at times. Drought in China, rain in Europe. Rain in Europe, drought in China.

Ivar Bardarson or Baardson in his *Description of Greenland* reported that by about 1342 the old sailing route along the 65 degrees North parallel to the Greenland Coast and then following the coast around to the colonies on the Western side, had been abandoned because of the increase of sea ice. A track going just a day and a night west from Iceland and thence southwest to avoid the ice was laid down instead.

Between the 1340s to 1360s more intensely cold winters and exceedingly cold summers forced the northern Norse settlers who were farming in Greenland to head south to the Eastern Settlement. With the abandonment of the Northern Settlements in the face of increasing cold, the walrus ivory trade collapsed which were used to pay tithes to the church of up to 1400 tons per year. Ties with Norway ended as ice conditions increased.

In India in 1342, there was a very severe famine in Delhi. Few of the inhabitants could obtain the necessities of life. This was during the reign of Sultan Muhammad Tughlaq.

In 1343 CE Asosan in Japan erupted with a VEI of 2.

On 25th November 1343 there was a tsunami in the Gulf of Naples that was caused by the collapse of the flank of the Stromboli volcano on Sicily in Italy.

In Norwich in East Anglia there was a very high wind by which the passage boat coming from Yarmouth sunk near Cautley and 38 people perished.

In 1344 CE all the rivers in Italy were frozen over.

In India in 1344-45, there was a famine that extended over the whole of Hindustan. It was very severe in Deccan. The Emperor Mohommed, it is said, was unable to procure the necessities for his household the situation was so bad. In 1344 CE there was a famine in the Deccan in southern India.

In 1345 CE Popocatepetl in Mexico erupted with a VEI of 2.

It was reported that in 1345 CE between Cathay (China) and Persia there rained a vast rain of fire: falling in flakes like snow and burning up mountains and plains and other lands, with men and women: and then arose a vast masses of smoke: whosoever beheld this dust died within the space of half a day.

There were disastrous crop failures between 1345 to 1348 CE in Europe. In some areas substantial levels of the population died.

In 1345 CE rains began in Italy in July and lasted 6 months. Famines then followed.

On 21st November 1245 in the Novgorod region in Russia, a southerly wind arose, with snow, and drove the ice into the Volkhov River, and carried away seven stays (of the great bridge).

In 1345, there was a drought in England. It was called "the dry summer". From March to the end of April, there was little to no rain. As a result, grain was very meagre the rest of the year.

Around 1345, a famine struck in and around the Delhi district of India. The famine was caused by a failure of the rains. Ibn Batuta, a native of Tangiers, Africa, was visiting the region at the time and wrote: "Distress was general, and the position of affairs very grave. One day I went out of the city to meet the wazir, and I saw three women who were cutting in pieces and eating the skin of a horse, which had been dead for some months. Skin was cooked and sold in the markets. When bullocks were slaughtered, crowds rushed forward to catch the blood, and consumed it for their sustenance. . . The famine being unendurable, the Sultan ordered provisions for six months to be distributed to all the population of Delhi. The judges, secretaries, and officers inspected all the streets and markets, and supplied to every person provisions for half a year." Other sources confirmed the relief was real.

In 1346 CE Asosan in Japan erupted with a VEI of 2.

In 1347 CE a large storm swept some 400 houses into the sea at Dunwich in East Anglia. Again? How many houses did Dunwich have or are we getting our dates mixed up?

The Black Death that struck between 1347 CE and 1352 CE reduced Europe's population by 25 million people from a population of 80 million. CO^2 may also have decreased after forests reclaimed agricultural lands after the pandemics occurred and there was sharp population decline. From a low in 1450 England's population would not return to its previous levels until the 1600s.

Ziegler refers that in 1347 in Cyprus: whilst the plague was just beginning a particularly severe earthquake came to complete the destruction. A tidal wave swept over large parts of the island.... A pestiferous wind spread so poisonous an odor that many…fell down suddenly and expired in dreadful agonies.

There was a great famine in 1347 in Germany. The misery in 1347 is described by Müllner in his Nuremberg Chronicle: "it was a very barren year. Everywhere the grapes and fruit froze. So, this year started the cruel mortality of almost the whole world was stricken until the fourth year. There was a great mass of locusts. Everywhere in the fields there were big piles of dead locusts. In some places, whirlwinds carried the dead locusts into the sea, where they washed ashore. The great stench from these rotting locusts poisoned the air. Others blamed the plague on the Jews and accused them of poisoning the wells and they were persecuted in a most violent manner. This mortality (great death) affected not only municipalities and large cities but also villages and caused many to become deserted. In many cities neither the council nor the court was held. Parents abandoned their children and children their parents."

In Italy in 1347, a dreadful famine swept away by absolute starvation vast numbers of the inhabitants. And in the following year a pestilence of a deadly nature swept the peninsula. "Such was the sufferings produced by these visitations that it was calculated that two-thirds of the whole population were destroyed."

On 25th January 1348 CE there was an ammonium spike in the atmosphere that indicated an atmospheric strike by a high energy body.

On 25th January 1348 CE it was recorded from numerous sources that there were more earthquakes than usual. The reports also mentioned that there were reports of a black comet seen before the earthquakes. There were also heavy mists and clouds, falling stars, blasts of hot wind, a column of fire, a violent earth tremor in Italy, a crescendo of calamity involving earthquakes, following which the plague arrived.

That same date at 11.00 pm there was an earthquake throughout Carinthia in Austria and Carniola in what is now Slovenia, which was so severe that everyone feared for their lives. There were repeated shocks and on one night there were twenty. Sixteen cities were destroyed, and their inhabitants killed as well as 36 mountain fortresses, and their inhabitants and more than 40,000 men were swallowed up or overwhelmed.

The 1348 Friuli earthquake, centered in the South Alpine region of Friuli, was felt across Europe on 25 January. The earthquake hit in the same year that the Great Plague ravaged Italy. According to contemporary sources, it caused considerable damage to structures; churches and houses collapsed, villages were destroyed and foul odors emanated from the earth. Striking in the early afternoon, the earthquake caused hundreds of casualties and destroyed numerous buildings. In Udine, the castle and the cathedral were severely damaged. In Carinthia, the town of Villach and numerous surrounding villages were largely destroyed by a major landslide followed by a flood of the Gail River. Even in Rome the earthquake allegedly took a toll: considerable damage was sustained by the Basilica of Santa Maria Maggiore; in the Torre delle Milizie, an upper floor crumbled, and the structure assumed the slight tilt it retains today. The sixth-century basilica of Santi Apostoli was so utterly ruined that it was left in an abandoned state for a generation. The earthquake coincided with the beginning of the Black Death in Europe; in contemporary minds the two disasters were connected, as acts of God, but accepted as something both tremendous and unexpected, and yet which also belonged to daily life. The historian of medicine A. G. Carmichael observes, "The earthquake of 25 January 1348 is likely to have fueled and focused specifically apocalyptical fears more than plague did." The only explicit reference to the earthquake as an omen of the end of the world comes in the chronicle of Giovanni Villani. Guglielmo Cortusi of Padua, as well as the bankers of Udine, saw it as a *memento mori* and a sign to repent, but not of imminent apocalypse. The earthquake figured in the diary of the German nun Christina Ebner, and was reported in numerous city and abbey chronicles, which have given modern historians opportunities of making the "Friuli event" one of the most thoroughly studied medieval earthquakes.

In 1348 CE in England and Wales, the land was so inundated with continuous rains that scarcely an article of food was raised. In England, there were violent rains from Midsummer to Christmas "so that there was not one day and night dry together." This wet season caused great floods, and a pestilence which raged a whole year. The earth was at the same time barren, and even the sea did not produce such plenty of fish as formerly. The mortality was so great that in the city of London, England, two hundred bodies were buried every day in the Charter-house-yard, besides those interred in other common burying places. This loss of life lasted from Candlemas to Easter. Would this then be 1348 to 1349?

Fourteen thousand people were carried off by a remarkable pestilence in Dublin, Ireland.

In 1348, there was a dry fog in England with earthquakes and volcanic eruptions. Where would English volcanic eruptions be as there are no volcanoes in Britain. There are earthquakes though. And we have had reports of volcanic eruptions before on several occasions. Except as I have stated before, there are no volcanoes in the British Isles or near them.

The summer of 1348 in southern France was remarkably hot producing a drought. The year 1348 was a year of the plague.

A severe winter struck Iceland and the sea was frozen around it.

The Black Death began at Cathay (China) in Asia, and in the neighborhood near the great sea. But whether it arose in India, Scythia, Tartary or Arabia, it went sweeping along through the Indians, Tartareans, Saracens, Turks, Syrians, Palestinians, Persians, Egyptians, Ethiopians, Africans, with the parts about Tunis or Trisibon. Then it went over all the Levant, through Mesopotamia, Chaldea, Cyprus, Gandy, Rhodes, and every island of the archipelago. Then it came to Greece and overran Europe. About the latter part of 1346 or the beginning of 1347, it reached Italy. On 28 September 1347, it landed on the English coast in Dorsetshire. In 1350 or 1351, it reached Scotland and Ireland. In 1350, it reached the Hungarians, Goths, Vandals, and the most northerly people. It had not fully finished its perambulation over the world before 1360 or 1362. If it was so favorable as to leave a third part of men alive in some few places, in others it took 15 out of every 16 people. In more, it utterly exterminated the human race. It laid waste some places as Arthemusia. In the Eastern parts died in one year 23,840,000 people.

The Venetians lost 100,000.

In Florence, Italy died in one year 60,000.

In Germany died 1,244,434.

In 1348, the pestilence in London, England was terrible, where 50,000 people perished. At Yarmouth 7,000 perished. In Norwich, England, 57,000 people died from the first of January to the first of July. In London, England from the first of February to the first of May, there were 2,000 deaths per week. From its landing place in Dorsetshire, it spread into Devon, and Somersetshire and Bristol, then to Gloucester, Oxford, and London.

On 19 April, it snowed and there was an earthquake. Villach, Austria was totally destroyed.

Also, in 1348 CE fire falling from heaven consumed the land of the Turks for sixteen days. What sort of meteorite fall was this?

The Dominican Friar Bartolomeo reported that in China in 1348 CE fire rained down from heaven in the form of snow (ash) which burnt mountains, the land and men. From this fire arose a pestilential smoke that killed all who smelt it within twelve hours, as well as those who only saw the poison of the pestilential smoke. What was this falling fire?

It was reported that Naples was destroyed on 25th January 1348. This was the Friuli Earthquake that was centered in the South Alpine region of Friuli in Italy and was felt across Europe. This was also the year that the Great Plague suddenly appeared and ravaged Italy. Numerous towns and villages were destroyed. There were numerous reports of pestilential gases appearing after the earthquakes across Europe, the Mediterranean and the Far East at this time that caused the mass deaths of thousands of people who died in dreadful agony. Striking in the early afternoon, the earthquake caused hundreds of casualties and destroyed numerous buildings. In Udine, the castle and the cathedral were severely damaged. In Carinthia, the town of Villach and numerous surrounding villages were largely destroyed by a major landslide followed by a flood of the Gail River. Even in Rome the earthquake allegedly took a toll: considerable damage was sustained by the Basilica of Santa Maria Maggiore; in the Torre delle Milizie, an upper floor crumbled, and the structure assumed the slight tilt it retains today. The sixth-century basilica of Santi Apostoli was so utterly ruined that it was left in an abandoned state for a generation.

In August 1348 a French record translated by Rosemary Horrox states that apparently in August 1348 a very large bright star was seen in the west over Paris. late in the day but before the sun set. "It was not as high in the heavens as the rest of the starts, on the contrary, it seemed rather near." As the sun sets the star appears to stay in one place and then once night had fallen this large and my implication unusual star sent out many separate beams of light, and after shooting out rays eastward over Paris it vanished totally: it was there one minute, gone the next".

Rosemary Horrox translated another document from the Paris medical facility which was written in October 1348 and refers to a deadly corruption of the air around us. "We believe that the present epidemic of plague has arisen from air corrupt in its substance…Also the sky has looked yellow and the air reddish because of the burnt vapors…and in particular the powerful earthquakes, have done universal harm and left a trail of corruption. There have been masses of dead fish, animals and other things along the seashore, and in many places, trees covered in dust…and all these things seem to have come from the corruption of the air and earth.

Hecker mentions a curious event in Cyprus on or around 1348 CE. On the Island of Cyprus, the plague from the east had already broken out; when an earthquake shook the foundations of the island, and was accompanied by so frightful a hurricane, that the inhabitants.. fled in dismay…The sea overflowed…Before the earthquake a pestiferous wind spread so poisonous an odor, that many… expired in dreadful agonies…and as at that time natural occurrences were transformed into miracles. It was reported that a fiery meteor, which descended on the earth far in the east had destroyed everything within a circumference of more than one hundred miles, infecting the air far and wide.

In regard to this event the Greek historian Nicephoros Gregoras wrote that while the plague was just beginning a particularly severe earthquake came to complete the work of destruction (in Cyprus) A tidal wave swept over large parts of the island, entirely destroying the fishing fleets and olive groves. a pestiferous wind spread so poisonous an odor that many, being overpowered by it, fell down suddenly and expired in dreadful agonies.

Interestingly enough comets have been found to contain poly-aromatic hydrocarbons which can be poisonous. Who knows what else they may contain? Fred Hoyle and N. C. Wickramasinghe have proposed that the great plagues that have attacked the earth have come with passing comets.

Prior to 1349 CE when the Black Death apparently arrived in Bergen on board a ship, in the central and southern areas of Norway, the medieval expansion continued and then stopped. In the great Hallingdal Valley the death rate amounted to ninety per cent of the population, with its through route, and about two-thirds along the pilgrim route to Trondheim through southern Sweden. There was no real recovery in Norway for two hundred years. The farms on the higher ground stood vacant for that long because any surviving occupants took advantage of the abandoned farms in the more productive lands in the valleys lower down.

On 2 January 1349, there was a flood of the River Ouse, which overflowed York, England, as far up as Micklegate. Micklegate Bar is the most westerly and main gate through York's walls. It was and is still used by royalty to enter the city of York. Historically it is where heads of traitors were displayed on spikes.

The winter of 1348-1349 was similar to the winter of 1323 CE .

The Baltic Sea was frozen over and passable from Stralsund to Denmark. Stralsund is located on the Baltic sea coast in northern Germany.

On 7 March 1349, an earthquake rocked northern Germany. There was a cessation of a two-year plague where one-fifth of the people died.

In 1349, the plague was very violent in Waghen in the East Riding of Yorkshire, England. The abbot and six monks died. It was also very violent at Beverley. It was also very violent at St. Alban's where the abbot, sub-prior, and many of the monks died. At Oxford, it was so dreadful that the colleges were closed, and there was scarce enough left in the city to bury the dead. The plague extended into Nottinghamshire and Derbyshire.

1350 CE

In 1350 CE Nevado del Ruiz in Colombia erupted with a VEI of 4 as did Cotopaxi in Ecuador which erupted with a VEI of 4. The two other VEI 4 eruptions were Mono-Inyo Craters in California and Kikhpinych in Kamchatka.

Other eruptions were Tungurahua with a VEI of 3 in Ecuador, and Rausadake in Japan with a VEI of 3. Zaozan in Japan had a VEI of 2, Taveuni in Fiji had a VEI of 2, and Mount Etna in Italy had a VEI of 2,

This is four volcanic winter causing eruptions in one year?

Was the Earth trying to blow itself up? Or was this still the **50 effect?

In 1350 CE in England, there was a drought that came after floods, storms and meteors. In 1350 in England, there were "floods, storms, tempests and fiery meteors in the air." There was a terrible thunder and lightning storm in England.

In 1350 in Barbary, North Africa, the grains exported from England caused a dearth in England. Barbary is now the region of Morocco, Algeria, Tunisia, and Libya. In 1350 there was a famine in Barbary, then in England on exporting corn (grains) thither. There was a great famine in Barbary and Morocco. Christian nations came to their relief transporting large quantities of corn (grain). This made the grain so cheap and plentiful in Barbary but left a famine at home in England. In England, this was followed by terrible inundations, storms, and tempests. These were succeeded by excessive drought and want of water. This led to the destruction of most animals and vegetables.

In 1350 the plague returned to Germany. The winter was cold and dry. A third of humanity died.

In 1350, there was a terrible disease throughout Europe called the Black Death. It killed men, horses, cattle, deer, bears, wolves, foxes, sheep, goats, hares, etc.

Before 1350 CE there were around 170,000 rural hamlets in Germany. After 1350 CE there were 130,000. These deserted settlements were called Wustungen.

In England the villages that did disappear by 1350 CE were the ones that that had declined most, on average by two thirds, in the years of famine earlier in the century.

A microfossil record of a sediment core from Disko Bay shows cooling in Western Greenland around 1350 CE.

By 1350 CE the smaller of the two Norse centers in Greenland, with only about 75 farms, the Vesterbygd, West Settlement, which was the more northerly of the two settlements in West Greenland, was wiped out by conflict or disease, possibly the plague. Some cattle and sheep were found wandering around unattended by any human owners when a ship visited the area from the other settlement.

On the coast of Norway, the population expanded and the country's economy expanded from 1350 CE to 1500 CE. This was possible by the increased cold-water outflow from the Arctic near Greenland that was compensated by the strengthening of the inflow of warm North Atlantic water with its fish stocks on the Norwegian side but by the seventeenth century the Norwegian fishery was devastated by the changing climatic situation.

In 1350 CE water levels at Jenny Lake in Grand Teton National Park, Wyoming, indicate that drowned trees that are now in 78 to 94 feet of water, indicate that they trees were growing from 1350 CE when there would have been no water where they had been growing.

In California this was contemporary with the water levels of the Sierra Lakes being reduced in this same period. This was only one of four major droughts caused by the winter jet stream over the northeastern Pacific, with its associated storm tracks staying well north of California and the Great Basin.

In the American Southwest there was a sudden decrease of building work using timber around 1350 CE. Had the population been reduced?

In 1352 CE Cotopaxi in Ecuador erupted with a VEI of 4.

In 1352 in England there was a drought with exceedingly dry summer. An ecclesiastical record from Winchester Bishopric, notes that there was a great heat…that lasted for the whole summer and meadow not mown due the heat, the drought in summer in Wargrave, Berkshire. It also mentions spring corn was short and grew badly due to the drought. This was in Farnham in Surry. In England in 1352, there was a dearth from a drought. The cattle died in the pastures for want of water and the Fens and marshes dried up. There was a way through them where none was before. This was a very dear year in England (everything was scarce and dear). About the Feast of All Saints, 1 November, came a tempest of wind stripping houses and churches, blowing down mills, and rooting up trees.

In 1352, the heat was excessive in Toscana (Tuscany), Italy.

The drought of the summer was on the Continent (Europe) so badly that many cattle perished on the field. The marshes and ponds were completely dry. Also, a great flood occurred on the Rhône River in the south of France.

In 1353 CE there was a long, cold, hard winter lasting from early December to mid-March in London and the south. This year was remarkable for scarcity of grain and provisions in England and France, occasioned by a great drought. It was called the "Dear Summer". Rye was brought out of Zealand in Denmark to support the poor, who otherwise must have perished for want of sustenance. Zealand is the largest island of Denmark.

In 1353, there was a great famine in England and France. From March to July 1353 in England, the country sustained a severe drought. Then famine ruled the land, and the peasantry, irritated by the attempted regulation of wages, in many places broke out in discontent. In Cheshire they rose in open revolt and attacked the servants of Prince Edward of Woodstock, Prince of Wales, Duke of Cornwall, and Prince of Aquitaine who were entrusted with supervising his interests.

In March 1353, there was the severest drought in Italy.

In England, in early spring there was a hurricane and from March to July came a scorching drought in England. This year produced a great dearth in England, but plenty of corn (grain) imported from Ireland settled it.

In Rome, Italy there was terrible thunder and lightning during the summer. At Cremona, Italy, there was a prodigious hailstorm. Each hailstone weighed from one to eight and a quarter pounds. This made a fearful slaughter of people and cattle.

In the middle of May 1353, a deep snow fell in Silesia and Poland after Gronau, Germany. It did not damage the fields or fruit gardens even though it lay on the ground for six days.

In 1354 CE Cotopaxi in Ecuador erupted with a VEI of 4, Popocatepetl in Mexico erupted with a VEI of 2 and Grimsvotn in Iceland also erupted.

In 1354, there was a great drought in Fukien, Honan, Hunan and Kwangsi provinces in China. The resulting famine was so severe that it led to cannibalism.

On 31 December 1355 there was a terrible flood in Germany that smashed dikes and dams and broke into local land drowning 2,000 people. This appears to be in what is now Northern Germany along the North Sea.

The Lisbon earthquake of 1356 was a violent earthquake that struck, among others, the Lisbon area in Portugal at the end of the day on August 24, 1356, a Wednesday, Saint Bartholomew's Day. It lasted a quarter of an hour and several buildings were destroyed, including the four ovens that baked bread in the city of Lisbon and the large bronze apples in the bell tower of the cathedral of Seville. The aftershocks were felt for a year.

The 1356 Basel earthquake is the most significant seismological event to have occurred in Central Europe in recorded history and had a moment magnitude in the range of 6.0–7.1. This earthquake, which occurred on 18 October 1356, is also known as the Sankt-Lukas-Tag Erdbeben (English: Earthquake of Saint Luke), as 18 October is the feast day of Saint Luke the Evangelist. The earthquake destroyed the city of Basel, Switzerland, near the southern end of the Upper Rhine Graben, and caused much destruction in a vast region extending from Paris to Prague. Though major earthquakes are common at the seismically active edges of tectonic plates in Turkey, Greece, and Italy, intraplate earthquakes are rare events in Central Europe. According to the Swiss Seismological Service, of more than 10,000 earthquakes in Switzerland over the past 800 years, only half a dozen of them have registered more than 6.0 on the Richter scale. A graben is an elongated block of the earth's crust lying between two faults and displaced downwards relative to the blocks on either side, as in a rift valley.

Shortly before the end of the year, there was a major earthquake in Switzerland. Twenty-four castles were totally destroyed. Was this the same earthquake that struck Basel?

In 1356 in Germany, the year began dry without snow. At Easter and Pentecost, there was a great snowfall. The autumn was cold but the winter was warm. At Christmas, the flowers were blooming and the trees gave forth buds. This was very warm unseasonal weather.

In 1357 CE Katla in Iceland erupted with a VEI of 4.

About the year 1357, there were earthquakes in Germany, at Mainz, in Hesse and Thuringia. There was major damage everywhere. The plague broke out again.

There was a great thunderstorm; the Igumen, head of the monastery of St. Nikola in Lyatka, perhaps Vyatk, now Kirov, Russia, was struck by lightning, and others and in Rogatitsa Street in Novgorod, Russia one person was struck dead, while other individuals by the mercy of God remained alive. This was during the same storm. Kirov is around 660 miles east of Novgorod.

In 1358 CE In Bologna, Italy, the snow was 10 Ellen (approximately 23 feet, 7 meters) deep. Other reports state that it was 10 Braccia (approximately 22 feet, 6.7 meters) deep. The average snowfall in Bologna is 55 centimeters.

In the regions around Metz, France in 1358, the great heat caused the vines to dry up and all the grapes to shrivel up. As a result, a glass of wine cost 5 sous (five centimes). After heavy rains, the Rhône and the Durance Rivers in France overflowed their banks in November 1358 and the floodwaters spread far into the countryside. A prodigious amount of snow fell in Provence, France during the winter of 1358. The harsh rains that succeeded the winter produced devastating floods. In France, the Rhône and the Durance rivers overflowed their banks and ravaged the surrounding countryside. The winter that followed was most severe. The river was filled with ice flows and famine followed.

In 1358 during the spring and summer, there was a great drought in Chihli, Shantung and Shensi provinces in China. The resulting famine was so severe that it led to cannibalism in Shantung and Shensi. Chihli was renamed in 1928 into Hebei.

The Black Death revisited England in 1358 and other places in 1359.

One hundred thousand people died in Florence, Italy between March and July. There were scarce 10 people out of 1,000 left alive in Italy.

In Numidia, North Africa, 800,000 people perished. Numidia is the Old Berber Kingdom roughly corresponding to modern Morocco, Algeria, and Tunisia.

In Greece, the living were insufficient to bury the dead.

A report from Paris in April 1360 CE stated that on Black Monday or Easter Monday when hailstones fell that killed both horses and men in the army of Edward III, from the extreme cold.

In 1360 CE El Chichon in Mexico erupted with a VEI of 5. This could cause a huge volcanic winter. Another eruption was Pacaya in Guatemala with a VEI of 3.

In 1360 CE King Edward III's ambition to conquer France during the Hundred Year's War was dealt a devastating blow by a tremendous hailstorm. By the spring of 1360, the English army had pillaged and burnt many of the suburbs around Paris before making camp outside Chartres. But on April 13 dark storm clouds billowed up and a fierce, bitterly cold wind blew. "A foul dark day of mist and hail, and so bitter cold, that sitting on horseback men died," described one chronicle. Thunder and lightning erupted. The storm unleashed a barrage of hailstones described as big as pigeon eggs, and even suits of armor appeared to give little protection. According to one estimate 1,000 men and 6,000 horses were killed in "such a tempest of thunder, lightning and hail that it seemed the world should have ended." This disaster became known as Black Monday. There is no more refined date for the El Chichon eruption than the year but El Chichon is a Plinian volcano that produces vast masses of sulfur and this is the primary cause of volcanic winters. Refer to the 1982 eruption.

In England in 1360, there was a great dearth this year and mortality of people. This was called the second plague because it was the second in the reign of King Edward III. There was a very great death of cattle and horses. 6,000 horses died in the army. Many houses were burnt by thunder and lightning. On 16 February there was a hurricane, the greatest. It did more damage than any within the memory of England. On 14 April there was a very bitter cold combined with mist and hail, which killed many people.

In 1361 CE Niigata-Yakeyama in Japan erupted with a VEI of 3.

On 3 August 1361, during the Shohei era, an 8.4 earthquake struck Nankaido, followed by a tsunami. A total of 660 deaths were reported. The earthquake struck Awa, Settsu, Kii, Yamato and Awaji provinces, Tokushima, Osaka, Wakayama and Nara Prefectures and Awaji Island. A tsunami hit Awa and Tos Provinces (Tokushima and Kochi Prefectures), in Kii Strait and in Osaka Bay. The hot spring of Yunomine, Kii (Tanabe) stopped. The port of Yuki, Awa (Minami) was destroyed and more than 1,700 houses were razed.

On the evening of 15 January 1362 in England, there began a very strong wind from the Southwest. It blew with such force so as to overthrow many strong and mighty buildings, towers, steeples, houses, and chimneys. This wind continued for six or seven days. Many edifices standing after the storm was over had been so shaken that they required restoration to prevent them from collapsing. This was followed by a very wet season; chiefly summer and harvest. Much corn (grain) and hay was lost or spoiled by the unseasonableness of the weather. There was great sickness in Britain for a year.

Also on 15 January 1362, a particularly dramatic first Mandränke (large drowning or Big Drowning of Men) occurred in the Netherlands, in consequence large parts of the outer North Frisian country perished. According to a Russian church clerk, the flood began on 15 January 1362, reached a climax on 16 January and ended a day after that. The chronicler Anton Heimreich reported that the western sea went 4 Ellen (9 feet, 3 meters) above the highest levels. A total of 21 levees were breached. Rungholt along with seven other parishes in the Edomsharde were destroyed and 7,600 people perished. The chronicles speak of a total of 100,000 deaths, a number that is certainly exaggerated. In some coastal sections of North Friesland before the tide came in, were part of the edge of the Geest. The Geest is a landform that is slightly raised above the surrounding countryside that occurs on the plains of Northern Germany, the northern Netherlands and Denmark. Tidal currents caused large bays to extend into wide estuaries. The present course of the German North Sea coast has been largely mapped out by the storm surge.

The Great Drowning, Grote Mandrenke, caused widespread severe damage across southeast Denmark and also along the East Coast. Sixty Danish parishes were swallowed up by the sea, with several thousand dead there. It suggests a rapidly deepening low moving swiftly across southern Britain and the southern North Sea with a high storm-surge event. This was also called the Saint Marcellus flood.

On the 15th January 1362 there was a severe gale from between south and west and lasted for a week. It affected large areas of southern Britain. A large number of buildings were blown down or damaged including St. Pancras Church, the church of Austin Friars in London, Norwich Cathedral and the original Abbey Gateway in St. Albans. There was also damage to shipping. The exceptionally severe gale caused great destruction with buildings, towers, windmills, all thrown down according to contemporary chronicles.

On 15th or 16th January 1362, the town of Rungholt and 28 churches, out of 59 at the time, with the associated villages were destroyed by a storm surge from the North Sea. Many lives were lost. This was on Nordstrand in Schleswig. Rungholt is now under the Wadden Sea. The storm surge created a great part of the Wadden Sea.

Some of the present islands of the German Bight were formed by coastal losses. There were great losses of land from Schleswig and the Danish Coast including many whole parishes. Over sixty parishes had been swallowed by the sea. There were 25,000 deaths.

Much of what was left of the town of Dunwich in East Anglia was destroyed in the Grote Mandrenke around 16th January.

In East Yorkshire on this same date the former great port Ravenser Odd, also called Ravensburgh, Ravensrodd or Ravenspur, was mostly destroyed along with several other towns in Holderness that disappeared into the sea. Ravenser Odd was at the mouth of the Humber.

Was El Chichon in 1362 in Mexico responsible for the Grote Mandrenke?

In 1362, a severe drought engulfed Honan (now Henan province) in central China at Loyang, Mêng-ching and Yen-shih. During the period between 27 March and 24 April, floods struck Fukien (now Fujian province) on the southeast coast of China at Kuang-tê.

On June 5th 1362 Oraefajokull in Iceland erupted with a VEI of 6. This is the highest VEI recorded in Iceland. Originally called Knappafellsjokull it erupted explosively, ejecting ten cubic kilometers of tephra. The wealthy district of Litlaherao was destroyed by floods, pyroclastic flows and ashfall. Sailors reported that pumice was in such abundance that ships could hardly make their way through it. Thick volcanic deposits obliterated farmland and ash travelled as far as Western Europe. More than 40 years passed before people settled the area again. This would have affected weather to a great extent.

During the time period between 22 July and 19 August, floods struck Hopei (now Hebei province) in northern China at Cho. Houses were flooded.153 In 1362, there was a great drought in Honan province in China. The resulting famine was so severe that it led to cannibalism.

Between 8 and 9 September 1362 a sea storm struck (Germany). Many who had survived the plague drowned. The waters of the western sea swept away 30 localities. The islands of Föhr and Sild (Sylt) were completely demolished.

In 1362, during the week of Easter, on April 17, a very severe frost killed the grape vines completely, along with walnut and other fruit trees in France at Tours, Angers, even in the Lorraine and beyond. These frosts, the humidity from the winter and almost continual rains combined to completely destroy the grapes, nuts and other fruits almost everywhere.

In 1362, there was a fearful mortality in London, England and Paris, France. In October, the Rhône River in France, flooded with such violence that the ramparts of Avignon, which had been raised shortly after the flood of 1226, were overthrown. The winter of 1362 was very wet in France.

1362 was a wet year in Britain.

The weather for late 1362 was cold or severe winters and frost from December to March in the Second Winter in 1363 in London and the southeast which was regarded as the worst of the two when taking the whole of Western Europe.

In 1363 in Mainz, Germany, there was a great plague that also killed the horses. The winter was so cold that even the birds froze.

The Winter of 1363-1364 CE was severe and struck Europe beginning on September 16th. The winter produced frost from September to April.

In 1364, the Rhône River at Arles, France froze to a considerable depth; loaded carts traveled on the ice. At Arles, the Rhône River divides in two major arms, forming the Camargue delta. This winter in 1364 was very severe, particularly in the north and the south of France where all the fruit trees tended to die. In Paris, the frost began on 6 December, and lasted 14 weeks. The snow was lying on the ground the whole time. As a result of this extreme cold, an extraordinary lack of meat soon followed.

In England, the frost lasted from mid-September until April.

In France, all the rivers froze in 1364. The frost accompanied by snow lasted until the end of March. The vines froze in several places into the roots. Very deep caves although they were protected by straw were not immune to the frost. Loaded carts crossed the Rhône River and the ice was in some places fifteen-feet thick.

In 1364, the tide of the River Humber in England rose to such an extraordinary height, that it overflowed its banks and inundated the adjacent countryside, destroying a once noted seaport called Ravenspur. Ravenspur was said to have been lost on 16th January 1362. This might be in regard to the remaining portion of Ravenspur.

In England, the frost was "very terrible" between the 16th of September to the 6th of April. The ground lay unplowed to the great loss of corn and fruit.

The summer of 1364 in southern France was remarkable due to excessive heat and extreme cold.

Was this the result of El Chichon in 1362 with a VEI of 5? Or Oraefajokull in Iceland in June 1362 which erupted with a VEI of 6? Are volcanic winter causing eruptions cumulative in their results? Was this the real start of the Little Ice Age?

In England in 1365 or 1366, there fell an abundance of rain in the time of hay harvest, whereby much hay and corn (grain) were lost. A great mortality of people followed. So many who went to bed well at night were found dead the next morning. Many of all ages and sexes died of smallpox.72 The year 1366 was a cold and wet year. The grain was expensive due to grasshoppers and mice feeding upon it. In Brunswick, Germany, a great mortality as the plague begins.1

In 1367, there was a "*morbida pestis*", a new type of plague in Mainz, Germany which struck people with a bloody cough followed by death in 3 days. The year was barren and everything was expensive. In the fall the plague struck Lübeck.

In 1367, there was a fearful mortality in London, England and Paris, France. Violent winds struck in December 1367, during the night of the Feast of Saint Lucia (13 December) in Flanders (now Belgium) and Brabant in Picardy in northern France, which had never seen the like. The winds came from the northwest. The ocean overflowed during the storm and swallowed several homes and villages on the banks of the sea.

In 1369 CE in England, there was a dearth. Wheat sold for 20s. per Quarter (quarter ton). This year began the next great plague called the third mortality. This was very great both for people and cattle, the like seldom heard of. The west country, as Oxford, was most afflicted by it. In 1369 in England, there was great pestilence among men and large animals. This was followed by inundations and extensive destruction of grain. Grain was very dear.

In 1369 during the period between 5 February and 6 May, a drought engulfed many regions of China.

In 1369 at Ho-t'ao om China a large star fell, started a fire and injured soldiers.

In 1372 CE Kolbeinsey Ridge in Iceland erupted with a VEI of 2.

The year 1372 was a hard winter. In Prussia and Thorn (now Torun, Poland), many people died from the cold. From 6 January, there was a lot of snow. This was followed by a dry autumn and a plentiful harvest.

In 1373 to 1372 CE from February on there was deep snow that laid upon the ground for seven weeks and on thawing occasioned a great flood. This was from Norwich Cathedral records.

In 1373 CE in the Novgorod in Russia, the Volkhov River flowed backwards for seven days owing to floods downriver.

The sea floods of 9 October 1373, affected mainly East Friesland in Germany. The storm surge was higher than the previous floods but the damage was not as great. This is because many people permanently abandoned the harbor towns and villages moving miles inland from the sea. For example, Westeel along Leybucht bay was abandoned. Many dikes failed. Norden and Marienhafe which is today kilometers from the seacoast, became port towns. Marienhafe is around six miles from the present coastline and Norden is around two miles. Friesland was also hit. Friesland and East Friesland were originally Frisia which crossed the Dutch and German border.

In 1374 the Rhine River in Europe flooded and is regarded as the worst flood in the last one thousand years. The Rhine River originates in the Swiss Alps and flows through Germany where from Andernach the flood-prone area widens in downstream direction until it becomes a river delta in the Netherlands. This was during the wettest winter of the last millennium. There were three large flood events around January 6th, January 25th and around 9th to 11th of February. The weather was warm so there were no ice jams. The flood around the 9th of February represents the largest flow by volume ever recorded by volume by observers at Cologne in Germany. It was possible to cross the walls of the city of Cologne by boat.

In Italy in 1374 and 1375, there was a famine.

The summer of 1374 in southern France was remarkably hot but also produced heavy rains in southern France.

In October 1374, flooding from the sea amid storms swamped several cities in the Netherlands. There was also sea floods damage in England as well.

Eruptions in 1375 CE were Lokon-Empung in Indonesia with a VEI of 3 and the Dieng Volcanic Complex also in Indonesia with a VEI of 3.

In 1375 CE in England, there was an excessive drought with heat.

In Italy in 1374 and 1375, there was a famine.

On 10 February 1375, the Weser River flooded and the water was in the cathedral in Minden, Germany.

On 13 November 1375, a storm damaged the steeple of the church of Saint Jakobi in Lübeck, Germany. The winter of 1375 was extremely cold in Germany. The Rhine River was frozen over for a quarter of the year and a four-week market was held on the ice.

On December 20th 1375 CE Asosan in Japan erupted with a VEI of 2. Other eruptions were Lokon-Empung in Indonesia with a VEI of 3 and the Dieng Volcanic Complex in Indonesia with a VEI of 3.

On June 20th 1376 CE Asosan in Japan erupted with a VEI of 2 and Kelud in Indonesia had a VEI of 3.

In 1376 CE in the Novgorod in Russia, for the second time in three years the Volkhov River flowed backwards seven days into lake Ilmen, owing to floods downstream.

On May 6th 1377 CE Asosan in Japan, as usual, erupted with a VEI of 2.

On 16th of November 1377 Friesland, Holland and Zealand in the Netherlands as well as Flanders in Belgium were hit by sea floodss from the North Sea. Thirty thousand people died. Holland is a province of the Netherlands.

Also, on 16th November 1377 a sea floods in northwest Germany caused 32 villages to be lost. Would this have been called the Saint Gertrude Flood?

In 1380 CE the Mono-Inyo Craters in California erupted with a VEI of 4. Also, in 1380 Melebingoy/ Mount Parker in the Philippines erupted with a VEI of 4 and Kirishimayama in Japan had a VEI of 2.

In 1381 CE on August 6th Mount Etna in Italy erupted with a VEI of 2.

The 1382 Dover Straits earthquake occurred at 15:00 on 21 May. It had an estimated magnitude of 6.0 M_s and a maximum felt intensity of VII–VIII on the Mercalli intensity scale. Based on contemporary reports of damage, the epicenter is thought to have been in the Strait of Dover. The earthquake caused widespread damage in south-eastern England and in the Low Countries. The earthquake interrupted a synod in London that convened in part to examine the religious writings of John Wycliffe, which became known as the Earthquake Synod. In England, the most severe damage was recorded in Canterbury, particularly to St Augustine's Abbey and Canterbury Cathedral, where the bell-tower was destroyed. The manor house and church at Hollingbourne, Kent were also badly damaged. This has been used to estimate intensity in the range VII–VIII. In London there was damage to St Paul's Cathedral and Westminster Abbey with an estimated intensity of VI–VII. In the Low Countries damage was reported in Ypres, Bruges, Liège and Ghent in Belgium.

In 1382 CE a massive storm surge in the North Sea and the English Channel killed 100,000 people along the Dutch and German coasts.

In 1384 CE Villarica in Chile erupted with a VEI of 2.

In 1384 CE during the summer there was possibly a major drought in the southeast of Britain. It was reported that streams and springs dried up and the deepest wells also failed. The dry episode lasted until early September.

In 1384, an unbearable dry heat reigned throughout France from the spring until the middle of August. The sources of water dried up during the summer of 1384 in France by the lack of rainfall and prolonged drought. In France, after a long drought and heat unbearable that extended until the middle of August 1384, there came heavy rains that lasted until March 1385. This caused the grapes to rot. On 23 April 1384, rye already had ears and was harvested in June. Towards the end of August, a severe frost began that lasted until the end of the year.

On 20 December 1385 there was a great earthquake throughout Teutschland (Germany).

During the year 1385, the winter was very cold in Northern Europe.

During the Winter of 1384-1385 CE the Rhine River, the Scheldt River and the Sea of Venice were frozen. The Scheldt River is a long river in northern France, western Belgium and the southwestern part of the Netherlands.

In 1385 CE Kelud in Indonesia erupted with a VEI of 3.

In 1387 CE on June 19th Asosan in Japan erupted with a VEI of 2.

By 1387 CE in Norway production and tax yields were from, in some districts, as little as 12 per cent to barely 70 per cent of what they had been around 1300 CE.

Villarica in Chile had a VEI of 2 in 1388 AD.

On October 16th 1388 CE Asosan in Japan erupted with a VEI of 2.

On December 1st 1389 CE Hekla erupted in Iceland with a VEI of 3.

In 1390 Asosan in Japan erupted again and had a VEI of 2.

The 1390s were regarded as a series of warm, dry summers.

In 1390 CE in England, there was a great famine arising from the scarcity of money to buy food. In England during the years 1390 to 1392, there was a great dearth from people hoarding up corn (grains). Also, in 1390 there was a dreadful pestilence in York, England.

The summer of 1390 in southern France was remarkably hot and produced heavy rains in southern France. The sea flooded beaches in France.

A strong wind blew across the world on Christmas Eve 1390.

Beginning on July 9 1391, the sun appeared to be obscured by certain thick and dreary clouds between that and the earth. The clouds rose daily for almost six weeks together. The north and east part of England were, at the same time, sorely afflicted with a pestilence. In a few weeks there died eleven thousand persons in the city of York. They were great, thick, dark clouds. At the same time there was a great mortality over all of England, especially in Norfolk. During both this and last year was great scarcity and dearth. The mortality was from an epidemic bloody flux, from eating large quantities of green fruits during harvest time. In 1391, there was a plague all over England.

In 1391, there was a high-water sea floods in Germany.

In 1391 during the period 2-31 July, floods struck Honan (now Henan province) in central China at Yüan-wu.

Also in 1391 in the vicinity of Shanghai, China, the sea suddenly overflowed and drowned 20,000 persons.

In 1392 CE Villarica in Chile erupted with a VEI of 2.

In 1392 CE since the water in tree sap acquires greater volume when it freezes in extreme cold, trees burst apart with a loud noise. In Strasbourg, France more fruit trees burst when the cold reaches -16° Reaumur (-20° C, -4° F). A great number of trees in France burst in the winter of 1392. In the year 1392 in France, the winter was severe. The winter of 1392 in France and in the North produced a very great cold. In 1392, during the marriage of Isabel of France who was 6 years old at the time, daughter of Charles VI, to King Richard II of England, there were wondrous winds that lasted for three months. The marriage ceremony took place at Calais, France. The persistent drought in the summer of 1392 dried up water sources and prevented the largest rivers in France from being navigable.

There was a famine in England that does not appear to be weather related. In England in 1392, there was a great scarcity for two years. People ate unripe fruit and suffered greatly from "Flux". The Corporation of London advanced money and corn (grain) to the poor at easy rates. One researcher attributes the cause of this famine to the hoarding of corn. In England during the years 1390 to 1392, there was a great dearth from hoarding up corn (grains). In 1392 there was a very bountiful good harvest. In England during 1392 and the previous two years, wheat sold from 16 shillings to 26 shillings per Quarter (quarter ton). After a long period of plenty came this dearth, chiefly in the center of England. This dearth came about not from want of corn (grain) but partly because the corn was hoarded and partly due to transportation. England had sufficient corn to serve its needs for five years. All the wool in England had been laid up for three years unsold. The parliament having prohibited its transportation and merchants would not buy it but at a very low price. For it was sold from 22d. to 3s. per stone. But in 1392 came both money and a plentiful harvest. The poor were relieved and the Nation was well stored. During the dearth, the poor suffered much in the bloody flux. In September, there were great and terrible thunder and lightning. In October, there were very great and long rains.

In 1392 in Germany, there was a late harvest, which came with an abundance not seen since in human history. The summer was hot. Beginning on 25 November, there were great snowfalls and many people met with an accident.

In 1393 CE there was a cold winter in Western Europe. This possibly affected parts of Britain as well.

In 1393 CE in France, the heat is so strong that the earth is burned and the rivers are dry.

In 1393 and 1394, the summers were dry and excessively hot in England.

In 1394 CE the medieval city Old Ostende in West Flanders in Belgium was lost to the sea during a sea floods on Saint Vincentius Night on 22nd January. The new city was built further back from the coast.

In 1394 CE in 1393 and 1394, the summers were dry and excessively hot in England.

During the months of December, January and February during the winter of 1394-95, the weather was extremely wet. Because of the rains, all the rivers of the kingdom of France overflowed their banks three times.

In 1395 CE Kelud in Indonesia erupted with a VEI of 2.

In 1396 in Holland (now the Netherlands), there was "another deluge," which formed the Marsdiep, separated the islands of Texel, Vlielandt, and Wieringen from the mainland, and submerged other districts. "This first raised the commerce of Amsterdam." The Marsdiep is the gap between Den Helder (on the mainland of Holland) and Texel (the largest Dutch Wadden Island). The Marsdiep connects the North Sea with the Wadden Sea. Texel is the large island in the Wadden Sea. Vlielandt or Vlielandt is one of the West Frisian Islands, lying in the Wadden Sea. It is the second island from the west in the chain, lying between Texel and Ter Schelling. Wieringen at that time was an island. Today, a large dike links Wieringen with Friesland.

In the English Channel, there was another great storm, on the occasion of the second Queen of Richard II landing. The second wife of Richard II brought a storm with her to the English Coast in which the King's baggage was lost, and many ships of the Fleet cast away. In England in 1396, in July and August, but especially in September, there were terrible hurricanes. These did great damage to churches and houses in many parts of the country.

There was in Paris, France at the end of October 1396, a terrible storm of wind, rain and thunder, which was felt throughout the north. The storm was so violent that it knocked down the tents of the royal camp. In Ardres, France at the end of October 1396, there came a terrible storm, mingled with torrents of rain. Then a north wind blew across continually with great fury for three months. This year was called "the year of the high wind". These winds intensified the night of November 17 for three hours. The Sea, the English Channel, overflowed this year. This indicates that there was a storm surge on the northern coast of France.

Beginning in 1396, there was an extreme famine in the Deccan region of India. "In 1396, the dreadful famine, distinguished from all others by the name of the 'Durga Dewee', commenced in Maharashtra. It lasted, according to the Hindoo legends for twelve years. At the end of that time, the periodical rains returned, but whole districts were entirely depopulated, and a very scanty tax revenue was obtained from the territory between the Godavery and Kistna (now Godavari and Krishna Rivers) for upwards of thirty years afterwards."

On February 17th 1397 CE Nasudake erupted in Japan with a VEI of 3.

In 1398 CE there was a cold winter in Western Europe. This possibly affected parts of Britain as well.

The winter of 1399/ 1400 was a very hard winter in Germany. The earth and sea froze hard in a short period of time. Not only did people use the Elbe River as a bridge to travel over the ice, but also went across the ice to Denmark.

In 1399-1400, the winter was very severe in Northern Europe and the frozen seas offered several armies an alternate route on the ice. "In 1400, the seas froze in Northern Europe". Fresh water freezes at 32 degrees Fahrenheit but seawater freezes at about 28.4 degrees Fahrenheit, because of the salt in it. When seawater freezes, however, the ice contains very little salt because only the water part freezes.

1400 CE

In 1400 CE Arenal in Costa Rica had a VEI of 4 and Zaozan in Japan erupted with a VEI of 3.

In the first decade of the 1400s summers were frequently and anomalously wet.

In various parts of Europe, the growing season shortened, perhaps by three weeks or more. The climate's accumulated warmth decreased, and the frequency of harvest failures increased. These were the dreaded "green years" when the crops fail to ripen in the north. As the climate deteriorated, barley, oats and rye were preferred over wheat except in the warmer parts of Europe. There were also many places where cereal ceased to be profitable and was given up in favor of sheep rearing to meet the increasing demand for wool.

In the 1400s the Caspian Sea was eight meters higher than now, (1984) after rising over the previous two to three centuries. There was a notably moister climate regime spreading over Central Asia. The Medieval Warm Period had been relatively dry in these regions.

As sea levels started falling there were more sandstorms in Europe and the British Isles. This would indicate that there was increasing ice buildup as the temperatures kept dropping.

In case you forgot, as climate gets colder, sea levels drop. As climate gets warmer, sea levels rise.

On Dartmoor in Devon are the ruins of Hound Tor or Hundatora that had existed since the Bronze Age. The ridges and furrows of medieval tilled fields at an altitude between 350 and 400 meters, 1150 to 1300 feet, did not grow anything after the beginning of the 1400s and the village was deserted. It was the same for the rest of Dartmoor. Hundatora is still the same now.

In the fifteenth century the rivers of England were generally deeper and bigger than in the late twentieth century.

In Europe there was evidence of increasing severity of windstorms and resulting sea floods and disaster by shifting sands in 50-60° North.

In England, Sweden and south Germany there was the main abandonment of small, unsuccessful settlements or villages in the fifteenth century that were deserted due to the lowering of the snowline and the crop growing areas.

In the fifteenth century there was turmoil in a period of climatic stress in Scotland.

In Denmark in the fifteenth century because of the deepening crisis in agriculture there was a drift to the towns and by the end of the century a more general emigration affecting the towns as well.

In the province of Skane in Sweden in the fifteenth century because of the deepening crisis in agriculture there was a drift to the towns and by the end of the century a more general emigration affecting the towns as well. This was the same as Denmark and probably Norway as well.

Much of the fifteenth century was comparatively dry in Indochina with shorter duration, intense monsoons which also reduced the amount of water that could be easily stored and used.

In the early fifteenth century Southeast Asia suffered from severe drought.

In Wales the port of Harlech on the west coast ceased being a port when it was permanently obliterated by sand dunes. These are the dunes of Morfa Dyffryn and due to them Harlech Castle which was built on a cliffside overlooking the Irish Sea is now half a mile, 800 meters, inland. As of 2023 Harlech Castle was now around one kilometer from the shore.

In 1400 sea floods enlarged the Zuyder Zee in the Netherlands and Amsterdam and Enkhuisen opened up as ports. Humberside in England was also hit by sea floods.

In 1400, a large storm surge called "the Frisian Flood" raged on the Frisian coast which is the coastal region along the southeastern corner of the North Sea that extends from The Netherlands to Denmark

On July 15th 1400 CE Dubbi in Eritrea erupted with a VEI of 2.

In 1401 in Germany, it began to rain on the Feast of Saint Gregory on March 12 until the feast of St. Lambert on September 17th. For almost a half year there was not a day that it did not rain. The winter grain was spoiled by the cold rain. The summer grain produced long stalks but little grain. Then came a famine and a miserable time.

The winter of 1402/ 1403 was similar to the winter of 1323 CE.

In 1402 the Baltic Sea was quite frozen over from Pomerania to Denmark. Pomerania was a historical region on the southern shore of the Baltic Sea formerly in Germany and now in Poland.

In Novgorod in Russia, during the winter of 1402/03 from St. Georgi's Day up to March, horses could travel over the Volkhov River on the ice.

In 1403, there was a huge snowfall in Silesia, particularly at Glatz, now Kłodzko, Poland, Patschkau, now Paczków, Poland, and Reiß, now Głuchołazy, Poland, so that high waters and floods destroyed many houses and tore through the city gates and took great multitudes of people away. The floods destroyed the bridge of Troppau, now Opava, in Czechia.

On 25 November 1403, there was a very large storm surge along the North Sea.

In 1404 CE on February11th Nasudake in Japan erupted with a VEI of 3.

In 1404 CE in England, there was an inundation from the sea and in France in June, a strong mist covered the whole country.

On November 19th 1404 there was a major storm surge affecting the North Sea. This also affected the English coasts. This was known in the Netherlands as the "First Saint Elisabeth Flood". Flanders and Zeeland were flooded in particular. Almost 3000 hectares of land was lost in West Flanders and Zeelandic Flanders. The water in the rivers was very high due to frequent rainfall. The dyke breaches and subsequent flooding caused extensive damage in Zeeland and Holland. Although the dikes in the area broke during this flood, it took decades before the entire area was under water and the Biesbock with its creeks and reeds was formed. There were great losses in Kent, England, in the Netherlands, in Zealand, in Flanders, now Belgium, etc. by breaking in of the waters that overflowed the sea banks, to the drowning and loss of much cattle, etc. Many people drowned in Kent in England.

In Autumn 1405 sea floods hit Denmark. Many thousands of people drowned and also in Autumn 1405 36,000 people died in Hamburg in Germany when a sea floods hit the city. This was probably the same sea floods.

In 1406 English visitors to a Royal Wedding in Denmark noted much sodden uncultivated land and that wheat was grown nowhere. These changes were more general in Jutland on the west coast rather than in the islands on the east coast. There was a graduation across the country with much less stress in the more sheltered districts of the islands of Fyn and Sjaelland farther east. In Denmark in the late 1300s and 1400s many farms were deserted, corn growing given up on and those farmhouses that were still maintained were now shared by several families.

In 1407, there was a dreadful pestilence in London, England, where 40,000 people perished.

The winter of 1407 to 1408 CE was recorded as being severe and affected most of Europe. It is regarded by climatologists as one of the most severe on record. The frost lasted for fifteen weeks and people were able to walk across the frozen Thames again. The winter was one of the most snowy, and was of outstanding duration.

In Europe, ice in the Baltic had allowed traffic between the Scandinavian nations, and wolves had passed over the ice from Norway to Denmark. The Baltic Sea was frozen over.

The Winter of 1407-1408 CE was similar to the winter of 1323 CE. The whole sea between the island of Gothland and Geland was frozen, and from Rostock in Germany to Gezoer, now Gedser in Denmark which is a distance of 25 miles. I cannot find a Geland in Latvia or in Sweden or in Google or any of my atlases including Medieval atlases. It must have disappeared or changed its name. There is the island of Oland 35 miles to the west of Gotland in the Baltic Sea.

Norway wolves reached Jutland in Denmark across ice rivers and Swiss lakes froze. The ice extends without interruptions from Norway to Denmark, so that the wolves invaded from the north into Jutland.

In the Baltic Provinces in 1408 the frost was very severe. Baltic Provinces in this time period were the Hanseatic League Baltic provinces of Estonia and Latvia.

In Europe the winter was also very severe.

In England, there was a long and severe winter, frost and snow in December 1407 and January, February and March in 1408. Thrushes, blackbirds and many thousands of smaller birds died of hunger and cold. In December 1407, began a frost of such violence and continuance, that the like was never heard of in England. It lasted fifteen weeks, and being accompanied with abundance of snow, it was greatly destructive to the smaller birds.

In 1408, the vines and fruit trees in France were killed by the extreme cold. Carts crossed the Seine River in France on the ice.

In 1408, the Danube River froze over its entire length in Eastern Europe.

The Maas (Meuse) River in Holland and France was frozen. The winter of 1408 began November 11 and did not end until late January. It froze all the rivers in France and destroyed the roots of vines and fruit trees. In Paris, France, the wagons rolled down the Seine River.

The thaw in France caused terrible devastation because of the flowing pieces of ice and because the rivers overflowing their banks. The first shock of the ice against the arch warned the inhabitants of many houses built along the shoreline to run for their safety. For when the ice flows broke, one could see icebergs 100 meters (330 feet) in length floating. The little-known wooden bridge at the Chatelet, and the bridge of St. Michael (then called the New Bridge) collapsed. The foundations of the large bridge mills were swept away. In many places similar misfortunes occurred.

In the year 1408 "The winter of this year, ruled strictly in Northern Europe to the banks of the Danube, and was the cruellest in 500 years. The winter was so long that it stopped by Feast of St. Martins (11 November) by the end of January, and so severe that the roots of the vines and fruit trees froze to death." "Since last Feast of St. Martins such a cold occurred, that no one could do business, and if I had just asked the clerk for an additional shovel of coal in order to protect the inkwell from freezing. So, the ink froze but always after two or three words with the pen so he couldn't keep records." The acute shortage of wood and bread was painfully felt. The mills were collectively still on the frozen river because of the frost.

In 1408 CE on February 24th Nasudake in Japan erupted with a VEI of 3 again.

Later that year on November 8th Mount Etna erupted in Italy with a VEI of 3.

In 1410 on March 15th Nasudake erupted again with a VEI of 3.

During the winter of 1409 to 1410 CE the tidal river Thames froze over for fourteen weeks.

In 1411 CE Kelud in Indonesia erupted with a VEI of 3.

In India, there was a great drought followed by a famine on the Ganges-Jumna Delta from 1412-1413.

In 1413 CE there was a famine in the Ganges, which is the river that separates India from Bangladesh, and Jumna Doab, the track of land between the Ganges river and the Jumna (now Yamuna) river in northern India.

In England on 12 October 1412, the "sea flooded thrice without ebbing." In England beginning on October 12, there were three floods in the River Thames, one upon another and no ebbing between. The likes of this event was never known before. This indicates that each tidal flood had not withdrawn or ebbed before the next tidal flood came along.

On August 18th 1413 CE there was a southerly storm of blown sand which obliterated the town of Forvie on the east coast of Scotland, north of Aberdeen. There would have been an unusually extreme low tide which would tie in with this event. The storm that did this lasted nine days.

In 1415 CE on May 21st Izu-Oshima erupted.

In 1416 CE Katla in Iceland erupted with a VEI of 4.

On September 2nd 1416 CE Izu-Oshima in Japan erupted with a VEI of 2.

During the Winter of 1419-1420 CE in 1420, the sea between Constantinople (Istanbul) and Iskodar (Üsküdar) Turkey was frozen and passable on ice.

In 1420, the severe winter increased the misery in France; a nation tattered by war, with its capital in the hands of the English. The famine was in Paris so great that the unfortunates spent their days searching for food. The wolves advanced to the suburbs of the city that was now like a vast wasteland.

On 31 August 1420, a huge earthquake shook what is now the Atacama Region of Chile. Landslides occurred along the coast and tsunamis affected not only Chile but also Hawaii and Japan.

Katla in Iceland erupted in 1421 with a VEI of 4.

On May 14th 1421 Izu-Oshima off the coast of Honshu in Japan, erupted with a VEI of 4.

On 17 November 1421 CE the sea broke in at the city of Dordt, Dordrecht, in the Netherlands and drowned 72 villages, and 100,000 people. It was a dreadful and most destructive inundation, overwhelming seventy-two villages, twenty of which were never recovered. The loss of life (nearly 100,000 persons on some authorities) and property was immense; many noble families were reduced almost to beggary. By this inundation the Biesbock, Biesbosch, was formed, and the town of Dordrecht separated from the mainland. The Biesbock is a freshwater tidal area and network of rivers and smaller and larger creeks with islands.

In 1421 CE there were notable storms and coastal flooding in the North Sea region. On November 18th the southwestern part of the Netherlands was inundated. This must have impacted the English Coast as well. This was called the Second Saint Elisabeth Flood. There was a storm tide in combination with extreme high-water in rivers due to heavy rains. Some sources say that there were between 10,000 and 100,000 deaths.

Due to sea floods the Zuyder Zee reached its greatest extent on 19th November 1421. The dikes had been damaged by a storm in mid-June 1421. This was the St. Elizabeth's Day flooding of the Grote Hollandse Ward. This is regarded as one of the twentieth worst floods in history. On the night of 18th to 19th November 1421 a heavy storm near the North Sea coast caused the dikes to break in a number of places and the lower-lying polder land was flooded. A number of villages were swallowed by the flood and lost, causing between 2,000 and 10,000 casualties. There was widespread devastation in Zeeland and Holland. The flood separated the two cities of Geertruidenberg and Dordrecht which had previously fought each other during the Hook and Cod (civil) wars. Most of the land remains flooded ever since that day.

On 19th November 1421 sea floods hit England.

In Novgorod in Russia, the Volkhov River flooded and washed away the great bridge, also the Neredich and the Zhilotug bridges. The Neredich and Zhilotug bridges were bridges over small tributaries of the Volkhov River in Novgorod. At Kolomentsa (the city of Kolmovo near Novgorod) it carried away the Church of the Holy Trinity, and in Shchilova, Sokolnitsa, and Radokovitsi Streets and in the Resurrection in the Lyudin quarter, service in the churches was performed only on raised platforms, and in the different quarters it washed away dwellings with all their stores; and it was so great that it poured out through the town gates to Rybniki, the Fisheries. On May 19, during Peter's Fast, there was a great storm by night in the skies; clouds came up from the south, and in the north thunder and fiery lightning came from the skies with frightful noise, and purple rain fell with stones and hail. During these two years, 1421 & 1422, there were great famine and plague, and three public graves were filled with the dead, one behind the altar in St. Sophia and two by the Nativity in the field. What were the stones that fell with the rain?

From 1421-1423 CE there was famine in Russia for three years which again led to cannibalism and an exodus to the West.

The winter of 1422 to 1423 CE was severe in Western Europe and possibly affected Britain as well.

In the Winter of 1421-1422 CE. "This year was the Seine, which was large, becomes quite firm." In 1492, the Seine River in France was swollen and then froze solid. In 1422, wine, verjuice (the juice of unripe grapes) and vinegar froze in the basement. The Seine River in Paris, France, whose waters were high, froze in less than three days as the cold grew sharply. Frost began on January 12th and there was still ice at Notre Dame in March. On 12 January 1422, there was the most severe cold that had ever been seen by man. It froze so terribly that in less than three days, the vinegar and wine in the cellars solidified into icicles hanging from the vaults of the cellar. The wells froze within four days. There was eighteen full days of this harsh cold. About a day or two before the severe cold started, there was a heavy snowfall (similar to the snowstorm that took place 30 years ago in the year 1392). Due to the severity of the frost and snow and because of the extreme cold, no one undertook to their jobs, but rather resorted to jumping, ball and other games to heat up. The cold was so intense that the ice in the courtyards, streets and near the fountain lasted until the Feast of the Annunciation (March 25). It was so cold that the ridges (combs) were frozen on the heads of roosters and hens.

Between September and November 1422 there was devastation by storm floods from the sea in Friesland and Holland Provinces, in the Netherlands.

In 1423, in what is now Germany the ice was thick enough to ride on from Lübeck to Prussia, and the Baltic Sea was covered with ice from Mecklenburg to Denmark. In 1423, the travelers went on the ice from Lübeck, Germany to Danzig (now Gdańsk, Poland). In the winter of 1423, the shores of the Baltic Sea from Lübeck to Danzig were frozen.

On 18th November 1424 CE there was the Third Saint Elisabeth Flood in the Netherlands. This is regarded as the greatest disaster of the 15th century and perhaps the greatest of the second half of the Middle Ages in the Netherlands. This mainly affected the southwest of the Netherlands. All the restoration work after the 1421 flood was destroyed. This flood wiped out the Groote or Hollandsche Waard embankment which had been rebuilt after the 1421 Saint Elizabeth Flood. This flood mainly affected the southwest of the Netherlands.

In 1424 another drought and epidemic of Plague caused many deaths in Russia.

In 1426 CE it was a dry year in London and the south of England.

In 1426 CE the winter was similar to the winter of 1323 CE. The ice was thick enough to ride on from Lübeck to Prussia again, and the Baltic Sea was covered with ice from Mecklenburg (part of Germany) to Denmark. In 1426, the ice bore riding upon it from Lübeck to Prussia.

In 1427 CE in England, it rained almost continually from Easter to Michaelmas, 29th September; hence dearth, famine and sickness. The sickness also came from the winter without cold. At the Feast of Saint Nicholas, 6 December, all vegetables still flourished. Next summer the plague raged. In 1427 in Great Britain, the "rain began on April 1, and did not cease till Hollontide. Hollantide, today celebrated as Halloween, was in earlier times celebrated on November 11." Old Halloween where children go out in the evening to sing songs and collect treats was on the feast day of St. Martin of Tours.

In Paris, France, the Seine River entirely covered the island "Ile de la Cité" of Notre Dame and the island "Île Saint-Louis" and rose on the embankment Saint-Paul to the height of the first floor of houses. The floods drowned the marshlands. The water level in the marshes rose over two feet.

The Catalan earthquake of 2 February 1428, known in Catalan as the *terratrèmol de la candelera* because it took place during Candlemas, struck the Principality of Catalonia, especially Roussillon, with an epicenter near Camprodon. The earthquake was one of a series of related seismic events that shook Catalonia in a single year. Beginning on 23 February 1427, tremors were felt in March, April, 15 May at Olot, June, and December. They caused relatively minor visible damage to property, notably to the monastery of Amer; but they probably caused severe weakening of building infrastructure. This would account for the massive and widespread destruction that accompanied the subsequent 1428 quake. Modern estimates of the intensity are VIII (*Damaging*) or IX (*Destructive*) on the Medvedev–Sponheuer–Karnik scale. scale. The ramparts of Prats-de-Mollo-la-Preste were destroyed. The clocktower of Arles-sur-Tech collapsed. The monastery of Fontclara at Banyuls-dels-Aspres was devastated. The damage sustained by the monastery of Saint-Martin-du-Canigou marked the commencement of its decline. The belltower and lantern tower of Sant Joan de les Abadesses fell down. The chapel at Núria was destroyed. The villages of Tortellà and Queralbs were entirely destroyed. Among the damaged structures were Santa Maria de Ripoll and Sant Llorenç prop Bagà. As far away as Perpignan and Barcelona the populace was gripped by panic. In the latter, the intensity was estimated at VI (*Strong*) or VII (*Very strong*). The rose window of the Gothic church of Santa Maria del Mar was destroyed. Robin de Molhet, lord of Peyrepertuse, who was travelling in his domains when the earthquake struck, quickly came to the aid of victims, which earned the recognition of Alfonso V of Aragon, who was away in Valencia at the time of the tremors. He was informed by the President of the Generalitat de Catalunya, Felip de Malla, in a letter. It is estimated that hundreds of people were killed in the disaster: two hundred are estimated at Camprodon, one to three hundred at Puigcerdà (due to the collapse of the church), twenty to thirty at Barcelona (in Santa Maria del Mar), and almost the entire population of Queralbs. The fallout lasted well over a year. The quake was probably the worst in the history of the Pyrenees though the first recorded only occurred in 1373. It remains to this day a point of reference for the study of seismic risk.

In 1428 CE it was a wet year in London and the south of England.

In 1430 CE Sheveluch erupted in Kamchatka with a VEI of 4 and Furnas with a VEI of 3 in Iceland. Also, in 1430 CE Katla in Iceland erupted with a VEI of 4. Were these combined eruptions capable of causing a volcanic winter? We will have to see.

In the 1430s the majority of winters, perhaps 7 or 8, contained several weeks of widespread severe weather. Not the paltry days we get at the end of the 20th Century and the beginning of the 21st Century. This was not matched until the 1690s in the depths of the Little Ice Age.

The coincidence of timing of waves of desertion over much of Europe points to a widespread and presumably external causes such as the behavior of the climate. The period when most desertions took place coincides with a well-documented period of frequent cold winters and wretched summers. The summers in the 1450s being most wretched as well as those in the late 1460s.

In Scotland in the 1430s bread had to be made from the bark of trees for want of grain there was such great famine.

In Norway the one thousand farms in Halogaland in north Norway were deserted and near the rich fishery districts in the Lofoten Islands up to 95 per cent of the farms had been abandoned, and elsewhere it was 65 per cent. The treeline was heading down.

In 1430 CE in Novgorod in Russia there was a great drought. In autumn the water level in the rivers and lakes was exceeding low; the soil and the forests burned, and there was very much smoke. Sometimes people could not see each other, and fishes and birds died from that smoke; the fish stank of the smoke, for two years.

In 1430, the vine and fruit trees in France were killed by the extreme cold.

The Danube was frozen again for two months.

The Seine River in France was crossable by pedestrians.

Travelers got on the ice between Denmark and Sweden.

In the year 1430, the winter was very strict in the north and the grape vine suffered greatly in Germany. After 1430 CE the upper limit of vineyard cultivation in Baden in southwest Germany was brought down by 220 meters. These height changes tend to verify the approximate magnitude of the change of summer temperatures.

In June 1430 CE it is now believed that a huge comet struck the ocean less than one hundred miles from the Chinese fleet of Zhou Man. Up to 173 ships have been counted as destroyed by this event. The comet incinerated many ships and hurled the blazing wrecks onto the South Island of New Zealand and the east coast of Australia as well as across the Pacific and Indian Oceans. Chinese and Mayan astronomers recorded the appearance of a large blue comet seen in *Canis Minor* for 26 days in June 1430. This date was compatible with the evidence. In November 2003 Dallas Abbott and her team announced where the comet crashed between Campbell Island and South Island of New Zealand. Deaths would have been in the multiple thousands. It was reported that around 1400 BCE two teams of geologists found sedimentary evidence in coastal marshes that the tsunami hit around 1400 CE. Researchers in Thailand and Indonesia wrote two articles in *Nature* about this. It has also been postulated that the Great Tsunami of December 2004 might have been triggered by a cosmic impact as well.

A particularly cold period struck Europe in the 1430s. There was damage to grain, vineyards, herbs and livestock. Food production fell and food prices went up.

The great Angkor State collapsed in Cambodia in 1431 AD. There had been a massive drought beforehand.

The winter of 1431 to 1432 CE was a severe winter in Western Europe and would have affected parts of Britain.

A severe winter began in Germany on 20 November 1431 and lasted until 4 March 1432. The rivers were frozen. "In 1432-1433, the Seine in France and all the rivers of Germany were frozen."

During the winter of 1432, the frost was very severe in England.

1433 was the year of the Great European Famine that lasted to 1438. It had been preceded by long cold winters and wet, cold springs. This was a period of great depopulation.

On May 10th 1434 CE Asosan in Japan erupted with a VEI of 2.

The winter of 1434 to 1435 CE was a severe winter. Perhaps the most severe of the last millennium. The Thames was frozen from below London Bridge to Gravesend. Sea-borne goods were landed at the mouth of the river and taken over the ice to London. There was a frost from the latter part of November to at least 14th February. This may have applied to a wide area of England.

The winter 1434-1435 was remarkable for the duration and severity of cold.

Winter in Flanders, now Belgium, lasted from the beginning of December to mid-March, and the thickness of the ice was more than an one Elle (~ 2.3 feet, 70 centimeters).

In Germany, many people died from the cold.

During the winter of 1433-1434 CE in Scotland there were reports of intense frost and that in England the Thames River was frozen sufficient to bear wagons. In London, England, the river was frozen below London Bridge to Gravesend from 24 November 1433 to 10 February 1434. Gravesend is a town in northwest Kent, England, on the south bank of the Thames, opposite Tilbury in Essex. Ships lying at the mouth of the River Thames could not come up the river. The River Thames was so strongly frozen, that all sorts of merchandizes and provisions brought into the mouth of the said river were unladen and brought by land to the city.

Was this all caused by the eruptions of Sheveluch in Kamchatka in 1430, as well as Furnas and Katla in Iceland in 1430? As well as the cosmic impact into the Pacific Ocean in 1430 as well.

In 1433, the River Thames and all other rivers of England and Scotland froze over; the Seine, Rhine and Danube rivers were closed to navigation early in December. The River Thames in England froze at Gravesend. This winter has been named in England, "the big chill"; the cold lasted from 24 November 1433 until 10 February 1434.

The Dardanelles and Hellespont in Turkey froze, as did many bays and inlets of the Mediterranean.

Ice formed in Algiers, in Algeria in North Africa.

The Strait of Gibraltar was almost impassable due to drift ice. Was it drifting from the rivers feeding into it or coming in from the Atlantic Ocean?

The frost began in Paris, France towards the end of December 1433, and continued during 3 months, less nine days. It recommenced towards the end of March and continued until the 17th of April.

The same year it snowed in the Netherlands for forty consecutive days. In 1434, it snowed in the Netherlands and in Paris, France for almost 40 days in a row.

During the winter of 1433-34, the frost began on 31 December 1433 and persisted for three months minus nine days. The frost reappeared at the end of March and continued until 17 April. The snow was higher than six feet in the streets of Carcassonne in the Languedoc region of southern France. Winter ruled in the city for three months.

In 1433, the winter was very severe again in Germany. There were severe frosts, when the large fowl of the air sought shelter in the towns of Germany. In 1434, all the rivers in Northern Europe and Germany froze.

On 25 April 1434 and the following night, there was such a heavy snowfall accompanied by extreme cold, that the greater part of the grape vines in Austria, Swabia, a region of Germany, and Hungary were destroyed.

In autumn 1436 CE a frost struck the crops during harvest throughout the entire Novgorod province in Russia. Also, during the autumn there were great floods. On a frosty night the ice carried away seven stays of the great bridge in Novgorod, and the little Zhilotug bridge was carried away.

During the summer of 1436 in southern France, humidity and heat competed.

In 1436 during the period between 7 April and 15 May, a drought engulfed China.

In 1436 because of famine and dearth many of the people left Novgorod and went to Germany. Frost killed the young crops.

On 31st October 1436 sea floods hit Schleswig-Holstein, Northwest Germany and Denmark, and North Friesland in the Netherlands. The islands of Nordstrand and Pellworm off the German coast were parted. They were originally part of the island of Strand off the coast of Schleswig.

The famine of 1437 CE in England was second only to the famine of 1315 to 1317 AD. In 1437 in England, there was scarcity. A bushel of wheat sold for 7 shillings being very dear (scarce), according to that time, so that the poor in Chester made bread of peas, veitches, and fern root.

In 1438 CE a gale on 23rd November did much damage in London and in England, there was a great frost that was unusually long. In England in 1438, came a great tempest, terrible winds and rains. As a result, came great scarcity of corn (grain), wine and bay salt. But the citizens of London, from the prudent care of their Lord Mayor, had a good supply of rye from Prussia. But the poor starved people in the country made bread of fern roots and the like. Wheat sold for 24s. a Quarter (quarter ton). In November began a terrible winter of frost and snow. In England in 1438 "In the 17th yeere of Henry the Sixt, by meanes of great tempests, immeasurable windes and raines, there arose such a scarcitie that wheat was sold in some places for 2 shillings 6 pence the bushell." (Another account places this event in 1439.) In England in 1439, during the 18th year of King Henry VI reign – "Wheat was sold at London for 3s. the bushell, mault at 13s. the quarter, and oates at 8d. the bushell, which caused men to eat beanes, peas, and barley, more than in an hundred years before: wherefore Stephen Browne, then maior (mayor), sent into Pruse (Prussia), and caused to be brought to London many ships laden with rye, which did much good; for bread-corne was so scarce in England that poor people made their breade of ferne rootes."

In 1438 CE at Smolensk in Russia it was reported that during a famine wild animals ate people and people ate people and small children. The famine was so severe that people died in the forests and along the roads.

1439 CE was regarded as a wet year in England and there was a famine from rains and tempests. Bread was made from roots.

In 1440 CE Katla in Iceland erupted with a VEI of 4.

In 1440 CE in England, there was a scarcity and in Scotland a famine. In England in 1440, there was a great scarcity and dearth of corn (grain). People were forced to make bread of beans, peas, barley and fern roots, etc.

In 1440 in Italy the weather held a southerly constitution, with great soaking rains that prevailed for a long time. The earth became a marsh and fruits abounded, then depopulating epidemics set in.

In 1441 CE Furnas in Iceland erupted with a VEI of 4.

In 1442 CE the rivers in the south of France froze. In the year 1442, the king spent the winter in Montauban in southwestern France, which was so severe that all flows such as rivers, and streams in that region were frozen. The troops held back in their quarters, because they could not move out in the severe weather. In the regions around Metz, France in 1442, there was great heat from April to June. It was so hot that several people worked the field without shirts, skirts or pants on and a portion of the wine was sour in the runners. The harvest began in Dijon, France on 13 September.

In Sweden there was a famine.

In Flanders a lot of the trees and fruits of the earth were frozen.

The winter of 1442 to 1443 CE was a cold winter in Western Europe and possibly parts of England.

In 1443 CE. Mahuika erupted in New Zealand. The Mahuika Crater is possibly twenty kilometers offshore. It is named after the Maori God of Fire. It is 12.4 miles long and 1.2 miles wide. It is to the south of the Snares. The crater was created around 1443 CE or 1491 CE. The tsunami that it caused was entered into Australian Aboriginal oral traditions.

In 1444 Sete Cidades in the Azores erupted with a VEI of 4.

Vulcano in Italy erupted on February 4th 1444 CE with a VEI of 3.

In March 1445 CE Margaret of Anjou, future queen-Consort of Henry VI, crossed the English Channel in horrendous weather. She crossed from Cherbourg in France to Portsmouth in England and as they approached the English coast, a storm blew up, probably a thunderstorm associated with a rapidly developing area of low pressure, and the ship wallowed in mountainous seas. The ship was guided with much difficulty through "The Needles Channel" to run before the southwest wind towards Portsmouth but the ship was dismasted, and it was beached near Porchester. The storm, the main area of low pressure, continued for several days, uprooting trees, killing cattle in the fields, causing rivers to overflow and flooding low lying ground. There was much damage to roofs and deaths were reported.

In 1445, during a thunderstorm, the steeple of St. Mary Redcliff in Bristol in southwestern England was thrown down, and the rest of the church was significantly damaged at St. Paul's Tide.

In 1445 bread was dear in Novgorod and not only this year but during ten whole years: one poltina or half a rouble for two korobyas (baskets); sometimes a little more, sometimes less; sometimes there was none to be bought anywhere. And amongst the Christians there was great grief and distress; only crying and sobbing were to be heard in the streets and marketplace, and many people fell down dead from hunger, children before their parents, fathers and mothers before their children; and many dispersed, some to Lithuania, others passed over to Latinism (Roman Rite), and others to the Besermeny (Muslims) and to the Jews, giving themselves to the traders for bread. The cause of this famine may have been related more to war than to poor weather.

Later in Novgorod on January 3 1446 there were heavy clouds with rain and the wheat and rye and corn (the autumn-sown crops) were beaten down altogether, both in the fields, and in the forests, all round the town for five versts (a verst is two-thirds of a mile, or 1067 metres) from the Volkhovets (river), and as far as the Msta river, for fifteen versts. The people bore into the town whatever they could gather up; and the townspeople collected to see this curious marvel, whence and how it came. This account might refer to a tornado.

In 1446 CE there were notable storms in the North Sea. There was coastal flooding and in April 1446 in particular, a North Sea storm occurred coupled with a significant tidal surge. Thousands died in coastal areas of the North Sea. In Dordrecht in South Holland in the Netherlands on 10th April 1446 sixteen parishes were devasted and over 100,000 lives were lost. This was caused by sea floods. There was also storm flood damage in northern Germany.

In 1448 CE the Thames flooded Poplar, Stepney and other places during March. It is not known if this was a storm surge or heavy rain or both.

In October 1449 CE a major storm flood affected the Dutch coast and possibly also the English coastline.

The winter of 1449-50 in France was very cold, very wet and very snowy. The winter began as early as October and the olive trees died.

1450 CE

In 1450 CE Pinatubo in the Philippines erupted with a VEI of 5 and Soputan and Kelud in Indonesia, both with a VEI of 3, erupted.

Around 1450 CE rainfall in the Lake Titicaca area in Bolivia reverted to the non-drought pre-1150 CE period when it was higher. The 300-year drought had ended. The Rio Catari, Katari River, people never returned to using cultivated raised fields on lake-edge wetlands. By this time, they were under the Inca Empire and went to rain-fed terrace agriculture on surrounding mountain slopes. This technique was very effective but not suitable for a dense population such as in Tiahuanaco which was nearby.

Sediment from Lake Edward on the border of the Democratic Republic of the Congo and Uganda in East Africa shows a pattern of droughts, from 1450 CE to 1750 CE.

By 1450 CE between twenty per cent and 60 per cent of all villages in the various parts of western, central and northern Europe had been abandoned by this date due to climatic reasons, not the Black Death.

In 1450 CE the Norse colonies in Greenland were deserted and there was no contact between the Norse and the Inuit as trading ceased between them. The fate of the remaining Norse is unknown.

In 1451 CE Kelud in Indonesia erupted again with a VEI of 3.

In 1452-1453 an unidentified volcano triggered the first large sulfate spike in the 1450s. This eruption caused a severe volcanic winter leading to one of the strongest cooling events in the Northern Hemisphere. This date also coincides with a substantial intensification of the Little Ice Age. 175 to 170 million tons of sulfuric acid were released. 36 to 96 cubic kilometers of tephra was ejected. The unknown volcano was in the New Hebrides Archipelago in Vanuatu in the Pacific. There is a possibility it was Kuwae which was a landmass that sank around this time. This was one of the largest volcanic events in the last two thousand years. The islands of Tongoa and Epi were its remnants. It is thought that the eruption of the Kuwae Caldera launched the Little Ice Age. On Epi Island, now fifty kilometers from the caldera, are the remains of a one-meter-thick pyroclastic flow. The eruption itself created a twelve-kilometer-wide submerged caldera. This was one of the seven biggest caldera forming events in the last ten thousand years.

In 1453 CE non-stop snow damaged wheat crops according to Ming Dynasty records. Other events were the Yellow Sea becoming ice-bound out to twenty kilometers from the shore.

In 1453 tree rings of bristle cone pines show frost damage from this period as do tree rings in Europe and China. Was this caused by the eruption of Kuwae in Vanuatu Around this time?

On 23 November 1454, during the Kyotoku era, an earthquake, possibly 8.4 or higher, shook the Kantō and Tōhoku regions of Japan at midnight, generating a tsunami that inundated 1.0-2.5 km of land, sweeping people away in Mutsu Province.

In 1454 CE as a result of the severe famine in Mexico during 1452-54, the Aztec king Moctezuma ordered his people to leave the city and search for food. Many parents sold their children in Totonacapan (currently Veracruz, Mexico) where grain was abundant. Girls fetched 400 ears of maize while boys fetched 500 ears each. Parents were able to buy back their child's freedom when situations improved.

During the Wars of the Roses, 1455 to 1485, rivers were much deeper and more navigable than now in England. They were high rivers that were much deeper than the present day (1974) and in the Medieval period were the norm.

On December 5, 1456, the largest earthquake to occur on the Italian Peninsula struck the Kingdom of Naples. The earthquake had an estimated moment magnitude of M_w 7.19–7.4, and was centered near the town of Pontelandolfo in the present-day Province of Benevento, southern Italy. Earning a level of XI (*Extreme*) on the Modified Mercalli intensity scale, the earthquake caused widespread destruction in central and southern Italy. Estimates of the death toll range greatly with as many as 70,000 deaths reported. It was followed by two strong M_w 7.0 and 6.0 earthquakes to the north on December 30. The earthquake sequence is considered the largest in Italian history, and one of the most studied. The December 5th shock struck at 23:00 local time, lasting approximately 150 seconds. Devastation was reported in five of the twenty regions of Italy: Abruzzo, Molise, Campania, Apulia, and Basilicata; whereas some damage occurred in Lazio and Calabria.) Weak aftershocks were documented following the December 5 earthquake. Contemporary sources made no distinction between effects of the December 5 and 30 earthquakes. Complete destruction occurred in a zone measuring 6,000 km² (2,300 sq miles). Whereas the total area affected was 18,000 km² (6,900 sq miles). The area of devastation was unusually large compared to most earthquakes in Italy; thought to be caused by the occurrence of multiple ruptures. The meizoseismal area stretched for nearly 180 km (110 mi), assigned X–XI (*Extreme*), where destruction of structures occurred. The unusually large area of the meizoseismal area is caused by multiple faults, separated by significant distances rupturing. The commune of Caramanico Terme experienced a maximum intensity of XI. Intensity IX–X was felt in the towns of Tocco da Casauria, Torre de' Passeri and Castiglione a Casauria. From the lower Aterno Valley (in the north), to Sulmona, and Navelli (in the southeast), the intensity was VIII–IX. Intensity VIII–IX was felt over an area that was 40 km (25 mi) wide. About 20 km (12 mi) away, the intensity gradually decreased to V.

The earthquake of December 30 struck at 21:30 which measured ~M_w 7.0 was not as severe in Naples. Regions closer to the epicenter reported in serious damage. Documentation of heavy destruction in Samnium and the Campania Plain might be due to conflicting reports of the previous event. Major damage occurred in Isernia. There was no damage in the areas between Castel di Sangro and Sulmona. Additional damage also occurred due to the aftershocks, which persisted up till early 1457. The aftershock sequence only ended in May 1457. Constant aftershocks occurred up till January 1457, and one caused significant damage in Naples between the night of January 8 and 9. Weak but frequent aftershocks continued to February. Three homes in Capua collapsed during an aftershock on 10 February. Four areas of extreme damage was identified by scientists at the National Institute of Geophysics and Volcanology; in the upper Pescara river valley (Tocco da Casauria, Popoli and Torre de' Passeri); northern Matese (Bojano and Isernia); Samnium and Irpinia; and the Monte Vulture region. A series of anomalous waves in the Bay of Naples also caused boats to crash. One small boat was completely destroyed though there was no casualty. There were also reports of a tsunami in the Gulf of Taranto, where it struck the Ionian coastline.

In the winter of 1457 to 1458 CE it was cold in Western Europe as well as parts of Britain. The winter of 1457-58 in Paris, France, was very severe. In the year 1457, the winter was severe and long lasting from the Feast of St. Martins (11 November) to 18 February. It froze so hard that you could travel on the Oise River and several other rivers on horseback and wagon. Lastly came heavy snowfalls, which came down in such massive quantities that when it thawed, it developed into such a flood as had not been seen in living memory and caused much damage. In Germany, the extreme cold froze the Danube River to a thickness that an army of 40,000 men were able to be encamped on the ice.

In 1458 an unknown volcano with equivalent power to Tambora in 1815 erupted. This was a VEI 7 eruption. This was found by studying ice cores. In the year following the eruption the study of tree-rings showed that there was a strong cooling that summer that registered a cooling of -1° Celsius and -0.4° Celsius the following year. Sulphate flux distribution in the ice cores suggests that the source volcano is in the low latitudes of the Southern Hemisphere. The main contenders are Kuwai in Vanuatu, Tofua in Tonga and Mount Reclus in Antarctica. It is interesting that there were records of volcanic ash raining from the sky in Europe and Eastern Asia. Was this possibly the same volcano that caused the volcanic winter in 1452-1453 also in the Northern Hemisphere? The eruption caused a massive sulfur spike and was the largest recorded in ice cores in the last 700 years.

The "Annals of Dublin" state that the River Liffey was entirely dry for the space of two minutes. This implies that there was a great drought for at least twelve months or perhaps longer. This was in 1459 CE.

The winter of 1459-1460 was intensely cold and similar to the winter of 1323 CE.

In 1459, the ice bore riding on from Lübeck, Germany to Prussia, and the Baltic Sea was covered with ice from Mecklenburg to Denmark again. The ice was thick enough that individuals could cross between Denmark to Lübeck, Germany. Horse passengers crossed from Denmark to Sweden.

In 1460 the Danube and the Rhine Rivers froze in Europe for several months. The Danube River froze for two months.

In 1460 the Rhône River in southern France froze for several months.

In Germany deer sought the towns for refuge from wolves. Packs of wolves came into the cities and attacked the people on the streets. The vineyards in Germany suffered a lot.

During the winter of 1459-60, both in Northern Europe and in Provence the winter was very cold. The Seine River in Paris, France came out of its banks and caused great devastation.

The Sporer Minimum, which was a sunspot minimum, lasted from 1460 CE to 1550 CE. Sunspot Minima generally coincide with minima in temperatures. Mimima is the plural of minimum.

The wet summer of 1460 CE was claimed to be one of the worst for one hundred years in the British Isles. In England, there was excessive rain during the summer. As a result, neither grass, corn (grain), nor fruit came to maturity or were fit to use. There were also greater inundations than had been for a hundred years before, which rapidly carried down mills and buildings, destroyed meadows and pastures, and made great destruction.

In the 1460s it had become recognized that the agricultural change in Norway had become permanent.

The Great Tempest destroyed the east window of Christchurch, Dublin, in Ireland, causing damage inside the cathedral in 1461 CE.

The winter of 1464 to 1465 CE was cold over Western Europe.

In Ireland during the winter and subsequent spring there was "exceeding great frost and snow and stormy weather so that no herb grew in the ground and no leaf budded on a tree until the Feast of St. Brendan (16th May), but a man, if he were the stronger, would forcibly carry away the food from the priest in church, even though he had the Sacred Body in his hands and stood clothed in Mass-vestments.

The winter of 1464 was very severe in Northern Europe.

In Flanders now Belgium no one had experienced a similar winter since the year 1408. It was cold from 10 December until 15 February without ceasing. One could travel across the frozen Schelde (Scheldt) River for a whole month.

In 1467, there was a great flood that overflowed the entire district of South Holland in Lincolnshire, England. The origin of the name Holland here is that from about 43 CE the sea moved out from the land here that led to an industry making salt out of seawater. Saxons and Vikings arrived, giving rise to some surviving place names like Holland which means High Land. The area reached its peak in the medieval period.

On October 20th 1468 CE there was a storm flood in the Netherlands known as the "Ursula Flood". This perhaps also affected English communities. This was supposed to have been far more forceful than the second Saint Elisabeth flood.

In the Winter of 1468-1469 CE due to the extreme cold, the wine in France was reduced to ice, and had to be cut with an axe. In France, the wines of the Duke of Burgundy froze in the casks. We distributed them piecemeal to the gentlemen. Many people died of exposure to the cold. Their extremities (hands, feet, etc.) were frozen.

In 1468 in Flanders the frost was very severe. It took the axe to break apart the frozen wine being distributed to troops in Flanders. The winter of 1468-69 is described by Philippe de Commines when he traveled to the land of Franchemont (near Liège, Belgium). "The largest cooling occurred between 14 and 17 November. Because of the great frost and intense cold most of the staff of the Duke of Burgundy had to walk to Franchemont. I saw some incredible effects of the cold. I found a gentleman whose foot was frozen, which he later could not move when it thawed. A page who had two fingers of his hand frozen. I saw a woman with her newborn child frozen to death. Three times I saw the wine, which they gave the Duke and his people broken with axe blows because the wine was frozen in the cask. This frozen wine was distributed to the people in a hat or basket. Hunger had us in great haste flee, after we spent eight days there".

The severity of the cold stretched up to Provence, in France where the grape vines suffered greatly.

Atitlan in Guatemala erupted with a VEI of 3 in 1469 AD.

On December 24th 1469 CE Miyakejima erupted in Japan with a VEI of 3.

On 1st November 1470 all the North Sea coasts were inundated by a sea floods and ten thousand people perished.

On November 3rd 1471 CE Aira in Japan erupted with a VEI of 5. This would cause a volcanic winter.

Between 1471 and 1476 CE Sakurajima on Kyushu in Japan erupted with a VEI of 5.

In May 1471 in Ireland, there was a great shower of hailstones, with thunder and lightning.

On November 30th, the Rhône River in France flooded. It destroyed two arches of the bridge Pont d'Avignon (Pont Saint-Bénezet) and part of the city walls in Limas in southeastern France.

In 1471, there was an awful pestilence in Oxford, England. In England, the winter was rigorous and weather stormy.

In the northeast coast of India, there was a famine in Orissa which is now Odisha on the east coast of India

On 6th January 1471 sea floods inundated Hamburg in Germany, Dithmarschen on the North Sea coast of Schleswig-Holstein and Friesland in the Netherlands. The water was an ell deeper than the 1412 floods. An ell was the unit of measurement in North Germany that was the length of the combined length of the forearm and the extended hand.

From 1470 CE or 1480 CE fishermen from Bristol in England were fishing on the Newfoundland Banks, which may have started because the fish stocks of the higher latitudes in the northeast Atlantic had deserted their former grounds as a result of the increasing spread of the Arctic cold water. The competition for fish was also complicated by Hanseatic competition in Iceland-Greenland waters. The climate was getting colder and colder.

In 1472, the plague swept away many of the inhabitants of Kingston-Upon-Hull, now frequently referred to as Hull.

On May 16th 1473 CE Asosan in Japan erupted with a VEI of 2.

In 1473 there was a most droughty summer and so hot that woods took fire. All the rivers dried up and the Danube River could be walked over in Hungary. This drought continued for 3 years.

The summer of 1473 was very hot in France. The heat lasted from June until December 1st and there was neither cold nor frost before Candlemas, 2 February 1474. In Dijon, France, the harvest began on 29th August. The heat around Metz, France, that year was so strong that on 1st May cherries were sold, and on the Feast of Saint Peters, June 29th, ripe grapes were sold. The harvest was over in August. Legumes could not be harvested due to the drought.

In England, there was a great drought and heat during 1473-75 after the two comets of 1472. Between 1473 to 1479 CE there were droughts with very hot summers in three successive years assumed to be applicable to south and central England. There were five fine summers in this seven-year period. The great comet of 1472 was visible from Christmas Day 1471 to 1 March 1472 (Julian Calendar), for a total of 59 days. The comet passed 0.07 AU from Earth on 22 January 1472, closer than any other great comet in modern times.

In 1475 CE towns about the Humber River in England were lost in a great sea floods. There was an inundation in England. The land near the mouth of the River Humber was swept away, and several villages were destroyed.

The severe cold of 1475 destroyed the olive trees of Languedoc, France.

In 1476 CE the Istula (Vistula River in Poland) flooded. There were locusts and a great inundation of the Istula (Vistula River in Poland).

The Rhine River froze.

On 16th October 1476 two parishes were destroyed in Schleswig-Holstein by a sea floods.

In 1476 in Germany, there was a normal snowfall. It began to snow on Christmas and continued daily until St. Dorothy's on February 6th and was fiercely cold and all the water and ponds froze until St. George's on April 23rd. The fish died. Many people froze.

At Breslau, now Wroclaw in Poland, many good-hearted citizens had baths and other rooms heated and furnished wherein many poor people preserved their lives.

In 1476, the plague raged again in Kingston-Upon-Hull, England.

The winter of 1476-1477 became progressively more severe; the ground was covered with snow. The cold was so great on Christmas Eve, more than four hundred men of the army of Charles the Bold, in Nancy in northeastern France died or had their feet frozen.

On 31st December 1476 a sea floods struck Friesland in the Netherlands.

The cold continued in January 1477. The snow, which fell in large flakes, obscured the day making it almost impossible for seeing far ahead.

In February 1477 the volcano Bardarbunga in northeastern Iceland erupted. Bardarbunga is underneath the glacier Vatnajokull which is Iceland's largest glacier. This was the largest known Icelandic eruption ever. It had a VEI of 6 and ejected ten cubic kilometers of tephra. This would have caused a volcanic winter.

In March 1477 CE Torfajokull in Iceland erupted with a VEI of 2.

In September 1477 CE a flood storm affected large areas of coastal Belgium, Germany and the Netherlands. This was known as the first Cosmos and Damianus Flood in Holland. There were thousands of deaths in the Netherlands and Germany.

On 13th October 1477 seventeen parishes were destroyed at the mouth of the Scheldt River in the Netherlands by sea floods.

On September 23rd 1478 CE Aira erupted in Japan with a VEI of 2.

In 1478, the plague again raged in Kingston-Upon-Hull, England. The plague raged so violently that there died in this town in a very short space of time 1,580 persons. All the churches, monasteries, priories, hospitals, schools, etc., were shut up and forsaken, and the streets were so little frequented that grass grew up in most parts of the town between the seams of the stones. The merchants forsook the port.

In 1478 there was a great plague in England, which destroyed more people than the wars waged during the prior 15 years.

In 1479 CE in St. Neots (Huntingdon in Cambridgeshire in eastern England), there was a hailstorm, "when the stones measured 18 inches round." In England in 1479 there were hailstorms in Huntingdonshire with stones 14 inches round.

Also, in 1480 CE the Seine River in France was frozen and the ice carried the weight of carts. The winter of 1480 in France did not begin until the day after Christmas. Then it froze very hard until 8 February 1481. The cold was so great that the rivers froze and carts crossed the Seine, the Marne, the Yonne rivers and all their tributaries. The cold continued after the thaw of February 8 until well into the month of May. The roots of trees were killed in several places. The winter of 1480 was severe and due to a large flood in Paris, France noteworthy

The winter of 1480 to 1481 CE was cold in Western Europe and possibly the British Isles.

The 1481 Rhodes earthquake occurred at 3:00 in the morning on 3 May 1481. It triggered a small tsunami, which caused local flooding. There were an estimated 30,000 fatalities. It was the largest of a series of earthquakes that affected Rhodes, starting on 15 March 1481, continuing until January 1482. Sources refer to destruction in Rhodes Town; the Palace of the Grand Master of the Knights of Rhodes was sufficiently damaged to require immediate rebuilding (Rhodes was at the time under siege by the Turks). The damage caused by the earthquakes led to a wave of rebuilding after 1481. Damage from the tsunami was said to be greater than from the earthquake.

The tsunami caused a large ship to break free from its moorings. It (or another ship) later sank with the loss of all its crew after running onto a reef. There were an estimated 30,000 fatalities. Reports of an earthquake felt an hour past noon on the seventeenth day of Muharram in Egypt may have been the effects of this earthquake. Shaking in Cairo was described as "severe", and minarets were observed swaying. A cresting on the Salihiyya Madrasa collapsed, killing two people. Reports of shaking also came from Fustat, Alexandria and Asia Minor. The tsunami appears to have been relatively minor, estimated at a maximum 1.8 meters. However, it was observed on the Levantine coasts and a tsunami sediment layer found at Dalaman, on the southwest coast of Turkey. Although the studies on sediment transport from tsunamis are limited, it is probable that the tsunami can be dated 1473 ± 46. The sediment found and studied appears to be consistent with the aforementioned tsunami.

On January 15th 1482 CE Mount St. Helens in Washington erupted with a VEI of 5. After this eruption ash and pumice piled six miles northeast of the volcano to a thickness of three feet. Fifty miles away the ash was two inches thick. Large pyroclastic flows and mudflows rushed down Mount St. Helen's west flanks and into the Kalama River Drainage system. This was exactly two years after the first eruption. This eruption was several times larger than the 1980 CE eruption.

1482 CE was recorded as one of the coldest winters in the Netherlands.

In October 1483 CE an extraordinary flood of the Severn near Worcester in England prevented the Duke of Buckingham from crossing to attack king Richard III. The Duke's army was dispersed, and he was taken and beheaded. There happened such a flood in Gloucestershire, England that all the country was overflowed by the River Severn; several persons were drowned in their beds. The River Severn overflowed for 10 days, and carried away men, women, and children in their beds; and covered the tops of many mountains. The waters settled on the lands, and the event was called the "Great Waters" for a hundred years thereafter.

In 1483, an epidemic first appeared in England called the "sweating sickness". This disease was particularly violent and in 24 hours the fate of the sufferer was decided for life or death. It chiefly attacked males, in the prime of life, and more especially the higher classes. This event occurred during the first year of King Richard III's reign.

On January 5th 1485 CE Asosan in Japan erupted with a VEI of 2.

In 1485, the disease "sweating sickness" returned to England. Many thousands of people died. In London, in one week, 2 mayors and 6 aldermen died.

In 1485 CE John Rous of Warwick listed 58 sites that were now abandoned in that county alone and had become depopulated in his life-time due to the changing climate.

In the year 1490, it was bitterly cold in Burgundy and the winter lasted 6 months. The winter was followed by a very great heatwave. The winter of 1490 was one of the harshest of which we had heard. The winter produced such a furious storm that the inhabitants of Marseille in southern France could not leave their houses for two months.

On 5 January 1491, the Paglia River as well as the Tiber River froze, so that people could cross it on foot for several days. Many keepers of cattle perished because they were victims of the weather. In the Winter of 1490-1491 CE a cold winter struck Florence, Italy on 10 January 1491. The River Arno froze. Then on 17 January freezing rain broke trees.

In March or April 1490 CE in Qingyang. Gansu Province, formerly Shaanxi Province, in China it was reported that ten thousand people were killed in the Quingyang Event with the deaths caused by a hail of falling stones. Stones fell like rain in the Ch'ing-yang Qingyang district. The larger ones were 4 to 5 catties (about 1.5 kg), and the smaller ones were 2 to 3 catties (about 1 kg). Numerous stones rained in Ch'ing-yang. Their sizes were all different. The larger ones were like goose's eggs and the smaller ones were like water-chestnuts. More than 10,000 people were struck dead. All of the people in the city fled to other places. This was in March-April 1490. The Chinese also reported the existence of a new comet that year as well. Were the comet and the fall of stones related? This could have been a meteor airburst or a meteor shower.

In 1490 in England there was a drought in London and the south.

In Ireland, there were great rain and floods all the summer; There was such a famine that it was called "The Dismal Year."

In England, there was considerable scarcity.

In Poland, there was a great dearth of cattle.

A letter from Pope Alexander VI in 1492 states "the church of Garda is situated at the ends of the earth in Greenland, and the people dwelling there are accustomed to live on dried fish and milk for lack of bread, wine and oil…shipping to that country is very infrequent because of the extensive freezing of the waters-no ship having been put into shore, it is believed for eighty years-or, if voyages happened to be made, it could have been, it is thought, only in the month of August…and… it is also said that no bishop or priest… has been in residence for eighty years or thereabouts.

Between 1492 CE and 1700 CE population fell by fifty million people in South and Central America. CO_2 may also have decreased after forests reclaimed agricultural lands after the pandemics occurred and there was sharp population decline.

On November 7th 1492 CE a great meteorite fell at Ensisheim in Alsace, now in France. It weighed 280 pounds. It was a stony meteorite and fell in a wheat field outside the walled town where it created a one-meter-deep hole upon impact. The meteorite, or what was left of it after it was souvenired by various people, was kept in a local church as an object of good luck and now resides in a local museum.

Also, in 1492 CE Columbus reported that a "a marvelous branch of fire" fell into the sea as he crossed the Atlantic.

In the Winter of 1493-1494 CE, the port of Genoa in northwestern Italy was frozen on December 25th and 26th. The winter of 1493-94 was remarkable for the severity of the cold, which was very severe in the south (Southern Europe). The lagoon and all the canals of Venice, Italy were also frozen so that pedestrians, wagons and horses could travel over the ice.

The Rhône River froze in 1493 in southern France.

During the second voyage of Christopher Columbus, on 16 July 1494, he encountered a violent hurricane at Cape Santa Cruz. Admiral Columbus remarked "that nothing but the service of God and the extension of the monarchy should induce him to expose himself to such danger".

In 1495, an Atlantic hurricane struck the West Indies causing a loss of life. "When the hurricane reached the harbor, it whirled the ships round as they lay at anchor, snapped their cables, and sank three of them with all who were on board." Before 10 March 1496 a hurricane wrecked the four ships of Juan Aguado as he was readying to return to Spain, along with two others in the harbor of La Isabela in the Dominican Republic.

In 1495 CE in India, there was a great dearth that occurred in Hindustan.

1497 CE there was an "intolerable famine throughout all Ireland and many people perished.

On 14th January 1497 18 parishes were destroyed by the sea in a sea floods in Holland in the Netherlands. Friesland in the Netherlands and Flanders in Belgium were also flooded by sea floods.

In 1498 CE there was a dry year in London and the south of England.

On 20 September 1498, during the Meio Era, a 7.5 earthquake occurred. The ports of Kii (Wakayama Prefecture) were damaged by a tsunami of several meters in height. Between 30 and 40 thousand deaths were estimated. The building around the great Buddha of Kamakura (altitude 7 meters (23 feet) was swept away by the tsunami.

In the Winter of 1498-1499 CE the frosts of the winter appeared in Hainaut in western Belgium in a very unusual form. On Christmas night there was a very heavy rain mixed with hail, the cold immediately formed a smooth ice flow. This was followed by so much snow, "that all, as the chronicler says, flowed together and with each other a mixed ice, hard as stone, formed." As the trees could not bear such a burden, "the branches broke with a crash." The branches that resisted caused by the wind a noise "like the rattle of a harness." This strange frost lasted twelve days, and when the thaw came, enormous pieces of ice fell from the church towers and damaged the ships and the chapels of the churches. The harvest of the apple and pear trees in the following autumn was very abundant, but there was a lack of food altogether, so that horses and cattle died of starvation. The farmers who had filled their barns with straw the previous year, had to remove it to give it to the animals to eat.

1500 CE

In 1500 CE Katla in Iceland erupted with a VEI of 4.

During the sixteenth and seventeenth centuries herring moved from the Norwegian coast to the North Sea. Herring normally inhabit waters with temperatures between 3° and 13°. In England the arrival of the herring produced a herring industry whilst the Norwegian herring industry collapsed due to the colder waters of the Norwegian Sea.

Whether this was produced by rather frequent anticyclones affecting the zone near latitudes 45° to 50° North and westerly winds over western Europe whereas in the previous century, like the period from 1550 CE to 1700 CE, was characterized by a remarkable frequency of anticyclones north of 60° North and winds from between northeast and southeast over Europe south of that latitude.

Between 1500 to 1550 CE there was a short warm period. This is confirmed by tree rings from California, and paleotemperatures indicated by isotope studies of calcite in a cave in a New Zealand.

In North Greenland and Iceland isotope measurements also indicate a warmer period.

In Somerset there were shallow lakes or meres in the sixteenth century CE which have since disappeared that had been there before in the centuries before Christ.

Also, in England through the period of 1500 CE to 1550 CE was milder, the Thames in London still froze over three times.

In Egypt the Nile floods were lower around 1500 CE than in the seventeenth, eighteenth and nineteenth centuries before falling again in the twentieth century.

There were substantial further losses of land to the North Sea by erosion of the low-lying coasts of the Netherlands, North Germany and Jutland in Denmark after the frequent high sea levels at the end of the medieval warm period.

The relatively warmer period from 1500 CE to 1550 CE caused an increase in population in Europe.

The surviving Greenland settlement named Osterbygd, East Settlement, where there were 225 farms survived until about 1500 CE, though in evident decline. The average stature of men in the graveyard at Herjolfnes was only 164 cm, 5 foot 5 inches., compared with 177 cm, 5 foot 10 inches in the early period of the settlement.

In Japan one of the greatest severity of winters was between 1500 to 1520.

In the sixteenth century mining operations in the Alps, which had reopened for a while in the High Middle Ages, were abandoned again as the cold and ice descended. In many places advancing glaciers had blocked roads and attacked mines and villages.

Analysis of Silver pine tree rings in New Zealand indicate that there was rapid cooling below average temperatures in New Zealand after 1500 CE.

In July 1500 an Atlantic hurricane struck the Bahamas causing a loss of life. Two caravels with all their crew were swallowed up by the storm.

In 1500 CE Mount Pinatubo erupted on Luzon in the Philippines with a VEI of 5. This was the Buag Eruptive Period. More than 10 cubic kilometers of material was ejected, and large parts of the surrounding areas were covered with pyroclastic flow deposits.

In 1502 CE in late winter or spring Prince Arthur, Prince of Wales at Ludlow in the Welsh Marches, contracted tuberculosis and died there allowing younger brother Henry VIII to succeed. The area is not normally exceptionally wet in a standard westerly climate so it suggests some abnormal synoptic pattern.

In 1502 CE in the Netherlands the winter was severe.

In Paris, France, the winter brought on a flood. The summer of 1502 in southern France was hot and dry. There was a drought in southern France in 1502.

A fleet of thirty-two vessels at Santo Domingo, Dominican Republic was to sail back to Spain. Christopher Columbus arrived and asked for shelter from the coming hurricane. Ovando denied his request. Columbus asked them to detain the fleet from returning to Spain because of the signs of an impending hurricane. Ovando did not believe him, and the sailors laughed at his prediction. Bovadilla, Roldan, and their party, were embarked; Guarionex was on board the Capitana, and 100,000 castellanos for the King, and another 100,000 castellanos belonging to the passengers, and the large lump of gold, which had been Garay's dish. The fleet sailed on 1 July 1502; and within twenty-four hours, twenty sail including the Capitana, with all on board, perished! During the time of Ferdinand and Isabella, the marc was a unit of weight equal to one-half pound. A marc of gold was equal to 50 castellanos. 100,000 castellanos would be equal to 1000 pounds of gold. Christopher Columbus arrived at St. Domingo in Hispaniola, now the Dominican Republic, towards the end of June 1502. The Admiral had the good fortune to get in a little creek in the island, where he weathered a very terrible storm, in which Bovadilla, his great enemy, and fourteen ships loaden, loaded, with treasure, and bound for Spain, perished. The enemies of Columbus having given the Court of Spain an ill opinion of him, it appears they employed Amerigo Vespucci, a Florentine, in the year 1497, to enlarge the discoveries of Columbus. The Americas were named after this lesser explorer. On 11-12 July 1502, a hurricane struck offshore the Dominican Republic causing approximately 500 deaths.

In 1503 CE there was a dry summer in London and the south.

The Po River in Italy was frozen and carried the weight of the army of Pope Julius II on the ice.

In 1504, a hurricane struck the north coast of Colombia causing 175 deaths.

In 1505 CE Atitlan in Guatemala erupted with a VEI of 3.

In February 1505 CE Asosan erupted in Japan with a VEI of 2.

The 1505 Lo Mustang earthquake occurred on 6 June 1505 and had an estimated magnitude between 8.2 and 8.8 making it one of the largest earthquakes in Nepalese history. The earthquake killed an approximate 30 percent of the Nepalese population at the time. The earthquake was located in northern Nepal, affected southern China, and northern India.

In 1505 because of a great famine in Hungary, parents killed and ate their own children and during the famine in Hungary in 1505 starving parents who butchered and ate their children were not punished.

In January 1506 CE there was a severe frost in England. The Thames froze throughout January and a horse and cart could cross the frozen river. On or around January 11th a major storm of wind affected at least the southern half of Britain and the southern North Sea. There was damage to St. Paul's in London and elsewhere. On 15 to 26 January 1506, there were tempests and hurricanes in England.

Also, in January 1506 CE the sea was frozen in the Mediterranean at Marseilles in the south of France. This implies that it must have been bitterly cold and persistently so since at least late December. It often needs some period of strong east wind as well to remove the heat from the water. This winter was extremely severe in Southern Europe. At Marseille, France, three feet of snow fell on the day of Epiphany, January 6th. The fruit trees died off. During the winter of 1506-1507 in southern France, there was severe cold, and large snowfalls. The chill of 1506-1507 completely froze the port of Marseilles, France and destroyed a large number of men and animals. On the Feast of the Epiphany (January 6th), 3.2 feet (974 millimeters) of snow fell on Marseilles. The mass of snow, fortunately protected trees and seeds from the cold.

In 1506, there was a great frost in England. The frozen River Thames at London bore the weight of carriages throughout January

On April 6th 1506 CE Asosan in Japan erupted with a VEI of 2.

Weather diaries from Eichstatt and Ingolstadt in Bavaria, in Germany, found no significant difference in winter temperatures, but summer temperatures were on average 7 to 8 per cent wetter and less warm.

In 1508 CE in Germany, there was a terrible hailstorm that destroyed trees, corn (grain), and (grape) vines, chiefly in the Duchies of Württemberg, Hohenberg, and Rottenburg on the Nickar, Neckar River. The hailstones were so large and tempestuous, that it broke windows and tiles of houses. At Stutgard, Stuttgart, Germany, a tempest arose and so great a flood of waters from the clouds that it filled the town. The city was in danger of perishing. Some men and oxen were lost and a part of the wall of the city was broken. Württemberg, Hohenberg, Rottenburg, and Stuttgart are located in southern Germany.

On 12-14 August 1508, an Atlantic hurricane struck the Dominican Republic causing a loss of life. "Many men were lost in this city and in the greater part of this island." The storm destroyed the entire population of Buenaventura (the date given is 3 August). On 3 August 1508, all the thatched houses in Santo Domingo, Dominican Republic, and several of those built with stone, every house in Bonaventura, and twenty sail of vessels, were destroyed by a hurricane. At first the gale blew from the north, and then shifted suddenly to the south.

The 1509 Constantinople earthquake or historically *Kıyamet-i Sugra* ('Minor Judgment Day') occurred in the Sea of Marmara on 10 September 1509 at about 22:00. The earthquake had an estimated magnitude of 7.2 ± 0.3 on the surface-wave magnitude scale. A tsunami and 45 days of aftershocks followed the earthquake. The death toll of this earthquake is poorly known; estimates range between 1,000 and 13,000 victims. The earthquake occurred in the northeast of the Sea of Marmara within the borders of the Ottoman Empire, and in the south of Prince's Islands, 29 km away from the capital Constantinople. It is thought that a fault ruptured between 70 km (43 mi) and 100 km (62 miles) from the Çınarcı Basin of the North Anatolian Fault Line to the Gulf of Izmit in the east of the Sea of Marmara. Major shocks occurred at half-hour intervals and were violent and protracted in nature, forcing residents to seek refuge in open parks and squares. Aftershocks were said to have continued for 18 days without causing any further damage but delayed reconstruction in some areas.

A tsunami is mentioned in some sources with a run-up of greater than 6.0 meters (19.7 feet) but discounted in others. The waves that surpassed the walls of the city and the Genoese Walls penetrated into the settlements. Especially in the Galata region, many houses were flooded. Seismologists and geologists believe that the tsunami observed in the Sea of Marmara was not only related to the earthquake, but also caused by seafloor landslides triggered by the earthquake. A turbidite bed whose deposition matches the date of the earthquake has been recognized in the Çınarcık Basin. Reports were sent to the capital that the earthquake caused damage even in Edirne, Çorlu, Gallipoli and Dimetoka, which were part of the Rumelia Province of the Empire.

On 26-27th September 1509 the provinces of Holland, Friesland and Zealand in the Netherlands were swamped by sea floods. This was the "Second Cosmas and Damianus flood."

Gamalama erupted with a VEI of 3 in Indonesia in 1510 AD.

On July 25th 1510 Hekla in Iceland erupted with a VEI of 4.

In 1510 CE in England there was excessive heat.

In Italy, there was a hailstorm "which destroyed all the fish, birds, and beasts of the country." Some of the hailstones weighed one hundred pounds.

The year 1510 in southern France was humid and rainy. There was a severe hailstorm in France in 1510. A black cloud came over the face of the heavens and darkened the air like night. In the midst of people's terror and astonishment, the most violent lightning and thunder burst from it, and hail began to fall. This increased in a most dreadful manner, and with a strong and suffocating smell, like burning brimstone. The hailstones were more like pebbles, their color bluish and their hardness like flint, till they softened to the wet. Some of these hailstones weighed a hundred pounds. This storm killed almost all the cattle, fowls and fish in the county and a vast number of the people. In Gulick and Juliers (now Jülich), there was such an extraordinary thunder and lightning storm. It struck all with a panic. A thunderbolt set fire to a magazine, which did great damage.

The territory of the Duchy of Jülich lies in present-day Germany (part of North Rhine-Westphalia) and in the present-day Netherlands (part of the Limburg province). Gulick is the Dutch word for Juliers.

In the winter of 1510 to 1511 CE there was a cold winter in Western Europe with this implying parts of Britain.

On 16th-17th January 1511 Friesland was flooded by a sea floods and in Northwest Germany the Jadebusen, or Jade Bight was enlarged by sea floods.

On September 1st 1511 CE Fujisan erupted in Japan.

On September 14th 1511 at Cremona in Lombardy in Italy, a monk was killed by a meteorite along with several birds and a sheep. Watch out for those falling stones. They are more common than you think.

On the 13th and 14th December 1511 CE a storm flood affected Dutch communities and also the English coastline.

In 1512 CE in Bologna, Italy, the snow fell so thick one could not see through it. This snow lasted until May.

In 1512, the grapevines froze to death during the summer in the countryside of Metz, France. The summer produced sinister cold.

In July 1513 CE there was a hot Wednesday, and several people were killed by the heat.

In 1513 CE a dearth, scarcity, of corn (grain), famine, rainy seasons and severe cold winters had afflicted Italy for two years and people were forced to eat uncommon and unwholesome food, and then in 1513, a contagious epidemic struck.

In 1513 there was a famine in England.

There was a severe winter in Western Europe in 1513 to 1514 CE and this included many parts of England. The Thames was frozen in January 1514. Carts crossed from Lambeth to Westminster which would imply extended periods of sub-zero temperatures together with persistent and perhaps strong easterly winds.

In the Winter of 1513-1514 CE the Meuse River froze over its whole course, and carts travelled from Liège, Belgium to Maastricht in southern Netherlands on the ice.

The winter of 1513-14 in Flanders was very severe. The loaded wagons traveled on the frozen rivers from Gorcum, now Gorkum, in western Netherlands to Cologne in west-central Germany on the ice.

In 1515, the frost in England was so intense that throughout January in London, carriages crossed over the River Thames on the ice from Lambeth to Westminster.

In 1515 CE on January 1st there was a most frightful and destructive storm in Denmark, which rooted up whole forests of trees, destroyed a great many houses, and blew down the steeple of the great church at Copenhagen. Many persons were killed.

"All Germany like a sea, and Cracovia, Kraków in southern Poland, flooded." The flood drowned many people. There was such a great flood in Germany that the country suffered much loss and looked like an island.

In July 1515, an Atlantic hurricane struck Puerto Rico causing the death of many Indians.

In 1515 during the period between 12 June and 10 July, a severe drought engulfed China. In 1515 during the summer, there was a great drought and dust storms in China.

In May 1516 during summertime in May of Jaijing 11th year, stars fell from the northwest direction, five to six-fold long, waving like snakes and dragons. They were as bright as lightning and disappeared in seconds. Many of them were recovered in 1958 when locals were looking for iron and the average weight was 150 to 1,500 kilograms. The stones were 92.35 per cent iron and 6.96 per cent nickel. Many of the small meteorites in valleys had been extremely weathered and had oxidized into limonite. This was near Nantan in China.

In 1516 CE it was hot and dry in London and the South. More generally, there was a drought with very little rain falling for nine months.

In December 1516 a major storm flood affected Dutch coastal communities and perhaps the English coast. The date given was the 27th, therefore "St. Steven's Flood".

In January 1517 CE a great frost started on the 12th. There was a severe winter across England and the Thames was frozen.

There was a very hot summer in London and the south of England.

In November 1517 CE a cold winter began in England. There were cases of frostbite and deaths by cold.

In February 1518 Hichijojima erupted in Japan with a VEI of 2.

In 1519 CE Colima erupted with a VEI of 3.

In 1519, an Atlantic hurricane struck near Jamaica causing a loss of life. Eighteen men from a caravel survived the hurricane.

In September 1519 CE Popocatepetl in Mexico erupted with a VEI of 3.

On December 31st 1521 Santa Ana in El Salvador erupted with a VEI of 3.

In the Netherlands on the 1st of November 1521, there was "a dire inundation of the sea, and 100,000 drowned." This overwhelmed 72 villages and drowned over 100,000 people and very many cattle.

In 1521, there was a great dearth and mortality in England. "Wheat sold in London for 20s. a Quarter (quarter ton)".

In India, there was a very general famine in Sind, now Sindh, now southeastern Pakistan. There was also a famine in the Bombay Presidency in India.

On February 15th 1522 CE Asosan erupted in Japan with a VEI of 2.

In 1522 CE the winter was severe in Europe.

In France, there was a great flood at Vivarais in Saint-Pierreville, France. In France in 1522, they began the harvest at Dijon on 5 September.

In Ireland in 1522, there was a great famine and at Limerick, Ireland, many thousands perished from the plague.

In Turkey from 1523-26, the rivers were greatly swollen, and pestilential diseases were prevalent.

In the year 1523 in France, the winter produced very severe winter storms. During the winter of 1523-24, the cold was felt in the autumn. In France the winter was severe and the snow began to fall on 2 November 1523. Due to the cold; the corn and vegetables froze in the fields. The lack of food continued until the next year's crops. By mid-August 1524, wheat and rye were still blooming and the other cereals were just as advanced. This made the food throughout 1524 very dear.

In England this winter began with heavy rain and strong winds and then a frost; so many people died from the cold, while others lost their toes. In England after long and great rains and winds, which had happened that season, followed so severe a frost that many died of the cold. Some lost toes or fingers, and many lost their nails.

In 1523, an Atlantic hurricane struck off the west coast of Florida in the United States. Two ships and their crews were lost.

In 1524 CE Momotombo with a VEI of 3 erupted in Nicaragua.

On April 30th 1524 CE Santa Ana erupted in El Salvador with a VEI of 3. That same day, April 30th 1524 CE, Fuego in Guatemala erupted with a VEI of 2.

In 1524 CE it was noted as a very hot summer in Dublin in Ireland.

In 1525 CE Mount St. Helens erupted in Washington.

Near the end of October 1525, a hurricane struck western Cuba causing approximately 72 deaths.

In June 1526, an Atlantic hurricane struck North Carolina in the United States and a Spanish brigantine was lost off Wilmington, North Carolina.

In October 1526, a violent hurricane did great damage at Española, Hispaniola which is now the island of the Dominican Republic and Haiti. The rivers overflowed their banks. No such storm had been seen in that island for many years.

In 1526, a meteor falls at Yung-chun in Fuhkien (now Fujian province) on the southeast coast of China, where it exploded and killed a number of people.

In 1527 Telica in Nicaragua erupted with a VEI of 3.

In May 1527 CE Asamayama in Japan erupted with a VEI of 2.

In October 1527, a hurricane struck Cuba causing between 60 and 70 deaths.

In November 1527, a hurricane struck the upper coast of Texas in the United States. Various accounts give differing fatality figures of 200, 191 and 162. The causalities were caused by the loss of the Spanish fleet.

In 1527 to 1528 CE there were the wettest pair of consecutive years in England since weather chronicles began. 1527 was wetter than 1528. In England, there was a great flood. In England, during the 18th year of Henry VIII, "In November 1526, December, and January 1527 fell such abundance of reine that thereof ensued great flouds, which destroyed corne-fields, pastures and beasts. Then was it drie until the 12th April; and from that time, it rained every day and night, till the 3rd June: whereby corne failed sore in the yeare falling." In 1527 in England from 1 November 1526 to 1 February 1527, there were continual rains; fearful floods; terrible destruction of corn (grain), cattle, and pastures. Then there was a drought to 12 April. Then daily rains till 3rd June. Hence there was a scarcity of corn (grain) in England and a dearth.

In England in 1527 during the 19th year of King Henry VIII reign – "Such scarcitie of bread was at London and throughout England that many dyed for want thereof. The King sent to the citie, of his owne provision, 600 quarters; the bread carts then coming from Stratford where nearly all the bakings were probably on account proximity to Epping Forest towards London, were met at the Mile End by a great number of citizens so that the maior and sheriffs were forced to goe and rescue the same, and see them brought to the market appointed, wheat being then at 15s. the quarter (quarter ton). But shortly after the merchants of the Stiliard, steelyard, brought from Danske (Danzig) such store of wheat and rye, that it was better cheape at London than in any other part of the Realme." The merchants of the steelyard were a famous guild of foreign merchants in England, connected with the Hanseatic League.

In 1528 CE in England, the last winter was wholly rainy and southerly. The spring was the same with very great and destructive inundations. In 1528 at Nottingham, England, it rained almost incessantly in deluges during the spring. This prevented the corn (grain) from being sown. As a result, there was an extensive crop failure during harvest. Grain was imported largely from Germany.

There was a great famine in Venice, Italy.

In 1528 and 1529, "sweating sickness" again plagued England.

In 1528 in Paris, France, the frost destroyed the wheat and vegetables.

In Augsburg in Bavaria in southwestern Germany on 19th July, there was a great hailstorm

In 1529 CE Telica in Nicaragua erupted again with a VEI of 4.

In northwestern Switzerland on June 13 or 14 1529, there was a great flood of the Rhine River at Basle, Basel.

In England on October 2 1529, there was a great flood of the River Thames.

In the year 1529, in France the winter was one of the most extraordinary that one has ever seen. Not only because there was no frost, but also because in the month of March the weather was as warm as during the Feast of Saint John, 24 June. As a result, the greater part of the rye in ears matured and was sold in Paris, France even before the new almonds of April. But the weather changed, and on 4 April a very strong cold struck. For a while it was feared that all the fruits of the country would be lost. Fortunately, rains soon came and beat back the effects of the frost so the harvest sustained no damage. The year 1529 in southern France was humid and rainy.

In 1528 and 1529, "sweating sickness" again plagued England.

On October 2nd 1529 CE the Thames was in flood.

In 1530 CE Soufriere in Guadeloupe in the Bahamas erupted with a VEI of 4.

Between 1530 and 1702 in Europe there was evidence of increasing severity of windstorms and resulting sea floods and disaster by shifting sands in 50-60° North.

The summers of the 1530s on the continent alternated in quality so strongly that graphs of the tree rings record from the oaks in Germany and vintage dates recorded in France and Switzerland produce a regular zig-zag pattern. This *Sagesignatur* is a prime example of a biennial, or alternate year cycle, which is present at times prominent in many series of climatic data. This suggests that the warmth of the 1500 to 1550 CE period did not match the levels of 1900 to 1950 CE in Europe.

On 31 August 1530, an Atlantic hurricane struck Puerto Rico. An uncounted number of deaths by drowning occurred. The inhabitants of San Juan, Puerto Rico was often called San Juan, were in great distress. The storms which had followed the hurricanes had made the rivers overflow their banks, and crops, trees, and herds had been washed away; so that the works at the gold mines, and other undertakings, were suspended.

A great sea floods, St. Felix's Flood, occurred on November 4 in the Netherlands. More than 100,000 people drowned. There was a general inundation by the failure of the dikes in the Netherlands. The was later called Evil Saturday or Kwade Zaterdag. The flood caused a lot of damage in Zeeland. All eighteen villages in South Beveland, North Beveland and East Watering were flooded. Zuid-Beveland could not be reclaimed from the sea and became known as the drowned land of Zuid-Beveland. Reimersaal, the third city of Zeeland was mainly wiped off the map by the flood and never recovered. On November 2nd 1532 another major flood in Zeeland destroyed most of the repair work.

Also in November 1530, a flood from the sea ravaged Calais (France), Antwerp (Belgium), Cluse, Gravesend (England), Mardyck (France), Dunkirk (France), Neuport (Newport, England) and almost all of Zealand (Denmark). Many towns disappeared. Flanders in Belgium and Zealand and Holland in the Netherlands were inundated by sea floods. 25 towns and many villages were lost.

On the night of the 14th of November, a storm surge flooded the coasts and estuaries in the southern parts of the English North Sea coast. This was particularly in Essex and Kent and also south Holland in the Netherlands, after three days of strong winds. A strong northerly wind was implied given the areas affected that was possibly aggravated by a secondary cyclonic center developing in the southern North Sea.

Between 1530 and 1575 CE the residents of Chamonix-Mont Blanc in France lamented the fact that the growing glaciers made the surroundings very cold. Complaints of physical destruction wrought by the glaciers only came later.

The earthquake of 26 January 1531 was accompanied by a tsunami in the Tagus River that destroyed ships in the port of Lisbon. The 1531 Lisbon earthquake occurred in the Kingdom of Portugal on the morning of 26 January 1531, between 4 and 5 o'clock. The earthquake and subsequent tsunami resulted in approximately 30,000 deaths. Despite its severity, the disaster was not widely documented until the rediscovery of contemporary records in the early 20th century.

On December 31st 1531 CE Fuego in Guatemala erupted with a VEI of 2.

In England the Isle of Thanet as well as the English coasts of Essex and Kent were also inundated in 1531.

On 2nd November 1532, villages were lost about the Scheldt River in the Netherlands. All coasts from Flanders in Belgium to Hamburg in North Germany were hit by sea floods. These sea floods were noted as being worse than the 1530 sea floods. There was a probable cosmic impact on the English coastline. This was the All Saints Flood. Several towns disappeared.

On November 15th 1532 Cotopaxi in Ecuador erupted with a VEI of 4.

On July 17th 1533 CE Asosan erupted in Japan with a VEI of 2.

In 1533, several Atlantic hurricanes struck Puerto Rico. "So many slaves were killed in the destruction of the twenty-sixth of July, August 23 and August 31".

In October 1533 CE Cotopaxi in Ecuador erupted with a VEI of 2.

In winter 1534 to 1535 frost lasted from November to February. The Thames was frozen below Gravesend which means that it was possibly also frozen upriver from this point. The river below Gravesend is at the head of the Thames Estuary so perhaps was only icy along the shoreline rather than being completely frozen all the way across. There was a great frost in England. It lasted from November through February. Goods were carried by land across Kent and Essex to London.

In 1534, there was a great flood in Poland. The flood began on 26 April. The winter produced very heavy snowfalls. In the spring, this was followed by heavy rainstorms that lasted for days, continuously day and night. On 8 May and three days after, heavy rain began falling in Poland and it seemed like there was so much water that it would cover the earth to the mountains. So, the greatest flood in Polish history began. The oldest inhabitants could not remember such a flood. Fields, lakes, gardens, houses and forests were all underwater. In Kraków, Poland, the water moved entire houses to the riverbanks.

On June 4th 1534 Cotopaxi in Ecuador erupted with a VEI of 4.

The great bridge between Kraków and Kazimierz was damaged and broken into three sections and then finally taken away by the waves to the suburb Grzegotki. The waters of the Vistula River finally began to recede and the river was back in its banks on 22 July. In the city of Kazimierz, the water destroyed several houses between the Monastery of Saint Catherine and Skalka. Kazimierz was a historical district of Kraków. The Dunajec River near the cities Nowy Targ and Sandomierz destroyed all the stonewalls, churches and windmills. Sandomierz is located in southeastern Poland. Nowy Targ is located in southern Poland. In the village of Szramowice, the water destroyed the church and seven houses. Szramowice is now Sromowce Niżne and is located in southern Poland. In the village of Trzemeśny, a church and many small lakes were destroyed. Trzemeśny is now Trzemeśnia and is located in southern Poland On 29 June, the San River flooded causing great destruction. Some villages were completely destroyed along with all their cattle, wheat, hay, windmills and wooden houses. On 15 August and the week after, rain again fell in Poland. By 24 August, the flooding was so great that the Vistula River again left its banks.

In 1535 CE Galeras Volcano in Colombia erupted with a VEI of 3.

In March 1535 CE Miyakejima erupted in Japan with a VEI of 2.

On March 22nd 1836 CE Mount Etna erupted in Italy with a VEI of 3.

The year 1536 produced many storms and tempests. Twenty-four ships were destroyed by one of these hurricanes on the coast of Provence, France.

In winter 1536 to 1537 CE in December and January the Thames was frozen in London. King Henry VIII rode on the ice-bound river from London, probably Whitehall, to Greenwich with his then wife Jane Seymour.

In March 1537 CE Mount Etna in Italy erupted with a VEI of 2.

In 1537 CE there was a wet summer in England. In 1537 in England, the summer was exceeding rainy. In December and January there was a great frost and the River Thames was frozen over.

In 1537, an Atlantic hurricane struck Puerto Rico and many slaves were drowned.

In 1537, a hurricane struck northwest Cuba and 2 ships were lost.

Eruptions in 1538 CE were Gamalama in Indonesia with a VEI of 3, Guagua Pichincha in Ecuador with a VEI of 3 and Villarica in Chile with a VEI of 3.

Between 1538 and 1541 CE the four years experienced drought with 1540 and 1541 particularly dry. In both of these latter years the Thames was so low that sea water extended above London Bridge, even at ebb tide in 1541. There were three successive fine and warm summers from 1538 to 1540. The weather in 1540 was so fine that picking cherries commenced before the end of May and grapes were ripe in July.

The summer of 1538 was scorching hot in Italy. The rivers dried up, and the air was filled with fiery meteors, so people felt earthquakes. Were these from the impacts? In the Kingdom of Naples, Italy, the Tyrrhenian Sea floor in a region of about eight miles (13 kilometers) was drained. In 1538, the sea by the Kingdom of Naples on the southern part of the Italian peninsula, was dry for eight miles together. What caused a sudden removal of the sea? And did it eventually return?

In Dijon, France, the grape harvest took place on 20th September.

On September 29th 1538 CE Campi Flegrei in Italy erupted with a VEI of 3.

In 1540 CE Aniakchak in Alaska erupted with a VEI of 4. Also, in 1540 Augustine in Alaska erupted with a VEI of 4.

In 1540 CE there was general warmth over Europe during the spring and summer. In England it was a hot summer with great heat and drought. There were also many deaths due to the Ague.

In 1540 there was so little water flowing in the Seine in Paris, France, that people were able to walk across.

In England, there was great heat and drought. After a calamitous year, there was an exceedingly early spring, which produced fine weather. The heat lasted from February through 19 September. During this time, it rained only 6 times and there was an early harvest in England. Cherries were ripe by the end of May. Grapes were ripe in July. The middle of the harvest was on 25 June. This year was remarkable for the abundance of corn (grain) and fruit. In England in the summer of 1540, there was an excessive drought. Wells, brooks and rivers were dried up. The River Thames was so low that the salt water flowed above London Bridge. During the end of summer there came a great mortality over the whole Nation because of an epidemic of pestilential ague and blood flux. But in other places, it was the hottest and healthiest year in the memory of man.

In Europe in 1540, the summer was much hotter and drier than in a large number of preceding years.

The drought in summer in Germany was so great that they suffered from a lack of the necessaries of life such as food and drink.

In Belgium, the grain and grape harvest was over at the beginning of August.

But in Dijon, France the grape harvest did not occur until 4 October. The price of corn (grain) in France went down to half. The summer of 1540 was unusually hot in France. the summer in France produced fine weather and heat that lasted from the month of February to the 19th of September. During all this period, it rained all but 6 times. At the end of May ripe cherries were eaten, and grapes in July; the 25th of June was the midst of (grain) harvests; and at the beginning of September, the vintage was at it height.

In Italy, after a drought of five months, a deadly heat wave occurred and the forests erupted in flames. In 1540 it did not rain over Italy with Rome dry for nine months. There were forest and city fires with many people dying of heat stroke and heart failure. The next warm summer of equal value was in 2003.

The glaciers of the Alps melted.

In 1540, the island of Sardinia in the Mediterranean Sea off the west-central coast of Italy,) was desolated by a famine.

From 1540 to 1543, there was a general famine in the Sind in what is now Pakistan.

In 1540, there was a famine in the Bombay Presidency in India.

In 1540 during the 7th moon in the vicinity of Shanghai, China, there was a roaring of the sea; a northeast wind; several myriads, a very great number, were drowned; it was a year of dearth; men and crops perished.

In Holland the year 1540 was called the "Big Sun Year". The lower part of the Rhine River from Cologne into the Netherlands was dry.

An expedition to Greenland in 1540 reported that the old Norse colony had died out.

In Central Europe there was a progressive cooling of the winters from the 1540s to the end of the century, which was repeated after the recovery to the 1620s and culminated in the very cold 1690s. Another recovery followed, but the winters of the 1750s to 1780s were on average cold.

In April 1541 CE Reventador erupted in Ecuador with a VEI of 3.

In July 1541 CE Mount Etna erupted in Italy with a VEI of 2.

In 1541 CE in France, it was extraordinarily hot.

In 1541, there was a remarkable drought in England. At Nottingham, almost all the small rivers dried up, and the River Trent was diminished to a staggering brook. The River Thames was so low that seawater, even at ebb, extended beyond London Bridge. Many cattle died for want of water, especially at Nottinghamshire, and many thousands of people died from grievous diarrhea and dysentery. It must have been quite extreme given that the previous year was notably dry. Cattle and other livestock were dying for lack of water and dysentery killed thousands of people.

In 1528, Cristóbal Guerra founded Nueva Cádiz on the island of Cubagua, the first Spanish settlement in Venezuela. Nueva Cádiz, with a population of 1,000 to 1,500, may have been destroyed by an earthquake followed by tsunami on 25 December 1541; it could also have been a major hurricane.

On April 29th 1542 CE Asosan in Japan erupted with a VEI of 2.

The summer of 1542 CE was wet in England after the drought of the previous two years.

In 1543 CE a great flood occurred on the Mississippi River in the United States on April 20. The river was 40 leagues (120 miles) wide. Explorer Hernando Desoto encountered a flood on the river near Memphis, Tennessee that extended over 40 days and likely extended to the lower reaches of the river. This was chronicled by Garcilaso de la Vega.

In England, there was a famine in 1543.

In France on September 6, there was a great flood, greater than any known in the memory of man (except for the Great Flood of Noah). The flood affected towns, cities and countryside and did incalculable damage to Vivarais and Dauphine. Vivarais is now called Ardèche and is a department in south-central France. Dauphine was a former province in southeastern France.)

The winter of 1543 to 1544 was cold in Western Europe and parts of Britain more likely. During the winter of 1543-44 in France, the cold froze the hogsheads of wine. It had to be cut with an axe. A hogsheads is a large barrels/casks that hold liquid measures, usually alcoholic, approximately 239 litres (63 gallons).) In France, the weather was so cold that the Provence wines ordained for the Army were frozen and cut with hatchets and carried away by the soldiers in baskets.

In England, wood, flesh (meat), and fish were very dear this winter. This is because the last summer was intemperate and rainy causing a great death of cattle. This winter the plague was in London, and the Terms were adjourned. It was a most rigorous frosty winter. In 1544, there was a plague at Canterbury, England.

In 1544, it was so cold in Paris, France, that wine froze, and it was sold in pieces by the pound.

In 1544, the cold was so severe in the Netherlands that wine was cut in blocks and sold by weight. In 1544, "The cold was so extraordinary that the wine froze in the casks. It had to be smashed with axes and was sold in pieces by the pound."

The winter of 1544 was remarkably severe all over Europe. In Flanders wine froze in casks and was sold in blocks by the pound weight. In Flanders, the wine in casks froze into solid lumps.

On July 28th 1544 CE Hiuchigarake in Japan erupted with a VEI of 2.

In southern France in November 1544 CE, the Rhône River produced flood disasters affecting Avignon, Arles, and Tarascon. In 1544, the Rhône River in southern France overflowed its banks on November 11. The waters knocked down a quarter mile (390 meters) of ramparts, city walls, of Avignon which is now part of southeastern France and the floodwaters covered the plain for eight days.

In June 1545 CE there was a fall of fist-sized hail in Lancashire.

On 25 July 1545, a tempest and hurricane struck Derbyshire, England.

On 20 August 1545, an Atlantic hurricane struck the Dominican Republic killing a large number of people.

The winter of 1545 to 1546 CE was cold in Western Europe and possibly parts of Britain.

A weather diary from Zurich in Switzerland recording data from 1546 CE to 1576 CE showed that the relative frequency of snow among the snowy and rainy days of winter was up 44 per cent up to 1563 CE and up 63 per cent from 1564 CE onwards to 1576 CE.

A report by Sebastian Munster, on a horseback journey, wrote that from the upper Rhone River Valley over the Furka Pass in Switzerland in August 1546, the Rhone where it emerged from the glacier front was about 200 meters across and ten to fifteen meters high. This was much further forwards than in the late twentieth century. The Rhone glacier had reached the broad valley bottom at the foot of the steep ascent to the Furka Pass but by no means achieving the size it had in the 18th Century.

On March 4th 1547 Hakusan in Japan erupted with a VEI of 3.

In 1547 CE there was intense frost at the end of the year in London and the south.

In 1548 Kelud in Indonesia erupted with a VEI of 3 and Merapi with a VEI of 3 also erupted in Indonesia.

In 1548 CE in Northern Europe, oxen drawn sledges traveled on the frozen sea from Rostock, Germany to Denmark.

In 1548, the winter was so severe that all the rivers in southern France froze. On 12, 13 and 14 November, the Rhône River flooded again in Avignon and Arles. The winter of 1548 was very severe all over France and all the rivers were frozen so that they could carry the weight of the heaviest wagon on the ice. The year 1548 was very rainy and accompanied by great floods in France.

The cold reigned from 1548 throughout Europe. The ice on most rivers in Europe was thick enough to bear heavy-laden wagons.

On 4 August, there was a great hailstorm at Louvain. There was another hailstorm on 5 September. Louvain is in central Belgium and on the 5th of September, there was another great hailstorm.

In England, it rained all summer in 1548.

1550 CE

In 1550 CE Michinmahuida in Chile had a VEI of 4, Katla in Iceland erupted with a VEI of 4, Witori in Papua New Guinea with a VEI of 4, and Maly Semyachik in Kamchatka with a VEI of 4. Other eruptions were Sheveluch with a VEI of 3 in Kamchatka, Oku Volcanic Field in Cameroon with a VEI of 3, Vulcano with a VEI of 3 in Italy, Kie Besi in Indonesia with a VEI of 3.

What sort of year was this? Another dating anomaly? A big year for volcanic eruptions though even if spread 25 years on either side of 1550.

In 1550, an Atlantic hurricane struck off the Florida Keys in the United States. A Spanish ship Vitacion, 200 ton, was lost during a hurricane near Havana, Cuba.

Winter temperatures between 1550 to 1590 in Europe were 1.3° colder than at the beginning of the century.

In 1550, there was a very great dearth in England. Wheat was sold for 16s. per bushel, which had sold at 10d. a little before.

Dukono in Indonesia erupted on November 20th 1550 CE with a VEI of 3.

On 18 December 1550 CE in England, "The Thames flowed thrice in nine hours."

In Scotland, great rivers in the middle of winter were dry, and in the summer so greatly flooded which carried downriver and drowned several villages and many feeding cattle from their pastures into the sea. Several whales came up the River Forth. There were many hailstorms with hailstones the size of pigeon eggs. The hailstones destroyed the corn (grain). In 1550, there was a general universal dearth and "sweating sickness" again reappeared in England. At Chester and York, it was accompanied by so great a dearth that wheat was 15 shillings per bushel.

Sea ice increased beyond its previous limits off Greenland and expeditions commonly had to make a wide sweep south of Cape Farewell to get round the ice to reach the west coast of Greenland. Conditions then deteriorated in Iceland too and the glaciers and ice sheets advanced.

Between 1550 CE and 1620 CE in Baden in southwest Germany the good wine years were rather under half of the frequency between 1480 CE and 1550.

The warm summers of the 1550s were dry in Switzerland followed by half a century of a predominance of wet summers, noticeably so from 1570s to the 1620s inclusive.

A one-hundred-year drought started in 1550 CE in Uganda. This had happened before in 1100 CE and would occur again in 1750.

In 1551, an Atlantic hurricane struck in the Gulf of Honduras and a ship with many persons were all drowned.

In December 1551 CE there was tidal flooding in the Thames as far up as Millwall. This would imply some sort of storm surge event.

There were Dutch coastal floods with a date of 19th of December 1551 being one of three significant storms floods to affect the Low Countries that winter.

In the Netherlands on 12 January there was a great flood. January and February 1552 CE were part of a notably stormy season. Two storms and floods were noted in the Dutch coastal communities in the months with the first being on 13th January and the second on 15th February.

During the winter of 1551-1552 on 13 January 1552, the sea broke in at Sandwich in Kent, England, and overflowed all the marshes thereabout, and drowned much cattle. On 12 & 13 January, there was a tempest of rain, snow, hail, rain, thunder and lightning. In 1552, there was a remarkable fall of red rain in England. "A heavy fall of rain, which lay on the grass as red as wine." In England in 1552 there was a drought in London and southwards.

In 1552 CE there were periods of major cyclonic storms and associated sea flooding affecting much of the North Sea. These sea floods were on the 15th and 18th to 25th of January.

The winter of 1552 was warm and dry in Italy.

On Friday 17 May 1552 between four and five clock in the afternoon, there was a particularly bad storm in Dordrecht, western Netherlands, which made the people flee in terror to their homes. For more than half an hour, giant hailstones fell. All the gardens were destroyed. Some hailstones weighed half a pound. On some stones "horns" and "crowns" were seen.

In Italy, the summer of 1552 was dry and burning hot and the drought lasted for five consecutive years.

The dry heat of 1552 in northern France consumed all the plants in June. In France, the harvest began at Dijon on 13 September.

In Budissina (Bautzen in eastern Germany) on 13 August, there was a great flood.

On October 7th 1552 CE Izu-Oshima erupted in Japan with a VEI of 3.

In 1553 Merapi in Indonesia erupted with a VEI of 3.

In 1553, a hurricane struck Texas in the United States. Sixteen ships of the New Spanish Fleet were struck by the hurricane and never heard from again.

In 1553, a hurricane struck the west coast of Florida in the United States causing approximately 700 deaths.

In 1554 in May Hakusan erupted in Japan with a VEI of 3.

Also, in May 1554 CE Hekla erupted in Iceland with a VEI of 2.

In November 1554, a hurricane struck Cuba. The admiral's ship was sunk. A small caravel was also sunk with all but two people drowning.

In 1554, an Atlantic hurricane struck the Mona Passage which is a strait that separates the islands of Hispaniola and Puerto Rico. A Spanish ship was wrecked by this hurricane.

In 1554, a hurricane struck offshore the southern coast of Texas in the United States. "Three ships from the New Spain fleet, the Santa Maria de Yciar, Espiritu Santo, and the San Esteban, were lost in a storm off what would later become Padre Island, Texas. A few survivors managed to escape in a small boat."

Around 1554, an Atlantic hurricane struck Bermuda. Of the other ships that had sailed with them (the San Miguel), weeks before, the leading ship from Veracruz sank in that hurricane that caught up with them in Bermuda. Twenty-five survived."

In London, England on the 1st of September 1555 CE, there was a great hailstorm and on 21 September, there were great floods on the River Thames.

On 30 September 1555, there was a notable inundation of the River Thames in London, England. It was caused by a great wind and rain.

In England, the year 1555 was a very wet, rainy, floody year. There was great scarcity. A fatal hot burning fever took hold in England. On the bare rocks on the seaside of Suffolk, grew of their own accord, a plentiful crop of peasons which is a type of bean. These were ripe in August. They grew where grass never grew before. These greatly relieved the poor, who carried them away in great quantities. As they gathered them, still more were coming on and others in blossom.

A plague infested Loughborough in Leicestershire, England beginning in 1555. It infested the city from Midsummer of 1555 to Midsummer of 1559. The plague went by several names including: Swat, New Acquaintance, Stoupe, Knave, and Know thy Master.

The summers of 1555 and 1556 in England were worse than the summers of 1550, 1551 and 1554 in which the harvests were mediocre or worse. In Colyton near Exeter in southwest England there was a decline in the population from 1550 onwards. There was a decline in the 1550s when deaths exceeded births for a number of years.

From the wetness and coldness of the season in 1555, several types of epidemics appeared in Paris, France.

In Western Europe the year 1555 was mostly excessively rainy.

In 1555, an Atlantic hurricane struck the Bahama Channel. The ship Capitana of the New Spanish Fleet was lost in the storm. The Old Bahama Channel is a strait off the northern coast of Cuba and the Sabana-Camagüey Archipelago and south of the Great Bahama Bank. It is approximately 100 miles (161 kilometers) long and 15 miles (24 kilometers) wide.

In September and November 1555 CE there were three Autumn storm floods affecting the Netherlands and possibly the English side of the North Sea as well.

The 1556 Shaanxi earthquake, known in Chinese colloquially by its regnal year as the Jiajing Great Earthquake or officially by its epicenter as the Hua County Earthquake, occurred in the early morning of 23 January 1556 in Huaxian, Shaanxi during the Ming dynasty. Most of the residents there lived in yaodongs—artificial caves in loess cliffs—which collapsed and buried alive those sleeping inside. Modern estimates put the direct deaths from the earthquake at over 100,000, while over 700,000 migrated away or died from famine and plagues, which summed up to a total loss of 830,000 people in Imperial records. It was the deadliest recorded earthquake in history, and in turn one of the deadliest natural disasters in Chinese history. More than 97 counties in the provinces of Shanxi, Henan, Gansu, Hebei, Shandong, Hubei, Hunan, Jiangsu, and Anhui were affected. Buildings were damaged slightly in the cities of Beijing, Chengdu, and Shanghai. An 840-kilometre-wide (520 miles) area was destroyed, and in some counties as much as 60% of the population was killed.

In 1556 CE Lyon, France experienced a drought. The rain stopped for four and a half months from March 26 until August The summer of 1556 was hot in southern France. In France the sources of water dried up. The grape harvest took place in Dijon, France on 5 September. Corn (grain) was in short supply this year.

In 1556 in Italy, the heat was excessive.

In Western Europe in 1556, there was a great famine not caused by weather. After a great scarcity of corn (grain), not from famine but because the rich corn-mongers had bought and hoarded it up, until it spoiled. This forced the poor to eat ox and swine dung.

In the Netherlands, a sudden and terrible plague broke out between Delph (Delft) and The Hague, which spread over the whole country in June.

In England in 1556 and 1557, there was a great scarcity of corn (grain) from the past great rains. All the corn was choaked and blasted, the harvest was excessively wet and rainy. Wheat sold for 55s. per Quarter (quarter ton), but a good and plentiful crop at harvest brought it to 4s. or 5s. In England from 1556 to 1558, there was a famine from great rains, bad and inconstant seasons, heat and long south winds. In 1556, there was a drought in England. The drought was so great that the springs dried up and wheat rose from 8 shillings to 53 shillings per quarter (quarter ton). At Chester, England the wheat was 16 shillings per bushel and barley was 12 shillings and very dear (scarce). In 1556 at Nottingham, England and throughout England, there was a plague and a dreadful famine.

In 1556, there was a famine in the Delhi District in India. During the first year of the reign of Akbar, there was a great scarcity in Hindustan. "In some districts, and especially in the provinces of Delhi, it reached a most alarming height. Though men could find money, they could not get sight of corn (grain). People were driven to the extremity of eating one another, and some formed themselves into parties to carry off the corpses for their food."

In 1558 CE Stromboli in Italy erupted with a VEI of 3.

In 1558, there were many tempests. On 9 January, a tempest struck Dover, England.

In London in 1558 CE there was a very hot summer in London and the south and in March 1558, there was a most destructive hurricane in England. In England, there was drought the whole year and hot. In 1558 in England, the River Thames went dry. Now there was another great scarcity of corn (grain) from want of workers to harvest it. Corn sold for 14s. per Quarter (quarter ton). In England, there was a cold winter with a north wind, a southerly rainy spring, and an excessively hot summer.

On 18 June, a tempest struck Calais, France and lasted for 5 days.

At harvest time, dysenteries broke out in France and Paris.

In Europe, all the spring, summer and at harvest time was hot and dry.

On 7 July 1558, there was a great thunderstorm (and possible tornado) within a mile of Nottingham, England. As it passed through two towns, it beat down all the houses and churches. The bells were cast to the outside of the churchyards, and some of the webs of lead were thrown 400 feet (122 meters) into the field like writhen leather. The water with the mud in the bottom was taken from the River Trent and carried a quarter mile (0.4 kilometers) and cast against trees. The trees were torn up by the roots and cast 12 score yards (720 feet, 220 meters) off. A child was taken from a man's hand and carried 2 spears length high and then let fall 200 feet (61 meters) off. The child died from the fall. Five or six men were slain during the storm during which hailstones fell measuring 15 inches (38 centimeters) in circumference. The hamlets of Sneinton and Gedling had all its houses and both their churches blown down.

On 15 July, a tempest struck France.

In July 1558, a storm of hail in Northamptonshire, England, when the stones measured 15 inches in circumference.

In 1558, there was a great mortality in England. At Nottingham much corn (grain) was lost in the fields for want of laborers. Many churches were shut up because the clergy were dead. East and West Retford suffered severely. In the small hamlet of West Retford between July and October, 82 people died.

In 1558, the spring, summer and fall were hot and dry in a large part of Europe. The grape harvest in Dijon, France began on 30 September.

On 20 August 1559, a hurricane struck offshore western Florida in the United States causing greater than 500 fatalities.

In 1559, a hurricane struck off the northwest Florida coast of the United States. Six Spanish ships were lost in the hurricane. On 19 September 1559, a hurricane struck off the coast of Florida in the United States. There was a great loss of life by a tempest from the north. There was a great loss of seamen, and passengers (less than 1500).

In 1560 Irazu in Costa Rica erupted with a VEI of 3.

Winter temperatures from 1560 CE in Central Europe were 1.3° C lower than in 1880 CE to 1930 CE or the first half of the sixteenth century.

On December 31st 1561 CE Gamalama in Indonesia erupted with a VEI of 2.

In March 1562 CE Asosan erupted in Mexico with a VEI of 2.

On September 21st 1562 CE Pico erupted in the Azores with a VEI of 2.

In the winter of 1562-1563, the River Scheldt froze near Antwerp, Belgium and the River Thames froze in London, England.

The winter of 1562-63 was severe in Flanders. In Antwerp, Belgium it began to freeze in mid-December. From the Feast of St. Stephen (26 December) to 5 January there was ice on the River Scheldt).

In London, England, it began on 21 December to freeze with such vehemence that by 1 January, people were crossing the River Thames on the ice; and the people amused themselves on the ice as on the mainland; but the frost was short-lived. The thaw began on 2 January and on the 5th there was no ice on the river to be seen.

During the night of 9 January 1563, a great tempest of wind and thunder at Leicester, England, which did great damage. In 1563, there were several great tempest of wind, thunder and lightning that struck England. These occurred on 9 January, in July, and on December 1-12.

On May 3rd 1563 Asosan erupted in Japan with a VEI of 2.

On 28th June 1563 and 26th July 1563 Agua de Pau on Sao Miguel Island in the Azores erupted. It had a VEI of 5. This as we know is volcanic winter material.

In January 1563 CE a tornado with possible wind speeds of 170 miles per hour hit Leicester in Leicestershire and caused considerable damage.

In 1563, a hurricane struck near Cape Canaveral, Florida in the United States causing 284 deaths.

From August to October 1563, five vessels went missing at the latitude of Bermuda in the Atlantic Ocean. This loss was likely attributed to storms/hurricanes.

On the 20th September 1564 CE there was a tidal flood in the Thames. This would imply a storm surge event, and quite a severe storm no doubt for this early in autumn.

In Liège, Belgium the frost lasted from 14 November 1564 to April 1565. On the river goods were sold from small stalls on the ice.

In 1564 CE on December 31st Gamonkara erupted in Indonesia with a VEI of 3 and also in December 1564 CE Asosan erupted in Japan with a VEI of 2.

In England, on 1 December 1564 began a frost, which froze the River Thames so as people went over on the ice. Boys and men played on it.

In England, from the 1st to the 12th of December 1563, there was greater thunder and lightning than any alive remembered. In London, England, famine and pestilence were said to have taken off 20,000 people.

Severe prolonged frost set in on the 7th December 1564. The court of Elizabeth II indulged in sports on the ice in the Thames River at Westminster. Football and other games were played on the ice.

In Paris, France, the winter lasted from 20 December 1564 to 24 March 1565.

In the Winter of 1564-1565 CE in England on the 21st of December, began a frost, which continued so extremely that on New Year's Eve people went over and along the Thames on the ice from London Bridge to Westminster. Some played football as boldly there as if it had been on the dry land; diverse of the court shot daily at pricks set up on the Thames; and the people, both men and women, went on the Thames in greater numbers than in any street of the city of London.

In 1564, there was a great frost in London, England at Christmas. "Surface of river Thames solid as a rock. The population left the streets to walk the whole distance from Westminster to London Bridge on the ice and the Queen Elizabeth was daily on the river. The frost broke up suddenly into fearful inundations, bearing down houses, bridges, and vessels, to destruction. At Chester the river Dee frozen over, so that people played football thereon."

In 1565 CE in August Pacaya erupted in Guatemala with a VEI of 3.

On 3 January 1565 it thawed and resulted in a very great flood. The thaw set in circa the 3rd January accompanied by a notable Thames flood. A notably unhealthy fog followed this thaw. The winter of 1564 to 1565 was notably severe as regards depths of cold and was amongst the top ten per cent of bitterly cold winters in the millennium. The Thames froze 20 to 22 times between 1564-65 and 1813-14.

On the 31st day of January 1565, at night, it began to thaw, and on the fifth day was no ice to be seen between London Bridge and Lambeth, which sudden thaw caused great floods and high waters, that bare down bridges and houses, and drowned many people.

On 2 February, there was a flood at Louvain, Leuven, in Flemish Brabant in what is now Belgium; wind blew the sea in. There was a great inundation from the sea.

In 1565, the Rhône River froze across its full width in Arles, France. In 1565 in France, the river remained frozen allowing carts to cross over the ice for two whole months. The cold killed the olive trees. In 1565, the Seine River in France froze since the start of January. Loaded carts travel on the ice on the River Meuse.

In 1565 in the Netherlands, the Scheldt River froze so as to bear laden wagons.

On 22 September 1565, an Atlantic hurricane struck Florida and the eastern coast of the United States. Many French vessels were lost at sea during the storm. Of the original 600 soldiers and sailors, 529 survivors could be accounted for.

On 24th December 1565 CE there was a Thames flood in London that was possible tidal and therefore storm surge related. On 16 July 1565 from nine at night to three in the morning, there was a great tempest of thunder, lightning and hail, in many place in England. In the morning of 24 December, there was a hurricane and west winds. The River Thames and the sea were both blown in with great damage done by both.

On December 31st 1565 CE Gamkonara in Indonesia erupted with a VEI of 3.

The winter of 1565-1566 CE in Russia saw widespread deaths and unburied corpses of people and animals. From the fifteenth century to the seventeenth century large areas of Russia were depopulated when the people left for other regions due to the climate changing.

On May 6th 1566 CE Kirishimayama erupted in Japan with a VEI of 3.

In summer and early autumn 1566 there was drought all summer and a harvest tide in London and south of it. In 1566 in England, the spring had great and almost continuous rains, with most frightful floods. The summer and harvest were droughty and clear. There was not one drop of rain the whole harvest.

In 1566 Guagua Pichincha in Ecuador erupted on October 17th with a VEI of 3 and Kirishimayama in Japan erupted again on October 31st with a VEI of 3.

On November 1st 1566 CE Mount Etna erupted in Italy with a VEI of 2.

At Commora, Comora or Komárom in northern Hungary on the border with Slovakia, broke out the Hungarian Fever, typhus, in the Emperor Maximilian's Army, just before he broke up the campaign against the Turks. The excessive spring rains had made them two months later in taking the field. It increased at Gewer and when his soldiers were disbanded, they carried the contagion along with them and dispersed it all over Europe, especially over Germany, Burgundy, Italy, Bohemia, now western Czechia, and Flanders; but chiefly in Vienna, Austria, through which most of them passed in their return home. They infected all houses there where they laid, and died so fast themselves, that the streets were covered with dead bodies. This increased the infection.

The rains of 1566 were stormy in southern France.

On 26 December 1566 in England, there was a great hailstorm.

In 1566 CE. Dithmarschen on the North Sea coast of Schleswig-Holstein in Germany was inundated by a sea floods.

There was a severe winter in 1567 CE in January. This was followed by a dry summer in England.

In 1567, off the coast of Dominica, six ships carrying 3 million pesos were wrecked in a storm (hurricane). The island natives killed all the survivors.

In 1567, a tempest and hurricane struck Paris, France.

Between 1568 and 1575 there was a massive famine in Poland, Lithuania and Russia.

In 1568 CE Savo erupted in the Solomon Islands with a VEI of 3.

In 1568 CE in England all spring there were continual rains. In London and the south of England there was an exceedingly hot drought. The summer was excessively hot, with a dearth of cattle.

In Rome, Italy, this year were such floods of the Tiber River, that they carried off and washed away, even to the foundation, a great part of the city, leaving very little behind. Besides the inestimable loss of the city and its great riches, and of innumerable cattle, 1,500 people drowned in it.

On the 18th of March 1568 there was a most dismal and destructive hurricane in England and the Netherlands. On 1 November 1568 the sea swelled excessively overflowing some banks, and broke down others, by a prodigious and unheard of deluge. It covered some islands of Zealand, a great part of the seacoast of Holland, and almost all Friesland. It was a foot higher than the deluge of 1528 which swallowed up 72 villages. Here was an incomparable loss of estates, but especially of men. In Friesland alone 20,000 people were drowned. Their bodies with the carcasses of cattle, household goods, etc. floated all over the fields, and sea being indistinguishable. People that had climbed to the tops of high hills and trees, when the flood just started were saved by boats.

In Italy in 1568 it was excessively hot and moist with a south wind.

From the 11th to the 21st of December 1568, the Rhône River in Western Europe was passable on the ice. The ice was sufficiently thick to support the weight of carts. On 11 December 1568, the carts crossed the Rhône River in southern France and the ice held until the 21st. It was reported that the winter in 1568 in Châtellerault in western France, was remarkable because of the snow and ice.

On 19 December 1568, unusually cold weather forced the Duke of Anjou, to break off his siege of Loudon in western France.

In England from 1568-1573, the years were mostly rainy, terrible, and tempestuous bad weather.

In 1569 CE in England and France, there were great floods.

On 13 January 1569, there was a great flood at Louvain, France with lightning.

In England on 13 July, there was a terrible storm of thunder and lightning with hail. In England from 1568-1573, the years were mostly rainy, terrible, and with tempestuous bad weather.

On 30 October 1569, a formidable hurricane (tornado) struck Ashley in Northamptonshire, England. It was 60 yards in breadth and spent itself in about 7 minutes. It was first observed assaulting a milkmaid, taking her pail and hat from off her head, and carrying her pail many scores of yards from her, where it lay undiscovered some days. Next it struck a yard in Westthorp (Westhorpe?) where it blew a wagon body off its axle, breaking the wheels and axle in pieces and blowing three of the wheels, so shattered, over a wall. Another wagon was driven into the side of a house. It blew the roof off the parsonage house.

In Italy in 1569 there was a great dearth of corn (grain) from excessive rains and mildew.

In 1570 CE Cayambe erupted in Ecuador with a VEI of 4 and Thera/Santorini erupted with a VEI of 3 in the Mediterranean.

In October and November 1570 CE a tidal flood affected the Thames estuary as far upriver as far as Erith. The tidal flood extended from the Humber to the Straits of Dover. This high tide was associated with severe gales and aggravated by heavy rainfall.

The 1570 Ferrara earthquake struck the Italian city of Ferrara on November 16 and 17, 1570. After the initial shocks, a sequence of aftershocks continued for four years, with over 2,000 in the period from November 1570 to February 1571. The same area was struck, centuries later, by another major earthquake of comparable intensity. The disaster destroyed half the city, permanently marked many of the buildings left standing, and directly contributed to – but was not the sole cause of – a long-term decline of the city lasting until the 19th century. The earthquake caused the first documented episode of soil liquefaction in the Po Valley, and one of the oldest occurrences of the event known outside of paleoseismology. Earthquake lights were seen above the city on November 15, 1570, the night before the first quake. Flames were reported to come out from the soil and rise into the air, probably small pockets of natural gas set free by cracks in the earth crust. The earthquake struck at dawn: three strong shocks hit the city on the first day; one – the strongest – the day after. The first strong shock struck at 9.30 (local time) November 16, 1570, its epicenter just a few kilometers under the city center. Six hundred pieces of stone masonry (mostly battlements, balconies and chimneys) are reported to have fallen, further damaging the flimsy stone and hay roofs. The following day the ground trembled again many times. At 8 pm a new powerful shock caused severe damage to walls and caused some buildings to sustain structural damage. Just four hours later, a new tremor caused new cracks and some collapse. At 3 am on November 17 the ground shook harder than ever; many buildings, damaged by the previous shocks, gave way and caved in. Many churches' facades, often built as self-standing walls rising well over the effective architecture, collapsed, including at the Duomo. Forty percent of the city buildings were damaged, including almost every public building. Some of them collapsed, and many churches sustained critical damage to pillars and main walls. Observers reported that the shallow bowl-shaped valley where Ferrara lies seemed to rise into a kind of hump, before coming back to its original profile. Damage to the city were assessed in over 300,000 scudi, a huge sum at the time. The event was a surprise to many scholars, since according to the then mainstream theory of natural philosophy, earthquakes were not meant to strike in winter or on flat land. Minor earthquakes had struck Ferrara in the past (events were recorded in 1222, 1504, 1511 and 1561, some of them causing little damage, and a stronger event in 1346). The exceptional length of the seismic swarm, unprecedented at the time in Ferrara, led some to believe it was a supernatural phenomenon. The earthquake's intensity has been assessed as VIII on the Mercalli intensity scale: only the 1346 event was similar in intensity, though minor urbanization led to less evident damage (but more victims), the other have been all marked as class VII or VI. Other seismic events would hit the city in 1695, 1787 (three shocks in ten days) and 1796.

On 10th March 1570, a great storm from the north generated high tides in England and also dropped significant snowfall and cold temperatures. "Septentrionis maxima Sævitia: Nivis flocci magni, ingens frigus. Maxime tumefcebat æstus Maris die & nocte, nam excurrebat in Agros late". (translation from Latin as written in Middle English: The greatest fury from the North: A great mass of snow and growing coldness. Very greatly swell the tide day and night, for they (tide water) were running widely in the fields.

The entire year of 1570 in northern France produced a suffocating heat. The rains of 1570 caused the rivers to overflow their banks in several parts of the kingdom of France.

Between September 1570 and January 1571, four ships were lost in a storm, hurricane, between Veracruz, Mexico and Spain.

In England on the 5th of October, there was an inundation from the sea. On the night of 5 October 1570, a great hurricane struck England. Near Rye in East Sussex in southern England, the sea broke in with a great flood and drowned a great marsh with herds of sheep, corn, etc. In Essex, Suffolk, and Norfolk were great losses. One by a tempest, wherein sheep, corn, cattle, houses, bridges, etc. were lost and carried down. On 5 and 15 October, there were tempest, hurricanes and rain in England. There was a dreadful hurricane that destroyed the Port in Liverpool, England.

Between 31st October and 2nd of November 1570 the great cities of the Netherlands were flooded by sea floods. Possibly as many as 400,000 people drowned. The flood extended from northern France to northwest Germany. This was the "All Saints" flood which was the greatest North Sea Storm flood after the 11th October 1250 event. There were coastline changes and cities were drowned on the continent. Many people were killed. It flooded the entire coast of the Netherlands and East Frisia and the effects were felt from Calais in Flanders to Jutland in Denmark and even Norway. Up to 25,000 deaths might have occurred. The Island of Wulpen in the North Sea was permanently lost to the sea. The flooding reached as far as the Alte Land on the Elbe River, the Vierlande near Hamburg in Germany and as far as Eiderstedt. Between the Ems and Weser Rivers around ten thousand people died, and entire villages disappeared. Tens of thousands were homeless.

On 1st November in the Netherlands there was an inundation of the sea. A strong northwest wind occurring during the high tides drove the sea with such violence against the dikes that several of them were broken down. The waters rushed in on every side, and rolling forward with resistless fury, swept away houses, trees, men and cattle, in one universal ruin. Entire villages were destroyed. The number of lives lost in Friesland alone was estimated at 20,000; and was very extensive in other provinces. "The damage to property incalculable."

The Spaniards (then at war with the Netherlands) imputed the flood, which occurred on All Saints' Day, to a vengeance of God upon the heresy of the land; the Netherlanders looked upon it as an omen portending some violent commotions.

Also on 1st November 1570, there was a dreadful flood at Antwerp, Belgium and on all the coast of the Netherlands that made infinite spoil.

On 1 November 1570 on All Saints Day, a great storm devastated large parts of the North Sea coast from France to Denmark. This great flood caused between 100,000-400,000 casualties due to drowning. In Holland 400,000 people drowned in 1570 CE. The storm called the All Saints Floods occurred on 11-12 November 1570. All Saints' Day is celebrated on 1st November in Western Christianity. This storm affected most of the North Sea between Britain and Denmark, and adjoining land areas. Presumably the Netherland was hit hardest. The cities Amsterdam, Muyden, Rotterdam and Dordrecht were all flooded. Somewhere between 100,000 and 400,000 persons were reported to have drowned. This represents an exceptionally high number of casualties, which should be seen in relation to the much smaller total population at that time.

In the village of Saint-Marceau in the Ardennes region in northern France, there were the most impetuous, terrible, horrible winds that were ever heard.

The summers of the 1570s in Europe were outstandingly cold.

In Switzerland the summers in the 1570s were wet until the 1620s.

From the end of November 1570 to the end of February 1571, the winter was so severe, that all the rivers, even those of Languedoc and Provence in the south of France, were so completely frozen, that they were passable on the ice with laden carts.

In Western Europe the winter lasted from the end of November 1570 to end of February 1571. The winter was so severe that rivers were frozen for three months so that wagons could drive over ice. The cold destroyed the fruit trees, even in the Languedoc, France, down to the root.

On the night of December 2 1570, the Rhône River at Lyon suddenly rose and overflowed its banks. No one living remembers such a sudden and substantial flood. In 1570 to 1571, the Rhône River in France froze. The grape vines and fruit trees in France were killed by the cold. The winter of 1570-71 froze the rivers in France for three months. The fruit trees were destroyed down to the roots.

The frost began in Flanders, now Belgium, on the eve of the Feast of St. Nicholas (December 5), and lasted until 10 March. Up to the very last days of winters, the Maas (Meuse) River, the Waal River and the Rhine River were still frozen.

From 1570 CE to 160 0CE there were great advances of the glaciers in Europe.

On 5 February 1571, there was a great flood at Louvain, France.

On 17 and 18 July 1571, a strong storm struck Germany causing damage to buildings.

In 1571 CE in Flanders, France and Germany in August, there were great floods.

From 15 to 23 August 1571, there was a very great flood in Flanders.

On 5 October 1571, a tremendous gale and flood struck England. Between Hummerston (Humberston) and Grimsby in Lincolnshire, twenty thousand cattle and sheep perished. Houses were blown down and bridges were washed away. Many ships were wrecked. The city of Bourne was flooded to the midway of the church's height. Boats rowed over St. Neot's Church walls. In Bedfordshire many trees were blown down. The gale was violent in Staffordshire, Oxfordshire, Buckinghamshire, Kent, Essex, etc., where great damage was done.

The tides of the River Humber flooded all the streets at Kingston-on-Hull to such a considerable height that the inhabitants were obligated to take refuge in their upper rooms.

On 17 December, there was a flood on the Rhine in Germany.

In 1571, the year was extremely intemperate in England with south winds, rain and fog. During this year and the several that followed, there was a great scarcity and dearth of salt, so that all fish and flesh (meat) were eaten unseasoned. In England, there was a southerly, rainy, cloudy, ugly harvest. In England from 1568-1573, the years were mostly rainy, terrible, and tempestuous bad weather.

In 1571, the San Ignacio, 300 tons, and Santa Maria de la Limpia Concepcion, 340 tons were lost off the Florida coast of the United States during a storm (hurricane). There were only a few survivors.

Between 1571 to 1572 marine flooding on the Lincolnshire coast between Boston and Grimsby resulted in the loss of "all the saltcotes where the best salt was made".

The winter of 1571-1572 CE was much moister, with either continual rains, wind or snow, to the middle of February, then came an intense cold with a continual north wind, and thick dark air to the Equinox (around March 20/21).

In 1572 CE Taal Volcano on Luzon in the Philippines erupted with a VEI of 3.

In the winter of 1572 to 1573 there was hard frost from 2nd November to 5th January and also a late spring in London and the south. There was also a cold winter in Western Europe. From November 1st cold winter began with deep snows and freezing rains until January 6th. Heavy snowfall fell on the Feast of the Second Epiphany in November. In England from 2 November to the Epiphany, 6 January, there was a hard frost, great and deep snow, with several rains which freezed as they fell (freezing rain), and therefore broke boughs of trees with the weight of the ice. The winds were north and west until after the Feast of the Ascension (generally in May or June). There was a very late spring. The weather continued daily to get worse and worse to the beginning of January.

On 2nd February 1573, Switzerland was so cold that Lake Constance froze for 60 days. Lake Constance is situated in Germany, Switzerland and Austria near the Alps.

In 1572, the winter was very severe in Flanders. The Maas (Meuse) River flooded and came out of its banks from the melting snows towards the end of February.

On 7 June 1573, there was a tempest and hurricane at Tocester (Towcester), in Northamptonshire, England. There was a sudden inundation of the riveret, small river, at Towcester.

In the Netherlands on the 1st of September, there was an inundation from the sea. Sea floods hit Friesland in the Netherlands, northwest Germany and Denmark. New bays formed in the northern part, and many were drowned.

In 1573 and 1574 in England, there was a famine.

In February 1574 Kirishimayama erupted in Japan with a VEI of 2.

In 1574 CE in Leyden in the northern Netherlands, a violent equinoctial gale broke through the dikes. By this means the city, then besieged by the Spaniards, was saved. An equinoctial gale is a gale or storm happening at or near the time of the equinox.

In Ireland in 1574, there was a shower of hail, which swept away good strong houses, and smothered whole flocks and herds.

In 1574, four ships sank during a bad storm, hurricane, in the Gulf of Mexico. Five perished from one vessel. The loss on the other vessels was unknown. These ships were part of a New Spanish Fleet that left Spain on 29 June 1574. Other researchers noted there was a violent hurricane on 27 August 1574 between Jamaica and Cuba.

In 1574, there was a great dearth in England without scarcity.

At 4 o'clock in the afternoon of 4 September 1574 there was a terrible storm of rain in London. In October and November, there was a great dearth there and some small plague. During the night of 18 November, a hurricane came out of the south. After harvest, the price of corn (grain) fell a little, but bay salt was dearer than ever was known. The spring was like summer and the summer was like spring. The whole harvest was like a bad winter, most rainy, cold and southerly. Most of the year, there was neither wind nor thunder. In 1574, there was a plague at Chester, England.

The winter of 1574-1575 CE in France was one of the most rigorous. In 1575 in northern France the weather was greatly inconsistent and produced an inequality of air.

Europe experienced a cold winter beginning in November 1574. The Rhine River froze and there was snow until April 1575.

In 1575 CE San Salvador erupted in El Salvador with a VEI of 3.

On September 8th 1575 CE Guagua Pichincha in Ecuador erupted with a VEI of 2.

The 1575 Valdivia earthquake occurred at 14:30 local time on December 16th 1575. It had an estimated magnitude of 8.5 on the surface wave magnitude scale and an estimated magnitude of 9.0+ on the Moment magnitude scale and led to the flood of Valdivia, Chile. Pedro Mariño de Lobera, who was Corregidor of Valdivia by that time, wrote that the waters of the river opened like the Red Sea, one part flowing upstream and one downstream. Mariño de Lobera also evacuated the city until the dam at Laguna de Anigua (nowadays Riñihue Lake) burst. At that moment he wrote that, while many native people died, no Spaniards did, as the settlement of Valdivia was moved temporarily away from the riverside. The effects of this earthquake are similar to the 1960 Valdivia earthquake, the largest ever recorded on earth, which also caused ensuing Riñihuazo flooding. These similarities show that large earthquakes have a pattern that spanned over several centuries.

The sea ice situation near Iceland grew rapidly worse between 1575 and 1600.

In the winter of 1575 to 1576 CE there was a cold winter in Europe. The Rhine River was frozen and there was great snow until April.

In 1576 CE Santa Ana erupted in El Salvador with a VEI of 3 and Colima in Mexico erupted with a VEI of 3.

On November 15th 1576 CE Asosan erupted in Japan with a VEI of 2.

In England on 17 March 1577 CE at Richmond in Yorkshire, there was a strange tempest, which overturned trees, cottages, barns, haystacks, and the church, with most frightful sights in the air.

On 4 August 1577, a violent thunderstorm struck England. It struck Bongay (Bungay) in the county of Suffolk with such violent rain, fearful flashes of lightning and terrible cracks of thunder that it struck fear into the hearts of those gathered in a church. A fearful flash of fire passed between two people (John Fuller and Adam Walker) as they were kneeling upon their knees, wringing the necks of them both at one instant clean backwards killing them.

A storm in Bliburrow, Bilsborrow, in Suffolk on 4 August between 9 and 10 in the forenoon. It rent (destroyed) the church and beat down the people in it. They were almost all smothered. On 4 August 1577, pestilence struck Bliborough (Bilborough in Suffolk, England. When the storm struck the church in Bliborough (Bilborough) where the lightning drave (drove) down about 20 people, a man and a boy died, and the others were scorched.

In 1577, pestilence struck Oxford, England. "During the Assizes, while the Court sat on the trial of a Popish bookseller, accused of circulating offensive pamphlets, a sudden sickness seized nearly the whole of the persons present, and within forty hours upwards of 300 died, among whom was the Lord Chief Baron of the Exchequer, the High Sheriff of the county, several justices of the Peace, and the chief of the Jurors."

In 1578 CE Momotombo erupted in Nicaragua with a VEI of 2.

On 17 March 1578, a tempest struck Yorkshire, England. On 4 August, a tempest and hurricane struck Suffolk, England.

1578 CE. Friesland in the Netherlands and northwest Germany were flooded by sea floods. Many lives were lost.

. In England on 4 February, it began to snow and continued to the 8th and was very deep. The north wind drove the snow into drifts, in which people and many cattle were lost.

There was frost on 10 February, then a thaw with continued rains a long time after. Hence such high waters and great floods as drowned marshes and low grounds. The River Thames so flooded Westminster Hall that fishes were left in it.

In England on the 14th of February 1579 four days of snow started with northerly wind and deep drifts. Many people and cattle were lost

On 24 April, there was another great and deep snow.

On 27 May, there was a great flood caused by rain in England.

In May 1579 snow was one foot deep in London after a five hour fall on the 4th.

Between August and November 1579, a storm (hurricane) sunk the Spanish Armada's 600-ton Almirante.

On September 27th 1579 CE Hakusan erupted in Japan with a VEI of 3.

In England in September and October 1579, there was a great inundation from the Sea by winds. On 14 October 1579 there was a great flood caused by the Sea.

In September and November 1579, great winds and raging floods carried down corn (grain), cottages, drowning pastures and cattle in many places in England.

Tempest in Hessen (Hesse) and Thuringia in central Germany did great damage; for hail as big as hen's eggs broke down the corn (grain) and grape vines; and floods did great harm to the grounds, people and cattle.

In 1579, plague struck Yarmouth (Great Yarmouth), England causing much havoc.

The year 1579 was so poorly regulated, especially on the side of Paris, France, that most grapes froze in clusters from excessive cold at the time of harvest.

In 1580 CE the Billy Mitchell pyroclastic shield volcano on the island of Bougainville in Papua New Guinea erupted. It is truncated by a two-kilometer-wide caldera filled by a crater lake. It has a twenty-year date variation. This was one of the largest volcanic eruptions in the history of Papua New Guinea so its effects could be from ten years either side of 1580. The VEI was 6. Fourteen cubic kilometers of tephra was ejected. If Billy Mitchell had erupted in 1570 allowing for the variation of the eruption date, then it may have been responsible for the extreme cold of the winters in the 1570s.

Also, in 1580 Sao Jorge in the Azores erupted with a VEI of 2 that same year. The ignimbrite eruption from the eruption extended 22 kilometers from the caldera to the coast.

From 1580 CE to 1700 CE was the time of the most sea ice and extreme cold in Iceland. This was especially so later on in the 1690s. This happened again in the late eighteenth and late nineteenth century.

By the 1580s the broad Denmark Strait between Iceland and Greenland was in several summers found entirely blocked by the pack-ice.

Up until the 1580s tree ring records show that there was a more genial climate in Lappland in the far north of Finland. This can be attributed to frequent anticyclones with some sunshine and some southerly winds. These are the same anticyclones that were responsible for the frequent northerly and easterly winds over much of Europe and North America at the time.

Sao Jorge in the Azores erupted on May 1st with a VEI of 3.

On August 11th 1580 CE Katla in Iceland erupted with a VEI of 4.

On December 7th 1580 Galeras Volcano in Colombia erupted with a VEI of 3.

In France in 1580 CE near Chamonix-Mont-Blanc the three great glaciers, the Argentiere, the Mer du Glace and Les Boussons were spreading through rifts in the mountains and descending almost to the valley floor. The Mer de Glace is the Sea of Ice and is presently 7.5 kilometers long and 200 meters or 660 feet deep. The maximum extent of the Mer de Glace was in the nineteenth century when it went down as far as the hamlet of Les Bois, and it was called Glacier des Bois. The glacier then was around three kilometers longer than in the present age.

In the 1580s the spread of Arctic ice to Iceland and polar water to the region of the Faeroes Islands, meant that the surface temperature of the North Atlantic between there and southeast Iceland became 5° C colder than in the late twentieth century. Consequently, there was a greatly strengthened thermal gradient between 50° and 61° to 65° North. This may have been the basis for the development of occasional cyclonic windstorms exceeding the severity of most of the worst storms of modern times. This is suggested that at a time of slightly lowered sea levels, as indicated by the first tide gauge, installed at Amsterdam in 1682 CE, and erosion and blowing sand. Water was being converted into ice in the Little Ice Age which therefore indicated the lowering of sea levels.

In 1581, Persia was desolated by famine and plague. Persia during this period of time, under the Safavid dynasty was at its height, and controlled all of modern Iran, Azerbaijan and Armenia, most of Iraq, Georgia, Afghanistan, and the Caucasus, as well as parts of Pakistan, Turkmenistan and Turkey.

On December 5th 1581 CE Fuego in Guatemala erupted with a VEI of 4.

On February 16th 1582 CE Asamayama erupted in Japan with a VEI of 2.

The following day on 17th February 1582 CE Asosan erupted in Japan with a VEI of 2.

In 1582 CE on June 5th Guagua Pichincha in Ecuador erupted with a VEI of 3.

On July 5th 1582 CE Asamayama in Japan erupted with a VEI of 2.

The winter temperatures in Denmark between 1582 and 1597 were 1.5° Celsius less than in 1886-1925. This was observed by Tycho Brahe and is lower than late twentieth century temperatures. Danish navy records as well as those from ships travelling from the Netherlands to southern Europe show that the wind was blowing from the east.

On July 15th 1583 CE Reykjanes in Iceland erupted with a VEI of 2.

On December 14th 1583 CE Asosan erupted in Japan with a VEI of 2.

In 1584 Merapi in Indonesia erupted with a VEI of 3.

In August 1584 Asosan in Japan erupted with a VEI of 2.

In 1585 Colima in Mexico erupted with a massive VEI of 4 on January 10th.

On January 15th 1585 CE Fuego in Guatemala erupted with a Vei of 2.

On May 19th 1585 CE La Palma in the Canary Islands erupted with a VEI of 2.

The 1585 Aleutian Islands earthquake is the presumed source of a tsunami along the Sanriku coast of Japan on 11 June 1585, known only from vague historical accounts and oral traditions. The event was initially misdated to 1586, which led to it being associated with the deadly earthquakes in Peru and Japan of that year. A megathrust earthquake on the Aleutian subduction zone in the North Pacific Ocean was hypothesized as the tsunami's source. Paleotsunami evidence from shoreline deposits and coral rocks in Hawaii suggest that the 1585 event was a large megathrust earthquake with a moment magnitude (M_w) as large as 9.25. At the same time, a number of Hawaiian natives died after their settlements were struck by a tsunami-like event described in oral traditions. Paleotsunami evidence was also found in the Hawaiian Islands corresponding to a large tsunami in the 16th century. Modelling of a magnitude 9.25 earthquake in the Aleutian Islands matched the descriptions and geological evidences in Japan and Hawaii.

In November 1585 CE Kirishimayama erupted in Japan with a VEI of 2.

Yakedake erupted in Japan with a VEI of 3 in December 1585.

In 1586 Kelud in Java in Indonesia erupted with a VEI of 5. This was the worst eruption of Mount Kelud killing over 10,000 people. Other eruptions that year were Merapi also in Indonesia with a VEI of 3, and Raung also in Indonesia with a VEI of 3.

Once again I ask about multiple volcanic eruptions and the weather.

The Tenshō earthquake occurred in Japan on January 18, 1586 at 23:00 local time. This earthquake had an estimated seismic magnitude (M_{JMA}) of 7.9, and an epicenter in Honshu's Chūbu region. It caused an estimated 8,000 fatalities and damaged 10,000 houses across the prefectures of Toyama, Hyōgo, Kyoto, Osaka, Nara, Mie, Aichi, Gifu, Fukui, Ishikawa and Shizuoka. Historical documentation of this earthquake was limited because it occurred during the Sengoku period. A tsunami was reported in Lake Biwa, Wakasa Bay and Ise Bay, however these may have been seiches as the rupture did not extend offshore. A run-up height of 3 meters (9.8 ft) was recorded at Ise Bay, while at Wakasa Bay, the tsunami was estimated to be 4–5 meters (13–16 ft). A wave was reported along the coast of Lake Biwa, slamming into homes and washing away many residents in Nagahama. The wave destroyed much of the city and flooded Nagahama Castle. There were over 8,000 fatalities in Ise Bay due to the tsunami. Multiple fatalities were recorded in Toyama Bay and along the Shō River. Analysis of sedimentary layers at Lake Suigetsu found no evidence of seawater entering the lake. A 2015 study found tsunami deposits in a paddy field in Ōi District, Fukui Prefecture dating to between the 14th and 16th centuries which corresponds to the event.

Banda Api in Indonesia erupted with a VEI of 3 on April 17th.

Fuego in Guatemala erupted with a VEI of 2 on June 3rd 1586.

In 1586 to 1587 CE there was a cold winter over Western Europe.

In 1586, the loss of two 120-ton Spanish ships was attributed to a hurricane in the Bahama Channel. Six or seven other ships were lost, including the San Juan of 120-tons. The Old Bahama Channel is a strait off the northern coast of Cuba and the Sabana-Camagüey Archipelago and south of the Great Bahama Bank. It is approximately 100 miles (161 kilometers) long and 15 miles (24 kilometers) wide.

In the end of September in England, there was a great destructive hurricane. This year till the next year's harvest, there was a great dearth in England. Wheat sold at 2l. 1s. 4d. per Quarter (quarter ton). Rye at 2l. 2s. 8d. and malt at 1l. 14s. 4d.

In 1586 there was a grievous dearth or famine in Hungary.

On April 7th 1588 CE Kirishimayama erupted in Japan with a VEI of 2.

On May 24th 1587 CE Kirishimayama erupted in Japan with a VEI of 2.

On July 24th 1587 CE Fuego erupted in Guatemala with a VEI of 2.

When the new English group of settlers arrived in 1587 the area around Roanoke in North Carolina was in severe drought that ran from 1587 to 1588.

In 1587 the Belgians groaned under a terrible plague and famine. For the inhabitants of great towns and villages in Flanders were either slain in war, dead of the plague or starved with hunger. All the country was waste, so that wolves and wild beasts stabled in the houses. These animals had become so numerous that they killed and tore in pieces, not only cattle, but also men, women and children. Dogs with hunger and madness ran up and down the country biting and killing cattle and one another. So great the desolation that neither fences nor walls were distinguishable from the rising ground. Nor could land be known by their owners. All were grown over with shrubs and bushes. Inconceivably great was the famine at Antwerp, Brussels, Bruges, etc. Honest, decent people begged from door to door in disguise. The vulgar and poor ate bones, excrements, etc.

In the Netherlands, and the United Provinces, their navigation and shipping saved them. Multitudes of people flocked thither.

In 1587 during the second moon, there was a fall of yellow sand in the vicinity of Shanghai, China. All who ate of the vegetables on which it fell, died. Generally, this is a marker for a deep magma volcanic event. Hydrogen fluoride is one of the gases released by massive flood volcanic eruptions. Fluorine is a pale-yellow gas that even in relatively low concentrations is very toxic. Fluorine attached itself to fine volcanic ash particles. This ash flung high into the atmosphere and spread by the prevailing winds eventually fell back to Earth and coated the skin of edible plants. Animals that ate these plants died. Even in areas that received as little as a millimeter of this ash (a fluorine content exceeding 250 parts-per-million (ppm)), poisoning occurred. Fluorine is very reactive and generally forms soluble fluorine salts. Rainfall dissolved and flushed these salts into rivers, streams and lakes, poisoning surface water supplies. We have a few volcanic eruptions in Japan to choose from that year.

In 1588 CE an Atlantic depression overwhelmed the Spanish Armada. Weather reports available from the Spanish Armada based on sixty days fixes the positions of the depression's centers with sufficient accuracy to indicate that their rates of travel on at least six occasions during that one summer corresponded to jet stream speeds at the limit of, or beyond, the maximum speeds in the late twentieth century.

In 1588, a hurricane struck Roanoke Island in North Carolina in the United States causing less than 116 deaths. Hunter hypothesizes that most of the settlers of Roanoke Island were killed by a hurricane. He indicates that of 116 people on the island in 1587, some returned to England before the storm and a few of the settlers survived the storm.)

At Grindelwald in Switzerland the lower glacier broke through its end-moraine. By 1600 the glaciers had thrust right forward to their most advanced position in recent centuries and destroyed several houses and barns. A terminal moraine, also called an end moraine, is a type of moraine that forms at the terminal (edge) of a glacier, marking its maximum advance.

In the Saas Valley in the Valais in southern Switzerland the Allalin Glacier created an ice dam in 1589. A lake called the Mattmarksee was formed. In September 1589 a flood resulted when the lake burst through the ice dam and caused disaster for the village's inhabitants. Half the fields were buried in debris and half the inhabitants were forced to emigrate. This happened again in 1633, 1680,1719, 1724, 1733, 1740, 1753, 1755,1764, 1766 and 1772. The ice-dammed Mattmarksee was a normal feature of the Saastal landscape from 1600 until the nineteenth century. The residual lake in 1920 had a maximum depth of four meters and held an estimated 560,000 m^3 of water compared to recorded 29-meter depth and $18.8 \times 10^6 \, m^3$ of water in 1834.

The cold winter of 1589 was so harsh that it completely froze the Rhône River. Mules, carriages, carts, all crossed in Tarascon in southern France like on a highway. Colonel Alfonsey even crossed the ice two or three times with the guns (cannons). Marshall Montmorency then crossed with his company of gendarmes.

In 1589, an Atlantic hurricane off Florida and the eastern coast of the United States caused a ship of the fleet commanded by Perez de Olesbal to be wrecked. Forty of her crew was rescued. On approximately 9 September 1589, four ships were struck by a hurricane and sunk in the Bahama Channel. Two of the ships were the Santa Catalina and the Jesus Maria, 350 and 400 tons respectively. The Old Bahama Channel is a strait off the northern coast of Cuba and the Sabana-Camagüey Archipelago and south of the Great Bahama Bank. It is approximately 100 miles (161 kilometers) long and 15 miles (24 kilometers) wide. In September 1589, a four-day storm (hurricane) struck the Bahama Channel. On the first day alone, the sea swallowed up a total of ten ships.

In the 1590s droughts and harsh winters led to famine in the Ottoman Empire.

In 1590 CE Reventador erupted in Ecuador with a VEI of 3.

On January 14th 1590 Colima erupted in Mexico with a VEI of 3.

On April 15th 1590 CE Asamayama in Japan erupted with a VEI of 2.

In 1590 CE in England, there was a drought all the year, and excessive heat.

A very strong heat and drought prevailed in 1590 in the temperate climate zones of Europe.

In Germany, there was a lack of hay, rowen and vegetables. Rowen is a second growth of grass or hay in one season. But wine was available. The heat wave caused numerous fires in Germany. In Thuringia, Germany, many cities and villages were destroyed by fires. In many places the forest fire started and burned, especially in the Bohemian mountains now in Czechia.

On 30 July, a fire was ignited in the vicinity of Vienna, Austria by the action of sunlight on the hay wagon, which then traveled into a dairy. Carts bringing hay home from the fields in Vienna, Austria were set on fire and burnt by the sun.

The grape harvest began in Dijon, France on 10 September, i.e. 14 days earlier than the mean. This is the earliest time since year 1556.

In England, there was a great drought through the whole year; so that corn was thin; wheat small; hay very little; herbs, peas and beans very few; little wine. Because of the dryness there were many fires in the Nation.

On 30 July, In England in September, there was hail, with thunder and snow.

In 1590 it rained all winter in Italy.

In early November 1590, a hurricane struck the Gulf of Mexico and more than 1,000 were killed.

In England there was a dry year in London and the south and was so great that a horseman could ride across the Thames at London Bridge. This was from 1590 to 1591. The River Trent was said to be almost dry. It had been very dry for the two years.

In 1590 during the 6th moon on the 18th day, in the vicinity of Shanghai, China, there was an unusual summer snowstorm. During the night, snow fell from midst of the moon, like the fine flowers of the willow, or shreds of silk. Taken in the hand, these snowflakes were all found to be hexagonal. This may be one of the earliest historical accounts of a diamond dust ice fall. Diamond dust is a ground-level cloud composed of tiny ice crystals. This meteorological phenomenon is also referred to simply as ice crystals.

Also, in 1590 in the vicinity of Shanghai, China, there was an overflow of the sea, destroying several thousand houses, drowning innumerable animals and more than 10,000 people. The survivors were attending to the interment of those who perished, when there was a false alarm that the Japanese were invading. This caused a panic and all fled to Shanghai city. In the process, several thousand people were trampled to death.

In 1591 Taal in the Philippines erupted with a VEI of 3.

When the Leaguers tried to attack the city of St. Denis, France (now a suburb of Paris) on January 3rd (the city walls protected by a broad moat); it was very cold and the moat was frozen to the ground and this allowed the attackers entry into the city.

On November 29th 1591 Asamayama in Japan erupted again with a VEI of 3.

During the winter in the year 1590-1591 in Provence, there was an abundance of snow, and the fruit trees were damaged by the cold.

In 1591, there was a drought in England. In spring, an uncommon drought struck Nottinghamshire. The summer produced strong west winds with little rainfall. The River Trent and other rivers were almost without water. The River Thames was so dried up that a man might ride over it on horseback near London Bridge.

In Italy in 1591, there was a famine. In 1591 a sore famine afflicted all Italy, till it was relieved with corn (grain) imported from Denmark, Holface, etc. Then it fell from 34 to 14 per measure. The dearth of Italy in 1591 forced multitudes to feed on herbs, roots and bread made of them; as of arum and earth nuts, fern roots; hence came the malignant fever in 1592.

On 17 July 1591, a fleet of seventy-seven sail sailing ships left Havannah (Havana, Cuba) for Spain: the smallest vessel in the fleet was 200 tons burthen, and the largest 1,000. About the 10th of August, in latitude 35°, in a gale of wind from the north, the general of the fleet, with 500 men on board, foundered; and three or four days afterwards, in another gale, five or six of the largest ships were lost with all their crews and the vice-admiral. About the end of August, in latitude 38°, they experienced another gale, during which twenty-two sail perished. Upon the 6th of September, the remaining forty-eight arrived within sight of Flores, where they were separated by another gale: so that of 123 sail ((sailing ships) that were expected in Spain this year from the West Indies, but twenty-five arrived. Seven were taken by the English off the Azores, and nineteen, with 2,600 men on board, were wrecked on the coast of New Spain, upon their voyage to the Havannah. The island of Flores is the westernmost point of the Azores Archipelago and of the European continent. New Spain is the Spanish colonial empire of North America, whose capital was Mexico City, Mexico. (From this account it appears approximately 91 Spanish ships were lost in storms and hurricanes in the Caribbean, Atlantic Ocean and off the Azores during the summer of 1591. The deaths of the crew and passengers might be around 13,000 individuals.

On 10 August 1591, another hurricane struck the Atlantic Ocean causing 501 deaths. In mid-August 1591 in the Atlantic Ocean, five or six of a group's largest ships and all their crews were lost.

Near the end of August 1591, twenty-two vessels perished in the Atlantic Ocean from a storm (hurricane). These ships were returning from Bermuda after 24 August. During the end of August 1591, "over a hundred ships, galleons and merchant ships were wrecked, their crews drowned, their riches lost" from a hurricane off the Azores.

In 1591, encountering storms (hurricane), "29 ships were lost, many on Florida's coast" of the United States.

In 1592 CE Asosan erupted in Japan with a VEI of 3 and Popocatepetl in Mexico erupted with a VEI of 3.

In 1592 the flooding of the great bend of the Niger River entered Timbuktu in Mali in Africa for the first time.

A plague struck Derby, England. It began in October 1592 and ended in October 1593. It was dispersed in every corner of the parish and there was not two houses together, next to each other, free of it.

In 1592 CE In England, there was an extreme drought and want of water and a great death of cattle from want of water. Springs and brooks were dried up. Horsemen could ride the River Thames at London. In June, the River Thames, in England was dry at London-bridge and many people passed and repassed on the riverbed. The River Trent and other rivers were almost dried up.

On 6 September 1592, the wind being southwest as it had been for 2 days before and very boisterous, the River Thames in London, England was made void of water, by forcing out the fresh water and keeping back the salt water. As a result, men in diverse places might go 200 paces over the dried riverbank and fling a stone to land on the opposite bank.

The winter was so cold in Austria that wolves entered Vienna and attacked men and beast.

In 1593 CE Raung volcano erupted in Java in Indonesia. It had a VEI of 5. Kelud in Indonesia had previously erupted with a VEI of 5 in 1585. Was there a continuous Southern Hemisphere cold spell due to these two events?

From 20th December 1593 to 3rd of January 1594 sea floods hit the Dutch and German coasts. Several bays were formed, and many lives were lost.

On 21 March 1594, there was a terrible tempest, hurricane and most destructive effects on trees and forest in England. On 11 April, there was an excessive rain, great floods and losses. In May and all summer and harvest time (except August), there were great rains and land floods. Corn (grain) was very dear.

In March or April 1594 on March 30th there was a gales and thousands of trees fell. The summer was unseasonable and there was extensive flooding of fields with the loss and spoiling of crops across England.

On 11 April, there was an excessive rain, great floods and losses.

In May 1594 in England, there were floods in Surrey, Cambridgeshire and Hertfordshire. On 11 May, there were great floods at Surry, England caused by rain and hail.

In 1594 CE in England during the 36th year of Queen Elizabeth's reign "In May fell many great showers of raine, but in June and July much more, for it commonly rained day and night till St. James's eve; and on St. James's day in the afternoon, it began again and continued for two days together. Notwithstanding there followed a fair harvest.

In May and all summer and harvest time (except August), there were great rains and land floods. Corn (grain) was very dear. In 1594, there was a dearth in England due to rain from the beginning of May to 25 July.

"But in September great raines raysed high waters, such as stayed the carriages, and bore down bridges, as at Cambridge, Ware, and elsewhere. Also, graine grew to be a great price – a bushel of wheat at 6s., 7s., 8s., etc., which dearth happened more through the merchants' overmuch transporting than the unseasonableness of the weather past."

In 1594 in England and Hungary, there was a famine.

During the siege of Paris, France, by King Henry IV, owing to the famine, bread which had been sold, while any remained for a crown a pound, was at last made from the bones of the charnel-house of the Holy Innocents (bones of dead children).

In the Winter of 1594-1595 CE Europe experienced a cold winter.

The Lagoons of Venice froze and didn't thaw until February 1595. The Po River in Italy was frozen and the Adriatic Sea at Venice froze. The Sea of Venice froze so that during three weeks no boats could be used. The Tiber River froze at Rome, Italy and men crossed it on the ice, a thing never known before or since.

In 1594, the sea froze on the coast of Marseille, France.

In Europe, the Rhine and Scheldt Rivers froze.

In 1594, the rivers of Northern Europe were frozen before Christmas.

The Kattegat froze, together with a large part of the Baltic Sea. The Kattegat covers 30,000 square kilometers and is between Denmark and Sweden.

The extreme cold of the winter of 1594-95 began on 23 December 1594.

The cold weather began again on 13 April 1595, which was as cold as Christmas, 1594. This period brought about many sudden deaths in Paris, France "particularly in young children and women".

The Rhine River in Germany, the Po River in Italy and the lagoons of Venice were all frozen.

During the winter of 1594-95, the severe cold broke on December 23. The extreme frost resumed April 13, 1595. On that day it froze as strongly as the day preceding Christmas.

In February 1595 food was scarce in Constantinople, the capital of the Ottoman Empire. It was a scarcity born of the evils of weather. Cold weather took a heavy toll on livestock and depleted animals which fell in large numbers to disease. In Anatolia people began to leave the countryside for food and crowded into towns which in turn helped spread epidemics.

On March 9th 1595 CE Nevado del Ruiz in Colombia erupted with a VEI of 4.

On June 1st 1595 Asamayama in Japan erupted with a VEI of 2.

In 1595 CE in Germany, there were considerable floods. The floods occurred in February and flooded all of Germany.

In 1595 in the Netherlands, Guelderland, the tract of the Rhine, Austria, Bohemia, Saxony, Silesia and other parts of Germany, were shocking and extraordinary floods, which overturned many villages, and made terrible slaughter of many cattle and people. Guelderland is now located in east-central Netherlands.

In 1595 and 1596 in England, there was great scarcity and dearth with profound shocking rains and great floods.

There raged a sore famine over all Italy, which reached Germany that forced people to eat uncommon and unwholesome food, such as green hedge crabs, mushrooms, dogs, cats, reptiles and etc.

In 1595, there was a plague at Canterbury in England.

In 1595-96 in Italy, Germany, etc. there was a famine. I could not find a definition of green hedge crabs anywhere.

There was so great a famine among the Turks in Hungary from 1595-97, that the Tartar women who followed the camp were forced to roast their own children and eat them.

There was a great dearth in England and Hungary during these three years.

In England in 1595, during the 36th year of Queen Elizabeth's reign. Since grain has lately been transported to foreign lands; grain in England has grown to exorbitant prices. In some parts of the realm; it has risen from 14s. to 4 marks the quarter ton. This is having a dire effect on the poor. And likewise, all other things made to sustain man have also increased in price, without conscious and reason. To remedy this condition, our merchants have imported much rye and wheat from Danshe, now Gdańsk in Poland. Because food was scarce, and even though the quality was not the best, yet it served our need in the extreme condition that we find ourselves in. Some apprentices and other young people about the city of London, without as much food as they are accustomed to, took butter from the market folks in Southwark, paying only 3d. when the owners could not afford to sell it under 5d. per pound. For this disorder, the said young men were punished on the 27th of June by whipping, setting on the pillory and long imprisonment.

In the Netherlands and Germany, there was a tempest and also one at Worcester, England with hail.

On September 7th 1595 CE Tinakula in the Solomon Islands erupted with a VEI of 3.

On November 22nd 1595 CE Miyakejima erupted in Japan with a VEI of 2.

In 1596 CE on May 1st Asamayama in Japan erupted with a VEI of 3.

In July 1596 CE there was a period of frequent and severe gales in Scotland that set in and lasted until August 16th. Many ships were lost on the east coast.

In 1596, there was a famine in Central India, which extended over the whole in Asia. In 1596 "there was a scarcity of rain throughout the whole of Hindustan, and a fearful famine raged continuously for three or four years. The king ordered that alms should be distributed in all the cities; and Nawab Sheikh Farid Bokhari being ordered to superintend and control the distribution, did all in his power to relieve the general distress of the people. Public tables were spread, and the army was increased in order to afford maintenance to the poor people. A kind of plague also added to the horrors of the period, and depopulated whole houses and cities, to say nothing of hamlets and villages. In consequence of the dearth of grain, and the necessities of ravenous hunger, men ate their own kind. The streets and roads were blocked up by dead bodies, and no assistance could be rendered for their removal." In 1596 CE during the 41st year of Akbar's reign, Akbar the Great, there was a famine in Delhi and its neighborhood in India. During this famine, Emperor Akbar sent officers in every direction to supply food every day to the poor and destitute and had also public tables opened at various centers of distress.

On January 3rd 1597 Hekla in Iceland erupted with a VEI of 5.

On January 17th 1597 Raung in Indonesia erupted with a VEI of 3.

In January 1597 CE Iwakisan erupted with Japan with a VEI of 2.

On April 17th 1597 CE Asamayama erupted with a VEI of 2.

In 1597 CE famine in Cumberland and Westmoreland in northwestern England was followed by plague in 1598 and 1599. Deaths were probably increased by malnutrition when poor peasants were forced off the land and into the towns in search of food. In these towns they crammed into crowded accommodation where diseases spread rapidly. Overall effects of ill-health and poor nutrition were shown in decreasing heights in people of some 2.5 inches between the High Middle Ages and 1700 AD.

In Norway there were some harvest failures at the end of the sixteenth century due to increasing cold and the lowering of the treeline.

The late sixteenth century through the first half of the seventeenth century was the driest era in the last five hundred years in the Mediterranean region. Drought destroyed crops and cold winters killed off livestock.

In Ottoman Bosnia wine production fell in the late sixteenth century where farmers shifted to growing plums for brandy as the grape harvest was as good as before.

In 1598 CE Banda Api in Indonesia erupted with a VEI of 3.

On May 13th 1598 CE Asamayama in Japan erupted with a Vei of 2.

Grimsvotn in Iceland erupted on November 7th 1598 with a VEI of 3.

In December 1598 CE Asosan, as usual, erupted in Japan.

By the late sixteenth century there was much aridity and the expansion of deserts in China. Cooling and overall aridity reduced the overall food production in the late Ming era.

The Winter of 1599-1600 CE in France was severe. The cold lasted from late November 1599 to late May 1600, with interruptions in the southern provinces of France. The cold was so great that it killed nearly all fruit trees and a large number of animals. The winter of 1599-1600 in southern France began in late November 1599 and lasted until the end of May. The winter killed a large number of cattle and almost all the fruit trees.

1600 CE

The 1600s in Europe were the coldest period of the Little Ice Age.

The most pronounced overall colder conditions in the little Ice Age were in the seventeenth century. This would be the coldest period of the Little Ice Age. And you thought that it was already cold.

A dearth of sunspots in the seventeenth century marked a period of markedly cooler climate during the height of the Little Ice Age.

Climate change during this phase of the Little Ice Age contributed significantly to the incidence of famine. There had been a renewed cooling phase from the late sixteenth to the seventeenth century that raised the risk of famine. It was not just cooling but variability that damaged crops.

Some reports state that in the sixteen hundreds the summer temperatures were cool with summer temperatures of 0.5° Celsius cooler than between 1961 and 1990.

If these were very minor temperature variations, then how did the ecotone move that had allowed grapes to be grown in southern Norway and England?

During the winter of 1600 in Lyon, France, many people lost limbs because they froze in the extreme cold.

On 14 April 1600, fell a great snow in England. The rest of April and all May were cold and dry. The late cold spring raised the price of all corn (grain).

On February 19th 1600 the volcano Huaynaputina in Peru erupted. This was the largest eruption recorded in South America. It caused a volcanic winter and temperatures in the northern hemisphere decreased. Cold waves hit parts of Europe, Asia and the Americas and the climate disruption may have played a role in the onset of the Little Ice Age. Floods, famines and social upheavals resulted. It had a VEI of 6 and 30 cubic kilometers of tephra was ejected. This eruption caused a volcanic winter. There was a series of worldwide cold winters and crop distribution. This led to massive famine around the world. It is recorded that in some cases people killed animals with their bare hands to preserve their furs for warmth.

Switzerland, Latvia and Estonia recorded exceptionally cold winters in 1600 to 1602.

Wine production collapsed in Germany and Peru in South America.

On February 22nd 1600 CE Iwakisan erupted in Japan with a VEI of 2.

Sometime in 1600 CE Suwanosejima in Japan erupted with a VEI of 4.

In Africa the severest phases of the Little Ice Age coincided with a time when the summer rains over West Africa were being held closer to the equator, rather than migrating seasonally to 15° to 20° North or beyond as in the twentieth century prior to the 1960s. This caused severe droughts in the Sahel zone at Timbuktu. In the central longitudes of Africa, Lake Chad (13°-14°North, 14°-17° East) was at a level of four meters higher than in the late 20th Century. There was mass migration of peoples to the moister south during the 17th and 18th Century.

In Norway the worst years for harvests were incidentally the best years for the herring fishery. The self-same conditions that produce harvest failures such as long-lasting harsh and stormy westerly and northerly winds drive the fish stocks of the Arctic Ocean, Barents Sea, in greater than usual numbers to the Norwegian coast. This happened between 1600 and 1602 CE.

A meteorite left a 116-meter diameter crater in Wabar. This was the site of a legendary lost city in the Empty Quarter of Saudi Arabia. There were also a 64-meter crater and an 11-meter crater. They had a depth of two meters and fell around 300 years ago.

The snowline was lower in Ethiopia in the 17th century.

There were numerous reports of glacial advances in the Alps in Europe. They were recorded as quite numerous. There was increased glacial advancement in Europe in the late 1600s to 1700 AD.

This would be the furthest extent of glaciation in Europe.

The seventeenth century was the coolest ever recorded for Japan, the same as in Europe.

Around 1600 CE near Chamonix in France. The glaciers of the Arve and other rivers had ruined much land in the parish of Chamonix, had destroyed twelve houses in the village of Chastelard, which later disappeared altogether, and that the village of Les Bois had been left uninhabited because of the glaciers. Les Bois and another hamlet also subsequently disappeared totally for the same reason. In 1600 the glaciers both on the French side and Italian side of Mont Blanc had advanced so far that the people of the valleys were in panic. Ice-dammed lakes which formed and burst as in Saastal, were the agency of much destruction. The hamlet of Les Tines was built over where Chastelard had been. La Mer de Glace was probably responsible as it had extended this far.

Asamayama in Japan erupted with a VEI of 3 on January 14th.

In late spring and all summer, it was very dry across East Anglia in England and the Low Countries or Netherlands.

On July 23rd 1600 Iwakisan erupted again in Japan with a VEI of 2.

One of the earliest famines in Russia of which there is any definite record was that of 1600, which continued for three years, with a death toll of 500,000 peasants. Cats, dogs, and rats were eaten; the strong overcame the weak, and in the shambles of the public markets human flesh was sold. Multitudes of the dead were found with their mouths stuffed with straw. In Russia in 1600, there was a famine and plague, of which 500,000 died and 30,000 in Livonia which is currently Latvia and Estonia.

We have met several Russian famines before around Novgorod.

On 12 September 1600, a hurricane struck offshore Mexico causing more than 60 deaths.

On 26-27 September 1600, a hurricane struck offshore Mexico causing between 103 and 940 deaths. Marx (1983) indicates that, in combination, the storms of 12 and 26-27 September 1600 caused about 1,000 deaths.

On 8 December 1600 at Venice, Italy, a strong southeast wind caused the highest tide. "Inundatio ventis 6 ped. temp. Sirocco." Translation from Latin: During the Sirocco, the wind caused a 6-foot surge/overflowing. The Sirocco is a Mediterranean wind that comes from the Sahara and reaches hurricane speeds in North Africa and Southern Europe.

In 1601, a hurricane struck Veracruz, Mexico causing 1,000 deaths.

In Ireland in 1601 to 1603, there was great scarcity and want and cannibalism was again reported.

After sunset on September 28th 1601 a meteorite was reported to have fallen in Hanau-Lichtenberg. This is a now extinct territory in Hesse-Darmstadt and Hesse-Cassel in Germany. In this period meteorites were recognized by the scientific class though the origins of meteors and meteorites was still regarded as terrestrial. They were regarded as ejecta from volcanoes or brought by hurricanes. No one dared even imagine that they had fallen from outer space. Not until the beginning of the nineteenth century after particularly heavy meteorite and meteor bombardments. The nineteenth century was a phenomenal period for meteoric impacts.

On 26 October 1601, there was a great tempest at Ostend, Belgium. The west-northwest winds caused much higher tides than usual.

Following the poor harvest of 1601 many peasants did not have enough seeds to sow the fields. The weather in 1603 was fine but many fields were empty and the famine in Russia intensified. During the two and a half years after the eruption of Huaynaputina, 127,000 bodies were buried in Moscow alone and one third of the Muscovite Tsardom perished from the famine. Many people were hurt by the famine from peasants to petty gentry and were forced to sell themselves into slavery. Others fled to the frontiers and joined Cossacks.

On January 23, 1603 at Besancon in Doubs, France. "In the year 1603, being in Besancon for the duties of my charge as Visitor to Sainte Claire monastery, it happened that on a Thursday, the 23rd day of January, between 7 and 8 P.M., we were told that all the people were assembling in the streets, terrified. I went out, and like the others I saw a great light in the air over the cathedral, covering the whole of Mount Saint Etienne with a round-shaped, heavy cloud, reddish in color, while all the air was clear and the sky so devoid of fog that the stars were seen shining brilliantly. "This light remained quasi-motionless over Mount Saint Etienne, and from there we saw it coming so low that it nearly touched the houses and lit up the nearby streets, but with a motion so slow that it was hardly noticeable, and it halted for at least a quarter of an hour over Saint Vincent Abbey, where some pieces of relics of two glorious Saints are kept. Then, escaping over the Grande place of Chammar to the Doubs river, it went away through the Grande Rue that goes to the bridge, and straight to the cathedral where it vanished, but as we said before, with such a slow motion that its travel lasted until 9:30 at night, which is to say at least two hours." This was not a meteor or bolide. What was it though? There were no balloons in this period, and they did not fly at night anyway.

People describe their encounters with the unknown, even with such things as auroras and meteors and meteorites, with their own cultural loading.

In July 1603 Mount Etna erupted in Italy with a VEI of 2.

In 1603, the winter in the south of France was very severe. Loaded carriages passed the Rhône River in southern France on the ice.

In 1603, the plague struck Boston in Lincolnshire, England. On 2 April it began at Aylesbury. In London, 30,578 people died of the plague. At Chester, 650 people perished.

In 1603, wars occasioned such a famine in Transylvania, now in Romania, that roots, herbs, and leaves of trees were people's usual food. Horses, dogs, cats, and rats were dainties to the poor. A mother ate six children and two men their own mother. Transylvania or the Principality of Transylvania, during this time period, was a semi-independent state, ruled primarily by Hungarian princes and included areas of modern-day Hungary, Romania and Serbia.

On August 8th 1603 a fireball appeared over Zurich. When I refer to fireball it usually means that it was visible across a whole region or a city. It was a common word for a meteor or bolide.

On September 19th 1603 a great bolide or fiery meteor exploded in the sky over Germany and Switzerland. It was seen to appear over Zurich. Two fireballs over the same place in less than two months? Does Zurich have a particularly good view of the Heavens?

On October 31st 1603 CE Grimsvotn erupted in Iceland with a VEI of 2.

Between 1603 and 1615 in New England in North America there were several winters of exceptional severity as far south as Virginia and in the interior of the country. Samuel de Champlain found bearing ice on the edge of one of the Great Lakes as late as June.

In 1604 CE on February 7th Iwakisan in Japan erupted with a VEI of 3.

On 1st March 1604, the wind was very great at west and northwest, with a furious tempest. The tide at Ostend, Belgium rising so high, as it had not done in forty years before.

In Pons in Lleida in Catalonia, Spain early in the morning of September 30 1604. In the early morning, already light, in the village of Pons or near it, close to Las Belianas, the whole diocese of Urgell, there were seen, in the air but very low, close to the earth, large squadrons of armed men that battled with great fury with the sound of weapons. And the first ones who saw it were some people who were working at a floodgate, and they went to report it in the village of Pons whence a great crowd of people came and saw the aforementioned omen. In the same way, early the same morning from Barcelona and above it, coming from the east and heading westward, there was seen to pass in the air a flock of birds such as starlings, which were of the thickness and blackness of crows, but with many feet and wings like a locusts. In the night those on the earth observed over the Monastery of St. Jerome of the Valley of Hebron great beams of fire like bars, very luminous and those on the sea saw them further away. In any case, everyone who saw them said they were in the direction of the north wind.

How well do humans interpret or normalize phenomena to describe them?

Are cultural loadings involved?

How much do our cultural leanings color our reports of meteoric and other celestial events?

This is 1600 CE. The dawn of the Age of Reason and greater education as well as interest in science and the logical and mechanical arts. This report did not sound like that.

The 1604 Arica earthquake is an earthquake that occurred at 1:30 pm on November 24, 1604, offshore from Arica in Chile. The estimated magnitude range is 8.0–8.5 M_s and possibly up to 9.0 M_w. It had a destructive tsunami that destroyed Arica and caused major damage at Arequipa. 1200 kilometers of coastline was affected by the tsunami. The recorded effects of this earthquake are very similar to those for the 1868 Arica event, suggesting a similar magnitude and rupture area of the megathrust between the subducting Nazca Plate and the overriding South American Plate. Tsunami deposits have been identified on the Chatham Islands that are likely to have been caused by a trans-Pacific tsunami caused by the 1604 earthquake. The tsunami was widespread and impacted many countries. The tsunami, along with the 1868 event, is considered one of "the greatest historical tsunami events along the Perú-Chile Trench". Tsunami run-up heights were estimated to be around 16 meters (52 feet) high. It was recorded along at least 1200 km (746 mi) and potentially up to 2,800 km (1740 mi) of coastline in South America between Lima and Concepción. In Oceania, the Chatham Islands near New Zealand have recorded what is very likely evidence of tsunami from this event as well

In 1605 CE Momotombo in Nicaragua erupted with a VEI of 4.

The 1605 Keichō earthquake occurred at about 20:00 local time on 3 February. It had an estimated magnitude of 7.9 on the surface wave magnitude scale and triggered a devastating tsunami that resulted in thousands of deaths in the Nankai and Tōkai regions of Japan. It is uncertain whether there were two separate earthquakes separated by a short time interval or a single event. It is referred to as a tsunami earthquake, in that the size of the tsunami greatly exceeds that expected from the magnitude of the earthquake. Or this report. On 3 February 1605, in the Keichō era, an 8.1 magnitude earthquake and tsunami struck Japan. A tsunami with a known maximum height of 30 meters (98 feet) was observed from the Bōsō Peninsula to the eastern part of Kyushu Island. The eastern part of the Bōsō Peninsula, Edo Bay (Tokyo Bay), Sagami and Tōtōmi Provinces (Kanagawa and Shizuoka Prefectures), and the southeastern coast of Tosa Province (Kōchi Prefecture) suffered particularly severely. 700 houses (41%) in Hiro, Kii (Hirogawa, Wakayama) were razed and 3,600 people drowned in Shishikui, Awa (Kaiyō, Tokushima) area. Wave heights reached 5 to 6 meters (16 to 20 feet) in Kannoura, Tosa (Tōyō, Kōchi) and 8 to 10 meters (26 to 33 feet) in Sakihama, Tosa (Muroto, Kōchi). 350 drowned in Kannoura and 60 at Sakihama. In total more than 5,000 people drowned.

On April 10th 1605 Iwakisan erupted in Japan with a VEI of 2.

In May 1605 Gamalama in Indonesia erupted with a VEI of 2.

The 1605 Qiongshan earthquake occurred on 13 July (33rd year of reign of Emperor Wanli, May 28th in the Chinese lunar calendar) that struck Hainan and the adjacent Guangdong province in China with an estimated magnitude of 7.5 M_s with a maximum felt intensity of X (*Extreme*) on the Modified Mercalli intensity scale. It caused widespread damage, including the subsidence of large areas of farmland, swamping many villages and several thousand people were killed. According to the Chinese records about the event, in the reign of the Wanli Emperor, "there was a thunderous sound, the public office collapsed, and the houses collapsed, and thousands of the dead were crushed in the county" and "the corpse is covered in pillows, bloody, touches the heart, and spit in the sky.

On October 27th 1605 Hichijojima in Japan erupted with a VEI of 2.

In December 1605 CE Asamayama erupted in Japan with a VEI of 2.

On January 23rd 1606 Hichijojima in Japan erupted with a VEI of 2.

In 1605, from a hurricane that struck Haiti, Dominican Republic and Cuba, three ships were lost but some men escaped.

In 1605, a hurricane struck the Caribbean island of Santa Margarita off the coast of Venezuela. Four galleons were lost.

In 1605, a hurricane struck offshore in Nicaragua causing 1,300 deaths. Marx (1983) is probably describing the same storm when indicating no survivors of 4 wrecks resulting from "a hurricane between Serrana and Serranilla banks" in 1605.

In May 1606 at Nijo Castle in Kyoto, Japan. Large numbers of "spinning fireballs" are reported over Kyoto, Japan. A "whirling, red wheel" hovered over Nijo Castle. The Samurai guards were put on alert. Some people interpret these as UFOs. They may well be meteoric phenomena with a dash of imagination.

On November 25th 1606 Colima in Mexico erupted with a VEI of 4.

On 20th January 1607 CE 2,000 people died around the Severn Estuary on the Bristol Channel. Lowlands on both sides of the estuary suffered inundation with the Somerset and Gwent levels suffering devastating results. It was thought that a severe gale from the west or southwest was responsible coupled with an astronomically high tide. The excess over prediction was 2.3 meters. As well as the cost in human life there was much damage and loss of housing and cattle, sheep and horses perished. There was also salt contamination of arable fields which would have rendered them useless for several years. Bristol and Barnstaple were badly affected.

A strong west wind brought in the sea into the River Severn with such violence; the water in several towns and villages ran higher than the housetops. 80 persons drowned as well as other damages. On 30 January 1606, there was a Great Flood. Hundreds of people drowned between Minehead and Slim bridge as salt water to a depth of 2 meters swept across the land both sides of the Severn estuary and the Bristol Channel. Strong westerly gales blew in the channel, most likely the offshoot of an extreme depression off southern Ireland. John Paul, Vicar of Almondsbury, described the incident thus..."But the year 1606, the fourth of King James, the river of Severn rose upon a sodden Tuesday morning, the 20th January, being the full prime day and highest tide after the change of the moon, by reason of a mighty strong western wind. So that Minehead to Slimbridge the low grounds along the River Severn were, that turning tide, overflown, and in Saltmarsh many houses overthrown, sundry Christians drowned, hundreds of rudder cattle and horses perished, and thousands of sheep and lambs lost. Unspeakable was the spoils and losses on both sides of the river." Low-lying land in Devon, Somerset, Gloucestershire and across South Wales was flooded. 2,000 people drowned, houses and villages were swept away and 200 square miles of farmland was inundated and livestock destroyed. The water went as far inland as Glastonbury Tor, 14 miles from the coast.

Great damage due to flooding was also recorded in East Anglian towns and villages, particularly across the fens.

On June 28th 1607 Mount Etna erupted in Italy with a VEI of 2.

In England, both the summers of 1607 and 1608 were dry and hot.

In 1607-1608 frosts killed many great trees, (Oaks, elms, ashes, etc.) splitting their trunks in England. This winter was considered to have been unmatched in England.

The winter of 1607 to 1608 was called the "Great Winter". Trees died due to the severity and strength of the frost. Ships were stranded by ice several miles out in the North Sea. In December there was a deep frost until midmonth, then a thaw until just before Christmas, then from the 21st of December an intense freeze until mid-January.

The Thames froze for two months during the winter of 1607-1608, sufficient to bear all sorts of sports, perambulations and even cooking. The frost lasted for two months. In Kenal in Westmoreland/Cumbria the severe weather lasted in parts of England until 20th February. England experienced a severe winter. The River Thames froze in December in London and there were frost fairs on the river until February. The winter of 1607-1608 in England had a great frost that lasted off and on for 7 weeks.

On the River Thames, London, England in 1608 from 10-15 January "the frost grew so extreme, as the ice became firme, and removed not, and then all sorts of men, women, and children, went boldly upon the ice in most parts; some shot at prickes; others bowled and danced, with other variable pastimes, by reason of which concourse of people, there were many that set up boothes and standings upon the ice, as fruit sellers, victuallers, that sold beere and wine, shoomakers, and a barber's tent, &c." In these tents were fires. The ice lasted till the afternoon of the 2nd of February, when " it was quite dissolved and clean gon." During the winter of 1607-08, there was a great frost and snow in England as observed in Alrewas in Staffordshire. It began on 5 December 1607 and continued until 14 February 1608. At this time all the rivers were frozen, and in most parts, they would bear the weight of horse, men and loaded wagons. Many of the mills were frozen up and could not grind any corn (grain). The cold did great damage to wheat, gresse (grass) and herbs. The River Thames was frozen over and the people crossed on the ice from Southwark to Lambeth. At York, the River Ouse was frozen over, and horses crossed the ice.

The winter of 1608 was long known as a "Great Winter". The cold reigned almost without cessation from 20 December 1607 until about mid-March 1608 in France, England, the Netherlands, Germany and Italy. The historians provided full detail about the effects of frost.

The winter of 1607 -1608 in eastern North America was notably severe. In Maine there were persistent northerly winds and such severe frosts that many people died both among the European and Native American populations. At Jamestown it was reported that the extraordinary frost in most of Europe was as extreme as in Virginia. Study of tree rings indicates that there was an enhanced further prevalence of northerly surface winds over the eastern and central parts of the continent. There was a general reduction of the westerlies, which were shifted to a lower latitude over the Pacific. Most of the continent was colder than in the late 20th Century. In the southwestern United States, there were some hints of warmth associated with a weak circulation of southerly winds.

In January 1608, the frost was very severe in London, England. The River Thames was frozen over. Wheat rose in the Windsor market from 36 shillings to 56 shillings per quarter (quarter ton).

In 1608 in Germany, the most rapid and deepest rivers are so cold and ice covered that loaded wagons drove over them.

The Scheldt River is frozen at Antwerp, Belgium and the Zuiderzee in the Netherlands froze.

All the rivers in France are frozen.

On 10 January 1608, it was so cold in the church of Saint-André-de-Arcs in Paris, France, that the wine froze in the chalice. "You had to bring a brazier (bucket of hot coals) to melt it." The bread served to King Henry IV of France on 23 January was frozen.

In the northern part of Europe, all the rivers were frozen. The ice was so thick in Flanders (now Belgium), that as the historian Mathieu says, "Antwerp (Belgium), when they saw the Schelde (Scheldt) River so frozen as in the year 1563, they set up several tents in which they feasted." Many people died in cities and in the countryside from this cold; while others remained paralyzed, and a large number had frozen hands and feet. The greater part of the young trees were destroyed, and a portion of the grape vines froze to the roots, and the cypresses, and many walnut trees were hit by the severe cold. England saw almost all its cattle destroyed.

In Scotland the Firth of Forth froze on January 20th 1608. In Scotland during January 1608 CE the Firth of Forth was frozen. In England the River Exe, south of Exeter, also had major ice formation by the latter part of January. In 1608 the sea froze as far as it ebbed on the east coast of Scotland and people walked out on the ice to ships in the Firth of Forth.

On 19 February 1608, the River Thames in England ebbed and flowed twice at noon.

In London, the River Thames was so frozen that loaded carriages went over the ice. Many birds were killed and a large portion of the plants was destroyed. The spring thaw caused great devastation. The ice from the rivers destroyed the ships, roads and bridges and the melting snow-swollen rivers flooded all the valleys.

Dams on the Loire River in France broke causing a second deluge, flooding neighboring lands.

In Italy, the waters of the Tiber River almost flooded Rome. These waters came down from the Apennines Mountains with such violence that several houses were thrown down and destroyed. In Padua, Italy, a tremendous amount of snow fell.

The spring in Italy was warm and moist. The harvest inconstant. Corn (grain) and grapes ill got; hence an epidemic.

In Danzig, now Gdańsk in Poland, the ditches were still frozen on May 15th.

In May 1608 between Angouleme and Cognac in Charente, France. "The day was calm and clear, and in an instant a large number of small, thick clouds appeared. They came down to the ground and turned into warriors. Their number was estimated between 10,000 and 12000, all handsome and tall, covered with blue armour, aligned behind deployed red and blue banners (…) This sight was such that peasants and even the nobility took alarm. They assembled in large number to observe these soldiers' progress; they noticed that when they came near a thick wood, to maintain their good order, they rose above it, only touching the leaves of the trees with the bottom of their feet, eventually walking on the ground again to a forest where they disappeared." "I have written this based on a manuscript report by the late M. Prevost, curate of Lussac les Eglises."

This was a very bloody period in European history so the current events would have a lot of influence on the observers. These apparent soldiers were not terrestrial and incidentally, we would meet Angouleme again. You will also notice that some places are more prone to meteorites and meteors than others as crazy as this sounds. How is this even possible when the Earth is whizzing through space at incredible speed at the same time that the Solar System rushes through space?

On July 18th 1608 CE Gamalama erupted in Indonesia with a VEI of 3.

In England, on 26 July 1608, there was a tempest of thunder, lightning and rain. In England, both the summers of 1607 and 1608 were dry and hot.

In Germany, the summer of 1608 was one of the hottest and it burned everything that had survived the great winter. Only cereals and the offspring of grape vines remain.

In Dijon, France, the grape harvest began on 1 October.

In 1608, a tremendous hurricane did incalculable damage at Beverley in Yorkshire, England.

In 1608 in France, the Loire River overflowed its banks and caused destruction of property.

During the winter of 1608-09 in Paris, France, many people lost limbs because they froze in the extreme cold. In 1608, the French historians cite great mortality of animals due to cold during this year.

In 1608 Samuel Champlain, the founder of Quebec in Canada, found bearing ice on the edges of Lake Superior in June, which is summer. How warm was the summer in the Great Lakes then?

On 23 November 1608, a major earthquake hit Sendai beach on the Sendai Plains in Japan, sweeping away and killing over 50 people.

In 1609 Banda Api in Indonesia erupted with a VEI of 3.

On April 5th 1609 Asamayama erupted in Japan with a VEI of 2.

On 4 August 1609, an Atlantic hurricane struck near the Bahamas, one ship was sunk. In 1609, a hurricane struck southeastern Bahamas causing 32 deaths.

In the Winter of 1609-1610 CE England experienced a severe winter beginning in October. The frost lasted 4 months and the River Thames was frozen. In England, the winter of 1609 was most rigorous hard frost from December to April. The River Thames became a highway. Birds and garden stuff were killed. In 1610, The River Thames in England was frozen and carried the weight of pedestrians. During the winter of 1609-10, the weather in England from December to April was very cold.

On February 6th 1610 Mount Etna in Italy erupted with a VEI of 2.

In 1610 CE there were four months of drought at Derby in England.

On 29th September, Michaelmas Day, 29th September, blizzards raged through Derbyshire and snowstorms were unparalleled in recent history. The hay harvest was severely affected in places.

In 1611 CE on April 15th Colima erupted in Mexico with a VEI of 3.

In 1611 CE it was a wet year in England with floods in January and February in the west of England with the Avon river system.

The 1611 Sanriku Earthquake occurred on December 2nd 1611, with an epicenter off the Sanriku coast in Iwate Prefecture in Japan. It triggered a devastating tsunami and a description of the event in an official diary from 1612 is probably the first recorded use of the term "tsunami". At about 10:30 on December 2, 1611 (Keichō 16, 10th month, 28th day), there was a severe earthquake, and at about 14:00 (local time), this was followed by a devastating tsunami. According to old documents, the earth shook violently three times. The tsunami reached its maximum estimated height of about 20 meters at Ōfunato, Iwate. The tsunami struck on the east coast of Sanriku from Sendai bay in the south to southeastern Hokkaido in the north, a greater length of coastline than was affected by the 1896 tsunami. According to old documents, 1,783 people were killed in the Sendai Domain, and over 3,000 horses and men in the Nanbu and Tsuguru domains. On the southern coast of Hokkaido, many Ainu were also drowned. Amongst the worst affected places was Ōtsuchi, with 800 deaths. There have been previous tsunamis listed though in Japan, so I am not sure what is meant by using the term tsunami. I will leave that to the academics.

In July 1612 at Fluelen in Uri, Switzerland. According to this broadsheet, some terrible and wondrous signs were seen in the heavens in Switzerland, on the 3rd, 4th, 5th, and 6th of July, 1612. They included three suns, three rainbows, a white cross, and two battling armies. The text in this broadsheet contains rhymed strophes. It describes the events rather superficially and repetitively addresses its Christian readers 'young and old' to repent their sins because wondrous signs from heavens are indicators of bad times, war, and menace.

You have to watch out for religious interpretation or misinterpretation as well with reports. Sometimes they were modified to suit a purpose.

In England in 1612 it was a hot dry summer.

On August 12th 1612 CE Asosan erupted in Japan with a VEI of 2.

On October 12th 1612 Katla in Iceland erupted with a VEI of 4.

On 31st October 1612 in North Friesland in the Netherlands, four parishes were flooded and new bays formed on the coast by sea floods.

Between 6th December and 13th December 1612 sea floods hit Nordstrand in Schleswig-Holstein. Floods remained all winter, and a new bay was formed. Nordstrand was an island up until this event and is now a peninsula. In Medieval times Nordstrand was originally part of the Island of Strand.

Also, on 6th December 1612 sea waves hit Flanders in Belgium and Holland in the Netherlands. Part of the city of Flushing, Vlissingen, on the former island of Walcheren, was lost.

In 1612 King James V1 of Scotland, who had acceded as King James 1st in England, used that power to create a plantation in Ireland of his Scottish subjects so as to relieve the suffering arising from the continual famines in Scotland and to reduce their frequency.

In 1613 CE on August 8th Asosan erupted with a VEI of 2.

In the Winter of 1613-1614 CE there was a severe winter at Boston, England.

In January 1614 at York, England, there was a heavy snow and 11 weeks of frost. Then the River Ouse overflowed and flooded the streets. The flood lasted 10 days and destroyed many bridges.

In 1614 on 19 February, fell such a storm of snow in the peak of Derbyshire, and over all the west of England, as was a full yard deep on the level. And because of such high winds, the snow was blown into vast snowdrifts. As a result, travelers on horseback or on foot went over hedges, fences, stonewalls, etc. The snow laid on the ground for a long time. It destroyed many cattle and sheep. A great scarcity of hay (immediately) followed. Corn (grain) next summer was very good and cheap.

The winter of 1613-1614 in Switzerland produced one of the three cases of longest snow cover.

Drought between 1614 and 1619 in China was so severe that the "History of Ming" described the land as burned.

On July 1st 1614 CE Mount Etna in Italy erupted with a VEI of 2.

In 1614 CE in Lincolnshire, England, the sea came 12 miles inland during a sea flood. In 1614 at York, England, after the great flood, there was a drought that continued until August. This caused a scarcity of hay and corn (grain). Hay sold for 30 to 40 shillings per load. At Leeds, it sold for 80 shillings.

Thunderstorms or rainstorms desolated Provence, France in 1614 and Southern France was very dry.

In the winter of 1614 to 1615 CE there was great snowfall over England. This snowstorm was regarded as of great severity and by February the Tay River was frozen over, such that foot and horse traffic could walk across it. In early March there was an enormous fall of snow and in Scotland this lasted for three days and there were many deaths of horses and men as people tried to move about. It was particularly bad across northern Scotland.

On 1 January 1615, a tempest of thunder and lightning struck Thuringia, Germany.

On 16 January 1615 at Youlgrave (Youlgreave) in Derbyshire, England, began the greatest snow which ever fell upon the earth in man's memory. It covered the earth 5 quarters deep upon the plane (on the level), and for heaps or drifts of snow, they were very deep, so that passengers, both horse and foot, passed over yates (gates), hedges and walls. It fell ten times, and the last was the greatest. The snow continued daily increasing until 12 March, when the sight of earth was no longer visible either upon hills or valleys. Then the snow began to decrease and was consumed little by little until 28 May. And then all the drifts were consumed, except for the one at Kinder Scout, which lay till Witson (Witsun) week (Pentecost, the seventh Sunday after Easter). (A quarter is a unit of measurement, usually for measuring cloth, equal to nine inches. Therefore 5 quarters would be equivalent to 45 inches.)

In 1615, there was a great flood at Boston, England and there was a great mortality among sheep, also at Boston, England.

During the winter of 1615, a very severe cold descended upon Germany, Hungary and the neighboring provinces on 20 January. This cold froze and damaged many grape vines and a significant amount of fruit trees.

On March 16th 1615 CE Banda Api in Indonesia erupted with a VEI of 3.

In 1615 CE the cod fishery at the Faeroe Islands began to fail and did so increasingly until there was no cod thereabouts for thirty years between 1675 and 1704. The water was getting colder and the cod were leaving for warmer temperatures.

The summer of 1615 was very dry throughout Europe and very hot. In the fields, everything was destroyed.

In Picardy in France a church was destroyed by lightning, which also killed several people. In Dijon, France the grape harvest was held on 21 September.

The drought was so great that in Germany more than 3,000 houses were consumed by fire.

On 12 September 1615, an Atlantic hurricane struck Puerto Rico causing some deaths.

In the Winter of 1615-1616 CE the Seine River in France froze in the beginning of the year; the ice takes place on 30 January.

The winter of 1616 brought cold weather to France. The cold was very severe on the royal army, which escorted the Queen of Poitiers to Tours, France. In Paris, the ice flow destroyed a support column of the bridge, Pont Saint-Michel.

On February 19th 1616 CE Mayon in the Philippines erupted with a VEI of 3.

On July 4th 1616 CE Galeras in Colombia erupted with a VEI of 3.

Two days later on July 6th 1616 CE Manam in the Philippines erupted again with a VEI of 2.

In 1617 CE Fuego in Guatemala erupted with a VEI of 3.

In England on 29 January 1617 CE, there was a tempest of thunder, lightning and earthquakes.

In Catalonia in northeastern Spain, there were great floods; 15,000 people perished. Other sources state 50,000 people.

The winter of 1617 (in Europe) passed without a significant frost.

On January 31st 1618 CE Iwakisan in Japan erupted with a VEI of 2.

After 1618 the population of Norway fell by twenty to thirty per cent over the next thirty years.

In the latter half of August 1618, a shower of stones fell in Styria in southern Austria. Styria in German is Steiermark. At Murakoz on the Styrian frontier three stones fell that were about one hundred pounds each.

In this period meteorites were regarded as the ejecta of volcanoes or created by lightning.....or to the peasants, signs from God. Usually not very good signs.

In 1618 CE in Europe, there were extraordinary tempests, inundations of rivers, eruptions of the sea, earthquakes, bloody rain, snow, hurricanes and meteors in the air.

On July 29th 1619 CE Grimsvotn erupted in Iceland with a VEI of 2.

In Frauenfeld on October 15th 1619 a fireball was seen in the sky. Frauenfeld is in the canton of Thurgau in Switzerland.

In 1619 in Prague in Bohemia, Czechoslovakia, now Czechia, a succession of fiery globes is observed. Some of them split into several parts or other globes. The report reads: "A strange and prodigious thing was seen in a village that is 6 leagues from Prague, the capital city of Bohemia. Never had we seen such a spectacular or frightful sign before. The inhabitants of the village were on guard as the country is full of soldiers, because of the partialities and differences that exist in the empire today. The village priest was with them at about 10:00 in the evening. He was praying, looking up at the sky, when suddenly he stopped, astounded by what he saw. He could see a globe that resembled the moon, but fiery. It divided into two parts, and one of the parts divided into four smaller globes. "The most amazing thing was that one of the globes disappeared, and in its place, we saw a bloody crucifix. These things stayed [in the sky] for a short time, and then disappeared gradually, finally vanishing into a big hole. Then we just saw a great globe which resembled the moon, as we had witnessed at the beginning. This whole process repeated three or four times, and then everything disappeared."

In 1619 at Fluelen in Uri, Switzerland. Herr Christophorus Schere, Prefect of Uri County, saw a bright, long object, fiery in color, near Fluelen, flying along Lake Uri: "As I was contemplating the serene sky by night, I saw a very bright dragon flying across from a cave in a great rock in the mount called Pilatus toward another cave, known as Flue, on the opposite side of the lake. "Its wings were agitated with much celerity; its body was long as well as its tail and neck. Its head was that of a serpent with teeth, and when it was flying, sparkles were coming out of it like the ones thrown by an incandescent iron when struck by smiths on an anvil. At first, I thought it was a meteor, but after observing more closely, (I saw) it was truly a dragon from the recognizable motion of the members. This I write to you with respect, that the existence of dragons in nature is not to be doubted anymore."

It was widely believed in this period that dragons lived in the rugged clefts and caves on the Pilatus, a mountain, above Lucerne in Switzerland. The dragons of the Pilatus were also called the Lindworms and there was a long history of them. But Lindworms were not reported to have flown.

In Geneva, Switzerland on the night of April 9 1620. "Two suns were seen, one red and the other one yellow, hitting against each other (…) Shortly afterwards there appeared a longish cloud, the size of an arm, coming from the direction of the sun, which stopped near the sun, and from that cloud came a large number of people dressed in black armed like men of war. Then arrived other clouds, yellow as saffron from which emerged some 'reverberations' (?) resembling tall wide hats, and the earth was seen all yellow and bloody. The sun became double, and it all ended with a rain of blood." Or this. Numerous local residents reported seeing an 'aerial battle'. First there were seen two suns, one red the other yellow. A little later there appeared a long cloud from this cloud a whole troop of black clad men emerged all were armed as if marching to war. Red clouds then appeared that resembled large hats and the whole earth sky became yellow and bloody, the whole affair ended in a bizarre rain of blood.

Meteorites, UFOs or hallucinations?

On June 3rd 1620 CE Asosan in Japan erupted with a VEI of 2.

A journal from a plantation in New England for December 1620 described "considering the weakness of our people, many of them growing ill with colds, for our former discoveries in frosts and storms, and the wading at Cape Cod had brought much weakness among us, which increased so every day more and more, and after was the cause of many of their deaths". The harsh winter was not the only source of woe for the Pilgrims. They had also arrived too late to plant crops. Only about half of those who had landed at Provincetown survived the first winter. In this period the climate of the Little Ice Age intensified the harsh and long winters. The weather was described in contemporary colonial journals etc. as an unaccustomed and severe climate of great extremes that they had never met before. They complained of cold and snow and drought.

In 1620 in Punjab in India, hot iron fell from the sky and burned grass. A dagger knife and two sabers were made from it. Meteorites were often used as a source of iron in the past and even recently in India. Even Inuit in Greenland used meteorites as sources of iron. This would now be classed as a siderite or ferrous meteorite that consists almost overwhelmingly of an iron-nickel alloy known as meteoric iron. Some of Tutankhamen's knives were partly made of meteoric iron.

The winter of 1620-1621 was very severe in the north and south of Europe.

During the winter of 1620-21, the Zuiderzee freezes up entirely.

A part of the Baltic Sea was covered with very thick ice.

In Italy, the ice on the lagoon in the Adriatic Sea held back the Venetian fleet.

The cold was very severe in Provence in France.

In 1620, there was a great snow in Scotland. There was a snowstorm lasting 13 days and known as the "thirteen days' drift" in Scotland, where on Eskdale Moor, out of 20,000 sheep, only 45 were left alive.

In England, there were fairs on the frozen River Thames in 1620.

The Venetian fleet was frozen in the ice of the lagoons of the Adriatic Sea.

Also, during the winter of 1620, the sea between Constantinople (Istanbul) and Iskodar (Üsküdar) Turkey was frozen and was passable on the ice.

On April 10th 1621 there was a fall of an iron meteorite in Jalandhar in India. It weighed 1,967 grams and was forged into sword blades. Jalandhar is in the Punjab. There had been a previous iron meteorite fall here in 1620…….or was there a year mix up?

On October 9 or 13, 1621 at Nimes in Gard. "Over the city of Nimes about 9 to 10 P.M., over the amphitheatre, was seen something like a great sun, very resplendent, which was surrounded by a number of other luminous torches. "It seemed to want to move straight towards the Roman Tower, over which appeared something like fiery chariots surrounded by very bright stars."

Were they all heading in the same direction?

In 1621 it was noted as being very dry in eastern Scotland but very cold and wet further south. Through autumn it changed to a lot of rain with a poor crop for the winter around harvest tide.

In the Winter of 1621-1622 CE in England, the frost was very severe from the 24th of November to the 7th of December.

In Italy, during the winter of 1621-22, the Venetian fleet was arrested by the ice in the lagoons of Venice." The Port of Venice was frozen. The Adriatic Sea froze from December to January.

The frost was very strong in the winter of 1621-22 in Flanders and northern France. The Dutch lost half their army due to the cold and hunger before Sluis in southwestern Netherlands.

The winter of 1621-22 was excessive in Europe.

In 1622, the rivers in Europe and the Zuyder Zee were frozen.

Ice covered the Hellespont in Turkey. (Hellespont is an ancient name of the narrow passage between the Aegean Sea and the Sea of Marmara. Today, it is known as Dardanelles.

In 1622 during the winter, it was so cold that all the rivers in Europe were frozen along with the Zuyder Zee (Zuiderzee) in the Netherlands.

On June 8 1622 CE Colima in Mexico erupted with a VEI of 4.

The 1622 North Guyuan earthquake struck Ningxia, China on 25 October with a magnitude of 7.0 M_s and a maximum Mercalli intensity of X (*Extreme*). It was the only recorded big earthquake in western China for 148 years, between 1561 and 1709. The earthquake occurred on the "rake of the Zhongwei-Tongxin fault", with a mid-seismogenic depth of about 15 kilometers (9.3 mi).

In 1622 and 1623, there was a famine in England. During the summer it was excessively wet and sultry hot.

In 1622, an Atlantic hurricane struck in the Bahama Channel. Two Spanish ships were lost in the hurricane. The Old Bahama Channel is a strait off the northern coast of Cuba and the Sabana Camagüey Archipelago and south of the Great Bahama Bank. It is approximately 100 miles (161 kilometers) long and 15 miles (24 kilometers) wide.)

In 1622, a hurricane struck off the coast of Florida in the United States. A Spanish ship Santa Ana Maria, 180-ton, was lost during the storm.

On 5 September 1622, a hurricane struck the Straits of Florida in the United States. More than 1,090 individuals were killed.

In 1622 the weather was inclement in the spring and summer across Britain but especially Scotland and the harvest was poor. In Scotland the harvest was stated to be catastrophic. There was great distress in the population of Scotland with death rates much higher than normal due to the famine prevalent at the time following the poor harvest being the second year in a row for this occurring.

On January 10th 1622 in the afternoon a large stone fell from the sky in Devonshire in England.

Also, in 1622 there was a meteoric fall at Tregnie, Tregony, in Cornwall. The stone weighed 1,250 grams. Cornwall and Devon are only seventy miles apart, which is not far for meteorites from the same shower to fall together in.

In 1622 and 1623, there was a famine in England and a plague in London, England.

In 1623 Pacaya erupted in Guatemala with a VEI of 3.

A stone was reported to have fallen from the sky on 10th January 1623. It was 23 pounds weight which sounded like the falling of a piece of ordinance. It resembled singed stone or half burnt lime. This was at the manor of Stretchleigh near Ermington in Devonshire, England. It fell near some men who were planting trees in an orchard and buried in the ground three feet deep. The stone was three feet long, two feet and a half in breadth and one and a half foot thick. This was an unclassified stony meteorite.

On 14 & 15 February, 13 March and 23 June 1623, tempests struck Strasburg (Strasburg might refer to Strasbourg, France; Straßburg, Austria; or Strasburg, Germany).

In Austria and Hungary, the Danube River greatly overflowed. On 12 and 18 February 1623, there were great floods on the Danube River.

On March 10th or 17th 1623 a fireball was seen over Zurich in Switzerland again.

On May 15th 1623 CE Zaozan erupted in Japan with a VEI of 3.

In England on 19 & 31 May and on 19 July, there were terrible storms of thunder and lightning.

In September 1623, a hurricane struck Cuba. One account lists 250 deaths, while another account cites 150.

On 19 September 1623, a hurricane struck St. Christopher's Island (St. Kitts) in the West Indies and destroyed the crop of tobacco. Was this the same hurricane as the one that hit Cuba?

On November 17th 1623 a meteor like a burning globe at sunset was seen over Germany. There was a fireball seen in the sky and detonations heard.

This winter of 1623-24 was very severe and as a result foiled the attack on Antwerp, Belgium by the army of the Prince of Orange. The winter produced tremendously heavy snowfalls and great disasters.

The winter lasted in England from mid-December to mid-January; and in Germany, the Danube River froze.

At Gierstedt in Anhalt, Germany at 8.00 pm on May 12 1624. From six to eight o'clock in the evening a multitude of men and chariots emerged from the clouds over this city in Germany. Gierstedt is not in Anhalt but actually in Thuringia next to Anhalt. Thuringia has come and gone over the centuries as a state but never changed as a region. It was possibly then in the principality of Anhalt.

In 1624 the city of Fez in northern Morocco was devastated by a major earthquake between three and four in the morning of 11 May. It had an estimated magnitude of 6.0 M_w and a maximum felt intensity of VIII–IX on the MSK scale. The earthquake caused severe damage in Fez and the surrounding area. It was felt as far away as Seville in southern Spain. Thousands of people died. Fez suffered the most severe effects of the earthquake. Very few buildings were unaffected, with many collapsing. The damage was greatest in the older part of the city, Fes el Bali. This difference in damage is reflected in the reported number of casualties, with over 2,500 dead in Fes el Bali compared to only eleven in the newer district of Fes el-Jdid and none in the Mellah. At Meknès, two towers were destroyed and two fatalities were reported, while at Sefrou four houses were ruined but there were no deaths.

In 1624 a fireball was seen over Tubingen in Baden-Wurttemberg, Germany.

Nikko-Shiranesan erupted in Japan with a VEI of 3 in 1625.

On September 2nd 1625 Katla in Iceland erupted with a VEI of 5. This is volcanic winter material.

In 1626 CE in March Vulcano erupted in Italy with a VEI of 3.

In 1626 CE in England, there were great hailstorms on the 29th of March and on the 25th to 30th of April.

In England on 6 June, there were great floods. There were tempests on 10 & 13-16 February, 20 March, 31 August, 1 September, 5 & 8-9 December, and 10 & 13-14 February.

In England, during the summer there was a drought and it was excessively hot.

In November, the weather was excessively cold.

December was mild soft warm weather, like a fine spring, yet it totally ceased and vanished.

In the summer of 1626, excessive heat was in England.

On 15 September 1626, a hurricane struck Puerto Rico causing 38 deaths.

The first grape harvest in Dijon, France was held on 1 October.

In 1627, there were tempests on 3-4 March, 13 & 27 October, and 17 December. Bohemia is in western Czechia.

On November 27th 1627 at Nice in France a stone weighing 59 pounds fell. If it was not meteoric, what was it doing in the sky? Remember, there were no aircraft in this period to drop stones from.

At 5.00 pm on April 9th 1628 a stone weighing 24 pounds fell near Hatford in Berkshire, England. Suddenly there was a hideous rumbling in the air following a strange peal of thunder. There were twenty discharges. Then there was the sound of a drum beating the retreat, amongst all of the angry peals shot off from heaven, at the end of each crack there was a hissing sound. One meteorite was seen at Bawlkin Green, one and a half miles from Hatford. The thunderbolt was seen by Mistress Green who caused it to be dug up. Other thunderstones were also found in the area. This was an unclassified stony meteorite.

In Austria on the Danube River on 10 September, "A cloud loaded with a sea of water burst and fell."

In Apulia (Naples), Italy, there were great floods and 16,000 souls were lost.

On October 26th 1628 Kirishimayama in Japan erupted with a VEI of 2.

Also, in October 1628 CE Sangay in Ecuador erupted with a VEI of 3.

The 1629 Banda Sea earthquake struck the Banda Sea, Indonesia on August 1. Its epicenter is believed to have been in the Seram Trough. A megathrust earthquake caused a 15 meter (49 foot) tsunami, which was recorded to have affected the Banda Islands about 30 minutes after the quake. The effects of the tsunami were reported as far as 230 kilometers (140 mi) away in Ambon. Many trees in the Banda Islands were reported to have been uprooted.

In 1629 a fireball was seen over Tubingen in Baden-Wurttemberg, Germany. Funny, Tubingen had one previously in 1624. Was this the result of bad handwriting?

There were outbreaks of plague in France in 1629-1630 AD.

There were outbreaks of plague in Italy in 1629-1630 AD.

In 1630 Raoul Island in the Kermadec Islands in New Zealand erupted with a VEI of 4.

On May 29th 1630 there were reports of a second sun over Magdeburg in Germany. Magdeburg had four fifths of its population of 25,000 people slaughtered only a few days before by Catholic forces during the Thirty Years War. This was a large city by the standards of the time being one of the largest cities in Germany. The city was burnt to the ground and only 200 of its 1,900 buildings survived. Some sources state that a meteorite or second sun crashed to earth here and caused the fire. Witnesses stated that it only took 3 or 3.5 hours to end up as embers and ashes.

In 1629 and 1630, there was a dearth in England. In London, they made bread from turneps (turnips).

On September 3rd 1630 CE Furnas in Iceland erupted with a VEI of 5. This as you know by now is volcanic winter material.

On 5 November 1630, Zealand (the largest island in Denmark) was completely overflowed with saltwater by the sea. On November 5th in the third year by tens {hence 30} after the sequemille (1600), Zealandia stands completely under salty waves.

On November 29th 1630 a 300-pound stone fell at Alsace in France.

In 1630-1631 CE a devastating drought afflicted the province of Gujarat in India and whole centers were depopulated. A Dutch merchant, returning from Swally (Suvali, India), reported that of 260 families only 11 had survived, while in Surat, India, a great and crowded city, he saw hardly a living soul, but at each street corner found piles of dead with none to bury them. Gujarat is a state in northwestern India on the Arabian Sea. Swally, Suvali and Surat are all located in Gujarat. In 1631 in India, there was a great drought and this drought extended through Asia.

In 1631 CE there was a very general and terrible famine in India. In this terrible famine, Shah Jahan opened a number of soup kitchens or alms-houses (langar), gave away a lac of Rupees in charity, and remitted taxes to the amount of nearly 70 lacs of Rupees (80 crores of dams) or one-eleventh part of the whole revenue.

In 1631, there was a great famine in the Deccan region in India. "There was a great deficiency of rain, and the drought was so intense that not a drop of dew could be found. The scarcity became so great that nothing but the herb bugloss (a coarse hairy plant of the borage family) was to be found in the shops of the bakers and druggists. The number of the dead exceeded all computation or estimate. Coffins and burial were not thought on in the Deccan. The distress compelled emigration to the north and east; but the poor wretches were reduced to such a state of weakness, that they did not accomplish the first stage. The towns and their environs, and the country, were strewed with human skulls and bones instead of seed. Men ate each other, parents devoured their children. Bakers ground up old bones, or whatever else they could get, and mixing the dust with a little wheat-flour, sold the cakes as valuable rarities to the wealthy. Human bodies dried in the sun were steeped in water and devoured by those who found them. Cities were depopulated by the death and emigration of the inhabitants. No such famine has been recorded in history. The Emperor (Shah Jehan) ordered distribution of provisions to be made in the cities and towns, especially at Burhanpore. Khandesh and Balaghat, with many other districts, were quite depopulated. Sultanpoor remained waste for forty years." The foregoing account refers especially to that part of India known as the Deccan; but other accounts indicate this famine was prevalent over all of India and also extended over the whole of Asia. In describing this famine, the Orissa Commissioners remarked, "that money could not purchase bread, and a prodigious mortality ensued. Disease followed famine, and death ravaged every corner of India."

The famine of 1631 due to the failure of the monsoon rains of 1629 and again in 1630 devastated all of monsoon Asia. Millions of cattle perished. Cholera epidemics carried away entire villages. Entire rural districts were depopulated as people moved elsewhere to get away from hunger and died by the roadside. Many areas did not recover for half of a century.

On 16 December 1631, at the Gulf of Volo off Greece, riding at anchor, about 10 o'clock at night, it began to rain sand and ashes. This continued until 2 o'clock the next morning. It was 2 inches thick on the deck of the ship. The crew cast it overboard using shovels, as they did snow the day before. There was no wind stirring when these ashes fell. This strange ash fell in other places. Ships coming from St. John D'Acre (Acre, Israel) over 100 leagues (300 miles, 483 kilometers) away also encountered this ash. Comparing the ash from these ships, they found that the ash were the same. This shower of ash was due to the eruption of Mount Vesuvius.

On December 16th 1631 Vesuvius erupted with a VEI of 5. This would have caused a volcanic winter and all VEI 5s can do this.

In December 1631 CE Asosan erupted in Japan with a VEI of 2.

Captain (Thomas) James wintered at Charleton (Charlton) Island at the southern end of James Bay, Canada during the winter of 1631-32. He was obliged to take harbor in the beginning of October 1631, the snow and ice began in that month, but the sea was not frozen close to the island until the middle of December. (On 29 November, the ship was deliberately sunk to keep it from being swept away or crushed by ice.) The cold was very intense until the middle of April. They endured great hardship in so long a winter, surrounded by a sea all covered with ice. On the

29th of April, it rained all day. On the 3rd of May, the snow was melted in many places on the island. On the 13th of May, the weather was warm in the daytime, but there was still frost in the night. On the 24th of May, the ice was consumed along the shore, and cracked all over the bay, and began to float by the ship. On the 30th of May, the water was clear of ice between the shore and the ship. On the 15th of June, the Sea was still frozen over, and the Bay full of ice. The 16th was very hot with thunder. On the 19th of June, they saw some open sea, and on the 20th all the ice was driven to the northward. The ship was refloated in June. The sea to the northward was full of floating ice until the 22nd of July.

On December 16th 1632 CE Banda Api in Indonesia erupted with a VEI of 3.

About midnight on September 27th 1632 an L chondrite weighing 1.04 kgs was found in a salt field near the seashore at Minamino in Japan. The owner of the field, Rokubei Murase saw the fall and recovered the meteorite soon afterwards. The stone was later presented to a Shinto shrine, Yobitsugi jinja, where it remains as a holy treasure to this day. A chondrite is a stony meteorite that has not been modified by melting or differentiation of the parent body. They generally come from the Asteroid Belt.

In Norway the worst years for harvests were incidentally the best years for the herring fishery. The self-same conditions that which produce harvest failures such as long-lasting harsh and stormy westerly and northerly winds drive the fish stocks of the Arctic Ocean, Barents Sea, in greater than usual numbers to the Norwegian coast. This happened in 1632 to 1634 CE.

The winter of 1632-1633 began early and was very hard. The Mercure de France reported that on 4 October 1632, the cold between Montpellier and Béziers in southern France was so severe that 16 Gardes du Corps (bodyguards) of Louis XIII, 8 of his Swiss and, 13 sutlers (civilian merchants who sell provisions to an army) died from the extreme cold.

In early 1633 CE a great storm hit the Scottish border region when vast numbers of sheep perished. There was also a severe frost. In Scotland snow lay around from December to March.

In England there was a major failure of the harvest.

1633 CE in Cork in southern Ireland, there was a "prodigious flood of the sea". The flood swept away some of the public buildings and bridges.

In 1634 CE Taal erupted in the Philippines with a VEI of 3.

In Summer 1634 loud thunder was heard during clear weather in Tobolsk in Tyumen Oblast, Russia. A cloud appeared and a stony meteorite broke the wooden cupola of the church of Dmitry Solunsky and fell inside the porch causing some damage.

On 5 October 1634, a hurricane struck western Cuba causing 40 deaths.

In Germany there was the Burchardi Flood on October 11-12 1634 in Germany and Denmark. The Burchardi flood (also known as the second Grote Mandrenke) was a storm tide that struck the North Sea coast of Schleswig-Holstein, North Frisia, Dithmarschen (in modern-day Germany) and southwest Jutland in modern-day Denmark on the night between 11 and 12 October 1634. Overrunning dikes, it shattered the coastline and caused thousands of deaths (8,000 to 15,000 people drowned) and catastrophic material damage. Much of the island of Strand washed away, forming the islands, Noordstrand, Pellworm, and several halligen or unprotected islands with no dikes of Sudfall and Nordstrandishmoor. The Nubbel and Nieland halligen were submerged into the sea. The former island of Strand was broken into four pieces. Six thousand people drowned, or two thirds of the entire population of the island. Before 1634 Strand had an area of 210 square miles. After the storm it was 31 square miles in area and composed of two islands, Nordstrand and Pellworm in Nordfriesland. These were also the highest recorded floods in southwestern Jutland. 50,000 livestock died due to 44 dike breaches and the water destroyed 1,300 houses and 30 mills. 17 churches were completely destroyed.

At 8.00 am on October 27th 1634 a shower of stones fell on Charollais, Charolais, France. This is now known as Charleroi and is in Belgium. The stones weighed up to eight pounds. Looks like a meteorite shower to me until we find another explanation.

In autumn and early winter in 1634 CE the far north of Scotland, along with Orkney and Shetland, were plagued by persistently stormy weather, often wet, such that great distress was caused due to famine. The harvest on Orkney was described as a famine.

On March 29th 1635 CE Gamalama in Indonesia erupted with a VEI of 2.

In 1635 a 350-gram meteorite fell on Calce in Veneto, Italy. This was after a blazing mass dashed across the sky.

A large stone fell near Verona in Italy in 1635. Calci is roughly 116 miles south of Verona. Were the Calce and Verona stones from the same shower?

On 14 August 1635, a great storm struck the early settlers in New England in what was to become the United States. The winds from the gale caused the tides to rise to height unknown before. At Boston, Massachusetts, the tides measured 20 feet (6 meters). The Narragansett Indians had to climb to the tops of trees to save themselves. Many failed to do so and were swallowed up by the surging waters. Trees were broken in two or torn up by the roots. The Indian corn, the main dependence of the colonists, was beaten down and destroyed. Houses were blown over. Several shipwrecks were caused by the storm, for there were at this time large immigration of settlers, and a number of ships were near the coast, having onboard many passengers and goods for New England. The ship Great Hope was driven aground near Charlestown, Massachusetts. The ship James with about one hundred passengers was almost driven onto the rocky shore of Piscataqua but the wind changed direction in the final minutes. The ship Angel Gabriel was dashed against the foam-covered rocks off Pemaquid Point in Maine.

On 24 August 1635, a hurricane struck the western Atlantic Ocean and the east coast of the United States causing 35 deaths.

Between 1635 CE and 1703 CE there were warmer conditions over the Pacific Region.

In 1635 at Port-Louis in Brittany. A 60-year-old man named Jean Le Guen, who lived in Riantec near Port-Louis, asserted that he had observed a procession of beings he took to be "angels" in the sky. They were going from Port-Louis to Caudan.

In the Winter of 1635-1636 CE in Western Europe, the carts drove on the ice on the Maas (Meuse) River.

During the winter of 1635-36 in Western Europe, the frost began in December 1635 and took a portion of January 1636.

On March 6th 1636 a large stone fell from the sky between Sagan and the village of Dubrow in Zielona Gora, Silesia, Poland. This was an unclassified stony meteorite.

On May 8th 1636 AM Hekla in Iceland erupted with a VEI of 3.

In Custrinensi Territory during December 1636 a great fiery globe making a terrific sound was seen. There was a Marquis of Custrinensi in seventeenth century eastern Europe. All sources are in Latin. There is a possibility that this was in Silesia. We will meet other localities that over the years have disappeared from the map or changed their names due to wars and conquest.

1636 was a warm year overall with a forward spring and a very hot summer. It was extremely dry. A drought was noted as having lasted from 1st March to well into September with completely rainless conditions. Trees by August were as if it were midwinter, given the loss of leaves. This was in England.

On November 29th 1636 at 10.00 am a stone weighing thirty-eight pounds fell on Mount Vaison in Provence-Alpes-Cote d'Azur, between Guilleaume and Provence in France. This was also an unclassified stony meteorite.

In East Friesland in the Netherlands on the 1st of September 1637 there were great floods.

In Friesland in the Netherlands on 10 September, there was a terrible storm of thunder and lightning.

In Custrinensi Territory during December 1636 a great fiery globe making a terrific sound was seen. There was a Marquis of Custrinensi in seventeenth century eastern Europe. All sources are in Latin. There is a possibility that this was in Silesia.

On July 1st 1637 CE Vesuvius in Italy erupted with a VEI of 2.

On September 29th 1637 Asosan in Japan erupted with a VEI of 2.

On October 3rd 1637 a fireball was seen over France.

In 1638 Raung in Indonesia erupted with a VEI of 4.

On February 24th 1638 CE Grimsvotn in Iceland erupted with a VEI of 2.

A series of mainshocks struck Calabria on March 27–28 and June 9, 1638. The first three earthquakes had moment magnitudes estimated to be M_w 6.6–7.1. On June 9, another mainshock estimated at M_w 6.7 struck the same region, causing further damage and casualties. The four earthquakes resulted in as many as 30,000 fatalities. The official number of deaths stood at 9,571, including 6,811 in Calabria Citeriore and 2,760 in Calabria Ulteriore. The total number of dead was likely higher as these figures did not account for the many deaths that occurred in the months following the first earthquake due to injuries and deprivation.

On March 27, at 22:00, the first and most destructive earthquake struck with an epicenter in the Savuto Valle or near the upper Crati River. It reached a maximum Modified Mercalli intensity level of XI (Extreme) in the heavily populated communes of Martirano, Rogliano, Santo Stefano di Rogliano, Grimaldi, Motta Santa Lucia, Marzi and Carpanzano. The earthquake destroyed much of the settlements in those towns. The town of Amantea suffered total damage, while minor damage was reported at Maratea and Reggio Calabria. According to Ettore Capecelatro, a jurist and official of the Kingdom of Naples, more than 10,000 homes were destroyed, while another 3,000 were rendered unsafe for habitation. Luca Cellesi, the bishop of the Roman Catholic Diocese of Martirano, was injured during the collapse of his castle in the town of Pedivigliano, where he reported that the population of his diocese fell from 12,000 to 6,500 after the quake. In Aiello Calabro, 408 homes were obliterated, and 655 residents were killed. At least 116 inhabitants were killed in Belsito, 234 in Grimaldi, 495 in Carpanzano, 229 in Conflenti, 173 in Malito, 532 in Motta Sta Lucia, 1200 in Nicastro, 102 in Piane Crati, 216 in Sambiase, 451 in Scigliano Diano and 126 in Feroleto. The town of Martirano was destroyed and 517 inhabitants were killed. Following the earthquake, the affected areas saw a decrease in population from migration. Many inhabitants of Motta Santa Lucia moved to Decollatura, and residents of Pedivigliano and Pittarella moved to Sila. Survivors from Scigliano and Carpanzano relocated to the Ionian coast and formed the communes of Mandatoriccio and Savelli. The two earthquakes on March 28 occurred in the southern tip of Calabria on Palm Sunday. One of the two shocks' epicenters was near Nicastro, where 3,000 people were killed. At least 600 of the total deaths in the city resulted from the collapse of a church. Many more residents were killed in Lamezia Terme, Falerna, Feroleto Antico and Sant'Eufemia Lamezia. The quake caused a destructive tsunami in the Gulf of Saint Euphemia. Damage at Sant'Eufemia Lamezia was so severe that the town was abandoned. A second shock occurred in the Serre Calabresi, causing fissures to appear in the ground. Sulfur and flames were reported emanating from the newly formed fissures. The earthquake was particularly destructive in Rosarno and Mileto, while the town centers of Borello, Briatico and Castelmonardo was destroyed. The June 9 nighttime earthquake registered a magnitude of 6.7 and a maximum Mercalli intensity of X. It affected the region of Sila, where six villages were destroyed. Extreme damage was reported in Catanzaro and Crotone, as well as in 13 other villages. It was preceded by two strong foreshocks in the early morning and afternoon, alerting many residents to stay outdoors. Despite the severity of damage, only 52 people were killed.

On July 3rd 1638 CE Sete Cidades erupted in the Azores with a VEI of 2.

At Cherbourg in Manche, France on October 21 1638. Four people were killed by a ball of fire while attending a local church, during a severe thunderstorm. The fireball was reported to have been more than six feet in diameter and entered the church through the ceiling window, destroying part of the roof. The ball smashed several of the pews and exploded, filling the church with a sulphurous odor. The minister of the church later explained the event as the work of the *Devil* who was upset with the good work, they were doing saving souls.

On October 21st 1638 there were tornadoes in Devon and Somerset in England. At Widecombe-in-the-Moor in Devon on the south-eastern flank of Dartmoor, a tornado struck the church of St Pancras with the utmost violence as a service had just begun. It was an afternoon service attended by 300 worshippers. A ball of fire moved through the church with a thunderous explosion. The roof and tower were wrecked, stone and masonry showered down both inside and outside the building. The tornado/ ball lightning killed and maimed scores of men and women, as well as a dog. People were snatched from the pews and whirled about. About sixty people were either killed or injured. It was all over in a few seconds.

In October 1638, an Atlantic hurricane struck off the southern coast of Puerto Rico. Two British ships were lost; two known survivors.

During the winter of 1638, the French galley ships were arrested by the ice at the Marseilles Sea. The winter of 1638 in southern France caused more damage than the winter of 1599-1600. The port of Marseille froze around the galleys.

In 1639 in China a large stone fell in a market. Tens of people were killed, and tens of houses destroyed. Exactly where is unknown. How big a stone was this?

In April 1639 in Yuan in Fengxian in Shansi Province, China. A red, white, yellow and blue "star" flew over a funeral, circling the village for a long time. The villagers were presenting their condolences to the family of Yuan Yingta, a minister of war under the Ming dynasty who had sacrificed himself on the battlefield while resisting the Man army. Suddenly a luminous object like a star, red, white, yellow and blue in color, flew over the funeral procession. This brilliant thing did not touch the ground, but it flew around the village for a long time, then rose up in the sky again. Its light was visible five kilometres away.

In Montbéliard in Doubs in east-central France on 21 June 1639, a cold wave struck that was so strong as in the full cold of winter. In 1639, almost no snow fell on the Alps. There was no rain in Provence, France. The Durance and other rivers dried up. The water level on the Rhône

River was down very low. The Durance is a major river in southeastern France. Its source is in the southwestern Alps.

In July 1639 in Santiago in Galicia, Spain. A short pamphlet published in Seville in Andalusia in 1639 titled "An Account of the Prodigious Visions of Armies of Men, Standards, Flags, Vessels, and Other Things, that Visibly have been Seen over a Long Time, near the Town of Santiago in Galicia, in the Fields of Lerida, since June to this Present Year of 1639", reported that "in Santiago three ships appeared in the air with the sound of drums and many people." Unfortunately, no more information was given. And these aerial phantom fleets would appear again.

In England in October, there were great floods. On 6, 24 and 27 December, there were tempests and hurricanes.

Eruptions that year were Chaiten in Chile which erupted with a VEI of 4, and Mauna Loa which erupted with a VEI of 3 in Hawaii.

in February 1640 CE Llaima in Chile erupted with a VEI of 4,

On April 4th 1640 a fireball was seen over Holland. Some years have barely anything.

On 8 April 1640 there was a sudden overflowing or inundation of the River Weland in Northamptonshire, England to an incredible height. It was called the Easter Flood. The water rose five feet, eight inches above the ground. At the same time there was a great and sudden overflowing of the River Nyne. This caused a great flood. The floodwater flowed into the lower rooms of the houses on both sides of Bridge Street in Peterborough. The rains that caused this flood did not fall on Peterborough but rather in the upper part of the county.

On July 31st 1640 Hokkaido-Komagatake in Japan erupted with a VEI of 5.

On 11 September 1640, a hurricane struck the western coast of Cuba. Thirty-six vessels were affected by the storm. Four ships were thrown on shore. Nearly all the sailors drowned, except 260 that were saved.

In Dresden in Saxony in Germany on the 23rd of September, there were great floods.

In England, on October 11 through 14, there was a most severe frost. It froze up all the rivers and brooks.

In England on 18 October, there was a terrible storm of thunder and lightning.

On December 26th 1640 CE Melebingoy/ Mount Parker in the Philippines erupted with a VEI of 5.

In December 1640 CE Awu in Indonesia erupted with a VEI of 3.

We had two category 5 eruptions in 1640 as well as 2X 4s. Do we add all these up to see what might happen or have happened? What happens when we have two VEI 5s?

During the 1640 drought in China people were reduced to eating bark and there were reports of people eating corpses.

During 1640-1655, there was a famine throughout India, principally felt in the Deccan in southern India and in Bengal. Today this is Bangladesh and the state of West Bengal in India.

In 1641 Kelud in Java in Indonesia erupted with a VEI of 4. Other eruptions were Galeras Volcano in Colombia with a VEI of 3 and Taal in the Philippines with a VEI of 3.

The 1641 Tabriz earthquake occurred on the night of February 5 in present-day East Azerbaijan province, Iran. The earthquake had an estimated surface-wave magnitude of 6.8 and an epicenter between Lake Urmia and the city of Tabriz. It was one of the most destructive earthquakes in the region, resulting in the loss of up to 30,000 lives. The earthquake occurred on a

Friday night. The communities of Khosrowshah, Osku and what is now present-day Azarshahr were completely devastated. Nearly all houses and public infrastructures, including historical monuments in Tabriz were razed to the ground. Many public baths and caravansaries were destroyed. A building in the city collapsed and buried many animals that were taking shelter from the winter season. Two important structures, the Masjid-i Ustad-Shagird and Arg of Tabriz suffered heavy damage. A large number of mosques experienced serious damage to their domes and minarets. The shock was also felt in Baghdad in Iraq. On Sahand, a rockslide was triggered, destroying a village and killing many people. Fissures appeared in the ground and erupted water. The earthquake was misdated to the years 1441, 1049, 1639, 1642, 1646, and 1651. Efforts to recover personal belongings and the dead continued for a month. Aftershocks were felt for six months. In the immediate aftermath of the earthquake, many survivors resided outside the ruins of their homes. Some residents returned to their homes but were killed due to collapses during the aftershocks. During the first two months after the earthquake, up to seven aftershocks were felt in a day.

At 9.00 pm on May 4, 1641 in Madrid, Spain. "The sky being very calm, without there being a single cloud in it, (Jose Pellicer reported) an extremely black and dark cloud, that approached from somewhere between the east and the north, dilated and narrow, crossing between the west and Midday, that was stationary for some time" – giving the impression that Pellicer may have seen something resembling the 'cloud cigars' dealt with in modern reports. Pellicer goes on to mention a burst of sound of unknown origin in the sky over Molina de Aragon. The people there heard "loud noises, bugles, drums, as if an invisible ferocious battle were happening but without anything to be seen."

If there are no UFOs though, then here we have an odd meteoric event.

In England, there were hailstorms on the 25th of June 1641 and the 14th and 19th of August with rain.

In England, there was a great hailstorm on 25 June 1641. There were further hailstorms in England on August 24 and 29 and rain. In 1641, the plague struck Stamford, England. Between 500 and 600 people died.

In Montbéliard in Doubs, France, on 27 July 1641 it froze. In Burgundy, France, the grape harvest began only on 3 October.

On September 14, 1641 at Akhaltsike, Georgia. Armenian chronicler Zacharia Sarcofag saw a strange phenomenon at sunset. The sky was not yet dark when suddenly "the ether on the eastern side was torn up and a big dark-blue light began to descend. Being wide and long, it came down approaching the Earth and it illuminated everything around, more brightly than the sun." The forward part of the light "revolved like a wheel, moving to the north, calmly and slowly emitting red and white light, and in front of the light, at a distance of an open hand, there was a star the size of Venus. The light was still visible until my father had sung, weeping, six sharakans, after which it moved away. Later we heard that people saw this miraculous light up to Akhaltsike." A sharakan is a brief prayer sung over two to three minutes, so the phenomenon would have lasted at least 15 minutes, according to researcher Mikhail Gershtein. It was not reported to have crashed or landed.

On 24 September 1641, an Atlantic hurricane struck from Hispaniola to Florida, in the United States. Eight ships were lost and many people perished. Many ships were lost in the

Bahama Channel. There were no survivors on four of the wrecked ships. There were some survivors on a fifth ship that was lost along the northeast coast of Florida in the United States.

Governor Winthrop in his journal mentioned that the frost was so great that the Boston Bay in Massachusetts, United States, was frozen over from the 18th of November to the 21st of December. The ice was so thick that horses and carts crossed over parts where ships had sailed. Loads of wood drawn by six oxen passed from Muddy River to Boston. It was frozen as far out to sea as one could discern. The great bay at Virginia was also frozen over, and all their great rivers.

In 1641 the Grand Canal in Shandong dried up. Cold also struck much of China including provinces in the south. An official describing the misery in Henen wrote "The people all have yellow jaws and swollen cheeks; their eyes are the color of pigs galls". Desperate people flocked to cities in search of food and there were reports of cannibalism. An account from Shanghai in China in 1641 stated that there was a massive drought as well as locusts, the price of millet had soared and the corpses of the starved lay in the streets. Grain reached three-tenths to four tenths of an ounce of silver per peck.

On April 6th 1642 CE Aira in Japan erupted with a VEI of 2.

During June 1642 burning fist-sized sulfur lumps fell on Loburg Castle in Brandenburg, Germany. Sulfur is not meant to come from the sky. It is generally a product of volcanoes. There were no volcanoes in Brandenburg and this sulfur is not the only sulfur fall ever recorded. You will meet it again and again…..and again. Even the Bible mentions sulfur falls, but under the name of brimstone. But where in the sky is there sulfur? And how could sulfur not burn up completely in our atmosphere? Was this connected to the ammonium increase in the atmosphere that same year that was recorded in Greenland?

At Aldeburgh in Suffolk, England around 5.00 pm to 6.00 pm on August 4th 1642. In the skies over Aldeburgh a large phantom battle occurred; gunfire was heard, and a large stone fell from the sky. What happened to this piece of rock is unclear. It was reported that the hot stone weighed four pounds and fell between Aldeburgh, or Aldborough and Woodbridge in Suffolk.

During 1642, there were three hurricanes in the West Indies. The second lasted twenty-four hours, during which, at St. Christopher's island (St. Kitts), twenty-three fully laden vessels were wrecked upon the coast. One of them belonged to the celebrated De Ruyter. The houses were all blown down, and the whole of the cotton and tobacco plants were destroyed. The salt lakes overflowed their banks and were for some time afterwards unproductive. Michiel Adriaenszoon de Ruyter was the most famous and one of the most skilled admirals in Dutch history. In September 1642, an Atlantic hurricane struck the Lesser Antilles. Men in 22 ships were drowned. In 1642, a hurricane destroyed all the houses on St. Kitts in the West Indies.

Greenland GRIP ice cores indicate a massive amount of ammonium entering the atmosphere in 1642 CE. This indicates that a cosmic body had passed through Earth's atmosphere. Had it landed or burnt out?

Also, in 1642 in the middle of December masses of iron fell at Ofen. The only Ofen that I can find was a free city of the Austrian Empire on the west bank of the Danube opposite Pest. Ofen was originally called Buda which means furnace. Pest also means furnace. Ofen is in Komarom-Esztergom, Hungary. This was possibly December 2nd or 12th.

In 1643 on March 31st Miyakejima in Japan erupted with a VEI of 3.

On April 20th 1643 Karkar Island in Papua New Guinea erupted with a VEI of 3.

On April 21st 1643, the following day, Manam Island in Papua New Guinea erupted with a VEI of 3. Manam is 75 miles northwest of Karkar in the Bismarck Sea on the north coast of Papua New Guinea.

On 23 January 1643, there was a terrible highwater flood at Friesland (now in the Netherlands) whereby much damage was done to the dikes and at the city of Gaes near Haerlingen (Harlingen, the Netherlands), the dead bodies streamed out of the earth as buried dead corpses floated to the surface. Google and none of my collection of atlases could find Gaes but it may well have disappeared in this flood. An alternate spelling was Maes and it could not be found either.

In England on 16 March and 3 May, there were terrible storms of thunder and lightning. The spring was very moist with almost constant rain.

In England, in 1642 and 1643, the summers were excessively hot.

In Italy in 1643, there was excessive heat.

In France, 1643 was a year of great scarcity of grain. In Dijon, France, the grape harvest began on 1 October.

Tree ring data showed that the drought in the final years of the Ming Dynasty may have been the worst since 500 CE. The frequency of drought in northern China was up to 76 per cent higher during the late Ming compared to its early years. Yields of key staple crops fell and the prices rose.

On February 20th 1644 CE Asamayama in Japan erupted with a VEI of 2 and four days later on February 24th 1644 CE Asamayama in Japan erupted again with a VEI of 2.

On 15 & 16 May 1644, there was a hailstorm in Staffordshire and Warwickshire, England. Some of the hailstones were the size of walnuts and others the size of half-crown pieces. "Maii 15 and 16, in divers places there fell great stormes of haile with haile stones of divers formes, some round as big as walnuts, and some flat as big as half-crown pieces; with thunder and lightning in three or four several places at one instant; the like seldom seen."

In Spain on 6 June, there were great floods.

in England on 16-18 June, there was a tempest with high winds.

In Montbéliard in Doubs, France in 1644, the heat for more than two weeks was so strong that the fish died in the rivers.

In August 1644, the plague was violent in Leeds in West Yorkshire, England.

In Dijon, France, the grape harvest began on 15 September.

On 1 October, there was a great flood again in Spain. In Spain and the Netherlands, there were considerable floods.

In October 1644, a great Atlantic hurricane struck the western coast of Cuba and the Straits of Florida causing approximately 1,500 deaths.

The Maunder Minimum ran from 1645 CE to 1715 CE. The Maunder Minimums are periods of little to no sunspot activity. Variations in sunspot cycles normally follow eleven-year cycles. The Maunder Minimum was a sunspot minimum. Sunspot Minima generally coincide with minima in temperature. Between 1645 and 1710 during the Maunder Minimum there was a reduction in solar activity and a rise in carbon-14 activity. This coincided during the height of the Little Ice Age. It appears that there were major fluctuations in global temperature and major changes in carbon 14 levels identified in tree rings. Long term changes in solar radiation may have a profound effect on terrestrial climate.

In 1646 CE in Friesland, in the Netherlands and Zealand, Denmark, there were great inundations. The sea drowned 110,000 people. The sea broke in at Dordrecht in the Netherlands and thereabouts and drowned 10,000 people. About Dullar in Friesland and Zealand, it drowned 100,000 people, and 300 villages, some of whose steeples and towers yet appear when the tide is out.

On March 15th 1646 a fireball was seen over Reutlingen in Baden-Wurttemberg.

In England, there were hailstorms on the 4th of May; 11th and 12th of July; and 17th of August.

On May 16th 1646 a there was a rain combined with sulfur at Hasnia in Italy. More brimstone?

Interestingly enough there was a report of hailstone-sized meteorites at Swaffham Prior, Cambridgeshire also on Thursday 16th May 1646 at 3 o'clock in the afternoon. After a peal of thunder with no rain two long bright flares of light appeared in a cloud that was hovering over the town. One disappeared straightaway and the other burst forth into a pyramid-shaped cloud that then split in two. On May 21st 1646 at Newmarket & Thetford. "Betwixt Newmarket and Thetford in the foresaid county of Suffolk, there was observed a pillar or a Cloud to ascend from the earth, with the bright hilts of a sword towards the bottom of it, which piller (sic) did ascend in a pyramidal form, and fashioned itself into the forme of a spire or broach Steeple, and there descended also out of the skye, the forme of a Pike or Lance, with a very sharp head or point (…) This continued for an hour and a half.

Suffolk has been busy for the last few years.

In May 1646 in The Hague, Netherlands. Strange people and animals appeared in the sky of this city. Coming from the SE a significant fleet of air ships with many sailors (occupants) on board approached this strange spectacle. A gigantic combat followed and at the time of the disappearance of the phenomenon a great cloud appeared where there was nothing before. Here are these aerial fleets again. Are we describing strange phenomena or misinterpreted meteorological phenomena?

Yes, don't worry. There will be many more meteoric reports as we continue on, but I am quite intrigued with these early reports, whatever they are.

In England on 26 June 1646, there was a terrible storm of thunder and lightning with rain.

On July 19th 1646 Kie Besi in Indonesia erupted with a VEI of 4.

In 1646, it was excessively hot in England. In 1646, the plague ravaged Bideford in Devon, England. It also struck Bingham in Nottinghamshire where it raged violently.

On October 2nd 1646 CE La Palma erupted in the Canary Islands with a VEI of 2.

On November 20th 1646 CE Mount Etna in Italy erupted with a VEI of 2.

In England, the weather was variable in 1647, but very rainy in 1648. In England from the harvest in 1647, both of the years (1647 & 1648) were southerly windy, and cold; all very rainy and floody. "This was a most exceedingly wet year; neither frost nor snow all the winter for more than six days in all. Cattle died everywhere of a murrain (cattle disease)."

On February 18th 1647 Asamayama erupted in Japan with a VEI of 2.

Also, on the night of February 18th 1647, a stone weighing half a hundredweight fell from the sky in the village of Polau near Zwickau in Saxony, Germany. Weird coincidence?

In August 1647 a stone fell from the sky near Stolzenau in Niedersachsen, Germany. Stolzenau and Zwickau are only one hundred and ninety miles apart. Stolzenau is almost on the border with Westphalia. Was earth running through a very long duration meteor shower with a limited width?

There was also a fall of stones in Westphalia in Germany in August 1647. Stolzenau is almost on the border with Westphalia. Was earth running through a very long duration meteor shower with a limited width?

Sometime in 1647 an eight-pound stone fell on a ship off Rochefort in Brittany in France. It was reported to have killed two men.

Interestingly enough these four sights are in a rough line travelling NE to SW over seven months. How long was the original meteor shower that it took seven months to cross over the earth, if that actually happened?

On January 8th or 10th 1648 a fireball was seen over Naples in Campania, Italy.

On March 15th 1648 a fireball was seen over Reutlingen in Baden-Wurttemberg in Germany.

On March 20th 1648 CE Asamayama erupted in Japan with a VEI of 2.

On June 15th 1648 CE Gamalama in Indonesia erupted with a VEI of 2.

On August 30th 1648 CE Asamayama erupted again with a VEI of 2.

In 1648 on a Dutch ship named "Malacca" in the East Indian Ocean two sailors were reported killed on board on route from Japan to Sicily by a meteorite. Other sources state that it was a volcanic bomb. This is our second ship death recorded that was actually noted. How many others might there have been in this period of little or no communication except over long time periods?

In England, there were very general floods in England, on 17 January 1649 (at Oxfordshire.

In England on 22nd January 1649, "Now was the Thames frozen over and horrid tempests frown'd." There was a great frost in England in January 1649. The River Thames was frozen over in London. On 22 January, there was a horrid tempest of wind.

In 1649 CE in France there was a flood. In January the Seine River in Paris, France, at the bridge "Pont de la Tournelle" reached a height of 7.65 meters (25.1 feet) above the zero mark (the low water mark of the year 1719). In 1649, continuous rain swelled the River Seine in Paris, France. The waters shook the small bridge "Pont Saint-Michel." In the living memory of the oldest bourgeois, they had not witnessed so great a rise of floodwaters in the Place de Grève (now called the Place de l' Hôtel de Ville) and the surrounding streets. The floodwater even overran the cemetery of St. John.

In February 1649 CE Nikko-Shiranesan erupted in Japan with a VEI of 3.

On May 11th 1649 a fireball was seen in Alsace. Alsace is now in France but has a history of moving from France to Germany and back again.

In England, there were very general floods in England (at Oxfordshire) on 17 June.

In July 1649 CE Asosan erupted in Japan with a VEI of 2.

On August 17th 1649 Asamayama erupted in Japan with a VEI of 2.

On 24 August 1649, there was a hailstorm at Peterborough in eastern England. Some of the hailstones were 9 inches in circumference.

On September 1st 1649 a meteor or fireball that seemed to move in bounds was seen over Hamburg in Germany.

In all September and October it was rainy and there were floods in England.

More of the Heligoland Archipelago in the North Sea was lost due to sea floods in November 1649. Heligoland is now 25 miles out to sea from the present coast and around 38 miles from the old coast.

Sometime in 1649 a purplish nebula entered Loughton Church in Cheshire, England, during a storm. Many people were killed who were inside the church. Was this a form of ball lightning?

We have met it in churches before.

I know, it is not a meteor or meteorite, but it is strange enough for this book.

The plague raged in Ireland and Shropshire in west Midlands England.

In 1649 and 1650, there was a famine in Scotland and the north of England from rains and wars.

In 1649 in Lancashire, England, there was a famine caused by the ravages of the armies. A plague followed.

1650 CE

From 1650 to 1710 temperatures across much of the Northern Hemisphere plunged when the Sun entered a quiet phase called the Maunder Minimum. Other sources say 1645.

Eruptions in 1650 CE were Sumaco in Ecuador with a VEI of 3, Paluweh in Indonesia with a VEI of 3, and Cameroon in Cameroon with a VEI of 3.

During the night of 18 January 1650 CE in England, there was a terrible storm. The cattle were so frightened that most of them broke out of the fields. Some in leaping broke their necks, others their legs. Some ran four miles off and when found were excessively hot.

In Leicester, England on the 29th of April, there was a hailstorm with thunder and lightning.

In Rome, Italy, during the summer of 1650, the heat was very strong and extremely dry. During the whole year, there was most excessive heat and drought, especially in the summer. After the harvest, the scorching heat was succeeded by very great rains and these were followed by a most rigorous cold.

On July 2nd 1650 CE Asamayama in Japan erupted with a VEI of 2.

On August 6th 1650 an iron meteorite fell at Dordrecht in South Holland. This is another iron meteorite which account for five per cent of all meteorite falls. They are believed to have come from the centers of now destroyed asteroids. Dordrecht cops it from all directions.

On September 27th 1650 CE Kolumbo, Kouloumbos, volcano on Santorini in the Mediterranean erupted with a VEI of 6. Sixty cubic kilometers of tephra was ejected. The volcano only appeared above the sea surface in 1649, the year before, and when it erupted it sent a tsunami that caused damage on islands up to 150 kilometers away. The volcano then sank again. The highest parts of the rim are now ten meters below sea level. Poisonous gases released during the eruption blinded or killed many Therans and their livestock. This new island had disappeared. The flames of the 1650 CE eruption were seen from Heraklion in Crete and tsunamis swept among the Turkish ships anchored on the island of Dia, close to the north coast of Crete.

On November 30th 1650 a fiery sword was seen in the sky about sunrise above Gloucestershire in England. To the untrained eye could this be a description of a comet? Ancient drawings of comets often resembled swords in the heavens.

In France, 1650 was noted for a great scarcity of corn; the price was three times higher than in the previous five years.

In 1650, an Atlantic hurricane struck St. Kitts in the West Indies. Twenty-eight ships were thrown on the roadstead of St. Christopher Island; the sailors drowned. During two different hurricanes, a total of twenty-eight merchantmen (merchant ships) were lost along with a great number of lives. St. Christopher island is commonly known as St. Kitts. A roadstead is a place outside a harbor where a ship can lie at anchor.

In 1650 and 1651 in Ireland, there was a famine throughout the country. There was the Siege of Limerick in western Ireland and Siege of Galway in west-central Ireland.

From the 1650s the population of Norway dropped by thirty per cent by the end of the century.

Between 1650 and 1850 CE the glaciers in the Alps regained an extent estimated in the Glockner region at about their Bronze Age minimum, when all the smaller ones had disappeared.

By 1650 woods were standing dead and then falling over and being covered in moss and then falling into the peat. This was in the county of Caithness. This area had been forested since Viking times.

On January 7th 1651 a fireball was seen over Switzerland.

In January 1651, the Seine River in Paris, France, at the bridge "Pont de la Tournelle" reached a height of 7.8 meters (25.6 feet) above the zero mark (the low water mark of the year 1719).

On February 18th 1651 Pacaya erupted in Guatemala with a VEI of 2.

On 23 February 1651 occurred St. Peter's high flood, whereby much damage was done to the dikes in Friesland (now in the Netherlands), Embderland, and elsewhere, and not far from Dockum (Dokkum in the Netherlands) by Oudt-woudumer-ziil, there was a breach of 42 roods (42 rods, 693 feet, 211 meters) long broken in the dike.

From 1st to 4th of March 1651 Holland, Friesland, Zealand in the Netherlands and Flanders in Belgium were hit by sea floods.

On 4 March, a great tide broke down St. Anthony's banks and overflowed all Dimermeer. There was significant damage in north Holland and Amsterdam.

On April 12th 1651 CE Asamayama in Japan erupted with a VEI of 2.

In England on 22 August, there was a terrible storm of thunder and lightning.

In Dorchester in southwestern England on the 23rd of August, there was a hailstorm with stones 7 inches in circumference. In England in 1651, it was very hot days at the time of harvest.

In England, the years 1651-54 produced scorching hot dry summers and dry years. In 1651 CE it was dry with a scorching summer especially in Kent and possibly Western Europe if this were the case.

Scotland was subject to greater dearth than the preceding year.

In Dijon, France, the grape harvest began on 22 September. This was another year in France where wheat was very scarce.

On December 22, 1651 at Almerdor, in the Netherlands, Dutch sailors saw a fleet of ships in the air, with many people and soldiers.

These are our mystery airships again. Misidentified meteoric phenomena? UFOs or just plain imagination?

In December 1651 sea floods hit North Friesland in the Netherlands and hundreds were drowned.

The thunderstorms of the year 1651 produced a great flood year in France. All the rivers overflowed their banks. In Provence, France on September 8th, the Durance River ascended to the gates of Avignon. In November at Grenoble, the Isère River overflowed bridges and fifty houses, drowned fifteen hundred beasts in the country and three hundred in the city. The flood left three or four feet of sand in the streets. The waters rose, they say, more than twenty feet above their usual height. Thunderstorms or rainstorms desolated Provence, France in 1651. The Seine River at Paris, France was so flooded that all houses near it were in danger, and great damage was done.

Two winters with the least Baltic ice in the port of Riga in Latvia occurred in 1651-52 and 1652-53. The winter of 1658-9 produced the opposite.

In 1652 CE Sheveluch erupted in Kamchatka with a VEI of 5. Other eruptions were Babuyan Claro in the Philippines with a VEI of 3 and Aogashima with a VEI of 3 in Japan.

On April 12th 1652 Asamayama in Japan erupted with a VEI of 2. This was the same date as the previous year. Do we have a mistake? Or a coincidence?

In May 1652 near Rome in Lazio, Italy. A single luminous object, 80 metres in size, was seen in the air. A mass of "gelatinous matter" then fell to the ground. Falls of jelly or gel? Was this a one off? Unfortunately, No! The Irish even had a name for it *pwdre ser* meaning Star Jelly. Wait until you meet New England in the United States in 1833.

On October 19th 1652 CE the Picos Fissural Volcanic System erupted in the Azores with a VEI of 2.

In England, the years 1651-54 produced scorching hot dry summers and dry years. In England, the summer was excessively hot and dry. England in 1652 saw a good harvest, particularly with fruit.

In 1652, there was a drought in Scotland. The warmth was very great, the summer being the driest ever known in Scotland. In 1652 the summer was noted for extraordinary drought across the whole of Scotland with high temperatures and a little rain. There was a great impact upon agricultural production, both good and bad. In 1653 the summer was one of great drought, and excessive heat across England. In Scotland the extended winter period was notably dry which would have been a disaster for autumn and winter-sown crops.

The summer of 1652 was very hot and very dry in Denmark and England. In Copenhagen, Denmark, the summer was excessively hot and dry.

At Dijon, France, the grape harvest began on 20 September. This was the third famine year of the grain.

Does this resemble our modern weather? As there is more documentation of events as we progress forward, the dates become closer and patterns appear to be more evident.

On 23-24 September 1652, an Atlantic hurricane struck the Leeward Islands. Three ships and crew went missing. In 1652, Prince Maurice was lost in a hurricane in the West Indies. Prince Maurice was a prince of the elector Palatine region of Europe. He was in a fleet under the command of his brother Prince Rupert.

The River Arve falls into the River Rhone, about 1000 paces beneath Lake Geneva, Switzerland. In December 1652, the River Arve swelled so that it not only overflowed its banks but also interrupted the course of the Rhone and forced it to re-enter in the Lake for a space of 14 hours.

On 13 July 1653, an Atlantic hurricane struck Barbados in the Lesser Antilles and St. Thomas in the Virgin Islands. One ship and crew was lost. At St. Vincent in the Caribbean Sea, there was "death of many savages".

In July 1653, it was so furiously hot in Poland, that in the regiment of foot (soldiers) which was the King's Guard, marching most of them barefooted upon sand, more than 100 fell down altogether disabled (heat stroke), whereof a dozen died outright, without any other sickness.

In Dijon, France in 1653, the grape harvest began on 11 September; 13 days earlier than the mean. In France the price of corn fell by half.

In 1653 CE on December 31st Gamalama in Indonesia erupted with a VEI of 3.

On February 25th 1654 CE Vesuvius, in Italy, erupted with a VEI of 3.

At 8.00 am on March 30th 1654 a shower of stones fell onto the island of Funen in Syddanmark, Denmark. These were stony meteorites.

In 1654 the summer in England was dry and scorching. In Scotland, from Edinburgh and Fife there was a great lack of water with wells drying up. There were lesser problems in the west of Scotland.

It was reported that a Franciscan monk was killed in Milan in Lombardy, Italy by a small stone that fell from the sky on September 4th 1654.

And you thought that fatal meteorite falls were rare?

We had already had reports of boat crew members being killed on two occasions as well as hundreds of people killed in a meteor shower in China.

In China where the climate was markedly colder than the late twentieth century, from 1654 to 1676, in Kiangsi Province, China. Orange and mandarin cultivation which had been practiced for centuries in Kiangsi was virtually abandoned after a succession of cold waves and severe frosts between 1654 to 1676. The trees were repeatedly killed by frost and the growers became fearful of attempting to replace them with new trees. It gets hot in Europe, and it gets cold in China. Like an amazing pendulum.

In this period in China the southwest monsoon was frequently interrupted by cold northerly winds.

During the Winter of 1654-1655 CE Mr. Fehre, Chief Secretary to Prince Radzivil, assures us, that in the war against the Muscovites and Cossacks in January 1655 at the siege of Bichow in White Russia (now Belarus), all their provisions of Spanish wines and peterfimen, and beer were in one night frozen upon the sledge, notwithstanding they were covered with straw. Insomuch that they were constrained to carry them into a stove to thaw them, which they could not do in two whole days, and were obligated to break the vessels, and put pieces of the ice wine into kettles, to thaw them over the fire in order to drink them. But he observed that the Hungarian wine resisted the cold better than the peterfimen. The scrue (screw lid) of a flagon of Aqua Vitæ (ethanol) being put to his mouth stuck close to his lips (froze to his lips) that he could not draw it off without drawing blood. The pool of the village (where they quartered) was so thoroughly frozen, that there was but very little water left between the ice and the bottom. Google would not tell me what peterfimen was. If the transcribing of the f was wrong, it could be petersimen that in Old Danish was "ot slags vin" which means "of kinds of wine".

In France during 1654-1676 CE the northern limit of vine growing in France was permanently shifted somewhat further south. The isotherm appeared to be permanently moving.

In 1654-1676 CE there was speculation in this period as to why the medieval vineyards were abandoned and decayed. There had been lamenting on the bad seasons for the vineyards in the fourteenth and fifteenth centuries by those who were attempting to grow grapes in England.

In 1655 CE Taranaki erupted in New Zealand with a VEI of 4.

In July 1655 CE Pacaya erupted in Guatemala with a VEI of 2.

On November 25th 1655 CE erupted with a VEI of 2.

In 1655 CE in Scotland there was a change in weather of excessive rainfall and snowfall. There were great problems for agriculture and therefore the food supply. Storms killed vast numbers of sheep and in the cold season the frost was severe enough to kill to kill broom and whins (gorse). A severe frost set in during February which lasted until April 15th.

In Paris, France, the Seine froze on 25 and 26 November 1655.

In 1655, the frost began on November 25 in northern France.

In the first days of December, it snowed. In 1655-56, the Seine River in France was frozen from the December 8th to the 18th. From 8 to 18 December the cold was the very great.

On December 10th 1656 Asamayama erupted in Japan with a VEI of 2.

In Scotland on December 10th the east coastal counties of Scotland, particularly Fife, the Lothians and the Firth of Forth estuary were affected by a severe gale from the east. This gale brought high winds, much snow, and the combination of snow and wind for many hours. It drove ships onshore, wrecked vessels in the harbors, and breached sea defenses, piers, harbor walls etc. many trees were also lost and buildings inland damaged or destroyed. For this implied wind direction there must have been a significant high block to the north and strong cyclogenesis to the south. In 1655, it was very cold in Scotland. The excessive snow and rain did great injury this winter. Cyclogenesis means the development or strengthening of an area of low pressure in the atmosphere, resulting in the formation of a cyclone.

From 18 to 28 December, the air was damp. The Seine froze again without interruption from 29 December to 28 January. On 29 December, the frost began again and lasted until 28 January 1656. A new frost occurred a few days later and the river again froze which lasted into March. This winter of 1655-56 in France and Germany was very severe. During the later frost, the cold was less severe than in December.

In Germany, the cold was so great that one could get in Wismar (Mecklenburg-Schwerin) onto the frozen Baltic Sea with a loaded four-horse wagon and travel a distance of 5-6 German miles, which has not been the case for many years. On the land, the wells were frozen to the bottom.

On the roads in Bohemia (now western Czechia), several people were found frozen to death.

In the winter and spring of 1655 to 1656 it was cold in Western Europe which implied as much for Britain.

In 1656 CE in Rome, Italy, there were floods and a famine.

There was a great drought in southern France in 1654-56. Rains were very rare.

In England beginning on 20 July 1656, there was two separate hailstorms in Norwich in quick succession. The following accounts were published. "The most Lamentable and Dreadful Thunder and Lightning in the County of Norfolk and the City of Norwich, on July 20, being the Lord's Day in the afternoon: the Whirlwind and thick darkness, and most prodigious hailstones, which being above 5 inches about, did so violently batter down the windows of the City, that three thousand pounds will hardly repair them. Divers (diverse) men and women struck dead. The firing of some towns, and whole fields of corn, by lightning, which also destroyed the birds of the air and the beasts of the field. Together with another most violent Storm, which happened on Saturday last, in the same County, for almost thirty miles together, performed the like terrible effects. Attested by ten thousand witnesses, who were either spectators, or partakers of the loss. Entered according to order, the 31 July, 1656".

Was the eruption of Taranaki in 1655 responsible for the odd weather?

The drought in England lasted to the spring. In England it was excessively rainy, unequal and southerly.

In July, there were so great rains, which caused the Danube River to flood over its banks. It broke down all the bridges and most of the mills. Many people were lost and a great number of cattle were carried away. Sixteen towns and villages were swept off by the irresistible torrents.

In Denmark, the summer was very unequal with heat, rain and south winds.

On 8 October, the River Thames ebbed and flowed thrice in three hours space.

In 1656, a hurricane struck Guadeloupe in the Lesser Antilles. Every vessel at anchor in the roads was wrecked and most of their crews drowned. The island of Guadeloupe in the Lesser Antilles in 1656 was desolated by a tremendous hurricane. Most of the houses were destroyed. All the domestic animals were killed. All the plantations were laid waste. Every vessel at anchor in the roads was wrecked, and most of their crews drowned.

There were great thunderstorm outbursts in 1657 that were similar to those of 1651 in France. These caused great floods. Camargue, France was buried by the Rhône River. The Camargue is the area where the two arms of the Rhône River form a delta in southern France.

In 1657, a hurricane struck off the Bahamas. Two salvage vessels were sunk in a storm off Gorda Cay.

On 2 August 1657 at Feversham in Kent, England, there was a very high spring tide. The interesting thing was the winds were at southeast, which deads (deadens, diminishes) the tide there.

On November 25th 1657 Asamayama erupted with a VEI of 2.

From 11th December 1657 to 21st March 1658 snow lay on the ground in England. This was one of the longest periods of snow lasting in England. There was a notably severe winter over Western Europe. In some parts of England, the frost lasted from 1st December to 10th March and ice was reported around the coasts. In Scotland the cold lingered through March into April.

In 1658, the bays and inlets of Northern Europe froze over early in December.

In England, there was a frost from 1st December to 10th March with a "north wind".. with a north wind even to January. During the winter of 1658, it was excessively cold in England and the price of wheat doubled. In England, the winter was severely cold. From 1 December 1657 to the Equinox (around March 20/21), the earth was covered with snow. There were north winds the whole time. It continued till 1 June like a winter.

During the winter of 1657-58 in France, an uninterrupted frost occurred from the 24th of December 1657 to the 8th of January 1658. Then the cold moderated. But then an extreme cold wave set in and the Seine River in France was entirely closed due to the ice. A slight thaw took place on 8 February, but the frost again recurred and continued from the 11th to the 18th of February 1658. In Provence many olive trees were lost to the cold. The winter in Paris, France: "it was cold, 24 December 1657 until 20 January 1658, but the cold at that time was not very sharp.

In the Winter of 1657-1658 CE King Charles X Gustav of Sweden was at war with Denmark. An intense cold wave descended on the Small Belt in the middle of December 1657 and it appeared that the Baltic Sea might freeze over. Charles X moved his army from Poland and approached Copenhagen from the south. He arrived at Haderslev in South Jutland on 28 January 1858. The cold on the night of 29 January was very severe. On the next morning, he gave the orders and his army crossed the frozen Small Belt on foot and invaded and conquered the island of Funen. He then traveled on the frozen Great Belt and leapfrogged through the islands of Langeland, Lolland, Falster, and finally his army reached Zealand on 11 February. The Small Belt is the strait between the Danish island of Funen and the Jutland Peninsula. The Great Belt is the strait between the main Danish islands of Zealand and Funen. In 1658, Charles X, King of Sweden, traversed the Little Belt with his army, artillery, caissons, baggage, etc. In 1658, Charles X of Sweden crossed the Little Belt over the ice from Holstein, Germany to Denmark, with his whole army. Charles X of Sweden crossed the strait to Denmark with his whole army, including the artillery, baggage and provision trains.

On 20 January 1658 in France, however; an unusually sharp violent north wind; very few people could remember to have seen such a piercing cold. Everything was frozen. This intense cold lasted until 26 January. On 27 January the weather turned somewhat milder and a hoped for thaw was in the air. But on 28 January, a very deep penetrating cold again reappeared and lasted until 8 February. On 9 and 10 February, the ice and the snow that had fallen in abundance began to melt. But at 2 o'clock in the morning on Monday the 11th, the wind again came from the north and northeast, and it froze the waters anew, and the frost was unusually severe. At sunrise no trace of the previous thaw could be seen. This severe cold lasted until 18 February. Finally on the 19th the winds changed to a northwesterly and then the winds began to blow from the west. The snow and ice again began to thaw and continued without interruption. On 21 February the ice broke, which completely covered the Seine River. On the 22nd the river began to swell. On the 27th and 28th the river came out of its banks and the inundation was greater than anyone could remember. From 6 o'clock in the evening of the 27 February until noon on the 28th, the water washed the walls of the church of St. Andre-des Arcs. One needed wooden planks to cross the street. At noon on the 28th, the water began to fall. Due to the cold several people were killed, others suffered with the loss of family members.

In 1658, the Seine River in France was completely frozen from the first days of January to the 21st. The rivers of Italy froze deep enough to bear the heaviest carts. The winter of 1658 in France destroyed the olive trees. The winter was accompanied by deep snows.

There was great snow in Rome, Italy on 27 February 1658.

In Italy, the rivers were frozen deep enough to carry the heaviest wagon. In Rome there was a tremendous amount of snow.

In England during the spring, the north wind and cold continued so rigorous and long, that farmers lost hope of their corn (grain) either growing or ripening.

In 1658 CE Merapi in Indonesia erupted with a VEI of 3.

On May 22 at Faversham (Kent), England, there were considerable floods.

On July 24th 1658 CE Asamayama in Japan erupted with a VEI of 2.

The summer of 1658 in England was remarkably warm, especially towards the end of the season.

In Modena in northern Italy, there was excessive heat and drought.

In Abdera in Thracia, there was an excessively hot summer. Abdera in Thracia was a city-state on the coast of Thrace, 17 kilometers east-northeast of the mouth of the Nestos, and almost opposite Thasos. The site now lies in the Xanthi peripheral unit of modern mainland Greece.

In Denmark and Copenhagen, there was drought and excessive heat.

From 1 August in England came such an excessive heat, as was truly uneasy.

On 22 August 1658 at Feversham in Kent, England, there was a very high tide in the afternoon, though the wind was southerly, and blew very stiff, which the seaman there wondered at.

In Europe, "the day that Oliver Cromwell died (3rd September), a storm so violent and terrible extended all over Europe." On 3 September 1658, the day that Cromwell died, there was a hurricane throughout Europe, which did very considerable damage. On 3 September 1658, a very alarming and destructive storm struck England in which many houses were blown down and others unroofed. Churches, steeples, and whole groves of trees were prostrated, and immense damage done to the shipping. Among a great many other vessels which were lost with most of their crews, were eight frigates and ships of the line, and two thousand officers and seamen perished. On 3 September 1658, there was a great gale in England. "It was such a night in London as had rarely been passed by dwellers in crowded streets. Trees were torn from their roots in the park, chimneys blown down, and houses unroofed in the city. Cromwell died that night." There was another great gale immediately before this one, which struck throughout Europe.

In Dijon, France, the harvest began on 30 September.

In October 1658 CE Hakusan in Japan erupted with a VEI of 2.

On November 3rd 1658 San Salvador in El Salvador erupted with a VEI of 3.

In 1658, there was a plague in London, England.

In 1658 in London, England, the tide in the River Thames ebbed and flowed twice in three hours.

In February 1659 Kirishimayama erupted in Japan with a VEI of 2.

In June 1659 CE Gamalama in Indonesia erupted with a VEI of 2.

At New-Market-Heath, in Suffolk in England at 10.00 pm in March 1659. It is observable that not long after this wonder from Heaven, had presented itself to the inhabitants of the Earth, but a bright star of a great magnitude, was seen glittering and sparkling up and down, even like unto that in the west; whose luster was so large an extent, and appeared so conspicuous, that although the air was dark, yet bright and serene were the glorious rays which streamed forth by its great splendor as if the Heavens received all its serene from its illustrious influences. This Star was seen for the space of two nights constantly about ten of the clock; yet in several forms; as sometimes like a fiery Dragon; sometimes like a flaming sword, and sometimes in an oval form casting forth a round flame as big as a bushel, and divers sparkling coals.

Meteorite or comet or UFO again? Take your pick.

On April 21st 1659 Hakusan erupted with a VEI of 2.

On July 24th 1659 Asamayama erupted in Japan with a VEI of 2.

In November 1659 CE Grimsvotn erupted in Iceland with a VEI of 2.

On November 11th 1659 CE Teon in Indonesia erupted with a VEI of 3.

On 8 December 1659, a gale struck England. A remarkably high wind, such as had never before been experienced in this country, did great damage to the houses in York.

In Leicester and Nottinghamshire. Starting at 1.00 P.M. people observed an object "in the perfect figure and form of a black coffin, with a fiery dart and a flaming sword flying to and again, backwards and forwards the head of the said coffin, which was with great wonder and admiration beheld by many hundreds of people." This was seen until 3:15 P.M., when it broke up with great brilliance. The five strange wonders, in the north and west of England as they were communicated to diverse honorable members of Parliament, from several country gentlemen and ministers, concerning the strange and prodigious flying in the air of a black coffin betwixt Leicester and Nottingham, on Sabbath day last a fortnight, with a flaming arrow, and a bloody sword, casting forth firearms of fire.

The Winter of 1659-1660 CE in Provence and Italy was very cold again. The olive trees were destroyed, almost completely.

In 1659, the frost was severe in England. The price of wheat doubled during the winter.

In 1659-1660, it was very cold in England, and the price of wheat doubled during the winter.

In 1660 two strata-volcanoes erupted on Long Island, also known as Pono, Arop and Ahrup in Madang Province in Papua New Guinea. Virtually all of the biota on the island was destroyed. The VEI was 6 and it would have caused a volcanic winter. 30 cubic kilometers of tephra was ejected. . The Long Island eruption deposited tephra across the New Guinea Highlands. This prompted legends of the "Time of Darkness". There were large pyroclastic flows from the eruption associated with a third phase of caldera collapse.

Other eruptions were Planchon-Peteroa in Chile with a VEI of 3, and Lewotolok in Indonesia with a VEI of 3

On February 23rd 1660 a fireball was seen over Wittenberg in Saxony-Anhalt in Germany.

In February 1660 CE Teon in Indonesia erupted with a VEI of 4.

At 8.00 pm on March 14th 1660 in London (Westminster) England, boat passengers saw a dark, then bright cloud dropping fire over Westminster. About 8 P.M. they observed "a white bright cloud which gave such a light that they could plainly see the windows of the Parliament House, and people walking to and fro upon Westminster Bridge". The cloud was seen to "drop down fire several times upon Westminster Hall and then it removed and (flew) over the Parliament Hall and did drop down fire upon that also several times".

On April 8th Asamayama in Japan erupted with a VEI of 2.

One evening in August 1660 at Statford Row, near London. The likeness of a "great ship" was seen in the air. It decreased in size and eventually disappeared. The worthy chronicler does not fail to inform us that "this is testified by an able Minister living not far from the place, who received the information from the spectators themselves".

One evening in September 1660 in London. "A gentleman of good quality and an Officer of Eminency in the late King's army and now a Justice of the Peace in the Country" reported seeing a bright light in the SW, along with six smaller ones. "Whilst he with several others, were with some admiration beholding them, they all fell down perpendicularly and vanished."

On the evening of October 3 1660 in Hull in England. The soldiers on guard at the South Blockhouse saw a large fiery object tapering off at one end and leaving a narrow stream behind. It was so brilliant that they could read fine print or take up a pin from the ground by its light. This object was in sight for half an hour. Someone who was approaching Hull that same night, coming from Lincolnshire, confirmed the first report: "He saw a very great light in the sky, whereby he could perfectly discern his way, though it was exceedingly dark." The whole relation-continues our chronicler-" is signified by letters from several eminent men in Hull who spoke with the eyewitnesses, as also by some inhabitants of London, who upon occasion have been at Hull since that time, and there from very good hands have received credible information concerning the premises."

On October 11 1660 in Hertford in England. A person of very good note and credit awoke at 4 am to see "a flashing like fire against his window and fearing some house near him had been on fire, he immediately arose and went to the window." He saw a large object with a circle around it, and two appendages above and below it, from which great flashes were indeed emitted. This object remained in view for several hours and was observed by others in the town.

Meteors generally only stay around for a few seconds at most…..but there might well be exceptions.

On the evening of October 17 1660 in Shenley in Hertfordshire. Five "exceedingly bright and glorious" naked men were seen in the evening sky. Were these bright men or bright stars or meteorites?

On October 27th 1660 Guagua Pichincha in Ecuador erupted with a VEI of 4.

On November 3rd 1660 Katla in Iceland erupted with a VEI of 4.

On the evening of November 30 1660 at Ilford in Essex. Very early in the morning two men saw a fiery cloud in the southwest. From under it appeared two bright objects as large as the moon, which began a dog fight in the atmosphere. One of them eventually grew dimmer while the other increased in size and remained in view for two hours, "a great part of which time they saw streaming from it streams of fire and streams of blood." It then diminished until it was no larger than an ordinary star.

On December 1 1660 at Houndsditch in London. At 5 am an inhabitant of Houndsditch saw an unexplained, bright object the size of the moon in the eastern sky.

In 1660 there was a sighting by two Dutch ships in the North Sea of an object moving slowly in the sky. It appeared to be made by two disks of different sizes. Who in the seventeenth century would even imagine a flying disc let alone two? There are no comparable aerial objects or creatures to compare or base flight on. But remember that in that same year aerial lights were seen that were also described as flying discs or flying dragons or flaming swords.

In Norway in 1660-1670 CE there was a great advance of glaciers.

There was a mild winter in England from 1660 to 1661 CE. Sameul Pepys, the diarist, mentions in late January that there had been a general lack of cold weather and that it was dusty which implied a warm and dry winter with plants well ahead for the season. In Central England the CET record shows an average at 5 degrees Celsius or roughly one and a quarter degrees Celsius above the all-series mean. The CET record is the longest instrumental record of temperature in the world and stands for Central England Temperature record.

In 1661 CE Ubinas in Peru erupted with a VEI of 3.

On 14th to 15th January 1661 sea floods hit North Friesland in the Netherlands and Hamburg in north Germany and many buildings and people were swept away.

On 18 February 1661, a great and dreadful storm of wind accompanied by thunder, lightning, hail and rain struck England. The damage was estimated at a little less than 2 million of money. Several people were killed by falling chimneys, houses, trees, barns, and windmills. Five or six were killed in London; one in Chiltenham (Cheltenham); one in Scaldwel (Scaldwell); one in Tewksbury (Tewkesbury); two near Elsbury; one at Northampton; one at Colchester; two near Ipswich; and 3 near Langton, to name a few. Many churches were damaged by the wind. This damage included broken windows and stonework. The lead was torn up. Pinnacles, spires and steeples were thrown down. In some cases, the timberwork in the church was broken and the pulpit and pews were damaged.

These included the churches at Tewksbury (Tewkesbury), Red Marly (Redmarley), Newin, Worcester, Hereford, Leighton Beau-defart, Eaton-Soken (Eaton Socon), Shenley, Waddon (Whaddon), Woolston (Little Woolstone), Finchinfield (Finchingfield), and Ipswich, to name a few. Many houses and buildings were blown down, and others extremely shattered and torn. Many barns were destroyed. These included 30 barns near Ipswich; an incredible number near Tewsbury (Tewkesbury); 11 barns at Twyning; 7 or 8 in Ashchurch; 5 in Lee; a great number at Norton; 140 in Worcestershire; 16 in Finchinfield (Finchingfield); and at least 15 at Wilchamsted (Wilshamstead), to name a few. Many trees were blown down. These included 1,300 trees in Bramiton Bryan Park; 600 in Hopton Park; and 3,000 oak trees in His Majesty's Forest of Dean, to name a few. Several persons lost whole orchards of trees in the counties of Gloucester, Hereford and Worcester. The winds were strong enough to take people up into the air. On the bridge near Wallingford House in London, several people were blown off and landed on top each other. In Herefordshire, a man was blown over a very high hedge. In a pond at James's Park, 2,000 fish were blown out and landed on the banks. Several people were blown into the air at Hereford. The rain that fell was as salty as brine.

In February 1661 at Darken in Surrey. A "discreet sober gentleman" saw a strange cloud in the evening sky, and two objects he compares to cathedrals or churches, "having upon it diverse goodly Pinnacles, and each of them a long streamer flying upwards upon it, and as he beheld it, he thought it grew up to a greater splendor and glory." The other object was darker. After a while, the large one emitted puffs of vapor and disappeared, while the smaller one grew and became brighter. The witness was called into his house and could not observe the end of the phenomenon.

On February 21, in Kent, England, there were considerable floods.

On March 20, 1661 at Canterbury in Kent. A very large "star" with an "opening" underneath, from which issued streams of fire was seen for thirty minutes.

The following day on March 21 1661 two armies were seen in the sky over London in England. Other witnesses saw auroral lights.

In England, there were great hailstorms on 11th April and 11th October.

On April 14th 1661 CE Asamayama in Japan erupted with a VEI of 2.

On April 23, 1661 at Bednall-Green, now Bethnall Green, in London. People saw a great pillar of fire with smaller objects (compared to "burning coals") within it, and at 10 o'clock that night "several persons near Piccadilly saw strange fiery clouds and other objects very terrible to the spectators, from some of whose mouths we received the information".

These aerial phenomena were being seen more often as the seventeenth, eighteenth and nineteenth centuries progressed. Was it from better observation techniques or increased events.

In April 1661 between Ilford and Romford in Essex. About 10 P.M. Captain Chelmsford, of Ipswich, and another man riding to London saw a fiery light with a green-white glow that changed direction. It approached at great speed, emitting light beams. When it was exactly overhead it suddenly changed direction again and disappeared at the horizon. Upon arriving in London, the two travellers had a notarial deed drawn up, recording their experience. A similar aerial mystery was seen here in November the previous year. Do some areas attract more phenomena than others?

In 1661, there was a drought in England. The River Derwent was so wonderfully dried up that in many places there was no water, and people might go over dryshod (without wetting the feet).

On 9th August 1661 a meteorite smashed through the roof of a house in China but there were no injuries. The meteorite broke the roof beam.

In 1661, upon Michaelmas day, 29th September, there was a great overflowing of the River Severn in England that drowned the low grounds lying by it.

On October 21st 1661 CE Asamayama erupted with a VEI of 2.

1661 CE was another mild winter but also a wet one in England. It was apparently wet over North and West Britain. The average CET mean was roughly 2°Celsius above the series average. There was no frost or seasonable cold.

In 1661 there was a famine in the Northwest Provinces and Punjab region in India. A famine struck India during the third year of the reign of Aurungzebe. "The rents of the husbandmen (farmers) and other taxes were remitted. The treasury of the Emperor was opened without limit; corn (grain) was bought in the provinces where the produce was least, conveyed to those in which it was most defective, and distributed to the people at reduced rates. The great economy of Aurungzebe, who allowed no expense for the luxury and ostentation of a court, and who managed with skill and vigilance the disbursements of the state, afforded him a resource for the wants of the people."

In 1661 CE there was a very general and terrible famine in India. Muhammad Amin Razwiny wrote: "Life was offered for a loaf, but none cared for it; rank was to be sold for a cake, but none cared for it. For a long time, dog's flesh was sold for goat's flesh and the pounded bones of the dead were mixed with flour and sold. Destitute at length reached such a pitch that men began to devour each other and the flesh of a son was preferred to his (son's) love. The numbers of the dying caused obstruction in the roads." Emperor Aurangzeb carried relief through every corner of his dominions. Whole provinces were delivered from impending destruction and many millions of lives were saved. He accomplished this by remitting taxes that were due. He expended immense sums out of the treasury in procuring grain from Bengal and the countries, which lie on the five branches of the Indus (which suffered less on account of the great rivers) and transported the grain by land and water to the interior provinces. The grain was purchased at any price, with public money; and it was resold at a very moderate rate.

On January 26th 1662 CE Kujusan erupted in Japan with a VEI of 2.

On the night of 17th to 18th February 1662 a major severe gale affected the southern half of Britain. Pepys wrote it was too dangerous to go out of doors with several people. 5 or 6 people were killed in London and several elsewhere across southern England. Houses were damaged and destroyed in London. There were also major falls of trees with above 1,000 oaks and as many beeches are blown down in the Forest of Dean. Also 57 elms were felled on an estate at Nettleton in Wiltshire. The amount of tree damage was such that a commission was set up to enquire into the state of English forests as these were important to the sustenance of the Royal Navy.

On 30 July 1662 there was a prodigious storm of hail at Ormskirk in West Lancashire, England. Hailstones were four inches about and more.

In the afternoon, on Macclesfield Forest, Cheshire, rose a pillar of smoke twenty yards broad and as high as a church steeple, which making a hideous noise, went along the ground for six or seven miles, leveling all before it. It threw down strong stone fences and carried the stones to a great distance from their former places. But falling on a moorish ground, it did little damage. Its noise frightened cattle; they ran out of its way and were saved. A cornfield it passed over was laid flat with the ground, as though it had been trodden with feet. It went through a wood and tore up 100 trees by the root. Coming into a mowed field with hay ready to be carried off, it swept all away so as scarce a handful was ever found. From this forest it went to Taxhall, then to Waily-bridge, and then to the Derbyshire Mountains, where it vanished. This is an interesting description of a tornado. Storms that produce tornados are very energetic and also produce hailstorms. The pillar of smoke is the funnel of debris brought up by the tornado. Tornados create loud sounds like freight trains. They either break large trees in half or uproot them

At noon on August 6 1662 in the High Tatras, at Mount Slavkovsky Stit in Slovakia. A luminous body hit the top of Slavkovsky Hill and smashed it into pieces. Next the UAP, Unidentified Aerial Phenomenon, landed at the Strba village and vanished. The event was witnessed by many people and later described in the Levoca town's chronicles. The UAP came from the Polish side of the border. On 9th August there was an earthquake and a major rockslide at the Lomnicky Peak and the altitude of Slavkovsky Stit, 2,453 meters, fell by three hundred meters. Something impacted with the mountain to reduce its height by three hundred meters! This must have been a big something and more than likely a meteorite.

In September 1662 CE Kirishimayama in Japan erupted with a VEI of 2.

In England, a very hard frost occurred on 28th November 1662.

In 1662, the winter produced a strong frost in England. In London, the River Thames was partially frozen over towards the end of November. In this frost, ice skates were introduced into England from the Netherlands. On 1st December, the king witnessed the performance of skating.

On December 10th 1662 there was a fireball seen in the sky and detonations heard at Novy-Ergi in Novgorod Oblast in Russia. This was followed by a shower of stones.

Meteorite falls are rare, are they?

In the Winter of 1662-1663 CE there was intense frost at Paris, France from the 5th of December 1662 to the 8th of March 1663. The Seine River in France froze in December 1662 completely. During this winter of 1662-63, which was very severe, the frost in Paris, France lasted from 5 December until 8 March. 6 In 1662 in northern France, there was a sustained frost from 5 December until 8 March. The cold moderated on three occasions.

Teon in Indonesia erupted with a VEI of 3 on January 18th 1663.

An earthquake was felt sharply in New England, though the date recorded for the event was 26 January 1663, as New England was using the Julian calendar at the time. A church record entry made by Reverend S. Danforth from Roxbury, Massachusetts (600 km from the CSZ) indicated the initial shock was felt around 6 pm that evening and several more shocks followed the next morning. On the shores of Massachusetts Bay, the tops of chimneys were broken on houses and pewter (a malleable metal alloy) was jarred from shelves. This level of damage is consistent with a modified Mercalli intensity of VI though this may have been because the early colonials had the capability of producing only relatively weak mortar. Using this MMI value and the distance from the epicenter one can estimate the magnitude of the earthquake using published intensity-attenuation relations. In a June 2011 report on the earthquake that was published in the *Bulletin* of the Seismological Society of America, John E. Ebel, a professor and researcher at Boston College, used these known relations that apply to earthquakes in northeastern North America and determined the magnitude to be 7.3 – 7.9. Great landslides along the Saint Lawrence, Saint-Maurice, and Batiscan Rivers made these rivers muddy after the shock, with the waters of the St. Lawrence being affected for up to one month. Near Trois-Rivières several waterfalls were transformed by these landslides, and one waterfall on the St. Maurice River near Les Grès was said to have been nearly leveled. At Saint-Jean-Vianney, Quebec, there was a large earthflow landslide in a sensitive clay, interpreted to have been caused by the 1663 earthquake. In 1971 this was the site of another much smaller earthflow that destroyed 41 houses and killed 31 people.

A severe frost occurred from the 28th of January to 11th of February 1663.

On February 5th 1663 along the St. Lawrence River in Canada many varieties of luminous phenomena preceded an earthquake. I include these luminous phenomena in this book as they are UAPs as well. As you read on you will find that UAPs are not rare during earthquakes either. They are called earthquake lights and are possibly releases of piezoelectricity or gases igniting.

Xxx 15.09.24

The 1663 Charlevoix earthquake occurred on February 5 in New France (now the Canadian province of Quebec), and was assessed to have a moment magnitude of between 7.3 and 7.9. The earthquake occurred at 5:30 p.m. local time and was estimated to have a maximum perceived intensity of X (*Extreme*) on the Mercalli intensity scale. The main shock epicenter is suggested to have occurred along the Saint Lawrence River, between the mouth of the Malbaie River on the north and the mouth of the Ouelle River on the south. A large portion of eastern North America felt the effects. Landslides and underwater sediment slumps were a primary characteristic of the event with much of the destruction occurring near the epicentral region of the St. Lawrence estuary and also in the area of the Saguenay Graben. The event occurred during the early European settlement of North America and some of the best recorded firsthand accounts were from Catholic missionaries that were working in the area. These records were scrutinized to help determine the scale of damage and estimate the magnitude of the quake in the absence of abundant records from that time period.

On February 5th 1663 along the St. Lawrence River in Canada many varieties of luminous phenomena preceded an earthquake. I include these luminous phenomena in this book as they are UAPs as well. As you read on you will find that UAPs are not rare during earthquakes either.

The 8th of February 1663 was a very hard frost in England.

On March 13th 1663 a fireball was seen over Malmoe in Sweden.

In April 1663 CE Unzendake in Japan erupted with a VEI of 2.

On May 14th 1663 a fireball was seen over Chiloe Island in Chile.

On August 15th 1663 there is a report of a huge fireball hovering over a lake for one and a half hours at Robozero in Russia. This is almost one and a half hours longer than the average meteor sighting.

On August 16th 1663 Toya in Japan erupted with a VEI of 5.

Popocatepetl in Mexico erupted with a VEI of 3 on October 13th 1663.

On December 7th 1663 the diarist Sameul Pepys noted "the greatest tide that ever was remembered in England to have been in this river, all Whitehall having been drowned". In December 1663 CE a flood driven by gales submerged Whitehall in London. It was produced by a high tide that was said not to have been exceeded for more than two hundred years.

On December 11th 1663 CE Unzendake erupted again with a VEI of 2.

Merapi in Indonesia erupted on December 31st 1663 with a VEI of 3.

In 1664 CE Pacaya in Guatemala erupted with a VEI of 3.

In 1664, a violent hurricane at Guadeloupe in the Lesser Antilles destroyed their potato crop.

San Martin in Mexico erupted on January 15th 1664 with a VEI of 3.

It was reported that a fireball had been seen in the sky over Saxony in Germany on April 8th 1664, the same day.

On August 3rd 1664 a meteor left a bright streak or train of light over Hungary.

On December 18th 1664 a fiery meteoritic phenomenon was seen in the sky over Croatia.

In the Winter of 1664-1665 CE in England there was a frost from 28th December to 7th February. In England in the latter end of 1664 began a most severe frost which continued to the latter end of March 1665. In 1664 in England until the beginning of March, there was a very violent frost that froze up all things from the beginning of winter.

The winter was very severe in France.

In Belgium there were very severe frosts and heavy snowfalls.

On 2 January 1665, the frost was so bitter in Poland that three soldiers died from the cold in passing a long ditch; and that diverse persons lost some of their limbs (to frostbite). The winter of January 1665 was similar to the winter of January 1655 in Poland. The winter in Poland was so severe that most of the wines froze and several people lost their limbs (due to severe frostbite), and others froze to death.

In January and February 1665, there were sharp frosts in England.

The 6th of February 1665 was described as "one of the coldest days, they all say, ever felt in England."

In 1665 CE in England, there were great flooding of rivers, very great floods, both from rains and from the Sea.

Was this extremely cold and volent weather caused by Popocatepetl in October 1663.

In England in February, there was a great tempest, accompanied by thunder, and lightning.

On July 2nd 1665 a great grey ball descended during a storm in Norfolk, England. It did great damage, killed one person and lamed others.

On July 2nd 1665 a great grey ball descended during a storm in Norfolk, England. It did great damage, killed one person and lamed others.

In England in 1665, the whole summer was very temperate; neither cold nor hot; dry nor rainy; but pleasant mild breezes which fanned the air and kept it healthy. Great plenty of all sorts of fruits, good and cheap.

In August 1665 plague struck London in England. The diarist Samuel Pepys recorded in his diary "This month ends, with great sadness upon the public through the greatness of the plague, everywhere through the kingdom almost. Every day sadder and sadder news of its increase. In the City died this week 7,496; and all of them, 6,102 of the plague."

In 1665, there was a great plague in England. One account gives the death toll at 68,000 people in London. Another account by Defoe reports the plague "May to July severe; August and September 8,000 persons died weekly; in the middle of September, 12,000 persons in one week, and 4,000 in one night; and in the whole 100,000 died." After an order to kill cats and dogs, it is said that 40,000 dogs and 200,000 cats were destroyed. In a third account in 1665-66, "in London 68,596 persons are said to have died of the plague". At Yarmouth, great havoc was made by the plague. The plague was very fatal at Derby. "The country people refused to bring their commodities to the marketplace, depositing them outside town; then retired to a distance till the buyer had deposited his money in a vessel filled with vinegar." At Winchester, the dead were carried out by cartloads at a time, and the plague was as bad as in London. At Eyam, 259 persons perished.

In 1665, there was a cattle plague in London, England.

On 25 October 1665, there was a great gale in London, England

Between the 16th and 17th November 1665, a storm did great damage to shipping all around the southern and eastern coast of England, with tidal surges reported at Yarmouth and inundation across the low-lying lands of Lincolnshire. The same storms also affected the Dutch significantly. There were further storms around 24th/25th with more significant damage, coastal flooding and shipwrecks, including Irish coast. There was a deep depression that brought the lowest barometric pressure ever measured in London, around 931 millibars.

On the 30th November 1665 English ships were trapped in the port of Hamburg in Germany. This was obviously an anticyclonic spell that had allowed continental temperatures to fall significantly.

In mid-December 1665 the River Thames in London was frozen over and blocked by ice by the end of the month.

On 17 December 1665, it was very cold in London, England.

As late as 1665 CE the total Norwegian grain harvest was reported to be only 67 to 70 per cent of what it had been in 1300 CE. In west Norway this was not exceeded until the middle of the eighteenth century. The total number of farms in 1665 was less than it was in 1300 CE and over the next one hundred years many farms were overrun by advancing glaciers and their land partly destroyed by avalanches, rockfalls, floods and landslides. On the Hardangar Vidda Plateau small new glaciers were formed, one of two which survive as dead ice today.

In 1665 and 1666 when London experienced its last episode of the plague and the city burnt down in September 1666, occurred in the middle of the coldest century of the last millennium. The two summers were very hot compared to normal.

In 1666 CE the Lassen Volcanic center in California erupted with a VEI of 3.

In 1666 CE there was a very great drought in England.

On 24 January 1666, there was a tempest at Hampshire in England with a thunderstorm at Andover.

In 1666 at Cransted in Kent, England, there was a shower of fishes. A great tempest of thunder and rain, and, although no ponds about, two acres were scattered over with whitings of the size of a man's little finger. This occurred on the Wednesday before Easter. Easter Sunday was April 25th.

On 10 May, there was a tempest of thunder and lightning at Oxford. On 12 May 1666 at Oxford there was a thunderstorm.

On July 2nd 1665 a great grey ball descended during a storm in Norfolk, England. It did great damage, killed one person and lamed others.

On 17 July, there was a hailstorm on the coast of Suffolk with some hailstones nine inches about. On 17 July 1666, a violent storm of hail fell on the coasts of Norfolk and Suffolk in east England. At North Yarmouth, the hailstones were comparatively small; but at Snape Bridge, one was taken up which measured a foot in circumference; at Seckford Hall, one which measured nine inches; and at Melton, one measured eight inches. At Friston Hall, one of these hailstones, being put into a balance, weighed two and a half ounces. At Aldborough, several of them were as large as turkeys' eggs. A carter (a person who hauls goods in a cart) had his head broken by hailstones even though he was wearing a stiff felt hat. In some places the wound bled, and in others, tumors arose. His horses were so pelted that they fled taking his cart with them. The hailstones were white, smooth without, and shining within.

In England on the 31st of July, there was a severe hailstorm and rain.

During the summer of 1666, there was an extreme drought in Somerset, England. In the moors (bogs) between Yeovil and Bridgewater, the dried pasture showed the outline of trees beneath. They were dug up and there was hundreds of oaks as black as ebony. The summer of 1666 was hot and dry in England.

In 1666, there was a plague at Sandwich and Stamford, England. At Stamford, upwards of 380 people died. In 1666, diarrhea prevailed in London, England and lasted until 1672.

In 1666, a hurricane struck Antigua in the West Indies. During the hurricane, two English warships were lost in the English harbor with a great loss of life. On 28 July 1666, Lord Willoughby fleet set sail with seventeen sails (sailing ships) and nearly 2000 troops (and took possession of St. Lucia.) On 2 August, his fleet was off Guadeloupe. On 4 August, three frigates and some smaller vessels were sent in, and destroyed the French ships in the Saints, Barbados. Symptoms of an approaching hurricane made Lord Willoughby extremely anxious for the return of the ships from the Saints; but the commanding officer's ship had suffered some damage and could not be refitted before night. At 6 p.m., the gale began from the north, and continued with great violence till midnight, when after a calm, which lasted for a quarter of an hour, it shifted suddenly to the east-southeast driving everything before it with irresistible violence. Every vessel and boat upon the coast of Guadeloupe was dashed to pieces. All the vessels in the Saints were driven on shore. The whole of Lord Willoughby's fleet, only two were ever heard of afterwards. An armée-en-flute of twenty-two guns got to Montserrat with only the stump of her mizenmast standing, and a fire ship got to Antigua, dismasted. The bottom of one ship was washed on shore at Cabsterre (Capesterre), Guadeloupe, and another at the Saints: the whole coast was covered with the wrecks of masts and yards) a figure from the stern of Lord Willoughby's ship was recognized among the ruins. The hurricane lasted twenty-four hours: houses and trees were blown down, and a great number of cattle killed. The sea rose and was driven to an unusual height. All the batteries, walls of six feet thickness, near the sea, were destroyed, and guns, fourteen pounders, were washed away.

The storm was felt at St. Christopher's Island (St. Kitts) and Martinico, (Martinique) but with less violence.

On 14-15 August 1666, a hurricane struck Guadeloupe in the Leeward Islands of the West Indies and Martinique in the eastern Caribbean. The hurricane caused approximately 2000 deaths. (Alexander (1902) notes, "17 sail with 2,000 troops...only two were ever heard of afterwards". Other references indicate that additional ships may have survived.

In summer 1666 summer temperatures in the south of England were 18° Celsius in July and 17° Celsius in August which indicated a rough anomaly of +2° Celsius. The heat, added to the dry weather, aided the extended risk of fire in populated areas. The climatological summer of 1666 was amongst the top ten or so warm summers in the CET series. CET is the longest record of temperature in the world and started in 1659. The letters CET stand for Central England Temperature.

In Dijon, France, the grape harvest began on 10 September; 14 days before the mean.

In England in 1666, there was a sundry of (many) tempest of thunder, lightning, rain, hail and wind.

Every month from November 1665 to September 1666 was dry. By August 1666 the River Thames at Oxford was reduced to a trickle. This drought contributed considerably to the Great Fire of London where many of the houses in London had a high proportion of timber, and old dry timber, in them.

In August and September 1666 there was a drought, and the prevailing east wind had dried the wooden houses of London so that they were like tinder. It was intensely hot and dry. There were east winds. The Great Fire of London occurred. The largest fire that ever occurred in London, England commenced on 2 September 1666 and continued for four days, and consumed thirteen thousand houses, eight-six churches and public buildings. St. Paul's Cathedral was among the number. The buildings were all destroyed on 400 streets.

In England in October, there were great floods on October 14 and 16.

In 1666, Lake Constance in Switzerland flooded.

In Lincolnshire, England on 13 October 1666, there was a dreadful storm of thunder, accompanied with hail, the stones as large as pigeon or even pullet eggs, followed by a storm or tempest, attended with a strange noise. It came with such violence and force, that at Welbourn, it leveled most of the houses to the ground. It broke down some trees and tore up other trees by the roots. It scattered abroad much corn and hay. One boy only was killed. It went on to Willingmore, where it overthrew some houses and killed two children in them. Thence it passed on and touched the skirts of Nanby and ruined a few houses. Keeping its course to the next town, where it dashed the church steeple in pieces, furiously damaging the church itself, both stone and timber work. It left little of either standing, only the body of the steeple. It threw down many trees and houses. It moved in a channel, not a great breadth. Otherwise, it would have ruined a great part of the country. It moved in a circle and looked like fire. It went through Nottinghamshire, where the hailstones were nine inches about. The whirlwind was about 60 yards broad. On Nottingham Forest, it broke down and tore up at least 1,000 trees, overthrew many windmills, overturned boats on the River Trent. In a village of 50 houses, it left only 7 standing. This was a description of a tornado.

There was an outbreak of the plague in Naples from 1666 to 1668 CE.

In the winter of 1666 to 1667 CE it was cold over Western Europe which implied parts of Britain. There was cold weather and hard frost in London on 31st December. At the end of December 1666, there was a very hard frost in London, England.

Ubinas in Peru erupted with a VEI of 3 in 1667 CE.

On the 1st January 1667 the Thames was covered in ice. The overall temperature figure for the three winter months of December, January and February shows an anomaly of -1.5° Celsius on the all-series mean. December was around one degree C below average, but January was bitterly cold with an approximate anomaly of -3° Celsius. February was about average, but this was followed by a very cold March. March was very cold with an anomaly of -3° Celsius which is perhaps one of the top 5 coldest Marches in the CET series. In the beginning of January 1667 in England, there was a hard frost from the 15th of February to the 19th of March.

In 1667, "On the 16th of March, a sharp northeast wind began to freeze very strongly, the sea that lies before Amsterdam; the Y on the 17th was solid; on the 18th we went from this city on the ice to North Holland; the Zuiderzee was completely frozen, and several ships were stuck in the middle of ice, which by the 1st of April is stopped." The "IJ" which sometimes shows up on old maps as "Y" or "Ye" is a river formerly a bay in the Dutch province of North Holland. So, the "Y" reference in this instance is the old Zuider Zee, that today in the area of Amsterdam is the LJmeer Lake. The winter in 1667 was very severe in the Netherlands, but extreme cold occurred late in the season, from 16 March to 1 April. 6

The 1667 Dubrovnik earthquake was one of the three most devastating earthquakes to hit what is now modern Croatia in the last 2,400 years, since records began. The earthquake occurred at around 8 in the morning on April 6, 1667. Survivors of the event witnessed a rumbling sound followed by a tremendous kick that rocked the city. It occurred at about 8:45 in the morning and lasted between 8 and 15 seconds. This event is thought to be the biggest one in the history of Dalmatia and practically defines seismic hazard in the coastal area of Croatia. The entire city was almost destroyed and around 3,000 to 5,000 people were killed. The city's Rector Simone Ghetaldi was killed and over three quarters of all public buildings were destroyed. At the time, Dubrovnik was the capital of the Republic of Ragusa. The earthquake marked the beginning of the end of the Republic. Many of the city buildings were reduced to rubble, with a majority of the ones that remained standing suffering significant damage. It is assumed that the large scale of destruction is due primarily to two factors: the previous earthquakes of 1520 and 1639, and the poor properties of the adhesives, prepared using brackish and sea water, used in the construction of the buildings. The Rector's Palace, the Major Council Hall and Sponza Palace all suffered severe damage. All the buildings lining the Stradun were destroyed, and passage through the street was blocked by rubble. Although the Franciscan Church and Monastery weren't destroyed by the earthquake, the subsequent fire destroyed much of the monastery, as well as its great library. Citizens of the city witnessed huge stones rolling down the hill of Srđ destroying everything in their way. Large cracks appeared in the land, and the city's water sources dried up. The dust created by the destroyed buildings was thick enough to obscure the sky. Later, a powerful tsunami devastated the port, flooding everything near the shore. Strong winds fueled fires from homes and bakeries, and the resulting blaze would not be extinguished for almost 20 days.

In 1667 CE in England, the air was cold and wet. Winds were from the north. Summer was very unequal.

In 1667, Nottingham and London, England were visited by the plague.

On 19 August 1667, a hurricane struck the island of Nevis in the Leeward Islands of the West Indies. Before the hurricane struck, there was a high mountain that was all green with trees. But afterwards, in most places it was bare. The wood lying in such a condition, with half trees, or stumps, or quarters, that one would think it almost incredible.

In Montbéliard in Doubs, France in 1667 the summer was very cold and dry. There was not a single month throughout the year in which it had not frozen. In Burgundy, France, the grape harvest began on 28 September.

On 1 September 1667, a tremendous hurricane desolated the island of St. Christopher (St. Kitts) in the West Indies. The hurricane blew with such violence, that all the houses and buildings were blown down. The inhabitants sought shelter from its fury by throwing themselves flat upon the ground in the fields. The French governor reported: "I hold myself obliged to inform you that this island is in the most deplorable state that can be imagined and that the inhabitants could not have suffered a greater loss or been more unfortunate except they had been taken by the English. There is not a house or sugar works standing, and they cannot hope to make any sugar for fifteen months to come. As for the manioc, which is the bread of the country, there is not one left, and they are more than a year in growing … I assure you that if peace is not made, or men-of-war sent to this country to facilitate the bringing of cassava from the other islands, that the inhabitants and troops will die of famine".

On 6 September 1667, an Atlantic hurricane struck Virginia in the United States. Virginia was originally called Wingandacoa. Buried in the ruins were much goods and many people and many lives were lost.

On September 23rd 1667 CE Shikotsu in Japan erupted with a VEI of 5. This was a volcanic winter eruption.

On November 15, 1667 at Mittelfischach in Baden-Wurttemberg, Germany. An engraving preserves the sighting of a "terrible sign of wonder" that took place during sunrise and was seen for several hours in the sky over the town of Mittelfischach. The image shows the sun shining through a break in the clouds while a group of people watch a formation of round lights. There is a scene of battle in the sky, and three crosses among dark nebulosities. The village is shown in detail to the left, with its church and a few houses.

The 1667 Shamakhi earthquake occurred on 25 November 1667 with an epicenter close to the city of Shamakhi, Azerbaijan (then part of Safavid Iran). It had an estimated surface wave magnitude of 6.9 and a maximum felt intensity of X (*Extreme*) on the Mercalli intensity scale. An estimated 80,000 people died. The city of Shamakhi is reported to have been completely destroyed. The city walls were said to have collapsed, as were the fortress and the congregational mosque. Baku was also affected with the wall of a palace reported collapsed. Landslides were mentioned in historical sources and some of the roads were so badly damaged that caravans had to find alternative routes. A death toll of 80,000 people is widely reported in sources but a lower value of 6,000–8,000 deaths has also been mentioned.

Shortly after 1667 CE after abandoned farms had been taken up again, troubles in Norway started again. The frequency of landslides, rockfalls, avalanches and flood damage increased dramatically and taxes had to be reduced. By the 1680s and 1690s these incidents had multiplied manifold and glaciers themselves were overrunning farmland.

In February 1668 CE Asosan erupted in Japan with a VEI of 2.

On June 19th or 21st 1668 after midnight a shower of stones, of which one weighed three hundred pounds and another two hundred pounds, fell near the village of Vigo, or Vago, seven miles from Verona in Veneto, Italy. These were H6 chondrites.

A major earthquake occurred during the rule of the Qing dynasty in Shandong Province on July 25, 1668. It had an estimated magnitude of M_s 8.5, making it the largest historical earthquake in East China, and one of the largest to occur on land. An estimated 43,000 to 50,000 people were killed, and its effects were widely felt. Its epicenter may have been located between Ju and Tancheng Counties, northeast of the prefecture-level city of Linyi in southern Shandong. The earthquake was felt in 379 counties, 29 of which experienced catastrophic damage. It also affected Jiangsu, Anhui, Zhejiang, Fujian, Jiangxi, Hubei, Henan, Hebei, Shanxi, Shaanxi, Liaoning, and Korea. There was a 1,000 kilometre (620 mile) radial zone of damage around Tancheng, Linyi and Ju County. It is considered one of the most destructive in Chinese history. The earthquake produced strong shaking assigned XII (*Extreme*) on the Modified Mercalli intensity scale, the most destructive shaking an earthquake could achieve. Seismic intensity VIII was over 16,800 square kilometers (6,500 sq mi) corresponding to an elliptical-shaped area along the fault zone.

Northern Anatolia in Turkey was struck by a large earthquake on 17 August 1668 in the late morning. It had an estimated magnitude in the range 7.8–8.0 M_s and the maximum felt intensity was IX on the Modified Mercalli intensity scale. The epicenter of the earthquake was on the southern shore of Ladik Lake. It caused widespread damage from as far west as Bolu and as far east as Erzincan, resulting in about 8,000 deaths. It is thought to be the most powerful earthquake in Turkey. The town of Bolu was reported to be almost completely destroyed by the earthquake, with 1800 fatalities. There was also severe damage further east along the fault, with another 6,000 reported casualties between Merzifon and Niksar. Some damage was also reported from as far east as Erzincan and at various locations along the Black Sea coast. The walls and towers of Samsun Castle were damaged and some parts of the structure "were demolished".

In August 1668 CE Asosan erupted again in Japan with a VEI of 2.

Also, in August 1668 CE Zaozan erupted in Japan with a VEI of 2.

As well in August 1668 CE Pacaya in Guatemala erupted with a VEI of 2.

During the winter of 1668-69, Captain (Zachariah) Gillam on the Nonsuch catch wintered in the southern end of the Hudson Bay in Canada. He anchored in Rupert River on the 29th of September 1668. On the 9th of December, the river was frozen up. In April 1669, the cold was almost over.

In 1669 CE Zaozan in Japan erupted with a VEI of 3.

In 1669 CE in England, the entire year was dry.

Mount Etna in Italy erupted on March 11th 1669 with a VEI of 3.

On April 15th 1669 CE Asamayama in Japan erupted with a VEI of 2.

On 20 June at Inspurg, there was a violent tempest of rain, hail, thunder and lightning. At Schwatz in northeastern Germany, the river overflowed, drowned all the neighboring fields and carried down 30 houses and drowned 200 people. Inspurg might possible be Insberg which is located south of Salzburg in the Austrian Alps.

In July at Holstein, Germany was a tempest with thunder and lightning, which so frightened the cattle, that many hundreds of them were lost. At Mecklenburg, Germany, there were many fires kindled by lightning in several parts of the country.

In the Netherlands in 1669, the spring and early summer by the continued influence of the north wind were exceptionally cold. The months of July, August and September influenced by a west wind were intolerably hot.

In Dijon, France, the grape harvest began on 11 September.

In England on 7, 8, 9, 12, and 20 August, there were severe storms of thunder and lightning. On 18 August 1669, there was a mighty torrent of water from Pendle Hill, which flooded the village of Worston in Lancashire, England. The furniture floated about in the houses.

On 17 August 1669, a hurricane struck a Caribbean Island near Nevis and Cuba causing 182 deaths.

On 1 September 1669 at Weymouth in Dorset, England, there was a very high tide. It was unusual because the weather was very calm and the little wind that was being at northeast, in the past contributed nothing at all to the tides in this haven.

A little before 23 September 1669 in the Atlantic Ocean, a hurricane struck and several Newfoundland ships were cast away by a storm. The news was reported by a vessel from Rochel.

On 23rd October 1669 on the east coast of Scotland there was a great storm of wind, rain and thunder and caused great losses on both land and sea. Ships were lost. Even in the harbor at Dundee. In the Firth of Tay some of the islands used for grazing cattle were submerged by the sea and all the beasts drowned. Trees were uprooted in many places.

On 30 October 1669, there was a frightful hurricane of whirlwind in Northamptonshire, England. On 30 October 1669, a tornado passed through Ashley (Ashley Green) in Northamptonshire, England. It was 60 yards wide and was on the ground for only around 6 minutes. It took a milkmaid's pail from off her head and carried it many score of yards distance. In one yard, it threw over a wagon breaking off the wheels and axles and blowing three of the wheels over a wall. It demolished the roof of the parsonage.

Before 9 December 1669, an Atlantic hurricane struck St. Kitts in the West Indies. During the storm, 25 merchant ships and others were cast away. (Another account relates to a 19 December 1670 account that a violent hurricane lasting eight hours struck St. Christopher's Island (St. Kitts) about the end of September last. (Thus, this event may have occurred in September 1669.)

The winter of 1669 to 1670 showed CET averages of -2° C to 2.5° C below modern day averages in Scotland. This was much colder than modern averages.

In 1670 CE eruptions were Galeras Volcano in Colombia with a VEI of 3, and Masaya in Nicaragua with a VEI of 3.

On April 26th 1670 CE Zaozan erupted in Japan with a VEI of 2.

On June 1st 1670 Aira in Japan erupted with a VEI of 2.

On 26 December 1669 in London, England, there was a great cold spell, freezing quickly for several days. After which there was a great snow. This cold spell was much colder than the winter of 1665 and the winter of 1666.

In 1670 between 25 January and 11 February, there was frost in England. In 1670, the frost in England was most intense this winter.

The Little and Great Belts were frozen, and many people perished. The Great Belt in Denmark (Danish: Storebælt) is a strait between the main Danish islands of Zealand (Sjælland) and Funen (Fyn). The Little Belt separates Fyn from Jylland. In 1670, sleighs traveled safely across the Little and Great Belts.

In 1670, the cold was intense throughout Europe. The Danube River was frozen so hard that it carried people, horses and wagons.

In Italy and France, there was severe cold. The extreme cold (in France) during January and February destroyed a large number of trees. The Academy of Sciences compares the cold from 1669-70 in northern France to the winters of 1608 and 1709. Rigor, in January and February, killed lots of trees.

In 1670 between 25 January and 11 February, there was frost in England. In 1670, the frost in England was most intense this winter and was intensely cold.

There was a flood in England on March 10.

On 7 October 1670, a hurricane drove all the fleet on shore in the harbor (Île à Vache, Haiti), except (the bloodthirsty pirate Henry) Moran's vessel, all of which, except three, were got off again and made serviceable.

In Bridgewater (Somerset), England, there were great floods on October 9 1670 CE.

On 13 October 1670, there was a very violent gale at Braybrook in Northamptonshire, England. It was only 6 yards wide. This most likely refers to a tornado event.

On 7th November 1670 a meteorite fell and broke the roof beam of a house in China.

In 1670 CE there were tremendous frosts and snow in Scotland.

The year 1670 was hot and dry in southern France.

In 1671 CE Iya erupted in Indonesia with a VEI of 3.

On February 27th 1671 at noon, stones weighing ten and nine pounds respectively fell at Oberkirch and Zusenhausen near Ortenau in Suabia, Swabia, now in Baden-Wurttemberg, Germany. They were unclassified stony meteorites.

On March 31 1671 at Schamaki near Erzerum in Georgia in the Caucasus Mountains. During an earthquake, vast numbers of fireballs fell from the sky. What connection is there between fireballs and meteors and meteorites and earthquakes? Are some earthquakes actually the result of impact events?

On August 18, 1671 at Regensburg in Bavaria, Germany. There were signs in the sky: An engraving shows an amazed crowd staring at ships in the sky, various mythical animals and armies arrayed for battle.

In August 1671 CE Pacaya in Guatemala erupted with a VEI of 2.

Around the 30th September 1671 hundreds of people in Flanders, Holland and Zealand were drowned in a sea floods.

On 9-11 December 1671, a storm of freezing rain struck Bristol, England causing vast destruction of trees about Bristol, Wells, Shepton-Mallet, Bath and Burton. There was no ice on any water but the rain froze as it fell. A branch from an ash tree weighed ¾ pounds (0.3 kilograms) had 16 pounds (7.3 kilograms) of ice on it. The ice being five inches (13 centimeters) in circumference on the branch. The trees along the highway from Bristol to Shepton were all thrown down. Also at Burton, the roads were all blocked up by fallen trees. The same ice storm struck Oxford. This weather was immediately followed by great heat. The bushes and the flowers were as forward as usually seen in April. An apple bloomed before Christmas. Some travelers were almost lost (died) from the coldness of the freezing air and freezing rain. All trees, young and old, on the highway from Bristol to Shepton, were so thrown down on both sides of the ways, that the road was unpassable (impassable).

Due to similar obstructions, the (mail/newspaper) carriers of Bruton were forced to return back. Some told me that riding on the snowy downs, they saw this freezing rain fall upon the snow, and immediately freeze to ice, without sinking at all into the snow, so that the snow was covered with ice all along, and had been dangerous, if the ice had been strong enough to bear them. Many travelers were stranded on the roads during their journey and were in great distress. On the 8th much snow had fallen. The incidences of freezing rain seemed to vary by elevation. The frost was very fierce and dangerous on the tops of hills and plains.

On 11 December 1671, one young man returned from a 5-mile journey, as he entered a warm room, he cried out in extreme torment in all parts of his body because of the unsufferable cold. After these frost were over, there was an excessive heat wave. As a result, an apple tree blossomed before Christmas. On New-Years-Tide, this apple tree bore apples as big as one's finger end. This freezing rain also affected the countryside around Oxford. And when the heat wave came, green apples were observed in diverse trees, particularly in the parish of Holywell. On 8 December (1671) fell a great snow. On 9 December, there was much rain, which swept off the snow. On 10 December, sudden fits of cold and warmth. Some travelers were almost lost by the freezing air and rain. Trees young and old, were torn and broken down; for the freezing rain falling on the freezing snow on the boughs, and presently turning to ice, broke down the trees. This frost was the same in Oxfordshire as in Somersetshire, a raining of ice or rain freezing as it fell and succeeded by the like heat. Great was the damage done to exotic plants by this and the frosts of 1683, 1684, 1709, 1716 and 1740.

On February 8, 1672 off Cherbourg in La Manche, France. Captain Isaac Guiton reports that a "star" came down; it split into two "ships", while a third one appeared later. The original reads: "An hour past midday, by the calmest weather in the world, appeared to us a star over our heads, about fifteen feet long. From there it went and fell to the north, leaving some smoke that formed into two ships, each with two lights and the mizzen and their large sails folded, both sailing into the south. The one on the north side was larger than the southernmost one. And as they sailed thus, they separated by about four feet, and another ship formed in the middle, seemingly bigger than the others, all black, and turning its bow to the north without any sails, yet equipped with its masts and ropes, as if resting at anchor. This seemed to us to take over half an hour. After which, they vanished to the south without leaving any trace…". This is the first mention of L'Astre Cherbourg or the Cherbourg Star. The solitary star would return to Cherbourg in 1822, twice, 1826, 1836,1838, 1841, 1848,1850, 1854 and 1905. And then it all stopped. Previously in 1638 a six-foot-wide fireball had crashed through the window of a church killing six people. UFO events were also seen in the late Twentieth Century in the same area.

On April 24th 1672 CE Fayal erupted in the Azores with a VEI of 2.

In May 1672, the drought lowered the water in the l'Yssel (sometimes called Gelderse l'Jssel River in eastern Netherlands) and the Rhine River in Germany. The river was fordable on one arm of the river at several locations. This allowed the army of Louis XIV to cross the river on June 5.

On July 12th 1672 CE Iwakisan erupted in Japan with a VEI of 2.

In 1672 CE on August 4th Merapi in Indonesia erupted with a VEI of 3.

In 1672, several great and violent rains fell in many parts of England in summer and harvest and washed away both corn (grain) and soil of many great fields. After this (winter), there were very long heats, causing excessive sweating both by day and night. Trees budded, flowers appeared as in April or May. On 2 September, there was shocking thunder and lightning at Leeds in West Yorkshire in north-central England. in England, there were tempests on 24-25 July, 2 September, 29 October, 19-20 December and 28 December. Some tempests produced great tides.

In December and winter of 1672-1673 there was a possible great storm after Christmas. There were widespread reports of damage due to low winds from the Channel Islands as far north as Richmond in Yorkshire and Dunfermline in Scotland and eastwards to the Low Countries. This seemed to be a vigorous, rapidly deepening depression crossing Northern Britain with a tightening gradient on its rearward flank. There was a great flood in Worcester on the River Severn and many shipwrecks attributed to stormy weather.

In Japan in 1672 flying luminous objects were seen during an earthquake.

On May 20th 1673 CE Gamkonora in Indonesia erupted with a VEI of 5.

In England, there were several tempests, which occurred on 16th February, May 25th (rain, thunder, lightning), June 23rd (rain, a spout), September 10-11th, and October 11th . In England, 1673 was a cold unseasonable bad year, and a very late lean harvest.

On August 12th 1673 CE Gamalama erupted in Indonesia with a VEI of 2.

At Hartshead in Yorkshire, England, an inundation struck on 11 September. In England in 1673, the year was cold and full of rough days. The harvest was late and the yield was poor.

In Dijon, France, the grape harvest only took place on 5 October.

In 1673, an Atlantic hurricane struck off the coast of Puerto Rico. A warship was wrecked but most of the (500) pirates made it ashore to Puerto Rico (alive).

In 1673 there was a fall of a stone at Dietling in Bavaria, Germany.

The 1674 Ambon earthquake occurred on February 17 between 19:30 and 20:00 local time in the Maluku Islands. The resulting tsunami reached heights of up to 100 meters (330 feet) on Ambon Island killing over 2,000 individuals. It was the first detailed documentation of a tsunami in Indonesia and the largest ever recorded in the country. The exact fault which produced the earthquake has never been determined, but geologists postulate either a local fault, or a larger thrust fault offshore. The extreme tsunami was likely the result of a submarine landslide. In an account by Georg Eberhard Rumphius, a German botanist, the earthquake occurred on a Saturday evening, at 7:30 pm local time, when locals on the islands were celebrating the Chinese Lunar New Year. The bells of the nearby Victoria Castle on Ambon Island began to clang by themselves. The earthquake was so strong as to knock people off their balance. Seventy-five stone buildings reportedly collapsed, killing 79 Chinese and five Europeans and injuring 35 others. Most of the injuries were fractured arms and legs. Among the casualties in the collapses were European settlers. Water began sprouting up from wells and the ground, some spurted upwards to a height of 20 feet (6.1 meters). Blue clay and sand also erupted from the ground. Many homes and roads on other parts of the island were cracked and severely damaged. Both Rumphius's wife and two daughters were killed during the earthquake after they were crushed by a falling wall. They were among the 31 Europeans who died in the earthquake and tsunami. Right after the earthquake, a large tsunami reportedly swept through the coast of the island. On the Hitu peninsula, the waves were thought to be as high as 100 meters (330 feet), nearly topping the coastal hills. Entire forests and plantations were uprooted and washed away. The tsunami was accompanied by a deafening noise. When it slammed into the coast, eyewitnesses described the flow as very dirty and foul smelling.

In the Winter of 1673-1674 CE the Zuiderzee is completely frozen in the Netherlands; "On 16 March we crossed it on foot, on horseback and sleigh on the ice between Stavoren and Enkhuizen in the Netherlands". Stavoren is a town in Friesland on the coast of then Zuyder/ Zuidersea, now the Ijsselmeer Lake. Enkhuizen is now in North Holland also but on the opposite side of the Ijsselmeer Lake. The winter of 1674 was remarkable in the Netherlands because of its severity and because of the late arrival in February.

In Europe near Marienburg in Borussia in what was Germany on 5 February 1674, there was a severe frost, which lasted to 25 March. Marienburg is now called Malbork in northern Poland. Borussia is a Latin name for Prussia.

On 4 April we skated on the sea at Haarlem, the Netherlands. Haarlem is in the peninsular region of North Holland.

In March 1674 there was a thirteen-day blizzard on the Scottish borders. Most of the sheep perished. The weather was extreme, and the CET shows an anomaly of -4° Celsius.

In July 1674 CE Pacaya in Guatemala erupted with a VEI of 2.

The summer of 1674 was one of the coldest on record with an average temperature of 13.7° Celsius. It was very wet which delayed the harvest.

On 1st August 1674, a great storm of whirlwinds (tornados), thunder, lightning and hailstones of prodigious bigness struck the Netherlands. At Amsterdam, many trees were torn up by the roots, ships sunk in the harbor, boats in the channels, houses beaten down, and several people were snatched from the ground as they walked the streets and thrown into the canals. The great and ancient Cathedral Church at Utrecht was torn in pieces by the violence of the storm and utterly destroyed. (This was actually the nave which had collapsed. The tower and the East End still remain) The vast pillars of stone that supported it were wreathed like a twisted club. Hardly any church or house in the town escaped the violence. France and Brussels (in Belgium) also suffered infinite damage from this storm.

On 10 August 1674, a hurricane struck Barbados in the Lesser Antilles blowing down 200 houses, and destroying the plantations, so that the inhabitants made but little sugar the two succeeding years. Eight ships were wrecked in the harbor, and 200 persons killed.

On October 6th 1674 two large stones fell from the sky in the Canton of Glarus in Switzerland.

Sometime in 1674 two Swedish sailors were killed on board a ship by a meteorite. Where is a mystery.

In 1675 CE Lewotobi in Indonesia erupted with a VEI of 3, as did Karangetang also in Indonesia with a VEI of 3.

On February 16th 1675 CE Asosan in Japan erupted with a VEI of 2.

In England on 24-25 May 1675 CE, there was a terrible storm of thunder and lightning.

In June 1675 CE Kujusan erupted in Japan with a VEI of 2.

In England, the summer was exceedingly rainy. The harvest was very unequal, like the months of March and April, sometimes clear; sometimes cloudy or rainy.

On 10 September 1675, a hurricane struck Barbados in the Lesser Antilles causing greater than 200 deaths.

In Burgundy, France in 1675, the grape harvest began on 14 October.

In November 1675, a storm so violent struck the Netherlands and caused several breaches in the great diques (dikes) near Enchusen and others between Amsterdam and Harlem (now Haarlem). Forty-six vessels were cast away at Texel and almost all the men drowned. These breaches caused a great inundation, which caused much damage. Many people, cattle and houses were lost. In November 1675 there were two sea floods that hit the coasts from Flanders to Schleswig-Holstein.

The winter of this year was not so severe. There was neither rain nor snow. A north wind in spring made intermittents very rife. Lightning can change the polarity of magnets. "Towards the year 1675 two English vessels went together on a voyage from London to Barbadoes (Barbados). Near the Bermudas, lightning broke the mast of one of them and tore the sails; the other received no damage. The captain of the second vessel having remarked that the first turned about and appeared to wish to return to England, asked the cause of this sudden determination, and learnt, not without astonishment that his companion thought he was following the route as at first. An attentive examination of the compasses of the vessel that had been struck showed them that the Fleur de Lys, or arrowhead, which usually stands on the compass-card for the north-pointing pole now indicated the south; so that the poles had been completely reversed by the lightning. It continued in this state during the whole remainder of the voyage."

In England there was a terrible storm of thunder and lightning on 26-27 December with hurricane.

Between 1675 and 1705 CE an area between Ireland and the Faeroe Islands in the Atlantic seems to have been about 5° Celsius cooler than the modern average in the late twentieth century. This was the year that the cod fishery collapsed and there were no cod seen around the Faeroes for the next thirty years. The cod, which thrives best in rather cold waters at between 4° C and 7° C has its kidneys fail at temperatures below 2° C and cannot venture into colder seas.

Between 1675 and 1677 it was reported that a stone fell from the sky into a fishing boat near Copinsha, or Copinsay, in the Orkney Islands of Scotland.

This would have been a surprise to the fishermen.

Did the boat sink?

How many boats or ships were sunk each year by meteorites before the advent of radio?

On January 24th 1676 a fireball was seen over Switzerland.

Twenty-eight days later on February 21st 1676 another fireball was seen over Switzerland.

Meteorites don't repeat locations you say?

Or do you just get a better view of the sky from Switzerland?

On March 21st 1676 a meteor larger than the moon was seen forty miles up at an estimated speed of 9,600 miles. This was over Italy.

On March 31st 1676 a great meteor was seen over Italy and Dalmatia. It was twice the size of the moon and went from east to west around 7.00 pm. It fell into the sea near Corsica with a detonation. It was seen from 174 miles high to 40 miles high. There was a hissing sound when it fell into the sea.

On April 8th 1676 a fireball was seen over Monte Pulciano in Tuscany, Italy.

The summer of 1676 was one of the 20th warmest summers across England and Wales with a CET value of 16.8° Celsius. June 1676 with a value of 18.0° C was the second warmest month in the CET series.

On September 22 1676 a fireball or meteor was seen over most of England.

In 1676 landslides and torrents destroyed farmland at Buar in Hardangar, below the Folgefonn ice sheet. This was in Norway.

In winter the Thames and the Derwent rivers were frozen. Huts to sell brandy were set up on the Thames. This was the first of three successive notably cold winters with considerable impact on wildlife and the populace. The winter of 1676 to 1677 had an anomaly on modern day winters of between -2° C and 3.5° C which is exceptionally cold.

On December 31st 1676 CE Gamalama erupted in Indonesia with a VEI of 2.

During the Winter of 1676-1677 CE the Seine River in France was frozen for thirty-five consecutive days from 9 December 1676 to 13 January 1677. Extreme cold reigned from 2 December 1676 to 13 January 1677 in northern France. The earth was covered with snow, and the river remained frozen thirty-five days. Then came wet weather.

In 1677 CE eruptions were Merapi in Indonesia with a VEI of 3, and Ubinas in Peru with a VEI of 3.

The Maas (Meuse) River remained frozen from Christmas until the 15th of January 1677. On the frozen river Meuse, they travelled from Christmas to 15 January with heavily laden wagons over the ice.

In February in England, we had a few mild frosts and frequent rains.

A few mild frosts and frequent rains prevailed in March in England. The sky was almost completely overcast.

The beginning of April in England was still cold and wet, but around the middle of the month, the temperature was mild, but soon afterwards came the cool weather again which held until 22 May.

On May 2nd 1677 El Misti erupted in Peru with a VEI of 2.

In Sweden (exact location not given) on the night of May 22-23 1677. On the night between the 22nd and 23rd of May, more than 200 people observed 'two armies' in the sky---one in the north, one in the south. Accompanying each army was a 'large star'. At first the army in the north drove the southern army back for some distance, but towards the end the struggle became more in favour of the latter, who finally drove the former away.

We are getting a good idea what these actually are.

On May 28th 1677 in the evening stones containing a considerable amount of copper fell in Ermendorf which is not far from Grossenhain in Saxony, Germany. Copper meteorites are not unknown, but they are rare. The most famous being the Eaton meteorite that fell in May 1931 in the United States.

July 1677 was very warm or even hot with a CET value of 18° C which was around 2° above the all series mean. This was after one of the coldest winters on record and before another winter record breaker in England.

On 4 November 1677, a low-intensity earthquake was felt in the area around the Bōsō Peninsula, but was followed by a large tsunami, which killed an estimated 569 people. The Bōsō Peninsula in Japan was struck by a major tsunami on 4 November 1677, caused by an earthquake at the southern end of the Japan Trench. It was felt onshore with only a maximum of 4 on the JMA intensity scale, but had an estimated magnitude of 8.3–8.6 M_w. The disparity between the maximum intensity and the magnitude estimated from the tsunami suggest that this was a tsunami earthquake. There are no records of significant damage caused by the shaking, but the resulting tsunami caused widespread damage and an estimated 569 people were killed.

On November 17th 1677 CE La Palma in the Canary Islands erupted with a VEI of 2.

On February 6th 1678 at Frankfurt-am-Maine in Hesse a fireball was seen in the sky.

On February 22nd 1678 Akita-Yakeyama in Japan erupted with a VEI of 2.

On March 1st 1678 CE Aira erupted in Japan with a VEI of 2.

Also, on March 1st 1678 CE Kirishimayama in Japan erupted with a VEI of 2.

In the beginning of July 1678, after some gentle rainy days, which had not swelled the waters of the River Garonne more than usual, one night the river swelled all at once so mightily, that all the bridges and mills above Toloufe (Toulouse in southwestern France) were carried away. In the plains which were below the town, the inhabitants who built in places which by long experience they had found safe enough from any former inundation, were by this surprised, some were drowned, together with their cattle, others only saved themselves by climbing trees or getting to the tops of houses. Others who were looking after their cattle in the field were warned by the horrible noise and furious torrents of water and fled but could not escape without being overtaken. At the exact same time the two rivers of Adour and Cave, which fall from the Pyrenean Hills (Pyrénées Mountains), as well as the Garonne, and some other little rivers of Gascoygne, which have their source in the plain, as the Gimone, the Save, and the Rat, overflowed in a similar manner and caused the same devastation. This flood was believed primarily due to the release of subterraneous waters. This observation was derived from the mineral content of the water and the formation of a new river.

Sometime in 1678 a shower of fire fell from the sky and burnt for half an hour in the streets in Sachsenhausen-Oranienburg in Brandenburg, Germany.

What causes fire showers?

Were these originally burning micrometeorites or meteoroids?

On August 19th 1678 CE Merapi erupted in Indonesia with a VEI of 3.

In August 1678 CE Pacaya erupted in Guatemala with a VEI of 2.

On September 2nd 1678 a fireball was seen over Glarus in Switzerland.

In 1680 CE Tangkoko-Duasudara in Indonesia erupted with a VEI of 5.

On January 16th 1680 there was a great body of fire followed by a whirlwind in Rostock in Siberia. Many houses were shattered. This is believed to be a meteorite impact. Where was Rostock and what is it called now if it still exists, who knows. None of my atlases or sources could find it in Siberia. There is a Rostock though on the Baltic Coast of the German state of Mecklenburg-Vorpommern in German.

In England on 11 February 1680, there was an irregular tide when the sea flowed three times in 5 hours.

On May 18th 1680 several stones fell near Gresham College in London, England. They were two and a half inches in diameter.

In May 1680 CE Krakatoa erupted in Indonesia with a VEI of 3.

On June 1st 1680 a fireball was seen over Leipzig in Saxony, Germany.

In Oxford, England in June 1680, there were great floods.

On 26 June 1680, there was a monstrous inundation near Londonderry, Ireland. It was believed that this inundation came from the release of subterraneous waters imbedded in the hills and from whence the waters gushed forth. Londonderry is located in Northern Ireland near Lough Foyle.)

In England, the summer was extremely hot and dry.

In Dijon, France, the grape harvest began on 9 September. This year was a good grain market in France.

In 1680, 1720, 1739 and 1740 (in Europe), storms of hail of one foot thickness fell.

In Breslaw (now Wrocław in southwestern Poland), there was great heat during the summer.

On 3 August 1680, a hurricane struck Martinique Island in the Caribbean Sea. During the violent hurricane over twenty large French ships and two English ships were totally lost in Cul-de-Sac Bay and the loss of life was great.

On 15 August 1680, an Atlantic hurricane struck the Dominican Republic. The storm submerged many vessels including 25 ships of France, causing the death of most. Several Spanish ships were also lost as well.

In England on 13 September, there was a terrible storm of thunder and lightning.

On December 17th 1680 a fireball was seen over Courland in Western Latvia.

Glaciers and glacier rivers destroyed meadows and pastures in Norway in the 1680s.

The Winter of 1680 /1681 CE in England was a long severe frost and an intense cold. The summer was excessively hot.

The winter was intensely cold in Europe.

The Little and Great Belts in Denmark were frozen, and many people perished. This year the cold was so severe as to split whole forests of oak trees.

In 1680 in southern France, the cold kills all the olive trees. The winter in 1680 was in Italy and Provence very severe. In Provence, the olive trees froze to death.

On February 10th 1681 there was a report of dragons and burning citadels in the sky over Hungary. Were these meteors? Were these an aurora? Auroras display dynamic patterns of brilliant lights that appear as curtains, rays, spirals, or dynamic flickers covering the entire sky. They normally appear in the Polar regions so would be a complete shock to people as far south as Hungary.

In July 1681 burning bituminous matter fell on a ship during a thunderstorm in Cape Cod in Massachusetts. How does bitumen come down from the sky? We will meet it again in the future as well. We have had many reports of this in the past but no scientific reason or cause.

Is it possible that some carbonaceous chondrites can become bituminous when exposed to extreme heat?

Carbonaceous chondrites contain a high proportion of carbon (up to 3%), which is in the form of graphite, carbonates and organic compounds, including amino acids. In addition, they contain water and minerals that have been modified by the influence of water.

In 1681, a hurricane struck the western Caribbean Sea. The loss of life was considerable from several ships.

In 1681, the island of Antigua in the Leeward Islands was desolated by a tremendous hurricane.

The winter of 1681 to 1682 CE in England was 3.5° C on average colder than modern temperatures.

On 26th January 1682 the coasts from Flanders to northwest Germany were hit by sea floods in which many thousands of people were drowned.

On 22 March 1682, the tide on the River Thames in London, England, ebbed and flowed three times in four hours.

On May 18th 1682 a fireball was seen over Glarus in Switzerland.

Also, in May 1682 fireballs were seen from many places in Germany.

On 6 June, at Tortorica in the Valley of Demana in Sicily, at 7 o'clock in the evening, there arose such a tempest of rain, thunder and lightning, which continued for 36 hours. At 1 o'clock the next morning, great torrents of water caused by these rains, fell down from the neighboring mountains with so great rapidity, that they carried down trees of extraordinary bulk, which demolished the walls and houses of the town. They overthrew St. Nicholas's Church, drowned the Archdeacon and many people with him. It left only fifty shattered houses, which fell soon after. It drowned 600 inhabitants, the rest were employed in their fields about their silk, fled to the mountains where they suffered much for want of provisions. The materials carried down by the flood, were so much, that they made a bank above the water, near two miles in length, near the mouth of the river, where the sea was deep before. Several other towns near were much damaged by it. Tortorica or Tortorici in Sicilian is located in the province of Messina in Sicily, a region of Italy.

In June 1682 a fireball was seen over Rochlitz in Saxony, Germany.

In 1682 CE on August 12th Vesuvius erupted in Italy with a VEI of 3.

In England, there were "Rain, hail, floods, all the summer." In 1682 there was a storm and flood at Brentford, in West London, England that did much damage. The sudden flood occasioned by the tempest was so great that the whole place was laid underwater. Boats rowed up and down the streets and several houses were carried away by the force of the current.

On September 1st 1682 CE Mount Etna erupted with a VEI of 2.

On November 30th 1682 a fireball was seen over Zurich.

On December 15th 1682 a fireball was again seen over Zurich.

On December 13th 1682 CE Sete Cidades in the Azores erupted with a VEI of 2.

In 1682 a fireball was seen over Niederelb in northern Germany. The Niederelb is a 108 kilometer section of the River Elbe, from Western Hamburg downstream to its mouth into the North Sea near Cuxhaven.

In 1683 CE Tangkoko-Duasudara in Indonesia erupted with a VEI of 3, Banda Api also erupted with a VEI of 3, and Serua erupted with a VEI of 3, all in Indonesia.

In June 1683 CE Asosan in Japan erupted with a VEI of 2.

On August 12th 1683 a fireball was seen over Zurich in Switzerland.

Three days later the Zurich fireball returned on August 15th 1683.

How could we have returning meteors or fireballs?

Was there a single stream that was not very wide that pinpointed over the Zurich area only as it passed?

In 1683 Ottoman forces advanced upon Vienna, the capital of Austria. They experienced cold and rainy conditions before suffering defeat. Swelling streams and rivers swept away bridges, mud made roads impassable and the carts in the wagon train frequently broke down. The ottoman cavalry also had to wait until grass was available for foraging, so the entire campaign slowed. The Ottoman army finally laid siege to Vienna in July, but a relief force broke the siege in September. The Ottoman forces gave up and returned home. The siege of Vienna was from July 17th- September 12th, 1683.

In 1683, a hurricane struck the east coast of Florida in the United States causing 496 deaths.

In England, August 1683 had one of the coldest summer temperatures on record with a temperature of 13.5° C with a reduction of 2.5° C on the all series mean or 3° C below modern values. October 1683 was decidedly colder with a CET average of 6.5° C or at least a 3.5° C average or -4° C on modern values. On the 4th and 5th 1683 an Arctic low came down lowering temperatures. This was one of the four or five coldest winters ever in the British Isles and parts of Europe and the coldest in the CET record. In most of England the ground was frozen to a depth of four feet yet in the southwest of England in Somerset there was no snow.

In England, there were fairs on the frozen River Thames in 1683-84. "In England on 9 September 1683, it was very rainy and then to the 16th, warm and pleasant, that night a great frost. This was the coldest winter in England, the longest hoar frost known in the memory of any living. In 1684, the River Thames at London, England froze eleven inches (28 centimeters) thick and was traversed by loaded wagons. "The people kept trades on the Thames as in a fair, till 4 February 1684. About forty coaches daily plied on the Thames as on drye land. Bought this book at a shop upon the ice in the middle of the Thames."

At the beginning of December, due to the extreme cold, there was fearful destruction of the trees and plants. Great oaks suffered; the bark was rifted (cracked) by the frost in the estates of Lords Weymouth, Chesterfield and Ferrars, and Sir W. Fermor. The figs killed to the ground. Elm, ash and walnuts cleft (split) by the frost, but not so much as oaks; the oaks in being cleft made a noise like a gun. Yew, holly and furze (gorse bush) in some places entirely killed and, in many places, lost their leaves. Rosemary, Laurustinus, Laurel, Arbutus, and Phyllyrea generally killed throughout the country. In dry mountainous places, trees escaped tolerably well. Firs and pines escaped the effects of the cold. There was great destruction of the herbs, plants, and flowers, except where covered with snow.

During the winter of 1683-1684, the River Thames in England was frozen below Gravesend. The frost in Britain lasted for 13 weeks.

Also, in 1683-1684 belts of sea ice five kilometers broad appeared along the Channel coasts of southeast England and France.

In 1683-1684 on the North Sea coast of the Netherlands the ice belt was 30 to 40 kilometers broad. Shipping was halted as in the Baltic which was also frozen.

On the 15th December 1683 there was a great frost in England and Central Europe.

The Thames was frozen down to London Bridge by 2nd January 1684 with booths on the ice by 27th January for more than a fortnight thereafter. Coaches were observed on the ice and the royal court of King Charles II visited the fair held on the frozen Thames. The Tees River in northeast England was also frozen amongst others. The frost was claimed to be the longest on record with the Thames being frozen for two months and the ice was eleven inches thick.

Sea ice was reported along the coast of southeast England and many harbors could not be used due to ice.

Some sources wrote that ice formed for a time between Dover and Calais with the two sides joined together. There were severe problems for shipping accessing such ports on either side of the North Sea.

Near Manchester the ground was frozen to a depth of 27 inches and in Somerset to more than four feet.

The winter of 1684 was incredibly severe. H. H. Lamb constructed a tentative mean seasonal pressure pattern with high pressure in the Faeroes area, an Arctic northerly from Spitzbergen to the Baltic, thence an anticyclonic east or northeasterly over northwest Europe and the British Isles. With both January, -3° C, and February, -1°, this is only one of four instances of successive sub-zero months.

Could the eruption of Tangkoko-Duasudara in Indonesia in 1680 with a VEI of 5 be partly responsible?

There were coaches on the ice on the River Thames (at London). There were shops on the River Thames until February. About 40 coaches plied for hire on the river daily. This was the longest frost on record, and the ice on the River Thames was 11 inches (28 centimeters) thick. Nearly all the birds perished. The frost lasted till 4 February.

On May 19th 1684 a fireball was seen over Annaberg in Saxony, Germany.

On November 13th 1684 a fireball was seen over Gottesgabe in Mecklenburg-Vorpommern, Germany.

On November 17th 1684 a fireball or meteor was seen over Brittany in France.

In England, on the 20th of December, 1684, a very violent frost began, which lasted to the 6th of February, in so great extremity, that the pools were frozen 18 inches (46 centimeters) thick at least, and the Thames was so frozen that a great street from the Temple to Southwark was built with shops, and all manner of things sold. Hackney coaches plied there as in the streets. There were also bull baiting, and a great many shows and tricks to be seen.

Smallpox raged in London. In 1684, the frost in England began at Christmas and lasted 91 days, and mortality increased. Coaches drove along the Thames, which was covered with ice 11 inches (28 centimeters) thick. Almost all the birds perished.

On 5th January 1685 trees split due to the cold. "How cold did it get during the winter the English Channel froze? The answer partly lies in the splitting of oak trees. The cold in England was so intense that the trunks of oak, ash, walnut, and other trees, were cleft asunder, so that they might be seen through; and the cracks were often attended with noises as loud as the firing of musketry. Both France and England reported that the trees split apart with the sound like a musket shot.

The temperature in England on 5 January 1684 fell to -40°F (40° C) or colder during this intense cold spell. Temperatures along coastal regions are somewhat moderated by the ocean temperature whereas temperatures inland will feel the full severity of an extreme temperature fall. Thus, in England, coastal regions such as London would not experience the same drop in temperatures as interior regions.

The winter of 1683-84 was so severe in England, where it could expunge (destroy) the more defensible and such as were enclosed, it has ravaged all that lay open, and were abroad, without any mercy. Many of the older trees, especially the oaks, were damaged by cleaving or splitting due to the severe frost. Many of these trees split apart. Oak trees split when the temperature reaches -7° C or 19.4° F. Nor has this damage been limited to only the standing timber, but to that which has been felled and seasoned, as Mr. Shish, the master-builder in his Majesty's Shipyard here, informed me.

Some of the splits in the trees were large enough for a man to see through it, and many times the cracks came with so great a noise, that as it was related from Needwood Forest, they made such a noise, that the keepers there thought that the deer were shot by the people of the country; and that in several parts they were heard as loud as guns, some having been cruelly frightened, especially in the evenings or nights, as they have passed within the hearing of this so unexpected and surprising a noise.

In England, there was a terrible frost of long continuance. "Many forest trees split. In the severe frost of 1683-84, not only oaks, but elms and ash of considerable bulk, and also walnut trees, were very much rent by the violence of the cold; oaks were most of all affected, and some split in such a manner as to be seen through, with a noise like the report of a gun. These clefts were not towards the same point of the compass."

During the winter of 1683 and 1708, coaches were driven over the ice on the River Thames in London, England and large fires were made on the ice. In England, the frost of 1683 and 1684 were both the severest. One lasted for 13 weeks.

In the winter of 1683-1684 all the French ports were closed for three or four weeks with the harbors being frozen over.

Krakatoa in Indonesia erupted on February 1 1684 with a VEI of 3.

Thirteen days later Izu-Oshima erupted in Japan on February 14th with a VEI of 3.

"On Candlemas Day (2 February) I went to Croydon market, and led my horse over the ice to the Horseferry from Westminster to Lambeth; as I came back I led him from Lambeth upon the middle of the Thames to Whitefriars' stairs, and so led him up by them. And this day an ox was roasted whole, over against Whitehall. King Charles and the Queen ate part of it. A whole street of booths, contiguous to each other, was built from the Temple Stairs to the barge-house in Southwark, which were inhabited by traders of all sorts, which usually frequent fairs and markets, as those who deal in earthenwares, brass, copper, tin, and iron, toys and trifles; and besides these, printers, bakers, cooks, butchers, barbers, coffee-men, and others, who were so frequented by the innumerable concourse of all degrees and qualities, that, by their own confession, they never met elsewhere the same advantages, every one being willing to say they did lay out such and such money on the river of Thames".

On 6th February the frost broke up. In the morning, I saw a coach and six horses driven from Whitehall almost to the bridge (London Bridge) yet by three o'clock that day, February the 6th, next to Southwark the ice was gone, so as boats did row to and fro, and the next day all the frost was gone.

During the Great Frost of 1683–84 in England solid ice was reported extending for miles off the coasts of the southern North Sea (England, France and the Low Countries), causing severe problems for shipping and preventing the use of many harbors. According to some sources, ice formed for a time between Dover (England) & Calais (France), with the two sides joined together. One of the ways that a river freezes occurs when great mass of broken chunks of ice flow downstream and then there is an extreme drop in temperature, which then freezes these icebergs together. I believe a similar process occurred in the English Channel. It is also credibly attested that vast solid cakes of ice, of some miles in circuit, breaking away from the eastern countries of Flanders and the Netherlands, &c. have been by the east and north-east winds driven upon the marine borders of Essex, Suffolk, and Norfolk, to their no small damage.

In 1684 there were belts of ice along the English and French coasts of the Channel and the coasts of Belgium and the Netherlands. The ice extended about five kilometers from the coast of England at Dymchurch in the Romney Marshes in Kent. The ice extended more than forty kilometers from the coast of the Netherlands.

The London Gazette reported that in Dover on February 1 1684: "This Road being almost clear of Ice, one of our Pacquet-Boats put to Sea yesterday with the Mails for Calais, though we cannot think they will be able to land them on that side; for from Dover Cliffs we can discern the Coast of France to be very full of ice. The Men on board the Dutch Doggers, which we told you in our last were put in here, reported that on the coast of Holland (now the Netherlands), and particularly off Sceveling, the Sea was frozen eight Leagues (24 nautical miles) from shore, and that in 16 fathom (96 feet deep) Water they had met with ice strong enough to bear, and that some of them had been upon it."

From a letter by Guillaume Fillastre, monk at Fécamp, France: "Some sailors from St. Valery en Caux, setting out to go fishing, were surrounded by ice nearly three leagues (9 nautical miles) out to sea, opposite the port of Veules, from which people could see them indicating by signs the danger they were in, but could not give them any help. In this extremity, they risked returning to land on foot, across the ice; which they achieved, happily, thanks to two planks which they placed one after the other as they advanced, to serve as a bridge over the icebergs, which were by no means neatly joined."

From "The World of Wonders: A Record of Things Wonderful in Nature, Science, and Art ..." (London, 1869), A private letter of the date of February the 9th of that year (1684), mentions the appearance of a great deal of ice in the Channel, adding that it was reported that the ice between Dover and Calais was within about a league of joining. On February 9 there was sea ice in the English Channel. The ice between Dover and Calais were "joined together".

The frost was also very severe in Northern Europe.

There was ice 27 inches (69 centimeters) thick in the harbor of Copenhagen in Denmark.

The winter of 1684 was excessively cold in northern France. Since the water in tree sap acquires greater volume when it freezes, in extreme cold, trees burst apart with a loud noise. In Strasbourg, France more fruit trees burst when the cold reaches -16° Reaumur (-20° C, -4° F). A great number of trees in France burst in the winter of 1683-84.

In 1684, the River Thames at London, England froze 11 inches thick. Loaded carts drove over it. During the big chill of 1683-1684 the ground was reported to have frozen to a depth of 50-90 centimeters in the southeast of England and near Manchester and in Somerset in southwest England. Wet ground was found to be frozen to a dept of four feet.

There was very severe cold in Paris, France from 11 to 17 January. During those seven days, the alcohol decreased in the bulb (alcohol thermometer) down to a point where it had not yet reached during other winters. The academics timed how long it would take wine to freeze in the open. It took 10-12 minutes time. There was an extraordinary amount of snow in the south (southern France).

The frost in February and March was so severe that one can almost cross all the rivers in Flanders (now Belgium) with carts. The winter of 1684 was also so cold in the northern France, but it was mild and dry in the south.

On the shores of England, France, Flanders and the Netherlands, the sea was frozen a few miles wide in such a way that for more than 14 days, boot packages could not enter the ports on or off. Most birds were killed; in the next summer we saw none. In the woods, many oak trees burst. The frost destroyed almost all the plants and the hopes of the peasants. As a result, in the main streets of London large piles of wood were lit so that the inhabitants who were forced to flee their homes after the Great Fire of London could warm up.

In 1684 CE in England, the spring was dry and cold and the summer was very hot and dry.

Jean-Dominique Cassini ranked the year 1684 among the warmest in an array spanning 82 years of great heat in Paris, France. Cassini developed a Fahrenheit thermometer, which he placed against the window of the tower northeast of the Observatory. He took his measurements between noon and three o'clock each day. The summer of 1684 produced sixty-eight days of a temperature of 77° F (25° C), sixteen days of a temperature of 87.8° F (31° C), and three days of a temperature of 95° F (35° C).79 The summer of 1684 was the first hot summer, over which we have thermometric data. In Paris, France there were: Hot days 68 days Very hot days 16 days Extremely hot days 3 days (It appears that hot days are defined as those with temperatures of 25° C and greater but less than 31° C, very hot days are those with temperatures 31° C or greater but less than 35° C, and extremely hot days are those with temperatures of 35° C or greater.) These peaks occurred on 10 July and on 4 & 8 August.

In July 1684 CE San Cristobal in Nicaragua erupted with a VEI of 2.

In England summer was preceded by a very harsh winter and a wet spring. The summer was hot and dry. In France, the drought was exceptionally severe. In Dijon, France, the grape harvest began on 4 September.

On November 5th 1684 CE Grimsvotn in Iceland erupted with a VEI of 2.

The winter of 1684 to 1685 though nothing as cold as the previous winter realized and anomaly of roughly -1° C and was 2.5° C below the all series mean and 4° C below the 1981-2010 Average. The Thames was frozen in London again.

1684 to 1685 was the most severe winter in Switzerland ever recorded.

In August 1685 CE Telica erupted in Nicaragua with a VEI of 2.

Also, in August 1685 CE San Cristobal in Nicaragua erupted with a VEI of 2.

In September 1685 CE Fuego in Guatemala erupted with a VEI of 2.

On October 3rd 1685 Vesuvius erupted in Italy with a VEI of 3.

On August 22nd 1685 a fireball or meteorite was seen over Germany.

From 1685 to 1688 a major drought hit India and caused intense famine.

From 1685 CE to 1704 CE the cod fishery in Iceland failed completely due to extremely cold water. The cod, which thrives best in rather cold waters at between 4° C and 7° C has its kidneys fail at temperatures below 2° C and cannot venture into colder seas.

In Norway the worst years for harvests were incidentally the best years for the herring fishery. The self-same conditions that which produce harvest failures such as long-lasting harsh and stormy westerly and northerly winds drive the fish stocks of the Arctic Ocean, Barents Sea, in greater than usual numbers to the Norwegian coast. This happened in 1685 to 1687 CE.

In the Faroe Islands in the Atlantic around this time the fishery failed for thirty years, and the cold water was extensive in that region.

In 1686 on March 26th Iwatesan erupted in Japan with a VEI of 3.

In Italy, during the years 1686-89 there was a great drought.

On 25 May 1686, there fell at Lille, a storm of prodigious hail, some stones above a pound weight. People broke one that had brown matter in it and threw it in the fire. It produced an explosion. The storm broke down trees and most glass windows and killed partridges and hares. Some of the hailstones were not only vastly large but appeared dusky in the center. Those that came down chimneys into fires, when the icy part was melted, and this brown substance was exposed to the fire, burst with a loud report (like a gunshot).

In June 1686, the inhabitants of Kettlewell and Starbotton in Craven, in the County of York, England, suffered a great loss by a sudden overflow of water. The towns are situated under a great hill on the east and west. The country is very mountainous and rocky. The descent of the rain after a thunderclap continued for 1½ hours with extraordinary violence. The rocks on the east side opened visibly, and the water they beheld thence into the air the height of an ordinary church steeple, so that the current of water came down the hill into the respective towns, as in one entire body, and with a breast (to rise over) as if it would have drowned the whole towns. Several houses were quite demolished, and not a stone left. Others graveled to the chamber windows. Some inhabitants were (permanently) driven from their houses. Currents of water ran through the houses. Mighty rocks descended from the mountains into the valley, and there they lay unmovable. Many fair meadows were covered with sand and stone. Household goods taken away into the great River Wharfe, and so lost, besides many quick goods (consumer goods). Many families were quite ruined.

The summer of 1686 was very hot in Paris, France. There was: Hot days 46 days Very hot days 8 days Extremely hot days 5 days (It appears that hot days are defined as those with temperatures of 25° C and greater but less than 31° C, very hot days are those with temperatures 31° C or greater but less than 35° C, and extremely hot days are those with temperatures of 35° C or greater. The peak temperatures occurred on 19-23 June.

On July 9th or 19th 1686 a fireball half the size of the moon remained motionless in the sky for seven and a half minutes over Leipzig in Saxony, Germany. It was estimated to be thirty miles high.

In Dijon, France, the grape harvest began on 4 September.

In September 1686 CE Gamalama in Indonesia erupted with a VEI of 2.

On November 12th 1686 the Saint Martin flood occurred in the Netherlands with 1586 deaths.

1686 CE. On 22nd October and 22-23rd November 1686 over 1,000 people were drowned during sea floods in Friesland and northwest Germany.

In England, the year 1687 was very rainy and the earth produced plenty of watery crude fruits. In summer the rivers were terribly flooded. Brooks overflowed their banks. Extraordinary tempests of rains demolished houses and buildings. Torrents carried along with them and drowned multitudes of people. At the time of ripe fruits were great swarms of gnats and insects. The year produced frequent tempests and hurricanes.

In 1687, there was a great estuary flood in the River Severn in Great Britain.

On 3 March 1687, there was a thunderstorm at Cloyne, Ireland. The next morning the barometer read 28.4 inches, the lowest that had been before seen there.

On March 26th 1687 Pacaya erupted in Guatemala with a VEI of 2.

On April 14th 1687 Iwatesan erupted in Japan with a VEI of 2.

In 1687 CE Gamalama in Indonesia erupted on May 10th 1687 with a VEI of 3.

On May 22nd 1687 a fireball was seen over Paris in Ile-de-France, France.

On May 26th 1687 a fireball was seen over Zurich and Berne in Switzerland.

Were these one of those pinpoint lines of fireballs?

On June 15th 1687 Serua erupted in Indonesia with a VEI of 3.

All summer was rainy in Germany.

In Italy, during the years 1686-89 there was a great drought.

The summer of 1687 in Paris, France was characterized by: Hot days 34 days Very hot days 6 days Extremely hot days 3 days (It appears that hot days are defined as those with temperatures of 25° C and greater but less than 31° C, very hot days are those with temperatures 31° C or greater but less than 35° C, and extremely hot days are those with temperatures of 35° C or greater.) The peak temperatures occurred on 29 June, 10 July and 16 August. In Dijon, France, the grape harvest began on 29 September.

In 1687 in England, the summer season was very dry and extremely hot and on 15 August, there was a waterspout seen at Hatfield in Yorkshire. On 15 August 1687 at Hatfield, Yorkshire, England, there was a land-spout (tornado), like those at sea (waterspout). There was an inundation not caused by weather, but rather a tsunami triggered by a massive earthquake.

On 10 October at 4 o'clock on Monday morning, there was a terrible shock of an earthquake, with a horrible roaring of the Sea at Lima in central Peru. Many houses fell and killed several people. At 5 o'clock a second shock and at 6 o'clock the greatest of all. The Sea bellowed, swelled and overflowed. The city was wholly overthrown. Several seaports were flooded. By the inundation, which carried off several ships nine miles into the land, much people and cattle drowned. At one place near the seaside were found 5,000 dead bodies and more were daily cast up so that at last the number of the dead was not known. The 1687 Peru earthquake occurred at 11:30 UTC on 20 October 1687. It had an estimated magnitude of 8.4–8.7 and caused severe damage to Lima, Callao and Ica. It triggered a tsunami and overall, about 5,000 people died. The port of Pisco was completely destroyed by the tsunami, with at least three ships being swept over the remains of the town. We have two dates for this event but they are close enough to not be a great problem.

On 4th and 5th December 1687, a great inundation happened in the River Liffey in Ireland from excessive rain and a violent storm. "By excessive rains, and a violent storm, there happened a great inundation in Dublin in east-central Ireland, which put the lower part of the city under water, up to the first floor; so that boats plyed in the streets. At which time Essex-bridge was broken down, when a coach and horses passing over it, fell into the river, where the coachman and one horse perished."

On the 18th December 1687 at Stockton-on-Tees a northeasterly storm and overland, sunk or grounded about 50 ships at Stockton in England. Around this time large areas around Doncaster in South Yorkshire were flooded. There was considerable damage to stored harvests. This appears to be a storm surge with a deep low crossing the southern North Sea coupled to a high inland rainfall. This led to a dramatic flood of the river valleys draining into the Humber estuary above Goole.

1688 was the coldest year of the 1680s and one of the six or seven coldest of the 17th Century and ran 8th of the coldest years in the CET series up to 2010. It had a mean value of 7.8° C.

On April 17th 1688 a fireball was seen over Heilbronn in Baden-Wurttemberg in Germany.

The 1688 Sannio earthquake occurred in the late afternoon of June 5 in the province of Benevento of southern Italy. The moment magnitude is estimated at 7.0, with a Mercalli intensity of XI. It severely damaged numerous towns in a vast area, completely destroying Cerreto Sannita and Guardia Sanframondi. The exact number of victims is unknown, although it is estimated to total approximately 10,000. It is among the most destructive earthquakes in the history of Italy. Three towns were completely destroyed by the earthquake: Cerreto Sannita, Guardia Sanframondi, and Civitella Licinio, a *frazione* of Cusano Mutri. In 20 more villages and towns near the epicenter, destruction was almost total. In an area of 50,000 square kilometers (19,000 sq mi), about 120 settlements took extensive damage, while about 40–50 other towns were damaged only slightly. Benevento was hit harshly by the earthquake, with over 80% of the buildings being significantly damaged or destroyed, including the Santa Sofia church. The poor quality of the buildings was a determining factor in the extent of the damages. Avellino and Naples also experienced extensive destruction, with several collapsed or damaged buildings; in Naples, more than thirty churches and religious buildings were severely damaged. In Cerreto Sannita only three houses remained standing, with the destruction also of a Poor Clares monastery and a Franciscan one. The exact number of deaths resulting from the earthquake is unknown, but it is estimated to total approximately 10,000. Cerreto Sannita, the hardest-hit town, lost half its inhabitants, 4,000 out of 8,000. In Civitella Licinio, which was completely destroyed, the only survivors were those who were working in the fields when the earthquake hit. In Guardia Sanframondi there were more than 1,000 victims. In Benevento 1367 people died out of a population of 7,500; the number of deaths was lower because many citizens were working in the countryside at the time. Hundreds of people died in San Lorenzello, due to a landslide. Several deaths were recorded also in Naples, Avellino and other towns.

On June 22nd 1688 CE Iwatesan erupted in Japan with a VEI of 2.

In June 1688 in Yunnan Province, China. A large yellow "umbrella-like" object rose from the ridge and came down again, with many lights: "In the year 27 under the reign of emperor Kangxi of the Qing dynasty, my brother in-law Bixilin went to his home in the mountains, 20 kilometres from the city of Kunmin. While staying there, he saw every day at noon, when the weather was clear, a large yellow cover like an umbrella that rose slowly above a ridge. This object threw such brilliant lights that he dared not look at it directly. It rose and got lost into the clouds. A little while later it would come down, always slowly, going up and down in the same way. At nightfall, the flying object lost its yellow color and turned paler and blurry. It disappeared completely when the sky was dark."

The summer of 1688 in Paris, France was characterized by: Hot days 40 days Very hot days 12 days Extremely hot days 1 day (It appears that hot days are defined as those with temperatures of 25° C and greater but less than 31° C, very hot days are those with temperatures 31° C or greater but less than 35° C, and extremely hot days are those with temperatures of 35° C or greater.) The peak temperature occurred on 9 September. In Dijon, France, the grape harvest began on 27 September.

In 1688, there was an epidemic fever in Dublin, Ireland that lasted from July to the middle of August and in London, England that lasted from May to June.

In Germany, spring and summer was most inconstant.

In 1688, John Clayton wrote to the Royal Society and described his stay in Virginia in the United States. He related the following accounts of thunder and lightning storms in Virginia. Dr. A. was smoking a pipe of tobacco and looking out the window when he was struck dead (by lightning), and immediately became so stiff that he did not fall, but stood leaning in the window, with the pipe in his mouth in the same posture he was in when he was struck. Lightning generally breaks in at the gable end of the houses, and often kills persons in, or near the chimney's range, darting most fiercely down the funnel of the chimney, more especially if there be a fire. Thunder split a mast of a boat at James Town (Jamestown). It is dangerous when it thunders standing in a narrow passage, or between two windows. Several people have been killed in the open fields. It is incredible to tell how it will strike large oaks, shatter and shiver them, sometimes twisting round a tree, sometimes as if it struck the tree backwards and forwards. I had noted a fine spreading oak in James Town Island, in the morning I saw it fair and flourishing, in the evening I observed all the bark of the body of the tree, as if it had been artificially peeled off, was orderly spread round the tree, in a ring, whole semi diameter was four yards, the tree in the center; all the body of the tree was shaken and split, but its boughs (branches) had all their bark on; few leaves were fallen, and those on the boughs as fresh as in the morning, but gradually afterwards withered, as on a tree that is fallen. I have seen several vast oaks and other timber trees twisted, as if it had been a small willow that a man had twisted with his hand, which I could suppose had been done by nothing but the thunder. I have been told by very serious planters, that 30 or 40 years since (1650-1660 CE.), when the country was not so open, the thunder was more fierce, and that sometimes after violent thunder and rain, the roads would seem to have perfect casts of brimstone; and frequently after much thunder and lightning for the air to have a perfect sulfurous smell. Lightning struck one of my large trees selectively peeling off the bark as it traveled down the length of the tree. When it was done, the tree had the appearance of a large candy cane.)

In 1688, there was a plague of cockchafers (also called May bug, mitchamador, billy witch, or spang beetle) in County Gallway (Galway) in Ireland. They covered the trees and clung to each other like a swarm of bees. Towards evening when they flew, they made a strange humming sound and darkened the air for 2 or 3 miles (3.2-4.8 kilometers) square. They ate up all the leaves off the trees for miles round making them as bare as in winter. Their grubs destroyed all the roots of the grass.

In Noremberg (Nuremberg in Bavaria in southern Germany), there was a rainy cold harvest.

In the winter of 1688 to 1689 from late December to early February there were extended periods of cold weather across England. It was noted as a severe winter. January was extremely cold with an anomaly of -2.5° Celsius. A frost fair was held on the Thames by 3rd January and the Thames was already full of ice such that boats could not navigate. By the 7th the Thames was almost frozen over which implies persistent sub-zero temperatures and strong east winds to allow ice to form sufficient thickness and stability.

The winter in 1688 was severely cold in Germany with great snow, followed by a sudden thaw and heat.

In Italy, during the years 1686-89 there was a great drought. At Modena and all over Italy, for three or four years previous, there had been an uncommon drought. During the drought there were plenty of provisions. But in 1689 about the vernal Equinox (around March 20/21), there fell great rains, which returned quickly after, rendering the whole spring frightful and good for nothing. The summer following was most rainy. About the Solstice and much more after all sorts of corn (grain) was wholly blasted and mildewed. But there were still hopes from the remains of the old store. At the beginning of September, and much more about the Equinox (around September 22/23), greater rains fell, which continued the whole month of October; so that it was with much labor and difficulty that the rivers were prevented from breaking down their banks and drowning the country. The last two months concluded the year pleasantly.

The summer of 1689 in Paris, France was characterized by: Hot days 27 days Very hot days 7 days Extremely hot days 1 day (It appears that hot days are defined as those with temperatures of 25° C and greater but less than 31° C, very hot days are those with temperatures 31° C or greater but less than 35° C, and extremely hot days are those with temperatures of 35° C or greater.) The peak temperature occurred on 10 August. In Dijon, France, the grape harvest began on 27 September. "You sing the vintage of Burgundy 27 September. You reap little wine, but it was excellent." In 1689, La Hire in France began taking observations using precipitation gauges. The year 1689 ranked as the driest year for next thirty years.

In 1689, there was a famine at Londonderry in what is now Northern Ireland. "The inhabitants glad to eat rats, tallow and hides."

On October 1st 1689 a brilliant bolide was seen in the sky over Boston in Suffolk County, Massachusetts. There was a detonation afterwards.

Beginning on 4 October 1689, there was rainfall at Bungay in Suffolk, England that lasted till noon on 10 October. The rain was continuous except for a few hours on the 6th. This caused a great flood that overflowed the lower part of Norwich and broke down the bridges at Bungay.

In 1689, a hurricane struck Jamaica. The hurricane was not very severe.

In 1689, an Atlantic hurricane struck Nevis. A dreadful mortality swept away one-half of the inhabitants of Nevis in the West Indies.

1690 CE seemed to be a very busy year. In 1690 CE Chikurachki in Kamchatka erupted with a VEI of 4. Other eruptions were Guntur in Indonesia with a VEI of 3, Banda Api in Indonesia with a VEI of 3, Colima in Mexico with a VEI of 3, and Koshelev in Kamchatka with a VEI of 3.

Six out of ten of the winters defined as severe in the CET series were between 1690 and 1699. This is a CET mean value temperature value for the months of December, January and February below 3° Celsius. In Northeast Scotland much outward migration of farming folk occurred after a series of bad harvests, with tales of mills falling into disuse with such recurring depth and persistence of cold in these winters, it was not surprising that the subsequent spring and summer seasons were also largely cold. All summers except for 1691 and 1699 were below the mean with two summers, 1694 and 1695, exceptionally cold. The mean value of the CET was 8.1° C which is about 2° C lower than modern times. In northern Britain there was a succession of wet springs and summer seasons, again compounding the problem of low temperature and poor growth of cereal and other crops. In Scotland in particular the oat harvest failed in seven out of eight years.

It was reported that several times between 1690 CE and 1728 CE that an Eskimo or Inuit was seen in his canoe in the Orkney Islands north of Scotland and the River Don near Aberdeen. In this period there were reports of sea ice and the situation in Scotland became serious. The cod fishery collapsed due to the fact that the ocean surface between Iceland and the Faeroe Islands, only a few hundred kilometers north of Scotland, was probably 5° colder than in the late twentieth century. The cod, which thrives best in rather cold waters at between 4° C and 7° C has its kidneys fail at temperatures below 2° C and cannot venture into colder seas.

In 1690 CE an awful snowstorm pounded Scotland. The storm lasted thirteen days and nights. During that time nine-tenths of the sheep were frozen to death, and many shepherds lost their lives.

In 1690 in Ireland, there was famine and disease.

On January 2nd 1690 an aerolite was seen in the sky over Jena in Thuringia, Germany.

In London, England on 11 January 1690: "This night there was a most extraordinary storme of wind, accompanied with snow and sharp weather; it did greate harme in many places, blowing down houses, trees, &c. killing many people. It began about 2 in the morning, and lasted till 5, being a kind of hurricane, which mariners observe have begun of late yeares to come Northward. This winter hath ben hitherto extremely wet, warm, and windy." In the winter of 1689 to 1690 and especially January on 11th 1690 there was a gale and storm of wind and snow and rain for three hours in England. Many were killed and there were reports of losses of shipping in the Tees estuary to significant losses in the waters of Kent and in the Thames Estuary. The winter up to this part had been extremely wet, warm and windy.

The January 1690 storm caused major losses to shipping interests on the French side of the Channel.

In France there was a flood. In March 1690, the Seine River in Paris, France, at the bridge "Pont de la Tournelle" reached a height of 7.5 meters (24.6 feet) above the zero mark (the low water mark of the year 1719). The summer of 1690 in Paris, France was characterized by: Hot days 34 days Very hot days 2 days Extremely hot days 1 day (It appears that hot days are defined as those with temperatures of 25° C and greater but less than 31° C, very hot days are those with temperatures 31° C or greater but less than 35° C, and extremely hot days are those with temperatures of 35° C or greater.) The peak temperature occurred on 31 July. In Dijon, France, the grape harvest began on 4 September. The summer in the area of Burgundy was very stormy. It produced a lot of wine of medium quality.

In 1690 in Italy, there was a famine from rains. In the beginning of 1690, the rains in Italy returned much severer than before, and were almost continual. The winter had been rainy and cloudy with some little cold and snow, which melted as it fell. The beginning of March was uncommonly dry and calm. But at the Equinox (around March 20/21), the heavens seemed to open their bosom and pour out their whole great reservoir of water. By one night's rain, all the country about Modena, Finlan, Ferraria, Mirandola, etc. were laid under water, deluged like a Sea. These cities standing up like little islands. This rainy weather continued the whole spring and summer, scarce one fair day. The wind was mostly from the north and cold. The mercury all the while stood higher in the barometer than ordinary in such a season. Frogs croaked over all the country. Fish was never more plentiful or freely eaten, from the scarcity of corn (grain).

In the beginning of June, mildew appeared on the corn again and increased to its total destruction both on low and high grounds. Of all the products of the earth, nuts alone escaped this plague. They were uncommonly good and plentiful. At the latter end of July, the rains stopped and we had two months very dry but cold weather. Near the end of September, the rains returned again but were moderate and useful. The last two months of the year were dry but moderately cold. Modena, Finlan, Ferraria, Mirandola are all in the Emélia–Româgna administrative region of northern Italy. Ferraria is now Ferrara.

In the Cairngorms in Scotland and elsewhere on the Scottish mountains there was permanent snow cover due to the cold.

In England the growing season shortened on the long-term average by about five weeks in comparison with the warmest decades of the 20th century. This reduced the yearly total reduction of summer warmth for the crops.

In 1690 CE a sea flood hit Zealand in the Denmark.

In the late 1690s glaciers were already overrunning farms in Iceland.

In 1690 when below the glacier Tverbreen in Jostedal in Norway. It could be seen high up in the mountains. Not many years after this the glacier had advanced about two hundred meters in only ten years and was threatening farmhouses in the valley. In Norway the frequency of landslides, rockfalls, avalanches and flood damage increased dramatically by the 1680s and 1690s. These incidents had multiplied manifold and glaciers themselves were overrunning farmland. These disasters and growing glaciers continued until around 1710. In the later decades of the seventeenth century farms were once again being abandoned in northern Norway.

The spring seasons of the 1690s appear outstandingly cold.

The summers of the 1690s in Switzerland were very wet and from 1690 CE to 1740 CE there was a massive increase in the size of glaciers in Europe.

In the late 1690s there were very high levels of the Nile River in Egypt that indicate that summer rains were heavy over Ethiopia.

In the 1690s northwest Siberia experienced great warmth, presumably due to southerly winds.

The winter of 1691 to 1692 was the sixth coldest year in the CET record. There was a mean value of 7.7° C around a 2° C anomaly compared to modern times. The winter had a CET anomaly of -2° C in January and with a value of 0.00° C in February led to a -4° C below the modern-day long-term average.

In 1691 CE Reventador erupted in Ecuador with a VEI of 3.

In April 1691 CE Asosan erupted in Japan with a VEI of 2.

In 1691 CE in Italy, it was hot and dry. The summer of 1691 in Italy was too hot and no rain.

The summer in Paris, France was characterized by: Hot days 44 days Very hot days 12 days Extremely hot days 5 days (It appears that hot days are defined as those with temperatures of 25° C and greater but less than 31° C, very hot days are those with temperatures 31° C or greater but less than 35° C, and extremely hot days are those with temperatures of 35° C or greater.) The peak temperatures occurred on 8, 9, 22, 23, and 28 August. In Dijon, France, the grape harvest began on 17 September; 10 days earlier than the average from the years 1689-1800. There was little wine but it was of good quality.

There was excessive heat and severe drought in Jamaica.

In England on 26 July, there was a terrible storm of thunder and lightning.

On 27 July 1691, a violent storm of thunder, lightning and rain struck in Everdon Field, near Daventry in Northamptonshire, England. Several people were at work reaping the corn (grain) in the fields when the storm appeared. The reapers, 20 in all, retreated for shelter to a quickset hedge (plant cuttings set in the ground to grow especially in a hedgerow) with a ditch by the side of it. Lightning killed 4 outright, 8 others were dangerously hurt. Of the rest, several were struck down, but recovered. One of those dangerously injured was a pregnant woman named Mary Bird. She had over her body nearly a hundred wounds, some as large as a man's hand, on each arm one, and one on each side of her belly. Out of most of her wounds came cores, some bigger, some less. The biggest were bigger than a walnut, dry and black like leather. She had two sores on the soles of her feet, but her shoes and stockings were not touched. She sat next to those that were killed. She was sensible of the stroke, and sensible that her husband looked pale, and then swooned away. She and her husband were both blooded (had blood drained), she within an hour after and her husband eight hours after, and they bled freely. Their legs were mightily swelled before they were carried out of the field. The woman was very sore and full of pain, so that she could hardly bear any clothes to touch her. She was three weeks ill before she could rise (to her feet). She continued ill for about a quarter of a year. No medicine used for burns did any good but occasioned some great torment to her. The first medicine that they perceived to do good to her was oil of St. John's Wort, and after the cores were coming out, the black salve (drawing salve). She went to full term and the child she bore had no marks or blemishes (and lived a healthy life).

In 1691 (in the Netherlands), there was a frosty dry winter; an excessive hot summer without rain. Winds were mostly east, northeast or north. The only stagnant water to be had were in the marshy countries, which was greedily drunk by thirsty parched laborers.

In 1691 in Italy, the year was as hot and dry as the previous two years were wet and rainy. The year began with a north wind and great frost. Roads were as dusty as in August. The summer was intensely hot.7

The Winter of 1691-1692 CE was awfully severe in Russia and Germany, and many people froze to death, and many cattle perished in their stalls.

Wolves came into Vienna, Austria and attacked men and women, owing to the intense cold and hunger. The wolves also attacked cattle.

All the canals of Venice, Italy were frozen.

The principal mouth of the Nile River in Egypt was blocked with frozen ice for a week.

In 1691 during the 12th moon, there was snow for four or five days in the vicinity of Shanghai, China. Men, horses, and animals froze to death. For half a month it was so cold that no one went abroad.

In 1692 CE in Noremberg (Nuremberg in Bavaria in southern Germany), the winter was very wet and cold. The harvest was very cloudy, rainy and cold.

In Italy) in 1692, the winter was exceedingly regular and agreeable to the climate. Spring, summer and harvest were the same. So was winter again neither too wet nor too dry; too hot nor too cold.

One evening in February 1692 during an episode of earthquakes, the village of Alari in Sicily seemed to be in flames for about six minutes. Were these earthquake lights?

On April 9th 1692 a fireball accompanied by detonation was seen over Temesvar in Hungary. This is now Timisaroa in Romania.

In the spring of 1692, a deluge, called the Great Flood, occurred at Delaware Falls (Trenton, New Jersey) in the United States. The first settlers of the Yorkshire tenth in West Jersey had built on the lowlands near the Falls and had been making improvements there nearly sixteen years. This flood, caused by the melting of the snow above, almost entirely demolished their settlement. The water rose to the upper stories of some of the houses, and many people were conveyed from their homes by canoes. Two persons, in a house were swept away by the torrent and were lost. Many cattle were drowned. The inhabitants, taught by experience the evils, of which the natives had forewarned them, fixed their habitations on higher grounds.

This year in Jamaica, the weather was very dry and hot in March, which was normally a very boisterous rainy month. From then until 7 June, it was excessively hot, calm and dry.

On 7 June 1692, an earthquake struck Jamaica. Within 2 minutes, most of the town of Port Royal was destroyed. The earth opened up and swallowed an abundance of houses and people. The water gushed out of the openings of the earth and people tumbled into it in heaps. Some people had the good fortune of catching hold of beams and rafters of houses, and these individuals were later saved by boats. Several ships were cast away in the harbor. The frigate Swan was carried away over the tops of sinking houses. Luckily the ship did not capsize and several hundred people retreated to this boat and were saved. Major Kelley, who was in the town at the time said that the earth opened and shut very quickly. He saw some people sink down to the middle and others sunk so low that just their heads were above ground and they were squeezed to death. The sky, which was clear before the earthquake, became in a minutes time as red and as hot as an oven. The fall of the mountains made a terrible crack, and at the same time dreadful noises were heard under the earth. The principal streets, which lay next to the key (quay), with large warehouses and stately brick buildings all sunk. But part of the town near the neck of the land, which ran into the sea, was left standing. At the end of this strip of land stood the castle, which was shattered but not demolished. Then the town was struck by a large tsunami. It drove most of the ships from their anchor. Then the sea immediately went out two or three hundred yards. It left the fish dry on the land. But the sea returned two minutes later and overflowed part of the shore. After the first great shock, as many people as were able got onboard the ships left in the harbor. They dared not venture back on shore for some weeks. The aftershocks still continued. It is estimated that 1,500 people died in the earthquake. And as many more by sickness from the noxious vapors that came out of the openings in the earth. The earthquake struck the entire island of Jamaica. Two mountains, which lay between St. Jago and Sixteen-Mile-Walk joined together and stopped the current of the river, so that it overflowed several woods and savannahs. On the north side of the island, over a thousand acres were sunk with houses and people inside them and a huge lake formed. This lake latter dried up, but there were no sign of the houses. At Yellows, a great mountain split and destroyed several plantations with people. One plantation was removed a mile from where it formerly lay. Houses were destroyed or damaged all over the island. It was estimated that 3,000 people were killed with those that were lost in Port Royal. The 1692 Jamaica earthquake struck Port Royal, Jamaica, on 7 June. A stopped pocket watch found in the harbor during a 1959 excavation indicated that it occurred around 11:43 AM local time. Known as the "storehouse and treasury of the West Indies" and as the "wickedest city in the world", Port Royal was, at the time, a key city in colonial Jamaica and one of the busiest and wealthiest ports in the Americas, as well as a common home port for many of the privateers and pirates operating on the Caribbean Sea. The 1692 earthquake caused most of the city to sink below sea level. About 2,000 people died as a result of the earthquake and the following tsunami, and another 3,000 people died in the following days due to injuries and disease. Two-thirds of the town, about 13 hectares (33 acres), sank into the sea immediately after the main shock. According to Robert Renny in his *An History of Jamaica* (1807): "All the wharves sunk at once, and in the space of two minutes, nine-tenths of the city were covered with water, which was raised to such a height, that it entered the uppermost rooms of the few houses which were left standing. The tops of the highest houses were visible in the water and surrounded by the masts of vessels, which had been sunk along with them."

In England, the summer was very rainy. In July of 1692, there were big floods in the north. In 1692, the summer in England was cold and there was a great deluge of rain until reaping (harvest) time.

The 1692 Salta earthquake took place in the Province of Salta, in the Republic of Argentina on 13 September at 11:00 a.m. local time. It registered 7.0 on the Richter magnitude scale and was located at a depth of 30 kilometers (19 mi). Aftershocks continued to be felt until 15 September. Salteño tradition has it that the number of victims was not higher because the earthquake occurred during the day and that the villagers were able to take measures to prevent greater damage. It is recounted that, in the middle of the chaos of the earthquake, while the houses were shaking and roofs were falling off, that the image of the Immaculate Conception (then called the Virgen del Milagro), then located in the Catedral de Salta, fell some three meters to the ground. Villagers, who had run to the church to pray, saw that the image was not only undamaged from the fall, but that it had fallen at the feet of the image of Christ. The villages interpreted that the image was interceding to Christ on behalf of the village. The following day the villagers paraded the image through the streets. The Salteños began venerating the image and praying for it to stop the earthquake. The tremors continued for two more days.

In Burgundy, France, the grape harvest did not begin until 9 October. It produced little wine and a great part of it was sour. The year was barren. There were rains and floods in the years 1692 in northern France. The cold humid period that lasted from 1692 CE to 1694 CE gave rise directly to famine. Some ten per cent of the population in Northern France died in the famine of 1693-1694. The rate of death was even higher in the Auvergne in the interior of the south of France.

Eruptions in 1693 were Teon in Indonesia with a VEI of 3.

Mount Etna in Italy erupted on January 9th.

A major earthquake on 9th January 1693 was followed on 11th January 1693 by the most powerful earthquake in Italian history. This was the same day as the first eruption that year of Mount Etna. The ensuing tsunami devastated the Ionian Sea Coast and the Strait of Messina between Italy and Sicily. It is unsure if the tsunami was caused by the earthquake, or a large underwater landslide triggered by the event. The 1693 Sicily earthquake struck parts of southern Italy near Sicily, Calabria, and Malta on 11 January at around 21:00 local time. This earthquake was preceded by a damaging foreshock on 9 January. The main quake had an estimated magnitude of 7.4 on the moment magnitude scale, the most powerful in Italian recorded history, and a maximum intensity of XI (*Extreme*) on the Mercalli intensity scale, destroying at least 70 towns and cities, seriously affecting an area of 5,600 square kilometers (2,200 sq mi) and causing the death of about 60,000 people. The earthquake was followed by tsunamis that devastated the coastal villages on the Ionian Sea and in the Straits of Messina. Almost two-thirds of the entire population of Catania in Sicily were killed. The epicenter of the disaster was probably close to the coast, possibly offshore, although the exact position remains unknown. The extent and degree of destruction caused by the earthquake resulted in the extensive rebuilding of the towns and cities of southeastern Sicily, particularly the Val di Noto, in a homogeneous late Baroque style, described as "the culmination and final flowering of Baroque art in Europe". The tsunami triggered by the earthquake affected most of the Ionian Sea coast of Sicily, about 230 kilometers (140 miles) in all. The first thing that was noted at all localities affected was a withdrawal of the sea. The strongest effects were concentrated around Augusta, where the initial withdrawal left the harbour dry, followed by a wave of at least 2.4 meters (7.9 feet) height, possibly as much as 8 meters (26 feet), that inundated part of the town. The maximum inundation of about 1.5 kilometers (0.93 mile) was recorded at Mascali.

On February 13th 1693 Hekla in Iceland erupted with a VEI of 4.

The spring was cold and there was almost constant rain, and north wind in Italy.

On 20 March 1693 at Oundle in Northamptonshire, East Midlands, England, there was a stormy day and a terrible tempest at night. There was great rain, winds from the southwest, and thunder with blue lightning, hail and rain most terrible. The lightning set fire to the steeple of Oundle.

Serua in Indonesia erupted with a VEI of 4 on June 4th 1693.

In Italy the summer was rainy and all corn mildewed. Harvest was intolerably hot and dry. There was excessive scorching heat and great drought. In Sicily, after the sun entered Virgo, the heat was great and at noon intolerable. On 1 August there was the most tempestuous day of hail, rain, and thunder. After that the earthquake struck. Another quake on the 11th desolated Sicily. Of the 254,936 inhabitants; 59,963 were swallowed up or killed. Sicily, late the most fruitful, rich and beautiful island in the world was left in rubbish and desolation. The summer of 1693 in Italy had excessive heat at the time of harvest.

In England, the heat was intense in September; and at noon it was unbearable.

The summer in Paris, France was characterized by: Hot days 33 days Very hot days 9 days (It appears that hot days are defined as those with temperatures of 25° C and greater but less than 31° C, very hot days are those with temperatures 31° C or greater but less than 35° C, and extremely hot days are those with temperatures of 35° C or greater.) The spring in Burgundy, France, was very cold and the grape harvest began on 27 September. It produced little wine, but the quality was good. In 1693 in France, there was an awful famine.

The summer of 1693 was dry in Finland and then a frost destroyed the meager harvest, especially the crop of barley. The harvests of 1693 and of 1694 were below normal.

In Germany during the beginning of the year it was very rainy; the latter part cold and frosty. The spring and summer were excessively hot.

In Britain and Ireland, October produced moderately warm weather, but there was some snow falling in the mountains and in the country. It turned suddenly extremely cold and was quickly followed by a hard frost for some few days at least. On 20 October 1693, there was a plague of locusts at Marthery in Pembrokeshire, Wales. A swarm of locust was seen in the air near Dôl-gelheu (Dolgellau) in Merionettshire (Merionethshire), Wales.

In 1693, there was an epidemic of the cold in London, England, which was very severe in October lasting 4 or 5 weeks.

There was also an epidemic of the cold that occurred in Dublin, Ireland in November where it was very severe for 4 or 5 weeks.

The epidemic of the cold was also present in France, the Netherlands and Flanders in Belgium.

On 19 October 1693 in Virginia in the United States there was a most violent storm, which stopped the course of ancient channels, and opened new ones, which never existed before. In 1693, there was such a dreadful storm in Virginia in the United States, that some rivers were stopped up and channels opened for others that were so large as to allow them to be navigated.

In 1693 and 1694, there were several occurrences of will-o'-the-wisps reported in Wales. About Christmas 1693 at Harlech in Merionydbshire (Merionethshire), sixteen ricks of hay and two barns which were filled with corn (grain) and other hay were set on fire by the "kindled exhalation", which were often seen to come from the sea and lasted at least a fortnight or three weeks. It annoyed the country, as well by poisoning the grass, as firing the hay, for the space of a mile or thereabouts. Those that saw the fire, say it was a blue weak flame, easily extinguished, and that it did not the least harm to any of the men, who attempted to save their hay, though they ventured close to or sometimes into the flame. All the damage sustained occurred constantly during the night. There were three small tenements in the same neighborhood (called Tydhin Sion Wyn (Tyddyn Sion Wyn)) whereof the grass was so infected, that it absolutely killed all manner of cattle that feed upon it. The grass was so infected these three years but not thoroughly fatal till this last year. As of August 1694, the strange fires continued there. It was observed to come from a place called Morva Bychan (Morfa Bychan) in Caernarvonshire (Caernarfonshire), about eight or nine miles off (over part of the sea). Cattle of all sorts, as well as sheep, goats, hogs, cows, and horses still continue to die. The place where it comes is both sandy and marshy.

In 1693, there was a dearness (scarcity) of all sorts of corn (grain) in England. Many poor people in Essex resorted to making bread from turnips.

In 1693 in the Free and Hanseatic City of Hamburg, Germany a very luminous, round "machine" with a sphere at its center, was seem crossing the sky. Does this sound familiar?

In December 1693 at Egryn in Merionethshire, Wales. A "fiery exhalation" came from the sea and set fire to the hay with "a blue weak flame." The fire, though easily extinguished, "did not the least harm to any of the men who interposed their endeavor to save the hay, though they ventured (perceiving it different from common fire) not only close to it, but sometimes into it." Egryn is on the coast around 18 kilometers (11 miles) south of Morfa Bychan.

In December 1693 CE Mount Etna erupted in Italy with a VEI of 3.

The winter of 1693/1694 in Italy was also very warm and dry. In Italy, the winter 1693 was cold and much snow (which is rare in Italy). In Italy, the winter was characterized by the most severe and scarcely to be paralleled cold frost and snow.

In the Winter of 1693-1694 CE in Germany and Italy, the frost was severe in November and December. In 1693 in Europe, there was great snowfalls and frost.

In Scotland there were seven years of harvest failure out of eight between 1693-1700. Was this due to a volcanic winter caused by the eruptions of 1693? The harvest, mainly oats, failed in seven of the eight years in all of the upland parishes of Scotland. This famine was so bad that the poorer sort of people frequented the churchyard to pull out nettles to eat and even fought over them which they greedily fed upon. This was from a parish record of Duthil and Rothiemurchus. Some sold their children into slavery.

In parishes all over Scotland from one third to two thirds of the population died of starvation. A greater number of deaths than the Black Death. Some whole villages and vast swathes of the countryside were depopulated at this time. About twenty per cent of the whole country was reduced to begging to survive from door to door. These were regarded as the ill years of King William's reign by the Jacobites. An indication of how cold it was in Scotland was the fact that several of the high-level tarns or lochans had ice on them all year round. A little lake in Straglash, Strathglass, at Glencannish, or Glencannich, had ice on it even in the warmest summer. These reports indicate that the required temperatures needed to be 1.5° to 2.0° C below twentieth century values averaged over the year. This was a lowering of twice to three times as great as that which has been substantiated in England from actual thermometer readings.

Eruptions in 1694 CE were Tangkoko-Duasudara in Indonesia with a VEI of 3, and Serua in Indonesia with a VEI of 3.

In the years 1694 to early 1697, cold winters and cool and wet springs and autumns led to extreme famine in northern Europe, particularly in Finland, Estonia, and Livonia. It is estimated that in Finland about 25–33% of the population perished, and in Estonia-Livonia about 20%. The famines to a lesser extent also affected Sweden (especially in the northern region), Norway, and northwestern Russia. The famine decimated the population of Finland and Estonia-Livonia either through prolonged starvation, epidemics and other diseases promoted by undernourishment, or the reliance on unwholesome or indigestible foods, and the contamination of water supplies.

In 1694 CE in Italy, it was hot and dry. There was a burning hot droughty summer in which five months passed without one shower of rain. Then came rain in October 1694 and the weather did not become fair again before April 1695.

In 1689, La Hire in France began taking observations using precipitation gauges. The rainfall in Paris in 1694 was 12.5 inches (318 millimeters). The year 1694 ranked just after 1689 as the driest year for thirty years.

On April 22nd 1694 a fiery exhalation rose up from the sea in Montgomeryshire in Wales. Grass was tainted and cattle were killed. The fire was a furlong broad and many miles in length burning all straw, hay, thatch and grass but doing no harm to trees, timber or any solid things, only firing barns or thatched houses. It left such a taint on the grass as to kill all of the cattle who ate of it. This had previously happened at Morfa Bychan and Egryn in December 1693, also in Wales. Was this a methane burst?

On May 29th 1694 CE Zaozan erupted in Japan with a VEI of 2.

On June 19th 1694 CE Iwakisan erupted in Japan with a VEI of 2.

On 20 June 1694, there was a hailstorm in Lohja and Siuntio, Finland (about 60° 10' N, 24° 10' E). Window glasses were shattered. Window glass was thick in those days, so the hailstones could not have been small. The hailstones destroyed the crop entirely on 3 farms, 2/3 of the crop on another farm, 1/2 of the crop on 2 other farms and 1/3 of the crop on another farm.

On July 4th 1694 Hokkaido-Komagatake erupted with a VEI of 4.

On 1 August 1694, a tornado struck Warrington in Northamptonshire, England. It carried 80 or 100 shocks of corn into the air out of sight, to a distance of one, four and five miles.

In 1694 at Topsham and Exeter, England, in Acremont Close, there was a waterspout. It lasted for 30 minutes and 3 or 4 wagonloads of corn (grain) were in the air at one time. On 7 August 1694 at Exeter, England, there was a landspout (tornado), like those at sea (waterspout).

The 1694 Irpinia–Basilicata earthquake occurred on 8 September 1694. It caused widespread damage in the Basilicata and Apulia regions of what was then the Kingdom of

Naples, resulting in more than 6,000 casualties. The earthquake occurred at 11:40 UTC and lasted between 30 and 60 seconds. There was serious damage to the area between Campania and Basilicata, with more than 30 municipalities being almost completely destroyed. These included Bisaccia. Sant'Angelo dei Lombardi, Bella and Muro Lucano. In Melfi, fifty buildings collapsed and the castle, cathedral, five monasteries and many churches were severely damaged. In Potenza, several buildings, the church and the Trinità Tower collapsed. The following number of casualties were reported, 700 at Calitri, 700 at Sant'Angelo dei Lombardi, 600 at Muro Lucano, 400 at Ruvo del Monte, 300 at Teora, 280 at Guardia Lombardi, 250 at Bella, 230 at Pescopagano, 190 at Cairano, 160 at Atella, 120 at Sant'Andrea di Conza and 100 at Tito.

On 27 September 1694, a hurricane struck offshore Barbados in the Lesser Antilles causing more than 1000 deaths.

Three weeks before 4 November 1694, a hurricane struck Barbados in the Lesser Antilles putting most of the ships in the road ashore.

A most severe sandstorm struck Scotland on 2 November 1694. The village of Culbin was covered over and lost for 230 years. Culbin is now Culbin Sands on the River Findhorn, on the southern bank of the Moray Firth in northern Scotland. From 1694-1699 in Scotland, there was a famine. Villages in northeast Scotland, near the Moray Coast, buried in sand due to a prolonged northerly or northwesterly gale. On November 1st there was a severe sandstorm that buried the village of Culbin. This was apparently one of the most fertile areas in Northern Scotland, but deforested Culbin Forest which resulted in destabilizing the land between Culbin and the mainland. This area was called Culbin Sands which referred to a large area of loose dune sand desert which is now the Culbin Forest. In its day the dune system was the largest in Britain. The town and port of Findhorn were also destroyed at this time.

In 1694 in England, there was a great dearth from rains, colds, frosts, snows; all bad weather.

Philippe de la Hire has the cold from the winter of 1694-95 in northern France as being among the most intense. The winter in 1695 was very harsh. La Hire's thermometer stood the whole time of the frost at 15° to 20° F (-12.1° to -8.5° C), except on 7 February, when it fell to 7° F (-17.9° C).

During the winter of 1694, there was ice in the Whangpu (Huangpu River), in China).

In 1694, the frost in England was so intense that many forest trees and oaks were split by the frost. During the winter of 1694-95 in London, England, the frost was of 7 weeks duration. England experienced a cold winter. There was continuous snow for 5 weeks.

Sea ice completely surrounded the whole island of Iceland.

In Bohemia (now western Czechia) during June, the summer was very cold and 3 intense frosts occurred leading to famine.

At Augsburg in Bavaria, Germany, from the middle of December 1694 to 11 March 1695, the wind was mostly east and exceedingly cold and cloudy. The harvest and beginning of winter were very wet. At Ulm in southern Germany, the winter of 1694/1695 was intensely cold and dry. The frost continued even to the spring then suddenly there was a cloudy, rainy thaw about the end of March.

In Finland, the beginning of 1696 was deceptively mild. The extraordinary snowy early winter of 1695-96 halted the forest work and traffic. But this was interrupted by a thaw in January. Spring came very early and the fields turned green.

1695 was regarded as a whole as one of the coldest years ever known in the British Isles. The value was 7.25° C on CET. Only 1740 was colder. The CET summer value was 13.25° C. The anomaly over the three summer months was -2.5° to -3.0° Celsius.

In 1695 when spring, summer and autumn temperatures were low and the summer months mostly about 2.0° or more below the 20th Century normal, the growing season was probably shortened by two months or more in England.

In Finland, the beginning of 1695 was the coldest winter since the year 1658. Wolves attacked people in their homes. Spring was late in coming and generally cold. Sowing of seeds could not be finished before midsummer, leaving too short a time for the crops to mature. The summer was also cold. The grains could not ripen before a killing frost came. The autumn was rainy and, in most regions, it was impossible to plant the seeds for the winter crop. In the region of Uusimaa, Finland, the rye did not blossom before the end of July and a frost in September destroyed the half mature grain.

On March 10th 1695 there was an inflamed cloud in the sky and sparks of flame fell for fifteen minutes at Chatillon-sur-Seine in Cote-d'Or, France.

The 1695 Linfen earthquake struck Shanxi Province in North China, Qing dynasty, on May 18. Occurring at a shallow depth within the continental crust, the surface-wave magnitude 7.8 earthquake had a maximum intensity of XI on the China seismic intensity scale and Mercalli intensity scale. This devastating earthquake affected over 120 counties across eight provinces of modern-day China. An estimated 52,600 people died in the earthquake, although the death toll may have been 176,365. At Linfen, where the maximum intensity was X–XI, many dwellings in villages were destroyed, killing their inhabitants. The shock knocked down structures in the city, including walls, temples, towers, homes, and warehouses. According to one account, more than 40,000 homes were destroyed. Conflagrations were started and the ground erupted out black-colored water, killing many. More than 60 villagers died at a settlement in Linfen when temples and homes collapsed. In another village, more than 80 of the 180 inhabitants died due to collapsing structures. The shock collapsed entire cave homes and buried many people. In Xinjian Village, Xiangling County, 7,000 people died. Many more livestock were also killed. In the aftermath, only a handful of homes, government buildings, temples and educational institutions survived. Linfen lost at least 28,000 residents or 16 percent of its population. The city accounted for 53 percent of the total death toll. Reports of damage also came from Hebei, Gansu, and Jiangsu provinces. An official report in 1875 stated that 52,600 people died, but a monument from the Yuan dynasty placed that figure at 176,365. Ground effects associated with the earthquake are still visible today. At Dongputou and Xiputou villages in the Xiangcheng District, a large valley splitting the villages into two. The Tongli canal, a source of water for agriculture in the region, was destroyed during the tremors.

On June 23rd 1695 CE Asamayama erupted in Japan with a VEI of 2.

Summer 1695 in Scotland was one of the first in a series of disastrous harvests in Scotland where famine ensued. In the cold-wet hunger years of 1695-99, Scotland lost between 5% and 15% of its people.

In Ireland, in the spring and summer of 1695, there were many stinking fogs in Limerick and Tipperary. What were stinking fogs? What made them stink?

In England in April 1695, the weather was extraordinarily fair for the most part and almost cloudless. May was remarkably wet, to the destruction of all fruits. All the dogdays of summer were exceedingly cold, like winter. The winter was warm and fair except two or three days of hard frost in the end of December.

On 24 July 1695, there was a violent thunderstorm with hail at Aberdeen, Scotland.

In Lappee (present-day Lappeenranta, Finland at 61° N, 27° 35' E) on Petersmas day in 1695, June 29th, there was an unusual hailstorm that quickly beat down the crops in all the fields.

The summer in Estonia in 1695 was very rainy. It rained from Johann's Day (June 24) till Michael's Day (September 29). As a result, the lowlands were flooded which destroyed the hay and crops.

In Italy, there were profound deluges in 1695. The Po River in northern Italy overran meadows, fields, and destroyed crops, leading to a severe famine in the area.

Lake Zurich in Switzerland, Lake Constance in Germany, Switzerland and Austria and Lake Neuchâtel in Romandy, Switzerland, froze completely and one could walk over them as one would travel over a bridge.

There were ice flows in the River Thames in England.

In 1695 a violent hurricane struck the Mauricius (Mauritius) Island in the Indian Ocean off the coast of Africa.

Rivers over a great part of Europe were in heavy floods in 1695-1697. Many of the rivers and lakes remained frozen for comparatively longer periods of time and didn't thaw until the late spring.

At Ulm, Germany, all August to 1 September, it was cold and rainy. September and October were very cloudy and excessively cold.

At Poson, Poznań, in Poland, the summer and harvest of 1695 was one continued winter of cold rain, raw frosts, mildew, etc. The summer began on 10 September and lasted till 10 December.

On September 22nd 1695 a violent storm affected the southern North Sea, the English Channel, Belgium, northeast France and the English coasts such as Norfolk, Suffolk, Essex, Kent and perhaps Sussex. Many ships were lost or damaged. Seventy coaling ships were beached with much loss of life as well as property on the East Anglian coast. Trees were felled in London and lives were lost elsewhere due to shipwrecks.

On 4 October 1695, a hurricane struck the Florida Keys in the United States. The hurricane caused the loss of a 933-ton warship offshore.

In October 1695, a hurricane struck offshore the Caribbean Island of Martinique causing greater than 600 deaths.

At Poson, (Poznań, Poland), after 10 December 1695, there came a great snow and a strong frost, which had no thaw or remission till 10 March 1696. All corn and herbs died and rotted under the snow.

In Estonia in 1695 there was famine.

In Norway the worst years for harvests were incidentally the best years for the herring fishery. The self-same conditions that which produce harvest failures such as long-lasting harsh and stormy westerly and northerly winds drive the fish stocks of the Arctic Ocean, Barents Sea, in greater than usual numbers to the Norwegian coast. This happened from 1695 to 1697 CE.

In 1695 the drift ice surrounded the whole of Iceland except for Snaefellsnaes on the west coast for many weeks. At most parts of the coast open water could not be seen from the highest mountains and merchant vessels could not make their way to the harbors for many months.

Also, in 1695 cod became scarce in the waters around the Shetland Islands.

As well in 1695 cod disappeared completely from the coast of Norway, except for a colony surviving in the inner part of Trondheim Fjord. The Arctic cold water had apparently spread across the whole Norwegian Sea. The sea conditions remained significantly colder than the late twentieth century until well after 1800.

In 1695 CE-1697 CE in Finland and Estonia there was famine.

The first part of the winter of 1695-96 was very cold and the snow was very high. But early in 1696, a thaw came. This pattern also happened in Sweden.

In Finland, the beginning of 1696 was deceptively mild. The extraordinary snowy early winter of 1695-96 halted the forest work and traffic. But this was interrupted by a thaw in January. Spring came very early and the fields turned green.

Early in l696, the cold in England, the Netherlands and northern Germany was excessive. Doctor Derham reported that at the Gresham College, London, England; the thermometer indicated a temperature equal to 1.6° F (-16.9° C). The average high temperature between December and February is 48° (9° C) and the average low temperature is 41° F (5° C).

On 26 January 1696, there was an intense frost in London, England. The temperature fell to 9 degrees below zero.

On 3 February 1696 on the Isle of Portland in the English Channel, there was a landslip (landslide). The great pier was demolished and much damage done, owing to excessive rain.

In the cold-wet hunger years of 1695-99, Scotland lost between 5% and 15% of its people.

In February, gooseberries in London begin to have a body. In March, dull, gloomy cold weather, blasting all the buds and ruining the spring.

In Finland on 7 March, winter struck back with a vengeance. Lakes and bays froze thick, so thick that people were able to drive across them (with wagons). When spring finally returned it was very late and summer was rainy. The crops were slow in ripening.

At Poson, (Poznań, Poland), the winter continued to 10 March 1696.

At Hildesheim in central Germany, up until 10 March there was warm moist winter weather. Following that was some weeks of severe winter weather.

Winter returned to Finland with a vengeance in March and it was impossible to sow the seeds until the end of May.

From Easter, Easter Sunday being April 12th, to 26 June, there were cold, wet excessive rains and great inundations. The rains rotted the hay. The spring till then was at a standstill.

In May, there was an extraordinary flood in England.

In June 1696, there was an inundation of the River Nyne in Northamptonshire, England.

In England from 26 June to 6 July, the weather was fair and then the rains returned. From 10 July, it rained incessantly 36 hours.

From 12 to 17 July, the weather was fair in England.

From the 17th of July to the 14th of August, both night and day, there were heavy showers daily. It laid all barley and oats in England.

On July 31st 1696 CE Vesuvius erupted in Italy with a VEI of 2.

During the night of 17/18 August 1696 a frost struck in England. In the morning the grains were coated with a thick layer of ice. A second frost occurred which finished off the rest of the crops. Estonia experienced a similar weather pattern in 1696.

To the 23rd of August, in England the weather was fair. The remainder of August was mostly rainy. To the end of the year, the weather was variable.

The heavy rains of summer ruined the crops in Finland. The harvest amounted to about one-fourth the seeds sown. Shortly after summer, there was no hay to be had for any price.

In Estonia in 1696, landlords could no longer feed their farmhands and servants and began dismissing them. Many of these recently unemployed along with destitute, hungry peasants turned to begging. Even some members of the nobility were reduced to this state.

In the autumn of 1696, the famine in Finland and Estonia became terrible. There was a pronounced rise in the death rates. "The peasants died like flies." Bodies of the dead were lying everywhere.

On 24 December, there were three tides in the River Thames in one day.

In 1696, it was a bad year for the crops and food was very dear (scarce) in England. In England, there was a great storm on the east coast; In England, 200 sail of colliers and some coasters were lost, with all their crews in a great storm, in the bay of Cromer, in Norfolk.

In 1696, a hurricane struck northwest Cuba. An unidentified ship was wrecked at Playa de Sabarimar, 7 leagues east of Havana in 35 feet of water during the storm.

The winter of 1696 was colder than had been known in New England in the United States, since the first arrival of the English. During a great part of the winter, sleighs and loaded sleds passed on the ice from Boston as far as Nantasket (Hull, Massachusetts). So great a scarcity of food, afterwards during the next year, had not been known; nor any grain ever been at a higher price.

Poson, (Poznań in Poland) went without rain in 1696; hence a great scarcity in 1697.

Rivers over a great part of Europe were in heavy floods in 1695-1697. Many of the rivers and lakes remained frozen for comparatively longer periods of time and didn't thaw until the late spring.

The winter of 1696-97 was extremely harsh in Finland and Estonia. The snow was very high so corpses were left unburied until springtime and then placed in mass graves. Cases of cannibalism were reported in Estonia.

In the winter of 1696 to 1697 the CET value was 1.3° C. this was more than -3° C on modern levels. The cold persisted through February and there was snow and soldiers in the armies and garrison towns were frozen to death at their posts.

In 1697 CE Klyuchevskoy in Kamchatka erupted with a VEI of 3.

In 1697 CE in what is now the United States, the winter was intensely cold in the American northeast. Boston harbor was frozen as far down as Nantucket. The Delaware River was closed with thick ice for more than three months so that sleighs and sleds passed from Trenton to Philadelphia, and from Philadelphia to Chester on the ice.

On January 13th 1697 at 5.00 pm hot stones fell at Pentolina, Manzano, Capraja and Paduel near Siena in Tuscany, Italy.

At Mansfeld in central Germany, January and February were intensely cold. March and part of April were unsettled, cloudy, snowy, rainy, frosty and clear. April the 1st and May began with hot summer weather, but followed by great storms of hail, especially the 21st, which did much damage. On the 27th sleet snow and an east wind to the end. Summer was often cold with frequent rains and very changeable winds. August was clear, but very cold. September 10th to October, great rains and shifting winds. November was cloudy and snowy. December was mild and rainy but ended cold.

In London, England from 15 January to 11 February, there was a hard frost with some small remissions.

In January, there was much snow in deep drifts. All January, there was ice upon the water, which on the 26th was eight inches thick (i.e. within 2 ½ times as thick as at any time on the Canal of St. James's Park in 1740). Yet on the 29th of January, there was lightning and five claps of thunder.

The winds were northeast almost the entire month of February with little sunshine, except for six days during the second week. On 14 February there was a great storm, and the lanes were blown up with snow several yards deep, that lasted the rest of the month. But the fields lay bare. (The winds blew the snow in from the flat fields into great snowdrifts.) On the 26th of February, the ice was four inches thick.

On 22 March 1697, the river near Noordwyck (now an abandoned town) on the Mauricius (Mauritius) Island (in the Indian Ocean off the coast of Africa) swelled in the space of 15 minutes to great heights. The sugar mill and sugar works and most of the ground were ruined. Most of the sugar canes were torn out by the roots by the violence of the currents. Then fifteen minutes later, the water was back to its normal level. The cause of this flood was unknown. There was only light rain at the time and no earthquakes.

On March the 24th and 26th, thunder and lightning, warm sunshine all day with sulfurous clouds and hot evenings.

From March to 11 April, there were cold northeasterly winds. The gooseberries not yet budded. On April the 11th, there was thunder followed by showers.

On 13 April, there was rain; and by 18 April, there were trees green with leaves, though no spring before. On April 25th, there were showers of fierce great hail with thunder and sunshine mixed. On April 27th, there was thunder and a storm of hail after. April was a cold month.

On 9 April 1697 in Wales, hailstones fell in Flintshire weighing 5 ounces.

In Cheshire and Lancashire, England on the 20th of April, "a storm of hail, which killed fowls and small animals, and knocked down horses and oxen; some of the stones weighing half a pound."

On April 22, it snowed hard from morning till noon, then a little sunshine; then snowed again very fast; then, sunshine followed with large hail (similar to the storm of April 1740).

On 29 April 1697, there was a thunderstorm, which produced great hail at Snowdon in Denbighshire, Wales. It also struck Flintshire, Chester, West Kirkby (West Kirby), Ormskirk, and Blackburn. The cloud was 2 miles (3.2 kilometers) wide and the length of the track 60 miles (97 kilometers) long. Hailstones of 5 ounces (142 grams) and of various shapes, broke nearly all the window, killed many fowls, and destroyed the green corn (grain). Some hailstones were 5 inches (13 centimeters) round. Scarcely any stones were as little as musket balls. But some were as large as hen's eggs and ½ pound (227 grams) in weight. Many sea fowl (sea birds) were killed. Poultry and sheep were also killed.2 In Cheshire and Lancashire, England on the 29th of April, "a storm of hail, which killed fowls and small animals, and knocked down horses and men; some of the stones weighing half a pound."

In 1697, a tremendous hailstorm struck the part of Denbighshire in northeast Wales bordering on the sea. All the windows on the weather side were broken by the hailstones. Poultry and lambs, together with a large mastiff (breed of dog) were killed. In the north part of Flintshire, several persons had their heads broken and were grievously bruised in their limbs. The main body of this hailstorm fell on Lancashire in northwest England, in a line from Ormskirk to Blackburn, on the borders of Yorkshire. The breadth of the storm cloud was about two miles, within which it did incredible damage, killing all descriptions of fowl and small creatures, and scarcely leaving a whole pane of glass in any of the windows where it passed. What was still worse, it ploughed up the earth and cut off the blade of the green corn (grain) utterly destroying them. The hailstones struck with sufficient force to bury themselves in the ground. These hailstones, some of which weighed five ounces, were of different forms, some round, others semispherical; some smooth, others embossed and crenulated, like the foot of a drinking glass, the ice being very transparent and hard; but a snowy kernel was in the midst of most of them, if not of all. The force of their fall showed that they descended from a great height.

On 29 April 1697 at Cheshire and Lancashire, England; on 4 May at Hertfordshire, England on 6 June at Monmouthshire, Wales on 9 June at Herefordshire, England fell shocking tempests of prodigious hail. The astronomer Halley communicated a paper to the Royal Society on this storm. Toward the end of April in 1697, a hailstorm struck Cheshire in northwest England. The storm was 2- miles wide and traveled a path for 60 miles before it was dissipated. The hailstones were as big as eggs and some were the size of a man's fist. They were pieces of clear, transparent, and very hard ice, with a white kernel in the middle, which seemed like a lump of snow. Some of these vast hailstones were smooth, others rough and sharp on the surface. They fell with a prodigious force, and killed fowls, lambs, and calves. They beat down the young crops of every kind. In some places where the wind drove them at a slanting angle, they plowed up the surface of the ground burying them an inch or two in depth. Trees were broken and shattered to pieces in many places. Houses were damaged. Many people who were outside during the storm were harmed. An extraordinary hailstorm struck Wales and England on 29 April 1697. The track of this storm was over 60 miles in length. The storm formed with southwest winds out of Carnarvanshire (Caernarfonshire), passing near Snowdon with a horrid black cloud attended with frequent lightning and thunder. It traveled as far as can be determined no further westward than Denbighshire, where it left St. Asaph to the right, and did much damage between it and the Sea, breaking all the windows on the weatherside, and killing poultry and lambs, and at Sir John Conway's at Desert, a stout dog. In the northern part of Flintshire, several people had their heads broke and were grievous bruised on their bodies. From Flintshire, Wales it crossed over the Arm of the Sea that comes up to Chester, England and was only felt in Cheshire, at the very northwest corner of the peninsula called Wiral, between Æfluaria of Chester and Leverpoole (Liverpool), at a town called West Kirby, where it hailed only 3 minutes. It was at Chester about 3 o'clock in the afternoon. The main body of the hail fell upon Lancashire, in a right line from Ormskirk to Blackborn (Blackburn), which is on the border of Yorkshire; but whether it crossed the ridge of hills into Yorkshire, we know not. The breadth of the cloud was about 2 miles within which it did incredible damage, killing all sorts of birds and small creatures. It scarcely left a whole pane of glass in any of the windows it passed. It plowed up the earth and cut off the blade of green corn (grain) so utterly destroying it. The hailstones buried themselves in the ground. On the bowling greens, where the earth was anything soft, they were defaced, so as to be rendered unserviceable for a time. Some of the hailstones weighed 5 ounces. They were of different forms. Some were round, some half round, some smooth, others embossed and crenulated, like the foot of a drinking glass, the ice very transparent and hard, but a snowy kernel was in the midst of most of them, if not all of them. Because of the force of the fall, the hailstones must have fallen from a great height. Near Bootle in Merseyside, England where the storm skirted, one of the hailstones measured 5 inches in circumference. At Bootle-Mill, the sea seemed to have risen to an extreme height and took on the appearance of a forest of woods. (As the hailstones struck with great force, they sent spouts of water up almost 5 feet high.) The hailstones were as big as Poot Eggs. Many seafowl and land fowl were killed by the hail. The storm was very violent at Linaker. It made holes in William Halsall's barns, broke branches off his apple trees, and made wounds in the green brow (hill) by his house. The holes were generally an inch to an inch and a half deep. William Halsall said that these hailstones fell so violently into the marl-pit (a pit from which marl is dug) beside his house that the spouts of water rose a yard and a half high. The hailstones were as big as duck eggs at Aughton Common and at Sephton (Sefton). One of the hailstones at Sephton weighed a full half pound. Two hailstones were weighed at Ormskirk. Each weighed ¾ of a pound. The hailstones at Ince (in Lancashire, England) varied between the size of duck eggs and goose eggs. In a little town next to the sea, they gathered birds killed by the hail by the bushels. On the seaside, at least seven varieties of dead birds were gathered including curlieu (curlew), sea-pye (oystercatcher, sea pie), sea swallow and gorre. At Bootle, a young woman who was running for shelter during the storm had her hat fall off and a hailstone hit her from behind the ear and made her tumble. A man was knocked off his horse. Another man pulled down his hat to save his face and a hailstone tore the brim from the crown, so far that he could put his hand through the hole. At Ormskirk, the hailstones rebounded 2 yards high. At Ince, two horses were knocked down in the plough and a man fell at the same time. At Crosby, some beasts were knocked down by the hail. Joseph Holland was found dead after the storm but it couldn't be ascertained whether his death was caused by hail or lightning. Two women were so badly beaten by the hail before they could find cover that the next morning, they could hardly turn them over in their beds.

From 29 April to 4 May, there was cloudless, intolerably, sultry, fainting, hot days. The heat was both day and night.

On the 30th of April, the first cuckow (observed). Gooseberries not yet blossomed. On 3 May, there was a great deep snow over all of England.

In the beginning of May in 1697, a hailstorm struck Herefordshire in the West Midlands region of England. In this terrible storm, not only fowls and young animals of all kinds were killed, but some of the larger ones. Some persons laboring in the fields, who could not reach shelter, suffered the same fate. Their bodies were black and blue, as if beaten to death with clubs. Oak trees were split in two. The branches of many other trees were torn down. The fields of rye were, in some places, cut down as if mowed off with a scythe. Many of the hailstones measured fourteen inches round.

On 4 May 1697, a hailstorm struck Hitchin, England. Hailstones were 7 or 8 inches (18-20 centimeters) about (in circumference). At Sir J. Spencer's, 7,000 quarries (a square or diamond shaped pane) of glass were broken. The hail split great trees and destroyed several hundred acres of wheat; and there were some stones 13 or 14 inches (33-36 centimeters) about. The ground was torn up, and there were at least 100,000 cartloads of hailstones. A southwest gale occurred at the same time. In Staffordshire, some of the hailstones were nearly 12 inches (30 centimeters) in circumference.

On Tuesday, 4 May 1697, at Hitchin in Hartfordshire (Hertfordshire), England, about 9 o'clock in the morning, it began to lightning and thunder extremely, some great showers intervening. This continued until around 2 o'clock in the afternoon, when a black cloud appeared and there was a sharp shower of hail. Some of the hailstones measured 7 and 8 inches in circumference. But the most extreme part of this storm fell at Offley, where a young man was killed, one of his eyes stuck out of his head; his body was all over black with bruises. Another person nearer to Offley escaped with his life but was much bruised. There was a house of Sir Joseph Spencer in which 7,000 quarries of glass were broken. The hail fell in such vast quantities, and so great, that it tore up the ground, split great oaks and other trees in great number. It cut down great fields of rye, as with a scythe, and destroyed several hundred acres of wheat, and barley. So much so that they ploughed it up and sowed the field with oats. The tempest was such when it fell that in 4 poles of land, from the hills near us, it carried away all the staples of the land, leaving nothing but chalk. The hail broke a vast number of pigeon's wings, crows, rooks, and other birds. The flood (of hail) came down, spreading 4 or 5 acres of land, (roaring) like the Bay of Biscay; and which is very strange, all this fell in the compass of one English mile. I was walking in my garden, which is very small, perhaps about 30 yards square, and before I could get out, it took me to my knees, and was through my house before I could get in, which I can modestly say was within a minutes time. It went through like a sea, carrying all wooden things like boats on the water. The greatest part of the town experienced this misfortune. The surprise was so great, that we had scarcely enough time to save our children and wives. There fell some hundred thousand cartloads of hail. I saw them 4 days later, and if the beds of hail had not been broken by people coming and going or trampled by horses, it might have laid until Michaelmas, 29th September. The hailstones were measured from one to thirteen and fourteen inches certain. Some people said they found hailstones that measured 17 and 18 inches. The hailstones had various shapes, some oval, others round, others picked, some flat. The damage done to our town was near 4,000 pounds.

In England from 4 to 25 May, it was cold. From 4 to 19 May, it was wet.

On 15 May, the woods were like winter in England.

On 19 May, it was a frosty night. The rest of May was fair and hot to the end, with a north wind.

In 1697, there was a northwest gale in Lancashire, England with hailstones 6-9 inches (15-23 centimeters) in circumference. Rooks and hares (birds and rabbits) killed, and vast quantities of glass broken.

On 6 June 1697, there was a hailstorm at Pont-y-pool (Pontypool) in Monmouthshire, Wales. The storm extended a mile and lasted half an hour. Some hailstones were eight inches (20 centimeters) about (in circumference) and very irregular. The hail broke all the beans and wheat.

On 6 June (9 June) 1697, there was a hailstorm at Westhide, near Hereford, England. The hail fell in so great quantity that it destroyed all the poultry, garden stuff, corn (grain), grass and windows. Some stones were 9 inches (23 centimeters) about (in circumference). In 1697, there was a hailstorm in Herefordshire, England. "Hail stones 13- and 14-inches round (33-36 centimeters)."

On 7 June 1697 in Charleville, Ireland near Limerick there had been a very wet spring, and as a result a bog moved over 40 acres of good ground, burying it sixteen feet (5 meters) deep.

June was seasonable enough in England. On 20 June, there was high winds and rain. On the 21st of June, there was excessive cold.

In June 1697 there was a severe flood caused by bog bursting near Charleville in County Cork, Ireland. The spring had been uncommonly wet in England and Ireland with frequent rain and hail. The peat bog had become so sodden and swollen that it exploded.

In 1697, it was a bad year for the crops and food was very dear (scarce) in England.

Excessive heat reigned again in July 1697 in northern France.

On 14th July 1697, in the region around the villages of Nummi and Pusula, Finland (at 60° 30' N, 24° E), there fell a "furious" shower of hail. The hailstones struck the cattle and made holes in the walls and roofs of houses. The crops on 19 farms was totally destroyed, 3 farms lost ½ of their crops, and another 3 farms lost 1/3 or their crops.151 In mid-July 1697 in Kalanti (present-day Uusikirkko, Finland (at 60° 50' N, 21° 35' E)), a hailstorm produced hailstones the size of hen's eggs causing great destruction on 5 farms. On 16 and 17 July, frost and mildew blasted the corn.

Universal rains during the summer of 1697, made all the rivers overflow in France. The rains lasted at least two months. The rain fell so hard for eight days from the Feast of Saint Peter (29 June), that in one night the Seine, the Loire and the Meuse rivers rose seven feet. The rivers continued to grow and overran their banks and flooded all the countryside, with the farmland, houses and their inhabitants.

At Poson, (Poznań, Poland), May and June most unequal, the heavens were terrible with clouds and cold rains. In July and August, the heat was excessive but often mixed with cold showers. Rivers over a great part of Europe were in heavy floods in 1695-1697. Many of the rivers and lakes remained frozen for comparatively longer periods of time and didn't thaw until the late spring.

In England August to the 10th, still calm; daily rain till the corn grew in the ear as it stood. 12 August frost to 10 September dry sun shiny weather, excellent harvest.

On September 15th 1697 Vesuvius erupted in Italy with a VEI of 3.

On 28 September, there was great hail in the night in England.

On 1st October 1697 all of the northwest German coast was hit by sea floods.

On 8 October, there was a great wind. October was a pleasant month in England.

On October 27th 1697 a round body of fire fell onto a powder magazine in Athlone in Ireland. The explosion wrecked the town.

In Autumn 1697 there was a sand-drift disaster on the island of North Uist in the outer Hebrides where the site was buried by sand.

There was also a great storm surge affecting both sides of the southern North Sea with it being most destructive on the continental shores.

On 25 November 1697 in London, England, there was ice three inches thick.

On December 12th and 15th, it was hot in England; the 12th, 18th, 19th and 20th, there was mist, hot and moist. From 10 to 30 December, it was as hot as in August; one could not bear the bedclothes. Yet there were frosts before and snow 12 inches deep.

In Finland in 1697, the famines, death and epidemics closely followed. This famine was so horrific that it brought on cases of cannibalism. In Ostrobothnia, Finland, "parents ate the corpses of their children, and children of their parents, brothers and sisters. In northern Karelia, Finland, court documents describe cases of cannibalism. In one township in Karelia, there were so many funerals that the church bell cracked. Storehouses and manor houses were plundered.

During the winter of 1697, there was a severe frost in England.

In the winter of 1697 to 1698 it was a severe winter. It was the tenth coldest in the 350-year series. There was an anomaly of -3.1° C when compared to the 1961-1990 average. The severe cold extended from England to Scotland. Rivers were frozen and there was frosty weather with heavy snow with deep drifts reported across the southeast of Britain. This would imply that the British Isles weather was dominated by a blocking high extending westwards over the country from Russia.

In 1698 CE Cotopaxi erupted in Ecuador with a VEI of 3.

In Scotland, they were reaping in January and beating the deep snow off it, as they reaped the poor green empty crop. Bread made from what was harvested would not stick together, but fell in pieces, and tasted sweet as if made of malt.

The frost, hail and snow persisted from January to May 1698 which was reputed to have been the coldest year between 1695 and 1742. The ice was eight inches on the coast of Suffolk. There was deep snow in England on 3rd May after a snowfall of up to six inches in Yorkshire on the first, and a keen frost, and the spring of 1698 was the most backward for 47 years. There was further snowfall on 13th May in London and Yorkshire with corn and fruit crops damaged. The spring of 1698 was regarded as the most backward for fifty years.

Ciremai in Indonesia erupted in Indonesia on February 3rd 1698 with a VEI of 3.

In London, England on 17 May 1698 CE, there was a great hailstorm.

On May 19th 1698 between 7.00 pm and 9.00 pm a large stone fell at Hindterschwendl in the Canton of Bern in Switzerland. This is also recorded as Walkringen in Bern as well as Waltringen in Berne.

In England on 31st May, the wheat was very low and there was cold weather.

On 3 June in England, it was cold with great lightning and thunder, loud and near, with fierce large hail three inches deep on the ground.

On 16 June in a warm rich soil, the first wheat ear was seen near London. This was the backwardest spring in 47 years.

In England in July the first part was wet. On the 9th there was a great deal of red lightning with unceasing thunder. On the evening of 15 July, there was a great rain. From 18 to 26 July, there was cloudless sunshine. There were no gooseberry tarts till July. On the 30th of July, the apple trees were in small blossoms as in the spring.

On 6 August 1698 in England, "Biggest raindrops known. There was one clap of thunder and then a shower of the biggest water droplets ever known. The most rain last four months known. Whole fields of corn (grain) were spoilt. On 13, 14 and 15 August there were frosts. The latter half of August was the most pleasant time in this year.

The first wheat cut in the middle of September, and much barley in swathe till December.

In the North much corn ungot at Christmas. The four last months had scarce two days together without rain (and with the exception of the period from 18 to 26 July) the wettest season known. Whole fields of corn (wheat) spoiled even in Kent; much more (spoilage) in the north. Horses were turned into (fed) the peas and barley. The earliest wheat not cut till the middle of September.

In Kent, September the 29th, barley standing uncut there; much lay in the swath till December. That which was brought in was soaked with wetness and almost useless.

In Scotland on 3 October, there was much lightning and pretty much thunder. On 15, 16 and 17 October, there were extreme cold nights with winds from the northnorthwest. On 30 October, there was a great deal of rain and snow with the winds from the northeast.

On 5 November 1698, there was a terrible flood, which destroyed a great part of St. Werburgh's Church in Derby, England.

In Scotland on 17 November, there was lightning and thunder.

In Scotland December was warm. On the 7th of December there was a hot steam. On 22 December, wheat was sown, which proved as forward in harvest as any. The seed time was so wet that there was hardly above half a crop sown this year.

On 22 December 1698, there was a thunderstorm at Warley Town near Halifax in West Yorkshire, England.

Much corn in the north of England was got at Christmas.

At Sepola in Corsica, France one night on December 24 1698. A local shepherd out in the fields observes from a distance a strange mist or fog suddenly take form above the small village totally covering it. Not thinking much about it, he prepares to go to church the next day. Arriving at the village he notices that the strange fog has already dissipated but finds the village strangely deserted. Unnerved and terrified he travels to the nearby village of Moltifao and alerts the local priests and Gendarmes (police) who form a search party that is unable to find any dead bodies or other villagers. They did locate a trail of footsteps that led to the local metal smith's residence. He is found hiding and terrified and tells a strange story of seeing a huge ball of fire rising from the ground near his hut. At the scene circular depression was found on the grass, and it is said that nothing grows there to this day. This sounds familiar as well but remember that we are in the seventeenth century.

At Tosa in Japan wheel-like luminous phenomena were seen in the sky during an earthquake.

In 1698 there was ice on the coast of England being 20 centimeters thick on the coast of Suffolk.

The year 1698 in England was the coldest year between 1695 and 1742. The winter of 1698-99 was extremely cold in London, England.

In 1699 CE Utara Wetar in Indonesia erupted with a VEI of 3.

In England in January, some of the days were perfectly warm. Although on some mornings there was frost.

In Scotland corn was reaped in January 1699, and the snow beaten off it. Bread made of it fell in pieces and tasted sweet like malt."

On the morning of January 5, 1699, a violent earthquake rocked the then Dutch East Indies city of Batavia on the island of Java, now known as the Indonesian capital city of Jakarta. Dutch accounts of the event described the earthquake as being "so heavy and strong" and beyond comparable to other known earthquakes. This event was so large that it was felt throughout west Java, and southern Sumatra. At least 28 people in Batavia were fatally wounded and 49 stone-constructed structures collapsed as a result of the earthquake. Some 21 homes and 29 barns were also obliterated. Almost every home situated in the city suffered some extent of damage. The earthquake uprooted and toppled many trees, blocking waterways and choking river systems. At Mount Salak, a volcano on the island of Java, the violent tremors triggered landslides and debris masses, blocking rivers and causing floods. The combination of earth and vegetation produced a debris flow that passed through the Ci Liwung river and into Batavia, where it entered the Java Sea. In the port settlement of Bantam, now Banten, the earthquake destroyed a storage warehouse and caused additional damages. In Lampung, Sumatra, the earthquake threw every home off its foundations. More than 100 people were reported killed in Lampung.

In Augsburg, Germany, in January, the winds were from the east or south. There were frequent snows, but they melted as they fell. But before the equinox (around March 20th or 21st) fell a great snow and the cold continued till May. The cold ended in long rains.

In Breslaw (now Wrocław, Poland), January was cloudy, rainy, windy and cold.

The summer in Paris, France was characterized by: Hot days 55 days Very hot days 5 days (It appears that hot days are defined as those with temperatures of 25° C and greater but less than 31° C, very hot days are those with temperatures 31° C or greater but less than 35° C, and extremely hot days are those with temperatures of 35° C or greater.) There were heavy rains in April and September.

On 30 January, the River Thames was full of ice.

In England terrible storms struck on 7 and 12 February.

In Breslaw (now Wrocław, Poland), the latter end of February was cloudy, rainy, windy and cold.

In Breslaw (now Wrocław, Poland), March began terrible with snow and hoar frost, till the milder spring came in. There was a famine at the time and many people were consuming unwholesome foods.

On 24 March in England, there was a storm of thunder and lightning, high winds and hail. There was another violent hailstorm on 30 March with loud thunder and yet very cold.

In England during the first half of April, the weather was very cold. People were forced to put on again their winter clothes, which they threw off in February. The last half of April had flying clouds and honey dews.

There were heavy rains in April and September in France.

England was hit with a stifling heat wave on 22 June 1699.

On June 29th 1699 Pacaya erupted in Guatemala with a VEI of 2.

For the past 9 years in England, June and July had been so cold that they were difficult to distinguish from the winters. But this year, 1699, produced one of the first of several hot summers. June and July were so hot that wheat began to be harvested on 1 August. June the 22nd and 23rd, it was sultry hot, like the summers of old. The 24th was sultry, and abundance of thunder, the sky being clear; only a few fleecy clouds, and sometimes a few small drops from one. It was intolerably hot to the end of June. The weather was kind to the wheat but not to oats and barley. Those crops were poor for want of rain. July was intolerably hot. There was little grass and no rain.

In Paris, the three summer months produced 130 mm (5.1 inches) of rain. In Burgundy the spring was late and wet.

In England on 11 August the nuts were full and on the 28th they fell out of the hulks.

There were hot days in August in England.

In Dijon, France, the grape harvest began on 5 September. It produced little wine but the quality was good. There were heavy rains in September.

September in England was mostly sultry hot, beyond what any month had been for nine summers before this. On 18 September, the sown wheat was already green on the ground.

In October in England the weather was warm, cloudless sunshine and very calm; as pleasant summer weather an in any month. The year 1699 was not only the hottest, but driest harvest of many years

In November and December in Scotland it was all like summer; warm pleasant sunshine.

On 26 November in Scotland there was snow yet it was warm.

On 30 November, the snow laid 8 or 9 inches deep in Scotland.

The middle two weeks of December in Scotland were perfectly warm.

In Scotland, some of their mosses took fire from small sparks and burnt till after Christmas.

In the upland parishes of Scotland more people died of starvation than of the Black Death. Failed harvest brought famine to Scotland in the 1690s. The population fell by fifteen per cent from a combination of death and emigration. The tacksmen with the excise who managed estates and collected rents, wrote "Many poor people were dying for want, and that the ground was manured in several places, and no Seed in the Countrey for sowing the same, which are two pregnant marks of famine". The privy council stated in 1698 "not only a Scarcity, but a purfeit famine, which is more sensible than ever was known in this nation".

The last half dozen years of William's reign had been the "dear" years of Scottish memory. There were six consecutive seasons of disastrous weather when the harvest would not ripen. The country had no means to food from abroad, so the people laid themselves down and died. Many parishes had been reduced to a half or third of their inhabitants. This was after 1692 as William of Orange died in 1702.

In the seventeenth century the valley bottoms in Western and Central Scotland in Ayrshire, Fife and elsewhere, were wet and ploughing had spread up the steep hillsides as the problem of draining the lower ground had not been tackled. This probably increased the vulnerability to the cold and wet of the community, to the cold, wet summers in the 1690s and in other bad years as far back as the 1580s. Parish records from Duthil and Rothiemurchus in Speyside, recalled that the poorer sort of people frequented the churchyard to pull a mess of nettles and frequently struggled and fought over it…being the earliest spring green which they greedily fed upon. So many families perished from want that for six miles in an inhabited tract there was not a smoke remaining. Nursing women were found dead upon the public roads and babes in the agonies of death sucking at their mothers' breasts. In the last decade of the seventeenth century, half the population of Scotland was supposed to have perished through famine.

During the late seventeenth century there was a renewed period of cold weather and drought that continued into the early eighteenth century. The deaths of people were enormous. It was not until 1850 that the Ottoman Empire regained the population that it previously had in 1590.

In 1699, a powerful cyclone struck the Sunderbans coast, Bangladesh causing 50,000 deaths.

Charleston, South Carolina in the United States was nearly depopulated by an awful tempest and inundation in 1699

Settlers in New England wrote that the climate had gotten colder. A 1699 almanac remarked "The seasons not as they used to be; the Summers turned into Winters; and the Winters embittered with hardships, which in the memory of many have not been known"

In the late 17th century annual mean temperatures in England were about 0.9° C (1.6°F) lower than between 1920 CE to 1960 CE. Between 1690 CE and 1699 CE the deficit was 1.5° C (2.7° F).

In 1699 at St. Didier near Avignon in Vaucluse, France. A priest saw a large light and three globes coming from the sky and merging together: "As I arrived near the oratory I saw the sky open, a great light appeared and soon I observed three globes of fire. The middle one was higher than the other two. I thought, 'here are the lights I have been told about.' Immediately I fell to my knees and thanked God for such a great marvel. At the same time, two more lights appeared, but a bit higher than the place where the chapel is located (…) The two globes merged with the middle one and vanished."

Continuation

As you can see, the climate has never been calm and smooth over long periods. We have volcanic winters and impact events as well as other unexpected climate affecting events. You will also see as you enter Volume 3 that we are getting better and more frequent data dates and more exact locations and themes. Volume 3 enters the new age of recorded climatology so that comparisons can be made with weather data as the new science developed.
Travel on to Volume 3.

The Bibliography is in Volume 4.

www.ingramcontent.com/pod-product-compliance
Lightning Source LLC
Chambersburg PA
CBHW062100220526
45471CB00010B/3545